Plant Microbe Symbiosis

Ajit Varma • Swati Tripathi • Ram Prasad
Editors

Plant Microbe Symbiosis

Springer

Editors
Ajit Varma
Amity Institute of Microbial Technology
Amity University
Noida, Uttar Pradesh, India

Swati Tripathi
Amity Institute of Microbial Technology
Amity University
Noida, Uttar Pradesh, India

Ram Prasad
Department of Botany
Mahatma Gandhi Central University
Motihari, Bihar, India

ISBN 978-3-030-36247-8 ISBN 978-3-030-36248-5 (eBook)
https://doi.org/10.1007/978-3-030-36248-5

This Springer imprint is published by the registered company Springer Nature Switzerland AG.
The registered company address is: Gewerbestrasse 11, 6330 Cham, Switzerland

Contents

Chapter 1
The Rhizobium–Plant Symbiosis: State of the Art

Nitin Kumar, Priyanshi Srivastava, Kanchan Vishwakarma, Rajesh Kumar, Hasmitha Kuppala, Sanjiv Kumar Maheshwari, and Siddharth Vats

Abstract Nitrogen is a vital element necessary for all living organisms (plants, microbes, and animals) for the production of nucleic acids, proteins, and other biomolecules in which nitrogen is needed. Nitrogen is the most abundant gas in the atmosphere of planet Earth, almost 79%. Even in its highest availability, living organisms cannot utilize this gaseous form, unless and until it is in fixed form, which is the reduced form, where it combines with hydrogen and forms ammonia. Photosynthetic plants use this fixed nitrogen to make organic matter, and the phytoproteins that are produced enter into the food chain. On death or during decomposition, microorganisms catabolize the proteins present in the body of dead organisms, fecal wastes, and other organic matter, releasing ammonium ions and forming the primary mechanism of the nitrogen cycle. Microbes exist everywhere, in soil, air, water, and even in extreme conditions, and they also need nutrients for their survival. Of all types of bacteria, some form the complex association known as symbiosis with other living organisms, which can be commensalism, parasitism, mutualism, predation, amensalism, or competition, proto-cooperation between bacteria and other organisms. Bacteria from the family *Rhizobiaceae* survive even nitrogen-limiting condition by symbiotic association with plants of the leguminous

N. Kumar · H. Kuppala
Department of Biotechnology, Periyar Maniammi Institute of Science and Technology, Thanjavur, Tamil Nadu, India

P. Srivastava · S. K. Maheshwari · S. Vats (✉)
Institute of Biosciences and Technology, Shri Ramswaroop Memorial University, Lucknow, Uttar Pradesh, India

K. Vishwakarma
Department of Biotechnology, Motilal Nehru National Institute of Technology, Allahabad, Uttar Pradesh, India

Present Address: Amity Institute of Microbial Technology, Amity University, Noida, Uttar Pradesh, India

R. Kumar
University Institute of Engineering and Technology, Kurukshetra University, Kurukshetra, Haryana, India

A. Varma et al. (eds.), *Plant Microbe Symbiosis*,
https://doi.org/10.1007/978-3-030-36248-5_1

family. This chapter discusses the whole mechanism involved in the symbiotic association between rhizobia and legumes.

1.1 Introduction

For most plants, the nitrogen present in the soil is not sufficient, but legumes are able to attract nitrogen through minerals and association with microbes that form a symbiotic relationship with the roots. Microbes provide support in gaining an association with the roots of legume plants: the relationship or association is termed as being symbiotic in type and the microbes are termed rhizobia. Such types of bacteria live in the soil and help increase the ability of crops to grow in association, improving nutrient uptake and transferring atmospheric nitrogen. The participating rhizobia may have a positive or negative effect on the plants, depending upon the species of the microbes and the environment in which the association occurs. Legumes are one of the most diverse and geographically widespread lineages of plants on Earth (Broughton et al. 2003). Legumes and rhizobia together fix atmospheric nitrogen, and the symbiosis also has a vital role in improving the organic fertility of soil as well as its economy (Jeffries et al. 2003). Nitrogen is an essential element of agricultural sustainability that involves the effective management of the soil. About 80% of biologically fixed nitrogen comes from syntheses formed between leguminous plants and species of *Rhizobium*, *Bradyhizobium*, *Sinorhizobium*, *Azorhizobium*, *Mesorhizobium*, and *Allorhizobium* (Vance 2001). Plants and microbes help each other: plants provide nutrients to the microbes and receive nitrogen in the reduced form from the microorganisms. Legume plants form a specialized atmosphere where rhizobia fix atmospheric nitrogen. These specialized plant structures, known as nodules, are generally established on the roots and at times on the stems of the plant (Kong et al. 2017). Microbes that are tolerant to stress have better nodulation ability and greater ability for nitrogen fixation of legumes to grow and survive under stressed conditions. Rhizobial populations vary in their tolerance to major environmental factors. In addition to nitrogen fixation, these beneficial microorganisms exhibit control activity as well: rhizobia are used as biofertilizers under severe conditions (Shiraishi et al. 2010). When rhizobia are not living symbiotically in root nodules they tend to live in a regular facultative manner. The symbiosis of legumes and *Rhizobium* spp. proceeds in a step-by-step manner that includes both the bacteria symbiont and host plant. First shown by Vincent in 1980, rhizobia attachment is followed by root hair curl formation, initiation of the meristem, invasion of bacteria into bacterial cells, and finally gene expression for nitrogen fixation and other important aspects of the plant–rhizobium interaction. Current studies on nodule formation are focused on different stages of nodule development, including genetic variability at all stages, cell biology, and biochemistry. Further research is focused toward identification of genes responsible for nodulation, host specificity, invasion, and other properties of rhizobia cells that are controlled by different genes. Genetics for plant–microbe specificity research has an exclusive and uniting role (Long 1984; Levy et al. 2018).

For the sustainable development of society, the rhizobium–plant symbiosis is being exploited. The increased use of chemical fertilizers has already started to show negative effects by degrading soil fertility, which is followed by severe health problems and environmental threats such as falling groundwater table, pesticide poisoning, soil erosion, waterlogging, and water contamination. For nutrient uptake by plants, plant–microbe interactions are important to help in solubilization, mobilization, and transformation of nutrients. These rhizospheric bacteria assist plant growth promotion by providing nutrients and by helping plants directly and indirectly in fighting against plant pathogens, bioaccumulation, improved soil structure, etc. (Bertrand et al. 2015), and for decades have been utilized for crop production (Davison 1988). These plant growth-promoting rhizospheric (PGPR) bacteria have also been exploited for polluted soil bioremediation by organic pollutant mineralization (Burd et al. 2000; Zhuang et al. 2007; Zaidi et al. 2008).

This chapter discusses the relationship between *Rhizobium* and legumes, and also the factors affecting its symbiosis. It also discusses the mechanism of their symbiosis and further continued with the discussion of biofertilizer application of rhizobia. The chapter ends with the discussion on the new aspects under the *Rhizobium*–legume symbiosis.

1.2 Legume–*Rhizobium* Symbiosis for Root Nodulation

Root nodule symbiosis helps the rhizobia fix nitrogen that is present in the atmosphere so that it is directly available for the growth of plants. Legume plants (Fabales) form symbiotic associations with single-celled microbes, known as rhizobia. In legumes, infection proceeds through intracellular and transcellular channels termed infection threads. At the same time, cells present in the cortex of the root are induced to divide, which helps to generate the tissues of the nodules. Some legumes are only infected through root hairs, and some legumes (such as *Neptunia*) can switch from root hair infections to epidermal crack infections in wet conditions. Rhizobia have defined a classic model of molecular crosstalk that is necessary for nodulation and initiation of symbiotic nitrogen fixation by rhizobia. Fixing of nitrogen does require specialized cells with organelles with cytoplasmic compartments termed symbiosomes. Within the symbiosome, the rhizobia ultimately differentiate into a specialized cell type called bacteroids and fix atmospheric nitrogen for the plant in exchange for sugars.

The structural arrangement of a root nodule (both anatomically and physiologically) shows an integration of metabolic and structural arrangements for the host and the microbe. Legumes emit flavonoids into the soil to attract rhizobia (Perret et al. 2000). The signals are received by the soil rhizobia that then emit a Nod factor in response, which is a complex oligosaccharide encoded by the Nod gene. The infection thread is formed by the curls of root hairs around rhizobia, which is triggered by receipt of the Nod factors in the legume root tip that helps to transport the rhizobia through the legume root. The signals that initiate the legume symbiosis are not simple. The flavonoids are not solely used for the attraction of rhizobia (Cesco et al. 2010). Other signaling molecules can also be released through legumes. According to some recent studies the lectins, carbohydrate-binding proteins, are also emitted by the plant and may be vital in the specificity observed in the symbiosis.

The symbiotic relationship between *Rhizobium* and legumes initially starts with two independently living organisms and leads to intimately dependent coexisting cells. *Rhizobium* has the ability to recognize some specific plants and stimulate the formation of root nodules in them to colonize tissues in the root. After transferring itself into the host cell, the *Rhizobium* cell becomes surrounded by the plant membrane, ensuring the supply of sanctuary and sugar to the *Rhizobium* cell in return for fixed atmospheric nitrogen supplied to the plant cell (Hirsch et al. 2001). The process that is responsible for fixing atmospheric nitrogen by microbes into ammonium so that it can be available for plants is known as biological nitrogen fixation (BNF). Every year nearly 200 million tons of nitrogen is fixed for nearly half of the terrestrial plants over the globe (Desbrosses and Stougaard 2011). The symbiosis is a complex process that is affected by various factors such as host plant species, *Rhizobium* species, and the terrestrial environment. The following subsections briefly discuss these attributes.

1.3 Diversity of Legumes Depending on *Rhizobium*

Arid regions have a wide distribution of wild legumes, either herbs or trees, which helps to maintain the soil fertility of these environments. The ability to fix nitrogen and tolerate extreme environmental conditions is found to be enhanced in wilder legume species when compared to crop legumes (Requena et al. 2001). The wild species of legume from arid environments bear a large pool of distinct rhizobia in their nodulated roots. Host specificity is observed to exist only in a few variants of rhizobia obtained from wild legumes, whereas the populations with a wide host range prevail (Zahran 2001). Phenotypic characterization and molecular techniques (plasmids, DNA–DNA hybridization, 16*S* rRNA, polysaccharides, protein profiles, etc.) allow us to classify the bacteria that were obtained from the root nodules of wild legumes into four genera: *Rhizobium*, *Bradyrhizobium*, *Mesorhizobium*, and *Sinorhizobium*. Inoculation of either wild or crop legumes or both with these rhizobia in reclaimed desert land cultivation is possible. A few recent studies have shown successfully working symbiotic relationships between the wild legume rhizobia and some grain legumes. Furthermore, when some N_2-fixing tree legumes (e.g., *Lablab*, *Leucaena*, *Sesbania*) were intercropped with forages, enhancement of biomass yield and quality of herb was observed. The wild legume rhizobia have also attracted the attention of biotechnologists in the past few years (Zahran 2001). These rhizobia may possess specific characteristic genes that can be passed to other rhizobia through various tools of genetic engineering to produce various economically important compounds. Therefore, these bacteria have been proved to be economically as well as environmentally important. Apart from root nodules, rhizobia can be found on the surface of roots, and the rhizospheric and non-rhizospheric part of the soil around the roots. The excretion of compounds called root exudates from the plants causes a rise in the rhizobia population in the rhizosphere, especially in legume plants. Moisture and temperature, salt content, and the alkalinity and acidity of the soil also decide rhizobial biodiversity in the soil.

1.4 Taxonomy and Host Specificity of *Rhizobium* Species

The three genera of rhizobial bacteria, *Rhizobium*, *Bradyrhizobium*, and *Azorhizobium*, have been classified with the agrobacteria and phyllobacteria into one family, the *Rhizobiaceae*, for many years (Jordan 1984). However, using different modern methods of bacterial systematics, such as numerical taxonomy, nucleic acid hybridization, and 16*S* rRNA analysis, a striking genetic diversity is observed within this family (Young et al. 1991). Therefore, it was established that the genera *Rhizobium* and *Bradyrhizobium* are only distantly related to each other. Close relatives that do not form symbiotic relationships with plants belong to these genera and are placed in different families. In the nomenclature of most microsymbionts, the species name usually suggests the host plant that can be invaded to cause nodulation. It also implies that symbiosis is a species-specific phenomenon.

It has been clearly understood that there is great variation in the extent of host specificity among the rhizobia (Young and Johnston 1989). Some strains of rhizobia have a very restricted range of hosts (*Rhizobium leguminosarum* bv. *trifolii*) whereas others, such as *Rhizobium* sp. strain NGR234, have a very substantial host range. With increasing experimental data, the complexity of the symbiotic relationship between legume species and rhizobia is being realized. *Rhizobium* sp. strain NGR234 alone is known to nodulate about 35 different genera of legumes as well as *Parasponia*, a non-legume (Lewin et al. 1987). An extensive study on the taxonomy of these bacteria has been performed recently. Several strains were found that were related to or classified into the existing species and genera of rhizobia, but other strains were not identified and had characters that were totally different from the existing one. This finding suggested that some new species and genera of root-nodulating bacteria need to be detailed. It has been observed recently that *Azorhizobium caulinodans* can also form a symbiont relationship with *Phaseolus vulgaris*, but further characterization of the genes involved in the formation of nodules by this isolate was not performed. Experiments conducted for the identification, cloning, and mutagenesis of the nodulation genes have led to the results that a single gene inactivation can greatly modify the range of hosts for a particular strain (Faucher et al. 1989). *Rhizobium leguminosarum* bv. *trifolii* and *Rhizobium leguminosarum* bv. *viciae* are found to have very similar symbiotic genes but vary greatly in the range of their hosts. Exchanging one gene of the microsymbiont may cause switching of the host range of the two biovars (Spaink et al. 1989). Analysis of plants to determine the phenotype of nodulation is also important, as it allows verifying the predicted host range based on the characterization of the molecular structure of nodulation signals. The root nodule rhizobia isolated from wild (naturally growing) legumes found in arid regions are very promising and multipurpose. Most have a wide host range, which gives these rhizobia an ecological advantage. As these rhizobia can form nodules with wild or crop legumes, they can act as a source of information to improve the symbiotic characters of other rhizobia genetically. Furthermore, symbiotic rhizobia isolated from naturally growing or wild legumes have more tolerance to ecological stress conditions (e.g., salinity, drought, high temperatures) than rhizobia obtained from cultivated legumes. Some rhizobia acquired from wild legumes were successful in establishing a functional symbiosis even under environmental stress conditions (van Rhijn and Vanderleyden 1995). The relevance of wild legume rhizobia is not limited to their ability to fix atmospheric nitrogen or to other soil activities that lead to the enhancement of soil fertility and plant productivity: some strains are of immense importance for biotechnological niches. These functions may include industrial production of polysaccharides, enzymes, and antibiotics. It can be confidently predicted that this research field will be the highlight of future biotechnological investigations.

1.5 Factors Affecting Legume–Rhizobium Symbiosis

The symbiotic relationship between rhizobia and plants is more sensitive to the stress imposed by higher salt concentrations than are independently living rhizobia (Zahran 1999). Various steps involved in the interaction of rhizobia and plant, the processes of nodule formation and metabolism, are adversely affected under salt stress, ultimately leading to decreased numbers of nodules (Singleton and Bohlool 1984). The complexity of rhizobial response and adaptation to salt stress indicates that many different physiological and biochemical processes affect the process of root colonization and early infection by the rhizobia (Nabizadeh et al. 2011). Nutrient imbalance, which is caused by the loss of control on nutrient uptake and transportation to different parts, leading to ion deficiency, is a major constraint imposed by salt stress. The major cations (Na^+, Ca^{2+}, Mg^{2+}, K^+) and anions (Cl^-, SO_4^{2-}, HCO^{3-}, CO_3^{2-}, and NO_3^-) accumulate in high quantities in the soil under salt stress. Salt injury occurs mainly because of the accumulation of Na^+ or Cl^- (or both) in transpiring leaves over the high limit, causing failure of the cellular mechanism to accumulate excess ions into the vacuole. These ions then spread rapidly in the cellular cytoplasm to inhibit enzymatic activity or are concentrated in the cell walls to dehydrate the cell (Munns 2002). The survival of rhizobia in soil is more affected by high temperatures than by low temperatures (Al-Falih 2002), affecting both partners in the symbiosis and all steps in the development of an efficient nitrogen fixation. Temperature affects infection of the root hairs and the differentiation of bacteriod, nodule structure, and legume root nodule functioning (Zahran 1999). The different strains and species of rhizobia differ in their optimum temperature over a range of 27–39 °C for growth in culture. The maximum temperatures are usually maintained up to 35–39 °C, but proliferation continues to occur up to 42 °C. Soil aggregates provide a survival advantage at high temperature to rhizobia in comparison to nonaggregated soil and seems to assist with dry rather than moist conditions (Zahran 1999). Water deficiency and other environmental stresses adversely affect N_2-fixing legumes, with drought as one of the major environmental factors affecting the productivity of plants (Zahran 1999). Drought commonly affects the osmotic balance of rhizobia causing osmotic stress, which consequently leads to morphological changes in the rhizobia, persistence and survival in soil, and root hair colonization and infection (Zahran 1999). The effort to develop stress-tolerant plants is very important to increasing crop productivity. Both biotic and abiotic stress-tolerant plants can be developed by using selective agents such as NaCl (for salt tolerance), polyethylene glycol (PEG), or mannitol (for drought tolerance). The selection of genetically stable somaclonal variations is used in regenerated plants to improve crop productivity (Manoj et al. 2011).

1.6 Mechanism Behind Root Nodulation

Nodulation occurs in three subfamilies of Leguminosae, namely, Mimosoideae, Papilionoideae (more advanced species), and Caesalpinioideae (less specialized and most primitive species). Also, the percent of nodulation varies in all these subfamilies. Caesalpinioideae is less specialized; it originated from the family Fabales and has developed a system for the synthesis of nodulation in legume plants. Plants from the Leguminosae family are very diverse habitat wise, both morphologically and ecologically (Trinick 1979). They are found from the Arctic region to tropical regions. There are many areas where nitrogen is present in sufficient amounts but these plants are still nodulated, so it can be inferred that this characteristic is not only a mere adaptation but instead the genetic outcome, as this peculiar characteristic is not found in other plants. Microbes of rhizobial family are put into the same group as that of agrobacteria and phyllobacteria. By the application of the new bacterial systematics approach based on 16*S* rRNA, nuclei acid hybridization, etc., *Bradyrhizobium* and *Rhizobium* are distantly related but still form nodules, whereas some of their closely related microbes do not (De Meyer et al. 2015). A list of the microbes that are involved in symbiotic association is given in Table 1.1.

Table 1.1 Microbes from rhizobium family and their host plant

Species no.	Microbial	Plant	Plant hormone role	Region of nodulation
1	*Rhizobium meliloti*	*Medicago*	Auxin, cytokinine, ethylene	Roots
2	*Rhizobium leguminosarum* biovar *viciae*	*Lathyrus*	Auxin, cytokinine, ethylene	Roots
3	*Rhizobium leguminosarum* biovar *trifolii*	*Trifolium* spp.	Auxin, cytokinine, ethylene	Roots
4	*Rhizobium leguminosarum* biovar *viciae*	*Lens* spp.	Auxin, cytokinine, ethylene	Roots
5	*Rhizobium leguminosarum* biovar *phaseoli*	*Phaseolus vulgaris*	Auxin, cytokinine, ethylene	Roots
6	*Rhizobium loti*	*Lotus* spp.	Auxin, cytokinine, ethylene	Roots
7	*Rhizobium meliloti*	*Melilotus*	Auxin, cytokinine, ethylene	Roots
8	*Rhizobium huakuii*	*Astragalus sinicus*	Auxin, cytokinine, ethylene	Roots
9	*Rhizobium cicero*	*Cicer arietinum*	Auxin, cytokinine, ethylene	Roots
10	*Rhizobium etli*	*Phaseolus vulgaris*	Auxin, cytokinine, ethylene	Roots
11	*Rhizobium leguminosarum* biovar *viciae*	*Vicia*	Auxin, cytokinine, ethylene	Roots
12	*Rhizobium* sp.	Tropical legumes such as *Parasponia* spp.	Auxin, cytokinine, ethylene	Roots
13	*Rhizobium meliloti*	*Trigonella* spp.	Auxin, cytokinine, ethylene	Roots
14	*Rhizobium galegai*	*Galega officinalis*	Auxin, cytokinine, ethylene	Roots
15	*Rhizobium fredii*	*Glycine max*	Auxin, cytokinine, ethylene	Roots
16	*Bradyrhizobium parasponia*	*Parasponia* spp.	Auxin, cytokinine, ethylene	Roots
17	*Rhizobium leguminosarum* biovar *viciae*	*Pisum*	Auxin, cytokinine, ethylene	Roots
18	*Rhizobium galegae*	*Galega orientalis*	Auxin, cytokinine, ethylene	Roots
19	*Rhizobium japonicum*	*Glycine max*	Auxin, cytokinine, ethylene	Roots

(continued)

Table 1.1 (continued)

Species no.	Microbial	Plant	Plant hormone role	Region of nodulation
20	*Bradyrhizobium japonicum*	*Glycine max*	Auxin, cytokinine, ethylene	Roots
21	*Bradyrhizobium japonicum*	*Glycine soja*	Auxin, cytokinine, ethylene	Roots
22	*Rhizobium fredii*	*Glycine soja*	Auxin, cytokinine, ethylene	Roots
23	*Rhizobium japonicum*	*Glycine soja*	Auxin, cytokinine, ethylene	Roots
24	*Bradyrhizobium elkanii*	*Glycine max*	Auxin, cytokinine, ethylene	Roots
25	*Azorhizobium caulinodans*	*Sesbania* spp.	Auxin, cytokinine, ethylene	Stem

1.7 Role of Nitrogen and Mechanism of Root Nodulation

Nitrogen is an important element necessary for all living organisms (plants, microbes, animals) for the production of nucleic acids, proteins, and other biomolecules where nitrogen is needed. Nitrogen is the most abundant gas in the atmosphere of planet Earth, almost 80%. Even after its highest availability living organisms cannot utilize this gaseous form, unless and until it is in fixed form, which is the reduced form where it combines with hydrogen, and forms ammonia. Photosynthetic plants use this fixed nitrogen to make organic matter and to make phytoproteins that then enter the food chain. At death or during decomposition, microorganisms catabolize the proteins present in the body of dead organisms, fecal wastes, and other organic matter and release ammonium ions to form the primary mechanism of the nitrogen cycle. Microbes exist everywhere, in soil, air, water, and even in extreme conditions (Nakagawa and Mino 2018). They also need nutrients for their survival. All types of bacteria form the complex association known as symbiosis, with other living organisms, which can be commensalism, parasitism, mutualism, predation, amensalism, and competition, proto-cooperation between bacteria and other organisms. Bacteria from the family *Rhizobiaceae* survive even in nitrogen-limiting conditions by the symbiotic association with plants belonging to the legumes family (Griesmann et al. 2018). Nitrogen-fixing root nodules are formed by the leguminous plants for utilizing nitrogen. Root infection in the legume plants is caused by rhizobia in a multistep process. *Rhizobium* and *Bradyrhizobium* respond by positive chemotaxis to flavonoids (Peck et al. 2006), dicarboxylic acids, and amino acids released by the plant roots (Dunn 2015). However, chemotaxis is not the main driving force: even if it is apparently not necessary for nodulation in mutants that are flagella deficient, it definitely has a role in rhizosphere establishment (Dunn 2015). Rhizobia mainly target sites that are prone to infections, such as the root surface of young growing root hairs. There are no anchors or loci for the attachment of rhizobia. In the past it was believed that attachment of *Bradyrhizobium* or *Rhizobium* spp. is host specific because of the binding of specific polysaccharide structures present on the surface of microbe to the lectins present in the plant roots (Smit et al. 1992). During the morphogenesis, root nodules need a coordinated control for spatiotemporal expression in bacterial genes and plant genes. Signals generated from plants and bacteria work in synergy to regulate the nodules on the plants (Crespi and Gálvez 2000). Some of the genes belonging to the plant that are induced during the morphogenesis and development of the root nodule are called nodulin. Nodulin genes have been studied and characterized, and their products have shown actions similar to that of the known regulators used by animals and plants in their signal transduction systems (Moling and Bisseling 2015). Model legumes were made the basis of a study for functional analysis of the molecular mechanism responsible for nodulation. Strategies such as mutagenesis and pharmacological approaches were used to understand the molecular mechanism for morphogenesis of this symbiont relationship (Foo 2017). There are various stages in the nodule development. The Nin transcription factor and G protein-mediated transduction have important roles in the early development of the nodule. Gene ENOD40 encodes for RNA, with short open reading frames, and is responsible for the formation of the primordium nodule. Transcription factors such as Krüppel (associated with vascular) and G proteins such as the Rab type function in differentiation and regulation and in the nitrogen-fixing zone and differentiation of the bacteroid. The root nodule is common in legumes (such as the pea, and pulses, which have a pod), and shows symbiotic association under nitrogen-limited conditions to nitrogen-fixing bacteria, called rhizobia (Suzaki et al. 2013). A general diagram depicting the formation of root nodulation in the legume plant is given in Fig. 1.1.

These bacteria attach to the roots of the plants, and in response to that the plant roots morphologically convert themselves into a gall-like structure and harbor the rhizobia. This gall-like structure also has a special type of phytochemical called leghemoglobin. Leghemoglobin creates a hypoxic condition that is perfect for *Rhizobium*, which is an anaerobic microbe. The main function of the rhizobia is to convert nitrogen gas (inert N_2) present in the air into ammonia (NH_3). By making

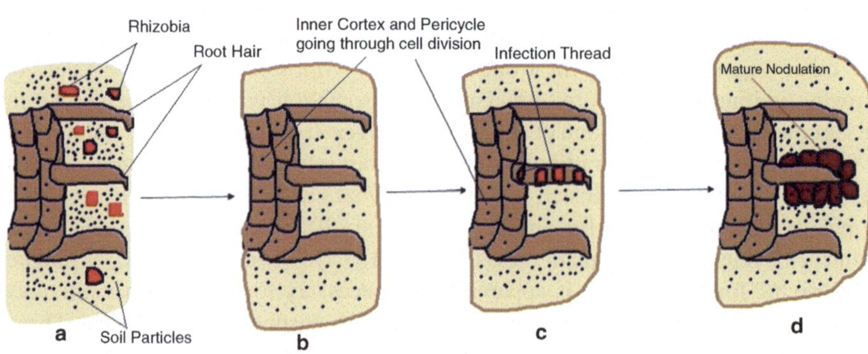

Fig. 1.1 Formation of root nodulation in legume plants

ammonia available to the plants, rhizobia help plants in synthesis of amino acids and nucleotides. Nodule formation is very common among plants of the Fabaceae by symbiotic association with rhizobia, which fix nitrogen gas and provide it to the plants for its function, receiving carbon-based energy in return. Nodulation is the differentiation of cortical cells of the roots to primordia (Oldroyd et al. 2011), which undergo organogenesis by the infection caused by rhizobia through the infection thread (Murray 2011). Auxin has a positive response in lateral root (LR) development whereas cytokinin (Marhavý et al. 2013) has a negative response in LR cell promotion and expression (Bielach et al. 2012). Plants undergo symbiotic relationships with microbes such as *Bradyrhizobium*, *Rhizobium*, and *Azorhizobium*, which are collectively termed rhizobia. These microbes cause development of the specialized organs termed nodules. These nodules act as a source for the conversion of nitrogen present in the air into ammonia, which acts as a nitrogen source in nitrogen-limited conditions. Nodulation is an interaction that is beneficial for both microbes and plants. Nodulation is a complex and sophisticated process leading to exchange of nutrients among microbes and plants. Of the many hormones involved in nodulation, cytokinin, auxin, and ethylene are the major components. All these hormones act differently, with positive and negative regulation effects. In the nodule primordium (NP), cytokinin and auxin have positive roles whereas ethylene has a negative role (Mortier et al. 2014). In the progression of infection threads (IT), auxin has a positive role whereas cytokinin and ethylene have a negative role (Murray et al. 2007). All these hormones act synergistically, and it is not possible to separate and analyze the individual action results of the individual hormones. Hormones meet, cross, and interact during the signaling pathways, and their interactions control the positioning of the nodule in the root.

1.8 Role of Ethylene in Preinfection Events

Interaction between leguminous plants and the rhizobium is initiated by release of flavonoids that are part of the plant chemotactic aspect. Rhizobia are attracted to the source of flavonoids, the root. In the microbe, flavonoids promotes its affinity for the Nod box by interacting with the NodD1 protein (Peck et al. 2006). This interaction leads to the synthesis of Nod factors (NF), which are recognized by the LysM receptor kinases: for example, Nod factor receptor r5 (NFR5) and Nod factor receptor r1 (NFR1). After proper binding, NFs initiate the common signaling transduction pathway (CSTP), which is called the common pathway because the same pathway is responsible for the symbioses in arbuscular mycorrhiza. The activated CSTP then initiates the programs responsible for nodule organogenesis, namely at the cortical and epidermal regions (Oldroyd 2013; Guinel and Geil 2002). Plants also respond to the presence of the rhizobia. Plants and microbes have a balanced act: the microbe does not cause strong infection nor do the plants show a high immune response. The microbe with its microbe-associated molecular pattern (MAMPs) and the plant with its immune system act in a fine balance. MAMPs cause activation of the immune response in plants, termed MAMP-triggered immune response (MTIR) (Gourion et al. 2015). The epidermis plasmalemma of the plants have receptors (FLS2 receptors, flagellin-sensing receptors) for flagellin-like molecules (Flg22) that are secreted by the microbes, similar to NFs. This binding activates calcium influx and the generation of free radicals such as reactive oxygen species (ROS), and, furthermore, upregulation of genes encoding ethylene response factor (ERFs), chitinases, and peroxidases. In *Lotus japonicus* plants, which also undergo nodulation, its transient defense response is dependent on LjNFR1 (*Lotus japonicus* nodulation factor receptor 1). After infection at the same site, other genes are also upregulated that are responsible for pathogenesis-related (PR) proteins and phytoalexin medicarpin (Breakspear et al. 2014). To meet the challenges posed by the plant immune system and its responses, rhizobia secrete exopolysaccharides (EPS), which chelate extracellular calcium ions, and liopoplysaccharides (LPS), to counter the attack of ROS and slow its production.

Similarly, plants balance out the MTI by effector-triggered immunity (ETI), activated by bacterial proteins such as NoP via type III effector molecules (T3SS) in the cytoplasm of the cell. These proteins slow the MTI. In response, the plant synthesizes leucine-rich repeat proteins that identify and recognize the rhizobial proteins (Gourion et al. 2015). It is important to note that defense-related hormones such as jasmonic acid, salicylic acid, and ethylene interact with such hormones as cytokinin, auxin, and DELLA proteins (Limpens et al. 2015). The simultaneous activation of the defense system and the symbiotic system is the outcome of crosstalk among proteins, hormones, and other phytochemicals. With time, the immune response of the plant becomes suppressed, allowing the bacteria to enter for rhizobial establishment (Gourion et al. 2015).

1.9 Organogenesis of the Nodule

The role of NFs in modulating the defense of the plants has not been fully understood until recently (Guinel 2015), but it is known that NFs have a dual role in cortical and epidermal functions. Two receptors are important in the functioning of NFs: LjNFR1 and LjNFR5, LjNRF1 being the important component of the entry receptor whereas LjNRF5 is the integral part of the signaling receptor. The entry receptor activates the epidermal program and the signaling receptor triggers the cortical program. The epidermal program is composed of steps related to rhizobial action, mainly root hair curling, infection thread formation, and progression in infection threads, whereas the cortical program is the key for all components related to nodule infrastructure. Both these programs are highly regulated and orchestrated. If not controlled, a pseudo-nodule can form in the absence of the bacteria, so that only the cortical program is activated without epidermal program activation. Within 24 h of infection by rhizobia, microbes activate scaffolding proteins (remorin and flotillins) responsible for the plasmalemma microdomain formation. Remorin protein includes

MtSYMREM1, a symbiotic remorin, and flotillins include MtFLOT2 and MtFLOT4 (Lefebvre et al. 2010; Haney and Long 2010).

MtSYMREM1 interacts with an ortholog of LjSYMRK (SYMbiosis receptor kinase) ortholog MtDMI2 (does not make infection 2). NSP2 with nodule inception (NIN) can upregulate the MtFLOT2 and MtFLOT4 for elongation in the infection thread. Exopolysaccharides are secreted by the bacteria to counter the concentration of calcium, which tries to prevent their cell entry, but plants also have EPR3, which can differentiate compatible and incompatible bacteria on the basis of EPS released by the bacteria (Kawaharada et al. 2015). On the entry of compatible bacteria, epidermal expression is activated and triggered by NF, causing curling in the root hairs. As is the epidermal program, the cortical program is required for nodule organogenesis, which is dependent on the dedifferentiation of the progenitor cells present in the nodule and targeted in the cortex by NF signals. These cells form nodule primordium (NP) and nodule meristem (NM). The nodule meristem grows in an outward direction whereas the infection thread grows inwardly (Guinel and Geil 2002). Various hormones such as cytokinin and auxins are important in nodule formation in legumes.

1.10 Genetic Basis of Phytohormones During Root Nodule Development

Phytohormones share in nodulation at the roots in legume plants because these are responsible for cell proliferation and differentiation (Durbak et al. 2012). Auxin and cytokinin have diverse functions, such as control in root nodulation and cell differentiation and proliferation. To elucidate the molecular genetic basis of cytokinin and auxin in nodule development, a recent study was carried out by Suzaki et al. (2013) on the model leguminous species, *Medicago truncatula* and *Lotus japonicus*. It was found that putative cytokinin signaling receptors *M. truncatula* cytokinin response 1 (CR1) and *Lotus* histidine kinase 1 (HK1) are important for nodulation by initiating the nodule primordia. Auxin is also responsible for the development of the nodule (Suzaki et al. 2013). Various pathways involved in nodulation are shown in Fig. 1.2.

1.11 Application of Rhizobia as Biofertilizers

Biofertilizers are substances containing living cells such as strains of microorganisms which, when supplied to a soil, promote the growth of the plant by increasing the supply or nutrient availability to the host. Biofertilizers are typically prepared as a carrier-based inoculant containing effective microorganisms that show a quality

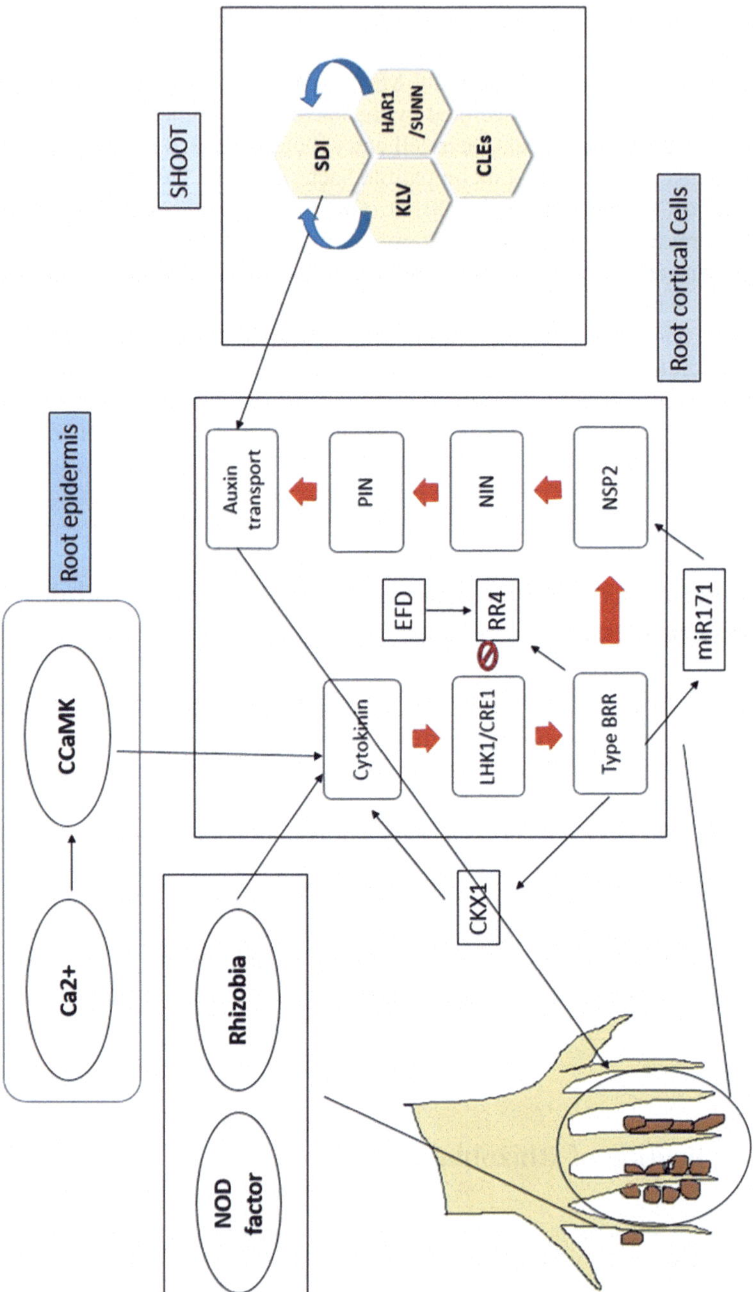

Fig. 1.2 Various chemical pathways involved in nodulation

Table 1.2 Different types of nitrogen fixation host groups by *Rhizobium* species

Host groups	*Rhizobium* species	Crops
Pea group	*Rhizobium leguminosarum*	Green pea, lentil, sweet pea, vetch
Soybean group	*R. japonicum*	Soybean
Lupine group	*R. lupine orinthopus*	Lupinus
Alfalfa group	*R. melliloti medicago trigonella*	*Melilotus*, alfalfa, fenugreek, sweet clover
Beans group	*R. phaseoli*	Phaseoli
Clover group	*R. trifoli*	Trifolium
Cowpea group	*Rhizobium* sp.	Moong, redgram, cowpea, peanut, kudzu
Cicer group (chickpea)	*Rhizobium* sp.	Bengal gram

relationship with the host plant, which simplifies handling a wide range of acceptance by the host.

Rhizobia have massive economic and agricultural value by providing the foundation of N_2 contribution to the soils of agricultural crops. In addition to N_2 fixation, many rhizobial strains exert plant growth-inducing traits such as the production of phytohormones, siderophores, and 1-aminocyclopropane-1-carboxylic acid (ACC) deaminase, as well as the solubilization of inorganic phosphate (Ahemad and Kibret 2014). These growth-promoting traits make the rhizobia valuable for both legumes and non-legumes. For improvement of plant growth, effective rhizobial strains have been screened and used as inoculants (Bloemberg and Lugtenberg 2001). The application of rhizobia as biofertilizers guarantees success in crop productivity and decreases the need for artificial fertilizers, which are costly and detrimental to the environment. A list of the microbes from *Rhizobium* species involved in nitrogen fixation is given in Table 1.2.

PGPRs might colonize the rhizosphere, the surface of the root, or even superficial intercellular spaces of plants (McCully 2001): these are the most efficient biofertilizers in accordance with the quantity of nitrogen fixation. The seven genera are highly specific for the formation of root nodules in legumes, denoted as cross-inoculation groups. A number of PGPRs have been used globally as biofertilizers, subsidizing growing crop yields and soil fertility, and therefore have the prospective to endorse sustainable agriculture and forestry (Khalid et al. 2009). Biofertilizers composed with solid carriers have a typical lifespan of 6 months. However, when the biofertilizer produced is liquid, nutrients and cell protectants can be supplemental so that the shelf life of the final product will be long term and the product will endure high temperature. As an outcome, liquid biofertilizers can persist in temperatures up to 55 °C, have a longer shelf life, of 12–24 months, with no contamination and loss of properties in storage up to 45 °C, and high commercial revenues with export potential and excessive enzymatic activity because contamination is nil. The only disadvantage of liquid preparations is that they are overpriced (Mahdi et al. 2010).

1.12 New Aspects of Plant–Rhizobia Symbiosis

The NF lipochitooligosaccharide (LCO) signals from plants secreted in the rhizospheric soil regions induce nodule formation by bacteria. Earlier, NFs were supposed to be specifically required for nodule formation. However, new studies have challenged this perspective, and place the earlier statistics in a new outlook, as signals related from NF apparently are part of a new and important idea of 'control of immunity' during symbiosis. One of the important studies has shown that mutant soybean with no NF receptor has been nodulated by *Bradyrhizobium elkanii* strains without the Nod genes that are accountable for the NF synthesis (Okazaki et al. 2013). More recent studies show the relationship between symbiosis development and suppression of plant innate immunity, although decrease in the level of immunity during symbiosis process can allow various pathogens to infect the host plant. Certainly, the capability to allow active decrease in the level of plant immunity while allowing infection by the symbiont may have been a specific cause for rhizobium host specificity. Nodulation changes the plant cell gene expression, which allows the symbiont to enter in a step-by-step manner (Irmer et al. 2015). A study on *Crotalaria* (Fabaceae) has shown that the changes in plant cell gene expression are related to pyrrolizidine alkaloid (PA) biosynthesis. PAs are among the chemicals that assist plant defense against plant-eating insects and are not involved in the process of symbiosis (Irmer et al. 2015). Synthesis of PAs by *Crotalaria* is detected only during nodule formation after rhizobial infection. The results of this study can be confirmed with the identification of homospermidine synthase (HSS) synthesis, which is the primary enzyme in the pathway of PAs biosynthesis, suggesting that it is not the microbiont but the plant which is synthesizing the PAs. HSS has been found mainly in nodules, which is also where the high concentration of PAs is found, which indicates that PA synthesis is controlled in the nodules only and PAs are the source for alkaloid transport to other parts of the plant. The connection between biosynthesis of nitrogen-containing alkaloids and nodulation in *Crotalaria* shows the outcome of rhizobia symbiosis on the plant defense system (Irmer et al. 2015). So, it is a practical challenge to have a non-legume crop plant achieve symbiosis with a nitrogen-fixing symbiont, and it has become a prime research interest to understand how leguminous plants tolerate the enormous rhizobial colonization. More research is needed for better understanding of plant innate immunity during nodulation. All these facts are telling us that the legume–*Rhizobium* association is not as simple as it seems but rather goes through cycles of complex mechanisms that are affected by various physical and biological factors. Thus, to study these factors is the new focus. New studies have improved understanding of the symbiosis process. As an example, Nod factors (NFs), earlier believed to be undeniably essential to plant–rhizobium symbiosis, have been reported as being not essential in some specific conditions. Similarly, an NF receptor earlier thought to only have a role in symbiosis has been shown as important in plant fights against pathogens. So, these new studies have shown the importance of the innate immunity of the plant–rhizobia symbiosis.

1.13 Conclusion

Plants and microbes help each other: plants provide nutrients to the microbes and acquire nitrogen in the reduced form from the microbes. Legume plants form a specialized atmosphere where rhizobia fix atmospheric nitrogen. These specialized plant structures, known as nodules, are generally established on the roots, although sometimes on the stems, of the plant. Formation of the plant and rhizobia symbiotic relationship involves complex communications and biochemical reactions. Although much research has been done, deeper study and further research should be focused toward identification of genes responsible for nodulation, host specificity, invasion, and other properties of the rhizobia cells that are controlled by different genes. Genetic research for plant–microbe specificity has an exclusive uniting role and, by enhancing understanding about all the genes, microbes, and metabolites involved, can help in utilizing this knowledge for sustainable development. The results can aid better understanding of the mechanisms as well as the biochemistry of nodule formation.

References

Ahemad M, Kibret M (2014) Mechanisms and applications of plant growth promoting rhizobacteria: current perspective. J King Saud Univ Sci 26(1):1–20

Al-Falih AMK (2002) Factors affecting the efficiency of symbiotic nitrogen fixation by *Rhizobium*. Pak J Biol Sci 5(11):1277–1293

Bertrand A, Dhont C, Bipfubusa M, Chalifour FP, Drouin P, Beauchamp CJ (2015) Improving salt stress responses of the symbiosis in alfalfa using salt-tolerant cultivar and rhizobial strain. Appl Soil Ecol 87:108–117

Bielach A, Podlešáková K, Marhavý P, Duclercq J, Cuesta C, Müller B et al (2012) Spatiotemporal regulation of lateral root organogenesis in *Arabidopsis* by cytokinin. Plant Cell 24:3967–3981

Bloemberg GV, Lugtenberg BJ (2001) Molecular basis of plant growth promotion and biocontrol by rhizobacteria. Curr Opin Plant Biol 4:343–350

Breakspear A, Liu C, Roy S, Stacey N, Rogers C, Trick M et al (2014) The root hair "infectome" of *Medicago truncatula* uncovers changes in cell cycle genes and reveals a requirement for auxin signaling in rhizobial infection. Plant Cell 26:4680–4701

Broughton WJ, Hernandez G, Blair M, Beebe S, Gepts P, Vanderleyden J (2003) Beans (*Phaseolus* spp.)–model food legumes. Plant Soil 252:55–128

Burd G, Dixon DG, Glick BR (2000) Plant growth promoting bacteria that decrease heavy metal toxicity in plants. Can J Microbiol 46:237–245

Cesco S, Neumann G, Tomasi N, Pinton R, Weisskopf L (2010) Release of plant-borne flavonoids into the rhizosphere and their role in plant nutrition. Plant Soil 329:1–25

Crespi M, Gálvez S (2000) Molecular mechanisms in root nodule development. J Plant Growth Regul 19:155–166

Davison J (1988) Plant beneficial bacteria. Biotechnology 6:282–286

De Meyer SE, De Beuf K, Vekeman B, Willems A (2015) A large diversity of non-rhizobial endophytes found in legume root nodules in Flanders (Belgium). Soil Biol Biochem 83:1–11

Desbrosses GJ, Stougaard J (2011) Root nodulation: a paradigm for how plant–microbe symbiosis influences host developmental pathways. Cell Host Microbe 10:348–358

Dunn MF (2015) Key roles of microsymbiont amino acid metabolism in rhizobia–legume interactions. Crit Rev Microbiol 41:411–451

Durbak A, Yao H, McSteen P (2012) Hormone signaling in plant development. Curr Opin Plant Biol 15:92–96

Faucher C, Camut H, De'narié J, Truchet G (1989) The nodH and nodQ host range genes of *Rhizobium meliloti* behave as avirulence genes in *R. leguminosarum* bv. *viciae* and determine changes in the production of plant-specific extracellular signals. Mol Plant-Microbe Interact 2:291–300

Foo E (2017) Role of plant hormones and small signalling molecules in nodulation under P stress. In: Legume nitrogen fixation in soils with low phosphorus availability. Springer, Cham, pp 153–167

Gourion B, Berrabah F, Ratet P, Stacey G (2015) *Rhizobium*–legume symbioses: the crucial role of plant immunity. Trends Plant Sci 20:186–194

Griesmann M, Chang Y, Liu X, Song Y, Haberer G, Crook MB, Billault-Penneteau B et al (2018) Phylogenomics reveals multiple losses of nitrogen-fixing root nodule symbiosis. Science: eaat1743. https://doi.org/10.1126/science.aat1743

Guinel FC (2015) Ethylene, a hormone at the center-stage of nodulation. Front Plant Sci 6:1121

Guinel FC, Geil RD (2002) A model for the development of the rhizobial and arbuscular mycorrhizal symbioses in legumes and its use to understand the roles of ethylene in the establishment of these two symbioses. Can J Bot 80:695–720

Haney CH, Long SR (2010) Plant flotillins are required for infection by nitrogen-fixing bacteria. Proc Natl Acad Sci USA 107:478–483

Hirsch AM, Lum MR, Downie JA (2001) What makes the rhizobia–legume symbiosis so special? Plant Physiol 127:1484–1492

Irmer S, Podzun N, Langel D, Heidemann F, Kaltenegger E, Schemmerling B, Geilfus C-M, Zörb C, Ober D (2015) New aspect of plant–rhizobia interaction: alkaloid biosynthesis in *Crotolaria* depends on nodulation. PNAS 112:4164–4169

Jeffries P, Gianinazzi S, Perotto S, Turnau K, Barea JM (2003) The contribution of arbuscular mycorrhizal fungi in sustainable maintenance of plant health and soil fertility. Biol Fertil Soils 37:1–16

Jordan DC (1984) Bradyrhizobium. In: Bergey's manual of systematic bacteriology, vol 1. Williams & Wilkins, Baltimore, pp 242–244

Kawaharada Y, Kelly S, Wibroe Nielsen M, Hjuler CT, Gysel K, Muszyñki A et al (2015) Receptor-mediated exopolysaccharide perception controls bacterial infection. Nature (Lond) 523:308–312

Khalid A, Arshad M, Shaharoona B et al (2009) Plant growth promoting Rhizobacteria and sustainable agriculture. In: Microbial strategies for crop improvement. Springer, Berlin, pp 133–160

Kong Z, Deng Z, Glick BR, Wei G, Chou M (2017) A nodule endophytic plant growth-promoting *Pseudomonas* and its effects on growth, nodulation and metal uptake in *Medicago lupulina* under copper stress. Ann Microbiol 67:49–58

Lefebvre B, Timmers T, Mbengue M, Moreau S, Hervé C, Tóth K et al (2010) A remorin protein interacts with symbiotic receptors and regulates bacterial infection. Proc Natl Acad Sci USA 107:2343–2348

Levy A, Gonzalez IS, Mittelviefhaus M, Clingenpeel S, Paredes SH, Miao J, Alvarez BR (2018) Genomic features of bacterial adaptation to plants. Nat Genet 50:138

Lewin A, Rosenberg C, Meyer ZAH, Wong CH, Nelson L, Manen JF, Stanley J, Downing DN, De'narié J, Broughton WJ (1987) Multiple host-specificity loci of the broad host-range *Rhizobium* sp. NGR234 selected using the widely compatible legume *Vigna unguiculata*. Plant Mol Biol 8:447–459

Limpens E, van Zeijl A, Geurts R (2015) Lipochitoologosaccharides modulate plant host immunity to enable endosymbiosis. Annu Rev Phytopathol 53:15.1–15.24. https://doi.org/10.1146/annurev-phyto-080614-120149

Long SR (1984) Nodulation genetics. In: Kosuge T, Nester EW (eds) Plant–microbe interactions. Macmillan, New York, pp 265–306

Mahdi SS, Hassan GI, Samoon SA et al (2010) Bio-fertilizers in organic agriculture. J Phytol 2:42–54

Manoj KR, Kalia RK, Singh R, Gangola MP, Dhawan AK (2011) Developing stress tolerant plants through in vitro selection: an overview of the recent progress. Environ Exp Bot 71:89–98

Marhavý P, Vanstraelen M, De Rybel B, Zhaojun D, Bennett MJ, Beeckman T et al (2013) Auxin reflux between the endodermis and pericycle promotes lateral root initiation. EMBO J 32:149–158

McCully ME (2001) Niches for bacterial endophytes in crop plants: a plant biologist's review. Aust J Plant Physiol 28:983–990

Moling S, Bisseling T (2015) Evolution of *Rhizobium* nodulation: from nodule-specific genes (nodulins) to recruitment of common processes. In: Biological nitrogen fixation, vol 2. Wiley, Hoboken, NJ, p 39

Mortier V, Wasson A, Jaworek P, De Keyser A, Decroos M, Holsters M et al (2014) Role of LONELY GUY genes in indeterminate nodulation on *Medicago truncatula*. New Phytol 202:582–593

Munns R (2002) Comparative physiology of salt and water stress. Plant Cell Environ 25:239–250

Murray JD (2011) Invasion by invitation: rhizobial infection in legumes. Mol Plant-Microbe Interact 24:631–639

Murray JD, Karas BJ, Sato S, Tabata S, Amyot L, Szczyglowski K (2007) A cytokinin perception mutant colonized by *Rhizobium* in the absence of nodule organogenesis. Science 315:101–104

Nabizadeh E, Jalilnejad N, Armakani M (2011) Effect of salinity on growth and nitrogen fixation of alfalfa (*Medicago sativa*). World Appl Sci J 13:1895–1900

Nakagawa S, Mino S (2018) Deep-sea vent extremophiles: cultivation, physiological characteristics, and ecological significance. In: Extremophiles. CRC, New York, pp 165–184

Okazaki S et al (2013) Hijacking of leguminous nodulation signalling by the rhizobial type III secretion system. Proc Natl Acad Sci USA 110:17131–17136

Oldroyd GED (2013) Speak, friend, and enter: signaling systems that promote beneficial associations in plants. Nat Rev 11:252–263

Oldroyd GE, Murray JD, Poole PS, Downie JA (2011) The rules of engagement in the legume–rhizobial symbiosis. Annu Rev Genet 45:119–144

Peck MC, Fisher RF, Long SR (2006) Diverse flavonoids stimulate NodD1 binding to nod gene promoters in *Sinorhizobium meliloti*. J Bacteriol 188:5417–5427

Perret X, Staehelin C, Broughton WJ (2000) Molecular basis of symbiotic promiscuity. Microbiol Mol Biol Rev 64:180–201

Requena N, Perez-Solis E, Azcón-Aguilar C, Jeffries P, Barea JM (2001) Management of indigenous plant–microbe symbioses aids restoration of desertified ecosystems. Appl Environ Microbiol 67:495–498

Shiraishi A, Matsushita N, Hougetsu T (2010) Nodulation in black locust by the gamma proteobacteria *Pseudomonas* sp. and the beta proteobacteria *Burkholderia* sp. Syst Appl Microbiol 33:269–274

Singleton PW, Bohlool BB (1984) Effect of salinity on nodule formation by soybean. Plant Physiol 74:72–76

Smit G, Swart S, Lugtenberg BJ, Kijne JW (1992) Molecular mechanisms of attachment of *Rhizobium* bacteria to plant roots. Mol Microbiol 6:2897–2903

Spaink HP, Weinman J, Djordjevic MA, Wijfelman CA, Okker JH, Lugtenberg BJJ (1989) Genetic analysis and cellular localization of the *Rhizobium* host specificity-determining NodE protein. EMBO J 8:2811–2818

Suzaki T, Ito M, Kawaguchi M (2013) Genetic basis of cytokinin and auxin functions during root nodule development. Front Plant Sci 4:42

Trinick MJ (1979) Structure of nitrogen-fixing nodules formed by *Rhizobium* on roots of *Parasponia andersonii* Planch. Can J Microbiol 25:565–578

van Rhijn P, Vanderleyden J (1995) The *Rhizobium*–plant symbiosis. Microbiol Rev 59:124–142
Vance CP (2001) Symbiotic nitrogen fixation and phosphorus acquisition. Plant nutrition in a world of declining renewable resources. Plant Physiol 127:390–397
Young JPW, Johnston AWB (1989) The evolution of specificity in the legume–*Rhizobium* symbiosis. Trends Ecol Evol 4:331–349
Young JPW, Downer HL, Eardly BD (1991) Phylogeny of the phototrophic *Rhizobium* strain Btail by polymerase chain reaction-based sequencing of the 16S rRNA gene segment. J Bacteriol 173:2271–2277
Zahran HH (1999) *Rhizobium*–legume symbiosis and nitrogen fixation under severe conditions and in an arid climate. Microbiol Mol Biol Rev 63(4):968–989
Zahran HH (2001) Rhizobia from wild legumes: diversity, taxonomy, ecology, nitrogen fixation and biotechnology. J Biotechnol 91(2-3):143–153
Zaidi S, Usmani S, Singh BR, Musarrat J (2008) Significance of *Bacillus subtilis* strains SJ-101 as a bioinoculant for concurrent plant growth promotion and nickel accumulation in *Brassica juncea*. Chemosphere 64:991–997
Zhuang XL, Chen J, Shim H, Bai Z (2007) New advances in plant growth-promoting rhizobacteria for bioremediation. Environ Int 33:406–413

Chapter 2
Diversity and Importance of the Relationship Between Arbuscular Mycorrhizal Fungi and Nitrogen-Fixing Bacteria in Tropical Agroforestry Systems in Mexico

Iván Oros-Ortega, Luis Alberto Lara-Pérez, Fernando Casanova-Lugo, Víctor Francisco Díaz-Echeverría, Gilberto Villanueva-López, Pablo J. Ramírez-Barajas, and William Cetzal-Ix

Abstract In Mexico, extensive production systems have caused a drastic reduction in tropical forests and in biological diversity. Most of the agroforestry systems (AFS) in Mexico use leguminous species that naturally associate with arbuscular mycorrhiza fungi (AMF) and bacterial nitrogen fixing that aid the uptake of N and P in poor soils of the tropics. The AMF and bacteria are predominant in tropical agroecosystems with wide ranges of hosts with potential to increase growth in forest species and in crop yield. Mexico is considered one of the countries with high diversity of plants within the countries of America with potentially high number of AMF species and bacteria in different SAF. Although we have considerable knowledge of the plants used in different AFS, the richness of soil microorganisms has received little attention in Mexico's tropics. Understanding of the structure and functional diversity of AMF and bacteria have allowed us to generate the bases for a sustainable AFS, increasing productivity and, at the same time, AFS work as reservoirs and biological corridors that could reduce degradation of forests.

I. Oros-Ortega (✉) · L. A. Lara-Pérez · F. Casanova-Lugo · V. F. Díaz-Echeverría · P. J. Ramírez-Barajas
Tecnológico Nacional de México, Instituto Tecnológico de la Zona Maya, Ejido Juan Sarabia, Quintana Roo, Mexico

G. Villanueva-López
El Colegio de la Frontera Sur, Villahermosa, Tabasco, Mexico

W. Cetzal-Ix
Tecnológico Nacional de México, Instituto Tecnológico de Chiná, Campeche, Mexico

A. Varma et al. (eds.), *Plant Microbe Symbiosis*,
https://doi.org/10.1007/978-3-030-36248-5_2

2.1 Introduction

On a global level, only 23% of the ecosystems can be considered intact while the remainder has been modified by human beings (Watson et al. 2016) due to the ever-increasing demand for food. Tilman et al. (2001) and Hansen et al. (2013) point that in little more than a decade, 2.3 million km^2 of forests would have been lost as a result of the increase in extensive agricultural production. These extensive systems of food production give rise to negative impacts on the environment, low levels of efficiency and rentability caused by the excessive use of fertilizers and pesticides, resulting in a significant impact on the biodiversity, alteration of the geochemical and hydrological cycles and the introduction of exotic species or plagues (Velásquez et al. 2002).

One strategy that can contribute to the implementation of sustainable models of food production is the diversification of crops through the establishment of agroforestry systems (Astier et al. 2017). The functionality of these AFS can be enhanced by means of a better understanding of the ecological interactions, taking into consideration that plant species establish a symbiotic association with soil microorganisms, such as arbuscular mycorrhizal fungi (AMF) and nitrogen-fixing bacteria which form highly functional structures in plant nutrition (Smith and Read 2008). These symbiotic organisms play an important role in soil structure, nutrient recycling, biological fixation of nitrogen and the transport of nutrients of difficult access for the plants (P, Zn, Cu, B, S), protection of the plant from pathogens and stress conditions (Smith and Read 2008; Goltapeh et al. 2008).

In numerous countries, there has been a significant advance in the understanding of the biodiversity and function of the microorganisms and plant species associated with AFS, employing fungi, bacteria and ecological principles to increase productivity. However, in the tropics of Mexico, information regarding the microorganisms and plants in AFS is limited if we consider the significantly larger biological, geographical, climatic, cultural and agricultural diversities in the area. With the intention of recovering degraded sites, while avoiding deforestation, in order to increase agricultural activity with sustainable practices, it is fundamental to have an understanding of the diversity of AMF and nitrogen-fixing bacteria and their potential effect on the development of leguminous plants and other tree species or shrubs with agroforestry potential.

2.2 Agroforestry Systems in the Mexican Tropics

On a global level, Mexico has an extraordinary biological richness, at a genetic level and in the variety of species and agroecosystems. It has been estimated that in any given group of 10 species existing in the world, one can be found in Mexico. For this reason, Mexico belongs to the group of the 12 most mega-diverse countries on the planet (CONABIO 2008; Sarukhán et al. 2009; Villanueva-López et al. 2019).

Together with Brazil, Colombia and Indonesia, Mexico is among the countries with the greatest richness of species, which includes between 60 and 70% of the diversity as well as a large number of endemic species (Rzedowski 1991; CONABIO 2008). In addition, Mexico also maintains high levels of diversity in the hyper-diverse taxonomic groups of microorganisms and arthropods (Llorente-Bousquets and Ocegueda 2008). This richness, among other elements, is due to several factors such as the geographical position occupied by Mexico, between two bio-geographical regions, (Nearctic and Neotropical), the diversity of environments, its rugged topography resulting from its geological history and the climatic variability manifested in the different regions (CONABIO 2008; Sarukhán et al. 2009; Rodríguez-Estrella et al. 2016). Mexico is also an important center of domestication and diversification of cultivated species, some of which are of global importance. More than 15% of the species consumed as food in the world originate in Mexico. By maintaining wild parents or ancestors, these crops can potentiate genetic diversity and thus improve food security (Sarukhán et al. 2009). This sophisticated and prolonged process of domestication and diversification has been possible thanks to the simultaneous development of the crops with their extensive biodiversity.

One worldwide strategy for the conservation of natural resources, biodiversity, genetic resources and ecological and evaluative processes has been through the determination of natural wild areas. In Mexico, these areas are identified as protected areas. However, given the size of their areas and interconnection (barely 13% of the national territory), they are still too small and isolated to safeguard stable, wild populations, ecosystems and the necessary processes for life and productivity on the planet (Torres-Orozco et al. 2015; Ferreira et al. 2018). In this sense, the AFS can serve as a bridge that connects, both functionally and structurally, areas destined for conservation, keeping in mind the integrity of the microorganism-soil-plant system and the sustainable agricultural productivity of large regions and rural sectors. Thus, there is a growing need to understand how to achieve greater yields in the agricultural production with fewer impacts, which requires quantitative evaluations of how the different production practices and the environmental variables affect yields (Tilman et al. 2001).

The trees used in the AFS belong to a wide diversity of native, functional groups, pioneer species and species of the original woodland flora (González-Valdivia et al. 2016), which can potentially host a good representativeness of the richness of species in the natural woodlands. In Mesoamerica, Mexico occupies the first place in richness of trees used in the AFS with dominance of the families: Fabaceae, Bignoniaceae, Malvaceae, Moraceae, Rubiaceae and Rutaceae (González-Valdivia et al. 2016). In the tropics of Mexico, multiple conformations of AFS have been identified with different degrees of complexity. One of the most studied AFS is probably that of coffee production, which is also one of the more complex systems, with respect to plant structure (Arias et al. 2012; Bertolini et al. 2018). For example, in the Peninsula of Yucatan a large diversity of tree species has been identified, among which the family Fabaceae is particularly noteworthy, given that it has approximately 225 species (Carnevali et al. 2010), which have been registered as alternatives for the transformation and improvement of agricultural systems. Among these species can be found *Leucaena leucocephala* (Lam.) de Wit, which has been studied under different agroforestry arrangements such as cultivation in alleys, fodder banks and improved fallow land. In fact, recently in Mexico, approximately 10,000 hectares of AFS, in the modality of silvopastoral systems, have been established in the states of Michoacan, Campeche, San Luis Potosi, Veracruz, Tamaulipas, Chiapas, Nayarit, Quintana Roo and Yucatan, among others, corresponding to the tropical belt of Mexico (Broom et al. 2013). These systems function under a rotational grazing system with the use of electric fencing in pastures cultivated with legumes associated with diverse tropical pasture such as *Panicum maximum* cv. Tanzania, *Cynodon plectostachyus* (K. Schum.) Pilg, and others grasses (López-Santiago et al. 2018).

In the humid and subhumid tropics of Mexico, the main tree species or bushy plant types preferred by the producers, due to their multiple uses as fodder, timber, fruit trees, honeybees, fuel or firewood, are the following: *Cedrela odorata* L., *Swietenia macrophylla* King, *Moringa oleifera* Lam. *Guazuma ulmifolia* Lam., *Piscidia piscipula* (L.) Sarg., *Gliricidia sepium* (Jacq.) Kunth ex Walp., *Lysiloma latisiliquum* (L.) Benth., *L. leucocephala* Wet. *Cocos nucifera* L., *Theobroma cacao* L., *Cordia alliodora* (Ruiz & Pav.) and *Tithonia diversifolia* (Hemsl.) A. Gray. (Fig. 2.1) (Villanueva-Partida et al. 2016, 2019).

2.3 Functionality of Soil Microorganisms in AFS

Soil microorganisms are one of the elements that contribute to the maintenance of plant biodiversity and to the functioning of ecosystems. Most of the trees and shrubs in the AFS present mixed association with nitrogen fixing bacteria and AMF, which facilitates the uptake of phosphorous and nitrogen, respectively. The family Fabaceae is predominant due to the large contribution of organic material and enrichment of the soils from the presence of nitrogen fixing bacteria (Barea et al. 2005). In order to increase the potential of these microorganisms, an adequate selection of the symbionts (fungi and plants) must be carried out. In addition, the tree species must be selected according to their functional compatibility, given that the species of AMF and bacteria respond in a differential manner within their host, and many plants are more susceptible than others to the mycorrhization and/or nodulation (van der Heijden et al. 1998; Smith and Read 2008; Sridhar and Bagyaraj 2017). The presence of bacteria and AMF in only one plant provides multifunctional benefits for the AFS, as it can result in a synergism that benefits nitrogen fixing. This is due to the fact that the nodulation requires high levels of P, which the mycorrhiza could translocate on arrival beyond the depletion zone of this element (Atangana et al. 2014).

In general, plant species present a category according to the dependence to associate with fungi in mycorrhizae: highly dependent, medium dependency or facultative and not mycorrhizal (Brundrett 2009; Atangana et al. 2014). The hyphae of the AMF function as an extension of the roots, which facilitates the translocation

Fig. 2.1 Schematic representation of some agroforestry systems characteristic to the Mexican tropics. (**a**) Grazing in *Cedrela odorata* plantations. (**b**) Grazing in *Cocos nucifera* plantations. (**c**) Living fence of *Gliricidia sepium*. (**d**) Plantations of *Theobroma cacao* in the shade. (**e**) Plantations of *Cordia alliodora* associated with banana. (**f**) Pastures in alleys of *Leucaena leucocephala*. (**g**) Fodder banks of *Tithonia diversifolia*. (**h**) Cultivation of corn in alleys of *L. leucocephala*. Photographs: G. Villanueva-López and F. Casanova-Lugo

of elements with poor mobility, such as phosphorous (P) and, in exchange, the fungus receives carbohydrates and lipids from the plant. In addition to these benefits, the mycorrhiza increases tolerance to saline and water stress, protects the root from pathogens, phytotoxic elements and heavy metals and also improves the structure while increasing the carbon input (C) to the soil (Smith and Read 2008). The mycorrhiza helps in the establishment and survival of plants in the field and has the potential to increase production. They are also essential for restoration and sustainable agricultural production. The majority of the trees and shrubs in the AFS are considered facultative and can include the presence of ectomycorrhiza (Brundrett 2009; Atangana et al. 2014). It has been considered that with a greater stratification of the AFS, the competition for light and nutrients is accentuated, which could reduce crop yield. However, the low selectivity of the AFS permits a food wide web of common hyphae that form a continuum in the roots with different plant species (Giovannetti et al. 2006). These networks of common hyphae would potentially allow interplant mobilization of carbon, water and nutrients as well as the suppression of non-mycorrhizal weeds (Cameron 2010). Although it has been demonstrated that the presence of trees or shrub species and the planting distance in the AFS considerably increases the harvest yields (Balakrishna et al. 2017), it is necessary to determine the influence and synergies of the AMF or nitrogen-fixing bacteria in the transfer of nutrients and carbon in the crops.

2.4 Soil Microorganisms in AFS in the Tropics of Mexico

2.4.1 Arbuscular Mycorrhizal Fungi

Within the tropical AFS, the AMF are the most important group due to their wide distribution, diversity and their capacity to improve the fitness of the plants (Cardoso and Kuyper 2006; Marinho et al. 2018; Prasad et al. 2017). The AMF belong to the subphylum of the Glomeromycota, which has approximately 316 species described at a global level (http://www.amf-phylogeny.com/). Fossil and molecular evidences suggest that this group of fungi has co-evolved with the plants for approximately 400 million years (Strullu-Derrien et al. 2014). Today, the AMF are widely distributed in most ecosystems and form mutualistic associations with bryophytes, ferns, lycopodium, gymnosperms and angiosperms (Wang and Qiu 2006; Brundrett 2009; Prasad et al. 2017; Varma et al. 2017). In Mexico, there are approximately 105 species of AMF registered, representing 36.3% of world richness (Alarcón et al. 2012). The main contributions to the taxonomic knowledge of the AMF derive from the herbaceous and fruit-producing plants in extensive agricultural systems; however, very few studies have focused on the agroforestry systems with only 21 species registered (Montaño et al. 2012).

In Mexico, the mycorrhiza were studied in different plant species; however, it has not been examined in the context of the AFS where a greater effort is required due to the richness, heterogeneity and spatiality of the trees in these systems (González-Valdivia et al. 2016; Villanueva-López et al. 2019; Villanueva-Partida et al. 2019). The diversity of AMF in the AFS will depend greatly on the diversity of the species integrated. The richness and abundance of plant species used in the AFS contain a greater richness of microorganisms in comparison with monoculture and a high representativeness of the conserved woodlands (Villanueva-López et al. 2019). Due to this influence, the AFS can be considered as reservoirs of AMF diversity, while allowing the connectivity of the landscape and habitat fragments that function as biological corridors of potential animal vectors of AMF, coming from adjacent, conserved environments (Janos 1996; Mangan and Adler 2000; Jose 2009). Given the traditional AFS can be simple or complex in their plant structures, comparative studies are required in order to determine the richness of organisms above and below the soil in these models of sustainable production. In the case of the tropics of Mexico, the most representative agroforestry system is the agrosilvopastoral, which includes mainly legumes, such as *G. ulmifolia*, *P. piscipula*, *G. sepium* and *L. latisiliquum* as well as forest trees such as *C. odorata* as living fences.

In this sense, we have addressed the study of the diversity of microorganisms associated with the legume and other tree species or tropical shrubs with agroforestry potential (unpublished data). We evaluate in study sites of the subhumid tropics the diversity of AMF spores in *L. leucocephala*, *L. latisiliquum*, *Cocos nucifera* and *Tithonia diversifolia* under natural and in silvopastoral systems (Fig. 2.2) as well as the level of colonization from *L. leucocephala* associated with *Cynodon plectostachyus* and *L. leucocephala* associated with *Panicum maximum*, adults and seedlings of *G. ulmifolia*, seedlings of *Moringa oleifera* at field capacity and at permanent wilting point (PWP), seedlings of *Cedrela odorata* at field capacity and at PWP, adults of *Swietenia macrophylla* and *Cocos nucifera*. (Fig. 2.3). It has been registered that the total colonization in roots by vesicles and arbuscules, at least on individuals of *L. leucocephala*, collected in natural sites, does not show significant differences in comparison with individuals of agroforestry systems. In seedlings of *Leucaena, Mahogany* and *C. odorata* in greenhouse cultivation inoculated with native AMF, we have observed high levels of total colonization, while colonization by AMF and dark septate fungi (DSF) has been found specifically in mahogany roots. The principal species of AMF with potential to integrate in agroforestry systems are *Acaulospora* spp., *Clareidoglomus* spp., *Diversispora* spp., *Glomus* spp., *Gigaspora* spp., *Rhizophagus* spp. and *Scutellospora* spp. (Fig. 2.2).

2.4.2 Nitrogen-Fixing Bacteria

The microbiota is one of the most important components in the maintenance of soil fertility (Sridhar and Bagyaraj 2017). In the AFS, the diversity of microorganisms includes the symbiotic nitrogen-fixing bacteria (Cardoso and Kuyper 2006; Atangana et al. 2014; Sridhar and Bagyaraj 2017). The nitrogen-fixing bacteria comprise three groups: (1) alpha and Betaproteobacteria (nodulated plants), (2) Actinobacteria (Frankiaceae) and (3) Cyanobacteria (nodulated plants) (Sridhar and Bagyaraj 2017) and are associated mainly with legume plants (Table 2.1). Each

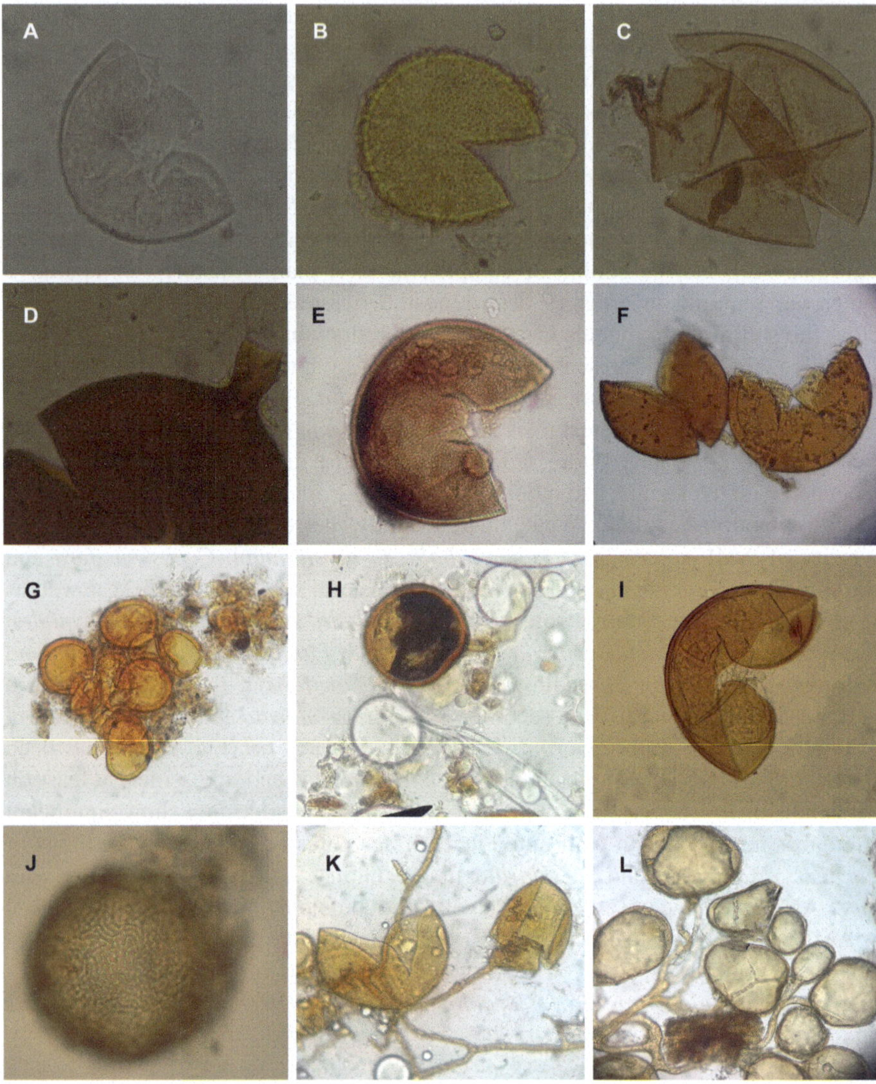

Fig. 2.2 AMF diversity in tree and shrub tropical species with agroforestry potential in the Mexican tropics. (**a–e**) *Leucaena leucocephala* + Grass – estrella + Grass – bombaza. (**f–h**) *Lysiloma latisiliquum*. (**i, j**) *Cocos nucifera*. (**k, l**) *Tithonia diversifolia*. (**a**) *Claroideoglomus*. (**b**) *Acaulospora*. (**e**) *Acaulospora scrobiculata*. (**g**) *Glomus* (**i**) *Gigaspora*. (**j**) *Acaulospora remhii*. (**k**) *Rhizophagus intraradices*. (**l**) *Rhizophagus irregularis*. Photographs: L. Lara Pérez, A. Salbador, L.F. Estrada, E. Mundo and J. L. Moen

year, the legumes of agricultural importance fix approximately 40–60 million metric tons of nitrogen while the legumes in natural ecosystems fix 3–5 million metric tons. Consequently, nitrogen fixing is a highly efficient process which requires only a tiny

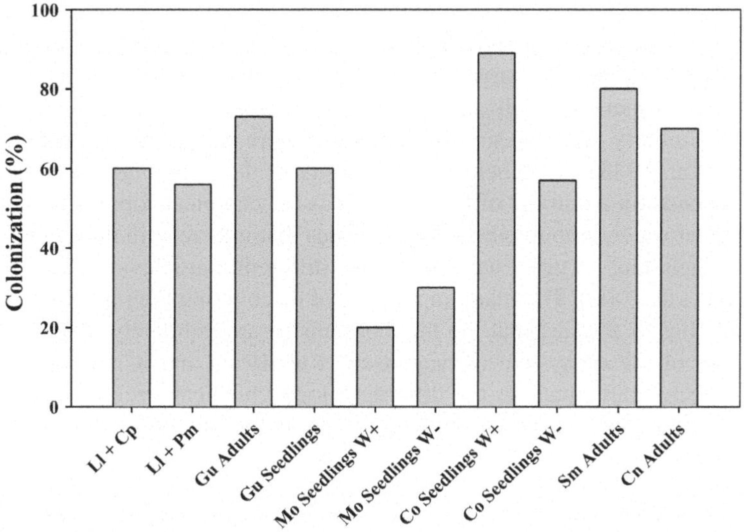

Fig. 2.3 Percentage of mycorrhizal colonization in roots of tropical tree/bush species under different growth conditions. Abbreviations: *Ll + Cp Leucaena leucocephala* associated with *Cynodon plectostachyus*, *Ll + Pm L. leucocephala* associated with *Panicum maximum*, *Gu Guazuma ulmifolia*, *Mo Moringa oleifera*, *W+* at field capacity, *W–* near at permanent wilting point, *Co Cedrela odorata*, *Sm Swietenia macrophylla*, *Cn Cocos nucifera*

Table 2.1 Principal characteristics of legume subfamilies (Sprent 2001)

	Caesalpinioideae	Mimosoideae	Papilionoideae
Number of genera (approximate total)	157	78	479
Confirmed as nodulant	8	42	297
Distribution	Mainly in the humid tropics	In tropical/subtropical areas, often in dry areas	From the tropics to the arctic, from dry areas to flood-prone areas
Growth habit	Mainly trees	Mainly trees and bushes	Trees, bushes and herbaceous
Nodulation	Rare, nodule structure is usually primitive	Common but with important exceptions	Very frequent with a few exceptions

amount of nitrogenase (an enzyme used by bacteria for the fixing of nitrogen) in order to carry out the process.

Phenotypes and the genetic diversity of various groups of fast-growing rhizobia have been described as associated with tree species such as *Acacia senegal* and *Prosopis chilensis* in some countries of Africa, where genetic similarity was found between the isolated strains of the two continents and strains belonging to the genera *Sinorhizobium* and *Azorhizobium*. Bryan (2000) mentions the existence of a very close relationship among the legumes and the dispersion of populations of rhizobia of tropical soils in symbiosis and free life (depending on the organic matter and the concentration of nitrogen in soil). This was demonstrated in studies carried out on the island of Maui where native strains of *Rhizobium* and *Bradyrhizobium* were isolated and associated with *L. leucocephala, Sesbania rostrata and Acacia mangium.* Similarly, other species of *Rhizobium* were isolated from shrub legumes (Woomer et al. 1988). Nonetheless, the diversity of the microorganisms present in the soil depends on a number of factors, such as the chemical composition, texture, availability of water, among other characteristics that intervene in the availability of nutrients, including oxygen tension in the different particles forming the soil (Mahmood et al. 2006). The main importance of discovering the microbial diversity in the soils lies in the fact that the microorganisms are intimately related with the productivity of the ecosystem (Zvyagintsev et al. 1991) and it is known that the microorganisms participate in the different biogeochemical cycles in the soil, or interact with the plants, promoting plant growth and increasing the productivity of the ecosystem.

Despite the great diversity of plants integrated in the AFS, studies of the Mexican tropics have focused only on a few species. For example, Roskoski (1986), determined nitrogen fixing in *G. sepium, L. leucocephala* and *Acacia pennatula* in association with *Rhizobium.* Similarly, by means of molecular biology, it was possible to identify the presence of the species *Sinorhizobium terangae*, *R. ettli* and two types of *R. tropici* in *G. sepium.* Another study carried out in Yucatan, on secondary vegetation, reported the species of *R. legominusarum*, *R. tropici*, *Allirhizobium spp.* and *Mesorhizobium* spp. associated with *L. leucocephala*, *G. sepium* and *Calliandra calothyrsus* (Bala et al. 2003). Other species of plants associated with AFS which have been addressed are *A. farnesina* and *A. tamentac*ea where the bacteria of the *Rhizobium* group have been identified (Martínez-Scott et al. 2002).

Nowadays, one of the main challenges is to measure and identify the diversity and composition of the communities of symbiotic bacteria in the AFS of the Mexican tropics, for example, the manner in which this was carried out in Mozambique, where they identified the symbiont bacteria in *Acacia xanthophloea* Benth., *Albizia versicolor* Welw. Ex Oliv. and *Faidherbia albida* (Delile) by sequencing the genes 16S rRNA, *glnII* and *recA* and found mainly in the classes *Bradyrhizobium*, *Mesorhizobium*, *Rhizobium* and species of *Ensifer.* In general, studies relating to symbiotic nitrogen-fixing bacteria are quite scarce and have focused on understanding the diversity and observing the effect of the inoculation on productivity in species of interest.

2.5 Conclusion and Future Prospects

Despite the fact that the AMF and the nitrogen-fixing bacteria are a functional, ecological group of importance for plant diversity, in the processes of establishment, productivity and forestry dynamics in Mexico, very few studies have been conducted on mycorrhizal status, diversity and distribution of these microorganisms. The knowledge of these interactions is of great value for the integration of new native species to the AFS. Unfortunately, there are areas and AFS where a deeper knowledge of the diversity is required as well as a coordination and systematization of the AMF and bacteria. Many of the studies of AMF diversity only identify the species at a genus level, perhaps due to the presence of species which have not yet been described. The integration of trap culture, and molecular studies could help in the correct identification of bacteria and AMF species that colonize the plants used in the AFS. The latest sequencing techniques offer a general panorama of the structure and diversity of functional groups of microorganisms and allow us to compare different scenarios and plant assemblage used in the AFS. With the correct identification of the symbionts and the establishment of pure cultures or in consortium, these are the basis for conducting physiological experiments in controlled environments and follow up in the field with the molecular identification of species. The compilation of this information would not only help us to understand the functional role of the microorganisms in the AFS but would also complement the strategies in agricultural production and the restoration of perturbed ecosystems. The results of this study of the AFS and the importance of their microbiological interactions can be directly incorporated to the strategies and public policies dealing with the interconnection of ecosystems, ecological integrity, food self-sufficiency and productivity. The mitigation of climate change must also be addressed in regions with high biological diversity and latent anthropogenic threats (hot spots), i.e. regions lacking alternatives to resolve the current problems.

References

Alarcón A, Hernández-Cuevas LV, Ferrera-Cerrato R, Franco-Ramírez A (2012) Diversity and agricultural applications of arbuscular mycorrhizal fungi in Mexico. J Biofertil Biopestici 3:115

Arias RM, Heredia G, Sosa VJ, Fuentes-Ramírez LE (2012) Diversity and abundance of arbuscular mycorrhizal fungi spores under different coffee production systems and in a tropical Montana cloud forest patch in Veracruz, Mexico. Agrofor Syst 85:179–193

Astier M, Argueta JQ, Orozco-Ramírez Q, González MV, Morales J, Gerritsen PR, Sánchez-Sánchez C (2017) Back to the roots: understanding current agroecological movement, science, and practice in Mexico. Agroecol Sust Food 41:329–348

Atangana A, Khasa D, Chang S, Degrande A (2014) Tropical agroforestry. Springer Science and Business Media, Dordrecht

Bala A, Ken PM, Giller E (2003) Distribution and diversity of rhizobia nodulating agroforestry legumes in soils from three continents in the tropics. Mol Ecol 12:917–930

Balakrishna AN, Lakshmipathy R, Bagyaraj DJ, Ashwin R (2017) Influence of alley copping system on AM fungi, microbial biomass C and yield of finger millet, peanut and pigeon pea. Agrofor Syst 91:487–493

Barea JM, Werner D, Azcon-Aguilar C, Azcon R (2005) Interactions of arbuscular mycorrhiza and nitrogen fixing simbiosis in sustainable agriculture. In: Werner D, Newton WE (eds) Agriculture, forestry, ecology and the environment. Kluwer, The Netherlands

Bertolini V, Montaño NM, Chimal-Sánchez E, Varela-Fregoso L, Gómez-Ruiz J, Martínez Vázquez JM (2018) Abundance and richness of arbuscular mycorrhizal fungi in coffee plantations from Soconusco, Chiapas. Mexico Revista de Biología Tropical 66:91–105

Broom DM, Galindo FA, Murgueitio E (2013) Sustainable, efficient livestock production with high biodiversity and good welfare for animals. Proc Biol Sci 280(1771):20132025. https://doi.org/10.1098/rspb.2013.2025

Brundrett MC (2009) Mycorrhizal associations and other means of nutrition of vascular plants: understanding the global diversity of host plants by resolving conflicting information and developing reliable means of diagnosis. Plant Soil 320:37–77

Bryan JA (2000) Nitrogen-fixing trees and shrubs: a basic resource of agroforestry. In: Ashton MS, Montagnini F (eds) The Silvicultural basis for agroforestry systems. CRC, Baton Rouge, LA, pp 41–60

Cameron DD (2010) Arbuscular mycorrhizal fungi as (agro) ecosystem engineers. Plant Soil 333:1–5

Cardoso IM, Kuyper TW (2006) Mycorrhizas and tropical soil fertility. Agric Ecosyst Environ 116:72–84

Carnevali G, Tapia-Muñoz JL, Duno de Stefano R, Ramírez-Morillo I (2010) Flora Ilustrada de la Península de Yucatán: Listado Florístico. Centro de Investigación Científica de Yucatán A.C., Mérida Yucatán, México. 328 p

CONABIO (2008) Capital Natural de México Vol I. Conocimiento actual de la Biodiversidad. Comisión Nacional para el conocimiento y Uso de la Biodiversidad, México

Ferreira AS, Peres CA, Bogoni JA, Cassano CR (2018) Use of agroecosystem matrix habitats by mammalian carnivores (Carnivora): a global-scale analysis. Mammal Rev 48:312–327

Giovannetti M, Avio L, Fortuna P, Pellegrino E, Sbrana C, Strani P (2006) At the root of the wood wide web: self recognition and nonself incompatibility in mycorrhizal networks. Plant Signal Behav 1:1–5

Goltapeh EM, Danesh YR, Prasad R, Varma A (2008) Mycorrhizal fungi: what we know and what should we know? In: Varma A (ed) Mycorrhiza, 3rd edn. Springer-Verlag, Berlin, pp 3–27

González-Valdivia NA, Casanova-Lugo F, Cetzal-Ix W (2016) Sistemas agroforestales y biodiversidad. Agroproductividad 9:56–60

Hansen MC, Potapov PV, Moore R, Hancher M, Turubanova SA, Tyukavina A et al (2013) High-resolution global maps of 21st-century forest cover change. Science 342:850–853

Janos DP (1996) Mycorrhizas, succession, and the rehabilitation of deforested lands in the humid tropics. In: Fungi and environmental change. Cambridge University Press, for British Mycological Society, Cambridge

Jose S (2009) Agroforestry for ecosystem services and environmental benefits: an overview. Agrofor Syst 76:1–10

Llorente-Bousquets J, Ocegueda S (2008) Estado del conocimiento de la biota. In: Conabio, Capital natural de México, vol. I: Conocimiento actual de la biodiversidad. Conabio, México, pp 283–322

López-Santiago JG, Casanova-Lugo F, Villanueva G, Díaz-Echeverría VF, Solorio-Sánchez FJ, Martínez-Zurimendi P et al (2018) Carbon storage in a silvopastoral system compared to that in a deciduous dry forest in Michoacán, Mexico. Agroforest Syst 93(1):199–211. https://doi.org/10.1007/s10457-018-0259-x

Mahmood KW, Yang N, Kidhwar Z, Rajputy A, Arijo A (2006) Study of cellulolytic soil fungi and two nova species and new medium. J Shejiang Uni Sci 7:459–466

Mangan SA, Adler GH (2000) Consumption of arbuscular mycorrhizal fungi by terrestrial and arboreal small mammals in a Panamanian cloud forest. J Mammal 81:563–570

Marinho F, da Silva IR, Oehl F, Maia LC (2018) Checklist of arbuscular mycorrhizal fungi in tropical forests. Sydowia 70:107

Martínez-Scott MM, Hernández-Hernández V, Palomo-Gil A, Vásquez-Arroyo J (2002) Diversidad genética de rhizobia asociada a cuatro leguminosas arbóreas del noreste de México. Rev Chapingo serie Zonas Aridas 3:9–18

Montaño NM, Alarcón A, Camargo-Ricalde SL, Hernández-Cuevas LV, Álvarez-Sánchez J, González-Chávez MC et al (2012) Research on arbuscular mycorrhizae in Mexico: an historical synthesis and future prospects. Symbiosis 57:111–126

Prasad R, Bhola D, Akdi K, Cruz C, Sairam KVSS, Tuteja N, Varma A (2017) Introduction to mycorrhiza: historical development. In: Varma A, Prasad R, Tuteja N (eds) Mycorrhiza. Springer, Cham, pp 1–7

Rodríguez-Estrella RB-M, del Val de Gortari E, Santos-Barrera G (2016) Impacto de las actividades humanas en la biodiversidad y en los ecosistemas. In: Balvanera P, Arias E, Rodríguez-Estrella R, Almeida L, Schmitter JJ (eds) Una mirada al conocimiento de los ecosistemas de México. Conacyt/UNAM, México D.F

Roskoski JP, Pepper I, Pardo E (1986) Inoculation of leguminous trees with rhizobia and VA mycorrhizal fungi. For Ecol Manag 16:57–68

Rzedowski J (1991) Diversidad y orígenes de la flora fanerogámica de México. Acta Bot Mex 14:3–21

Sarukhán J et al (2009) Capital natural de México. Síntesis: conocimiento actual, evaluación y perspectivas de sustentabilidad. Comisión Nacional para el Conocimiento y Uso de la Biodiversidad, México

Smith SE, Read DJ (2008) Mycorrhizal symbiosis, 3rd edn. Academic, New York, p 605

Sprent JI (2001) Nodulation in legumes. Royal Botanical Gardens, Kew, p 146

Sridhar KR, Bagyaraj DJ (2017) Microbial biodiversity in agroforestry systems. In: Dagar J, Tewari V (eds) Agroforestry. Springer, Singapore

Strullu-Derrien C, Kenrick P, Pressel S, Duckett JG, Rioult JP, Strullu DG (2014) Fungal associations in Horneophyton ligneri from the Rhynie Chert (c. 407 million year old) closely resemble those in extant lower land plants: novel insights into ancestral plant–fungus symbioses. New Phytol 203:964–979

Tilman D, Fargione J, Wolff B, D'Antonio C, Dobson A, Howarth R et al (2001) Forecasting agriculturally driven global environmental change. Science 292:281–284

Torres-Orozco D, Jiménez-Sierra CL, Sosa RJ, Cortés-Calva P, Solís-Cámara AB, Iñiguez-Dávalos LI, Ortega-Rubio A (2015) La Importancia de las Áreas Naturales Protegidas en Nuestro País. In: AM O–R, Pinkus-Rendón J, Espitia-Moreno IC (eds) Las Áreas Naturales Protegidas y la Investigación Científica en México. Centro de Investigaciones Biológicas del Noroeste S. C., La Paz B. C. S., Universidad Autónoma de Yucatán, Mérida, Yucatán y Universidad Michoacana de San Nicolás de Hidalgo, Morelia, Michoacán, México, p 572

Van der Heijden MG, Klironomos JN, Ursic M, Moutoglis P, Streitwolf-Engel R, Boller T et al (1998) Mycorrhizal fungal diversity determines plant biodiversity, ecosystem variability and productivity. Nature 396:69

Varma A, Prasad R, Tuteja N (2017) Mycorrhiza: function, diversity and state-of-art. Springer, Cham. ISBN 978-3-319-53064-2. http://www.springer.com/us/book/9783319530635

Velásquez A, Mas JF, Díaz-Gallegos JR, Mayorga R, Alcántara C, Castro R, Fernández T, Bovvo G, Palacio JL (2002) Patrones y tasas de cambio de uso del suelo en México. Gaceta Ecol 62:21–37

Villanueva-López G, Lara-Pérez LA, Oros-Ortega I, Ramírez-Barajas PJ, Casanova-Lugo F, Ramos-Reyes R, Aryal DR (2019) Diversity of soil macro-arthropods correlates to the richness of plant species in traditional agroforestry systems in the humid tropics of Mexico. Agric Ecosyst Environ 286:106658

Villanueva-Partida CR, Casanova-Lugo F, Villanueva-López G, González-Valdivia N, Oros-Ortega I, Díaz-Echeverría V (2016) Influence of the density of scattered trees in pastures on the structure and species composition of tree and grass cover in southern Tabasco, Mexico. Agric Ecosyst Environ 232:1–8

Villanueva-Partida CR, Casanova-Lugo F, González-Valdivia NA, Villanueva-López G, Oros-Ortega I, Cetzal-Ix W, Basu SK (2019) Traditional uses of dispersed trees in the pastures of the mountainous region of Tabasco, Mexico. Agroforest Syst 93(2):383–394

Wang B, Qiu YL (2006) Phylogenetic distribution and evolution of mycorrhizas in land plants. Mycorrhiza 16:299–363

Watson JEM, Shanahan DF, Di Marco M, Allan J, Laurance WF, Sanderson EW, Mackey B, Venter O (2016) Catastrophic declines in wilderness areas undermine global environment targets. Curr Biol 26:2929–2934

Woomer P, Singleton P, Bohlool B (1988) Ecological indicators of native rhizobia in tropical soils. Appl Environ Microbiol 54:1112–1116

Zvyagintsev D, Kurakov A, Filip Z (1991). Microbial diversity of forest, field and polluted by lead soddy-podzolic soils. Simposium 11. Moscow University Press, Moscow

Chapter 3
Nitrogen Fixation in a Legume-Rhizobium Symbiosis: The Roots of a Success Story

Sahana Basu and Gautam Kumar

Abstract Legume-rhizobium symbiosis is an exquisite mutualistic interaction responsible for nitrogen (N_2) fixation in the terrestrial ecosystems. In this symbiosis, specialized root nodules are formed on host plants, where the reduction of atmospheric nitrogen to ammonia takes place that can be readily assimilated by the host plants. Compatibility of the host plant and colonizing bacteria is required to establish a successful symbiosis. Therefore, understanding the complex mechanisms for recognition of suitable host and colonizing rhizobacteria is a considerable challenge. Metagenomic studies have revealed the involvement of diverse molecular mechanisms regulating the symbiotic specificity and N_2 fixation in legume-rhizobium symbiosis. Abiotic stresses, including salinity, drought, high temperature, soil acidity, and soil nutrient deficiency, are major constraints for biological N_2 fixation in legume-rhizobium symbiosis. Several symbiotic associations exhibit tolerance to these extreme environmental conditions by sustaining their N_2-fixing potential. The present chapter includes an updated study of the symbiotic specificity and N_2 fixation in the legume-rhizobium symbiosis.

3.1 Introduction

Plants require the external supply of inorganic elements for synthesizing organic materials. Nitrogen is one of the major nutrient elements involved in the plant growth and development. Being the chief nitrogen sources, the availability of nitrate and ammonia is frequently limited. Nitrogen present in the Earth's crust is taken up by the plants through direct or indirect absorption in the form of nitrate (NO_3^-) or ammonium ions (NH_4^+). However, the atmospheric nitrogen uptake in plants is mediated by the biological nitrogen fixation (BNF)—the reduction of gaseous

---

S. Basu
Department of Biotechnology, Assam University, Silchar, Assam, India

G. Kumar (✉)
Department of Life Science, Central University of South Bihar, Gaya, Bihar, India
e-mail: gautam@cub.ac.in

A. Varma et al. (eds.), *Plant Microbe Symbiosis*,
https://doi.org/10.1007/978-3-030-36248-5_3

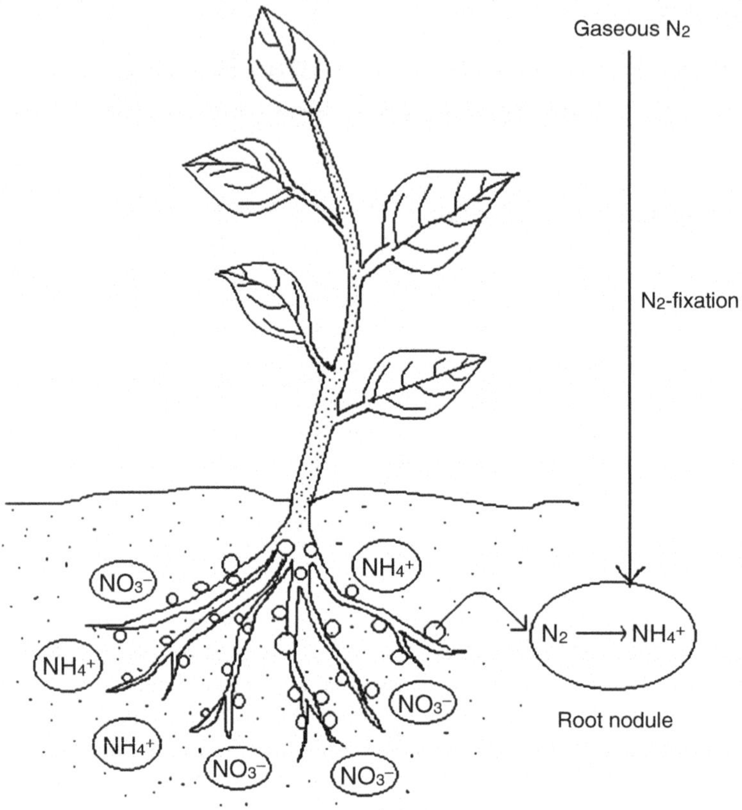

Fig. 3.1 Nitrogen uptake in terrestrial plants

nitrogen (N_2) to ammonia (NH_3) that is readily assimilated by the plants (Fig. 3.1). Nitrogen fixation is carried out by the prokaryotes, including bacteria and cyanobacteria present in free-living or symbiotic form.

Plant species of the Cucurbitaceae, Fabaceae (Leguminosae), Fagaceae, and Rosaceae families establish the nitrogen-fixing symbiosis with bacteria through the development of root nodules (Udvardi and Poole 2013; Fujita et al. 2014). The legume-rhizobium symbiosis contributes most of the ammonium to the plants, including the members of Leguminosae plant family and soil bacteria *Rhizobium*, *Azorhizobium*, *Bradyrhizobium*, *Sinorhizobium*, and *Photorhizobium* (collectively known as rhizobia) (Newton 2000; Table 3.1). Rhizobia are phylogenetically distinct α- and β-proteobacteria, involved in the symbiotic N_2 fixation. Rhizobia stimulate the formation of nodule (root or stem) plant organs for the N_2 fixation and assimilation.

Table 3.1 Symbiotic N_2-fixing rhizobia with their respective leguminous host plants

Rhizobial symbionts	Leguminous host plants
Azorhizobium caulinodans	*Sesbania rostrata*
Bradyrhizobium japonicum	*Arachis* sp., *Cajanus* sp., *Glycine max*, *Vigna radiata*, *Stylosanthes guianensis*
Rhizobium leguminosarum bv. *viciae*	*Cicer* sp., *Leucaena leucocephala*, *Lathyrus* sp., *Lens esculenta*, *Pisum sativum*, *Vicia faba*
Rhizobium leguminosarum bv. *phaseoli*	*Phaseolus vulgaris*
Rhizobium leguminosarum bv. *trifolii*	*Trifolium subterraneum*
Rhizobium meliloti	*Medicago* sp., *Melilotus* sp.
Rhizobium etli	*Phaseolus vulgaris*
Rhizobium tropici	*Phaseolus vulgaris*, *Leucaena leucocephala*, *Medicago* sp., *Macroptilium atropurpureum*
Mesorhizobium loti	*Lotus japonicus*, *Lupinus* sp.
Photorhizobium sp.	*Aeschynomene* sp.
Sinorhizobium fredii	*Glycine max*, *Leucaena leucocephala*, *Medicago sativa*

3.2 Root Nodule

3.2.1 Definition and Types

Rhizobial colonization of host plant root cells induces the development of specialized plant organs regulating the mutual symbiosis which are called nodules (Fig. 3.2a) (Oldroyd and Downie 2008). The nodules contain a hemeprotein called leghemoglobin (lb) that regulates the oxygen concentration of the N_2-fixing cell. The lb is responsible for the pink color of the nodule. Legume nodules are divided into two types—determinate nodule and indeterminate nodule (Nap and Bisseling 1990). Determinate nodules are formed on the tropical legumes, such as soybean (*Glycine max*), kidney bean (*Phaseolus vulgaris*), peanut (*Arachis hypogaea*), and southern pea (*Vigna unguiculata*), by *Rhizobium* and *Bradyrhizobium*. These nodules arise just below the epidermis with a transient meristem. The infected cells show synchronous differentiation, and the globose nodules contain homogeneous population of N_2-fixing cells. The uninfected cells are dispersed throughout the nodule, involved in assimilating NH_4^+ as ureides (allantoin, allantoic acid, and citrulline). In contrast, the indeterminate nodules are formed on the temperate legumes, such as pea (*Pisum sativum*), clover (*Trifolium repens*), broad bean (*Vicia faba*), and lentil (*Lens esculenta*), typically by *Rhizobium*. These nodules arise near the endodermis and nodule vasculature connecting with the vascular system of root. These appear as cylindrical nodules with persistent meristem having longitudinal gradient of differentiation. The uninfected cells of these nodules assimilate NH_4^+ as amides (asparagine, glutamine).

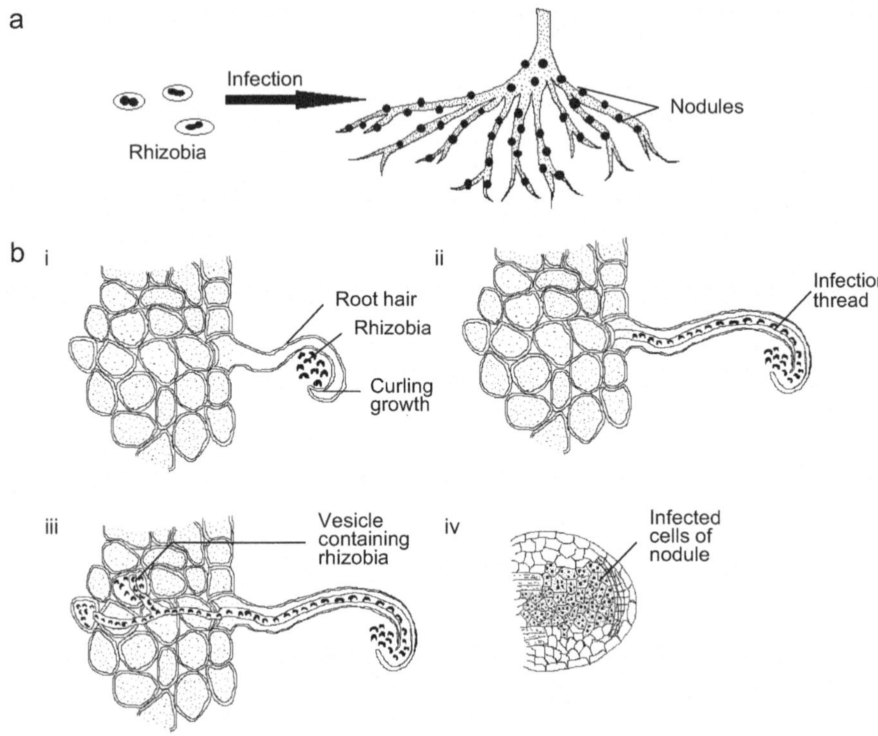

Fig. 3.2 Nodulation in legume-rhizobium symbiosis. (**a**) Nodules on the leguminous host plant root. (**b**) Development of root nodule in leguminous plants. (i) Rhizobia colonize the root hairs of leguminous plants in response to chemical attractant released by the host plants. Plant root hair shows curling growth in response to bacterial factors. (ii) Cell wall of root hair is degraded forming infection thread from root cell's Golgi vesicles that reaches end of the cell. The membrane of infection thread fuses with root hair cell plasma membrane. (iii) Rhizobia are released into cortical cells forming new infection thread that joins the previous one. The infection thread elongates, branches, and reaches the target cells. (iv) Membrane of host plant surrounds the bacterial cells (peribacteroid membrane), and the cells are released into the cytosol

3.2.2 Nodule Formation

Nodulation of plant roots involves three different stages—(1) preinfection, (2) infection, and (3) nodule organogenesis (Fig. 3.2b).

Development of root nodules is mediated by specific gene products, including the transcription factors playing the key role (Diedhiou and Diouf 2018).

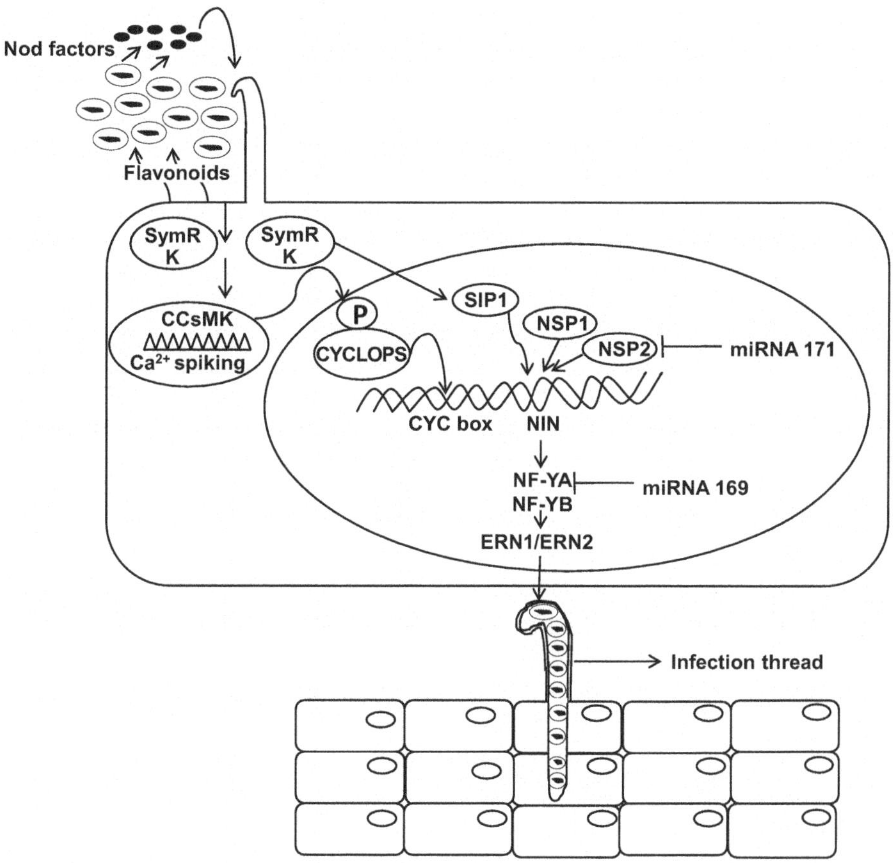

Fig. 3.3 Signaling cascade of events occurring during nodulation in legume-rhizobium symbiosis. The calmodulin-dependent protein kinase CCaMK/DMI3 induces the nuclear calcium spiking. Phosphorylation of the CYCLOPS with CCaMK/DMI3 promotes gene expression initiating the nodulation. The NIN expression is controlled by the SIP1 and NSP1/NSP2. The miRNA171 and miRNA169 regulate the TFs NSP2 and NF-YA, respectively

3.2.2.1 Preinfection Stage

In response to the chemical attractants, such as flavonoids and betaines, secreted by the young leguminous plant roots, specific rhizobia bind the root hairs (Fig. 3.3). The signaling compounds produced by the rhizobacteria involved in the recognition of compatible plant symbionts are called Nod factors (NFs) (Oldroyd and Downie 2008). The nod genes identified in *R. meliloti* and *R. leguminosarum* bv. *viciae* and *trifolii* have been broadly categorized into four classes—nodD; nodA, nodB, and nodC (common nod genes); hsn (host-specific nod genes); and other nod genes. Mutations of the nod genes have been reported to block the nodulation or alter the host range (Perret et al. 2000). Bacterial nod genes (NodD) perceive plant signals of the flavonoid family. Additional nod gene regulators include NolA and NodV/NodW (two-component system) in *Bradyrhizobium japonicum* and NolR in *Rhizobium* and *Sinorhizobium* spp. The nod genes trigger the biosynthesis of lipo-chitooligosaccharides (LCOs) (Masson-Boivin et al. 2009). Higher concentrations of jasmonate, vanillin, betaine, and xanthone (alternative plant compounds) have also been reported to trigger nod gene expression (Cooper 2007). The LCOs induce the nuclear calcium spiking in the root cells of the host plant (Granqvist et al. 2015). A calcium and calmodulin-dependent protein kinase (CCaMK/DMI3) decodes the calcium oscillations. Phosphorylation of the CYCLOPS transcription factor (TF) with CCaMK/DMI3 promotes gene expression and initiates the nodulation (Singh et al. 2014). The SymRK interacting protein (SIP1) binds to the AT-rich region of the promoter leading to the gene expression. SIP1 is also involved in the initial communication between rhizobia and plant root cell (Zhu et al. 2008). At the early stages of infection, calcium oscillations induce the NODULE INCEPTION (NIN) factor that initiates the bacterial infection in the root epidermis (Madsen et al. 2010). NIN also activates the expression of NF-YA1 (NF-YA1/NF-YA2 and NSP1/NSP2 complex), subsequently regulating ENOD11expression by direct activation of the ERN1 transcription (Laporte et al. 2013; Laloum et al. 2014). ERN1 activates ENOD11 expression, and NSP1 also shows direct interaction with the ENOD11 promoter in the presence of NSP2 associated with CCamK/DMI3 (Hirsch et al. 2009; Cerri et al. 2012).

3.2.2.2 Infection Stage

At this stage, the root hair wall of plant is digested by the bacteria, and infection thread (IT) is formed, through which bacteria are transferred to the cortical cells of plants. The bacteria are differentiated into bacteroids and endocytosed into symbiosomes. In legume symbiosis, plant cortical cell division is induced by *Rhizobium* forming the final nodule primordium. The CYCLOPS, a IPD3 orthologue, has been found to control the formation of IT and bacteroid release in the nodule (Ovchinnikova et al. 2011). The CYCLOPS after phosphorylation binds to the CYC-box promoter and induces the NIN expression. This consequently activates the pectate lyase gene involved cell wall degradation of host plant (Xie et al. 2012). Products of NF-YA1 and NF-YA2 are involved in cortical cell divisions (Soyano et al. 2013). Rhizobial infection and nodule organogenesis have been reported to be induced by the ERN1/ERN2 (Cerri et al. 2016). Therefore, the CYCLOPS TF has been recognized to control the ERN1 and NIN, but during early phase of symbiosis, these TFs play separate roles. Two antagonist miRNAs, miRNA172 and miRNA156, control different phases of the symbiosis (Lelandais-Briere et al. 2016). The ARF16 (Auxin Response Factor 16) has been observed to be induced at the early stages of the infection, depicting the necessity of auxin signaling initiation of IT (Laplaze et al. 2015). The miRNA390 is also involved in auxin signaling pathway by regulating ARFs through the formation of transacting small interfering (tasi) RNAs (Lelandais-Briere et al. 2016). Cytokinin also plays an important role in nodule initiation at the root cortex by regulating miRNA171h/MtNSP2 through the cytokinin-dependent CRE1 pathway (Suzaki et al. 2013; Lelandais-Briere et al. 2016).

3.2.2.3 Nodule Organogenesis

Nodule primordia, with a peripheral vascular system and infected cells, are induced by *Rhizobium* through the cortical cell division (Pawlowski and Sprent 2007). Involvement of various TFs has been studied in different legume-rhizobium symbiosis. The NF stimulates the expression of NF-YA1 and NF-YA2 in the epidermal cells (Laloum et al. 2014). The SIN1 (Scarecrow-like 13 Involved in Nodulation) reduces the expression of NF-YA1 and G2-M cell cycle genes CYCLIN B and Cell Division Cycle2 (Battaglia et al. 2014). The NFY TF promotes cortical cell division by developing lateral root and nodule (Baudin et al. 2015). The expression of TF Mszpt2–1 gene encoding a Kruppel-like zinc finger in vascular bundles of roots and nodules is required for developing the central nitrogen-fixing zone (Frugier et al. 2000; Diedhiou et al. 2014). During early nodule organogenesis, the miR171h has been reported to target NSP2 (Hofferek et al. 2014). In response to auxin, miRNA160 and miRNA167 regulate the pathways of root development by targeting ARF10, ARF16, and ARF17 and ARF6 and ARF8, respectively (Lelandais-Briere et al. 2016). The miRNA167 and miRNA393 cause reduced auxin sensitivity and decrease the number of nodules (Mao et al. 2013). The miR393j-3p also restricts the development of nodule by repressing the nodulin gene ENOD93 (Early Nodulin) (Yan et al. 2015).

3.3 N_2 Fixation in a Legume-Rhizobium Symbiosis

The N_2 fixation in legume-rhizobium symbiosis is carried by the enzyme dinitrogenase (EC 1.18.2.1). It is a multimeric protein complex made up of two proteins of different size—molybdoferredoxin (Mo-Fe) protein and azoferredoxin (Fe) protein. Nitrogenase catalyzes the reduction of atmospheric N_2 to NH_3.

$$N_2 + 8e^- + 8H^+ + 16\ ATP = 2NH_3 + H_2 + 16ADP + 16Pi$$

The Mo-nitrogenase requires high energy (16 mol ATP) for reducing each mole of N_2. Moreover, the enzyme is extremely oxygen sensitive, whereas the symbiotic rhizobacteria are strictly aerobic. The photosynthetic derivatives provided by the legume host plant in the form of abundant carbohydrate and citric acid cycle

Table 3.2 Functions of different nif genes involved in the nitrogen fixation in the legume-rhizobium symbiosis

Nif genes	Functions
NifA	Regulatory gene
NifB	Synthesizes a Fe-S containing precursor of FeMo-co
nifDK	Codes for the a and b subunits of dinitrogenase
nifEN	Encodes the molecular scaffold for assembly of the FeMo cofactor
nifH	Codes for dinitrogenase reductase
nifM	Helps in NifH maturation
nifQ	Incorporates Mo into FeMo-co
NifX	Substitutes the NifY for FeMo-co binding function

intermediates accomplish the energy requirement, and root nodules provide the anoxygenic environment required for N_2 fixation. The synthesis, processing, and assembly of the nitrogenase complex are carried by the nif genes. The number of nif genes varies with the physiology of the colonizing bacterium. Based on the previous studies, 15 nif genes have been reported in *A. caulinodans* (Lee et al. 2008) and *B. japonicum* (Kaneko et al. 2002), and 8 and 9 nif genes have been found in *R. leguminosarum* bv. *viciae* (Young et al. 2006) and *S. meliloti* (Galibert et al. 2001), respectively. Functions of different nif genes are presented in Table 3.2.

The nifA, nifB, nifDK, and nifEN are the core nif genes (Masson-Boivin et al. 2009). In rhizobia, the NifJ and NifF electron transfer proteins are missing and replaced by the fixABCX gene products. Similarly, the nifS and nifU genes are substituted with the icsS and iscA (housekeeping paralogs), respectively. NifY is replaced by the NifX (absent in *R. leguminosarum* bv. *viciae*) but is not involved in stabilizing apo-NifDK. The nifW and nifZ genes are also absent in rhizobia. Therefore, the nitrogenase assembly machinery in *S. meliloti* and *R. leguminosarum* bv. *viciae* is trimmed. It was previously suggested that nif gene number might vary according to physiology of rhizobacteria (Rubio and Ludden 2008). Alternatively, unidentified proteins might replace the missing nif products.

Rhizobial nif genes are regulated by the NifA protein. The synthesis and transcriptional activity of NifA in rhizobia are restricted by oxygen due to the extended cysteine-rich domain, but the transcriptional regulation varies in different rhizobia. The nifA expression of *A. caulinodans*, *B. japonicum*, and *S. meliloti* is carried out by FixLJ contrasting with *R. leguminosarum* bv. *viciae* that lacks the fixLJ (Fischer 1994). In this two-component regulatory system, FixL is the O_2-binding heme-based sensor. Although rhizobia exhibit plasticity for the nif gene composition and regulation, N_2 fixation occurs under apparent physiological condition.

3.4 Effect of Abiotic Stress on Legume-Rhizobium Symbioses and N_2 Fixation

Several environmental conditions including salinity, drought, high temperature, low pH, and soil nutrient deficiency are limiting factors to the legume-rhizobium symbioses (Zahran 1999).

3.4.1 Salinity Stress

Salinity is one of the most detrimental abiotic stresses leading to marked changes in the plant growth pattern and decreasing their productivity (Basu et al. 2017; Kumar et al. 2009). Salt tolerance in leguminous plants greatly varies with genotypes and developmental stages (Cordovilla et al. 1995a, b). Several studies have revealed that leguminous plants including *Glycine max*, *Phaseolus vulgaris*, and *Vicia faba* are more tolerant toward salinity than *Pisum sativum*. Salt-tolerant lines of *V. faba* have been reported to sustain the nitrogen fixation under salt stress (Cordovilla et al. 1995a). The legume-rhizobium symbioses and nodulation are more salt sensitive than rhizobia (Zahran 1991). The early stages of legume-rhizobium symbioses are inhibited by salt stress. In soybean-*Bradyrhizobium japonicum* symbioses, salinity has been reported to lead to root hair deformation and complete suppression of nodulation (Tu 1981). Salt stress also reduced bacterial colonization in faba bean (*Vicia faba*) (Cordovilla et al. 1995b). Reduction in nodular respiration and lb production under salt stress affects the nodulation and nitrogen fixation in leguminous plants (Delgado et al. 1994; Walsh 1995). Salinity-induced toxicity also disturbs the microbial populations of soil, severely affecting the cellular ultrastructure through distortion of cell envelope and cytoplasm (Zahran et al. 1997). Increased osmotic stress (0.2–1.44 MPa) has been found to affect the extracellular and capsular polysaccharide (LPS) synthesis of rhizobia, impairing the legume-rhizobium interaction (Breedveld et al. 1991). Being more salt-tolerant than the leguminous host plants, the symbiotic bacteria (*Rhizobium*, *Bradyrhizobium*) exhibit variable level of salt tolerance. *Rhizobium meliloti* and *Rhizobium leguminosarum* (Breedveld et al. 1991) have been found to be more salt-tolerant than the other rhizobia strains (Mohammad et al. 1991). In contrast, the growth of *R. meliloti* has been reported to be inhibited by magnesium (Mg^{2+}), depicting $MgCl_2$ to be more toxic (Jian et al. 1993).

Rhizobacteria adapt the saline environment by the intracellular accumulation of osmolytes (Smith et al. 1994a). Hypersalinity increases the intracellular free glutamate and/or K^+ levels in different rhizobia cells (*R. meliloti*, *R. fredii*, *Sinorhizobium fredii*), restricting the Mg^{2+} flux during osmotic shock (Susheng et al. 1993; Fujihara and Yoneyama 1994). Some of the rhizobia cells (*R. meliloti*) accumulate N-acetylglutaminylglutamine amide that acts as an osmolyte under salt stress (Smith et al. 1994b). Accumulation of disaccharide trehalose is also reported in *R.*

leguminosarum and peanut rhizobia growing under hypersaline condition (Ghittoni and Bueno 1996). Moreover, glycine betaine also acts as an osmoprotectant in salt-tolerant strains of *R. meliloti* (Smith et al. 1988). In *Sinorhizobium meliloti*, disaccharides (sucrose and ectoine) are used as osmoprotectants (Gouffi et al. 1999). Ectoine acts as another osmoprotectant in growth improvement of *R. meliloti* and plays an important role in stimulating the synthesis of endogenous osmolytes (glutamate and trehalose) (Talibort et al. 1994). Accumulation of organic osmolytes (amino acids) and the inorganic minerals (Na^+, K^+, Mg^{2+}) is found to increase in rhizobial cells under salt stress that play an important role in osmoregulation (Zahran et al. 1997). Polyamine content is also increased in salt-tolerant rhizobial strains (*R. fredii*) to balance the intracellular pH and maintain the ionic homeostasis under salinity stress (Fujihara and Yoneyama 1993). Salt stress leads to the production of osmotic shock proteins in rhizobial cells (Zahran et al. 1994). Effective legume-rhizobium symbioses in the saline environment require the selection of salt-tolerant rhizobia (Zahran 1991), as salt-tolerant rhizobia strains enhance nodulation and develop effective N_2 fixation under salt stress (Zou et al. 1995). Genetic structure of the colonizing bacteria can also be changed, since some root nodule bacteria exhibit DNA-DNA hybridization in the salt-affected soils (Zahran 1992). Tolerance of host plant is also a significant determinant in forming successful symbiosis (Craig et al. 1991). Therefore, the salt-tolerant legume host plant genotypes are required to be selected and matched with tolerant rhizobia strain for effective symbiotic association under salt stress (Cordovilla et al. 1995a).

3.4.2 Drought Stress

Drought occurring due to deficiency of soil moisture is a major constraint for plant growth and productivity (Basu et al. 2017). The distribution and survival of soil microbiome also depend on the variation in soil water content (Orchard and Cook 1983). The severity of the drought regulates the growth of the microorganisms, thereby affecting the legume-rhizobium symbioses (Williams and De Mallorca 1984). Though rhizobial populations occur in extreme arid environment and exhibit effective nodulation, the population densities have been observed to be declined under drought stress (Tate 1995). Distribution of *R. leguminosarum* in a sand and silt loam soil has been observed to be influenced by the initial soil moisture (Postma et al. 1989). Moreover, drought has also been reported to cease the movement of *R. trifolii* (Hamdi 1970) and strains of *B. japonicum* (Wadisirisuk et al. 1989). Morphological changes of the rhizobia are the foremost response under drought resulting in reduced infection and host plant nodulation, as observed in *R. meliloti* (Busse and Bottomley 1989) and mesquite rhizobium (Shoushtari and Pepper 1985). Water scarcity has been found to cause significant decrease in the number of infection threads inside root hairs, inhibiting the nodulation in *T. subterraneum* (Worrall and Roughley 1976) and *Vicia faba* (Zahran and Sprent 1986). Drought also hindered the *R. japonicum* infection in soybean (Hunt et al. 1981).

Soil moisture deficiency acts as a constraint for the nodulation and also the symbiotic N_2 fixation in leguminous plants (Albrecht et al. 1994). Drought tolerance in legume varies with developmental stages, with vegetative stage being more susceptible (Pena-Cabriales and Castellanos 1993). On the other hand, drought sensitivity of the microbial population also varies with the rhizobial strains (Busse and Bottomley 1989). Therefore, for successful legume-rhizobium interaction and enhanced N_2 fixation, drought-tolerant rhizobial strains need to be selected within their legume host range. Furthermore, the legume host plant is more sensitive to drought as compared to their bacterial partner. Different genotypes of *Vigna radiata* (Rai and Prasad 1983) and *Trifolium repens* (Robin et al. 1989) show differential N_2 fixation under drought stress, showing the significant role of N_2 fixation in improving soil fertility. Therefore, the soil moisture content also requires to be optimized for improved host plant growth (Tate 1995). In contrast, some leguminous plants including *M. sativa* (Keck et al. 1984), *Arachis hypogaea* (Venkateswarlu et al. 1989), *Desmodium intortum* (Ahmed and Quilt, 1980), and *Cyamopsis tetragonoloba* (Venkateswarlu et al. 1983) exhibit extreme drought tolerance. Drought-tolerant legumes maintain their cell turgidity through enhanced osmotic adjustment, which has been considered as an indicator for selecting legume host under drought stress (Ford 1984). Accumulation of osmotically active solutes like proline has been reported in different legumes, including *Phaseolus vulgaris* (Kapuya et al. 1985) and *Glycine max* (Fukutoku and Yamada, 1982). Accumulation of free amino acids and pinitol (o-methylinositol) (low-molecular-weight solute) under drought stress has also been reported in some of the tropical legumes (Ford 1984). The involvement of K^+ in minimizing the drought effect on N_2 fixation of *P. vulgaris* and *V. faba* has also been reported (Sangakkara et al. 1996).

3.4.3 Heat Stress

Elevated soil temperature is one of the major constraints for N_2 fixation of legumes in tropical and subtropical regions (Michiels et al. 1994). High temperature at the rooting zone disturbs the bacterial infection and N_2 fixation in legumes (Kishinevsky et al. 1992; Hungria and Franco 1993). Root hair infection, bacterial differentiation, and nodule structure are greatly affected by elevated temperature, consequently delaying or inhibiting the nodulation (Graham 1992). Temperatures between 42 and 45 °C inhibited soybean nodulation and N_2 fixation (LaFavre and Eaglesham 1986), whereas 35 °C resulted in small nodule formation with low specific nitrogenase activity in bean (Piha and Munnus 1987). Both heat-tolerant and heat-sensitive strains of *Rhizobium* produce heat shock proteins under heat stress (Michiels et al. 1994). High temperature between 35 and 40 °C has also been reported to change the LPS mobility pattern of some rhizobial strains (Zahran et al. 1994).

3.4.4 Soil Acidity/Low pH

Soil acidity caused by extreme low pH is one of the serious problems for productivity of legume plants (Graham 1992; Correa and Barneix 1997). Successful legume-rhizobium symbiosis requires neutral or slightly acidic soil (Bordeleau and Prevost 1994). Acidic soil affects survival of *Rhizobium* and nodulation, thereby limiting symbiotic N_2 fixation (Munns 1986). Nodulation of the legume host plants is disturbed under acidic soil conditions. Leguminous plants, including *M. murex*, *M. truncatula*, and *T. subterranean*, exhibit low pH tolerance as indicated by dry matter yield (Evans et al. 1990). Increased level of inoculation has also been found to enhance nodulation under acidic conditions (Pijnenborg et al. 1991). Rhizobial colonization to root hairs of the host plant and rhizobial multiplication are affected by soil acidity causing reduced nodulation (Taylor et al. 1991). Low pH also affects the rhizobial competitiveness for nodule sites (Vargas and Graham 1989). Symbiotic efficiency of rhizobia under acidic conditions varies with their strains. Several studies suggested that acid tolerance of only one symbiotic partners leads to successful nodulation under acidic soil (Vargas and Graham 1988). Heavy metal activity (Al) is also associated with the acidic soil with pH of 5.0 that inhibits nodulation (Bordeleau and Prevost 1994). Rhizobia exhibit differential responses to Al toxicity under low soil pH. Several *Rhizobium* (Vargas and Graham 1988) and *Bradyrhizobium* (Graham 1992) strains are Al-tolerant at low pH. Leguminous plants show marked variation in Al and Mn tolerance. Nodulation shows more Al sensitivity as compared to that of the host plants (Graham 1992).

Selection of low pH-tolerant legume host plants can minimize the effects of acidic conditions. Several acid-tolerant genotypes of lentil (*Lens culinaris*) have been identified to produce excess aspartic, citric, glucenic, malic, and succinic acids in root exudate (Rai 1992). The Al-tolerant plants exude higher organic acids and ligands forming stable chelates with Al and thereby reducing the toxicity (Foy and Lee 1987). Additionally, soil acidity has been treated with lime and superphosphate by decreasing the aluminum (Al) and manganese (Mn) concentrations in soil (Peoples et al. 1995). Subsoil acidity has also been treated with application of coal-derived calcium fulvate (Van Der Watt et al. 1991) that has increased the pH more than gypsum, $CaCO_3$, Ca-EDTA, or $Ca(OH)_2$. Decrease in nodulation and N_2 fixation has been observed after soil treatment with bicarbonate (Tang and Thomson 1996) or carbonate (Tang et al. 1998). Amelioration of soil acidity by lime and other substances including carbonate must be optimized to avoid the inhibition of legume-rhizobium symbiosis (Bordeleau and Prevost 1994).

3.4.5 Soil Nutrient Deficiency

Soil nutrient deficiency is another detrimental abiotic stress for crop productivity often accompanied by the salinity or soil acidity. Plant growth can be protected by eliminating the toxicity generated by increased sodium (Na^+) and chloride (Cl^-) ions under salt stress, which can be counterbalanced with potassium (K^+) and calcium (Ca^{2+}) ions (Glass 1983). Similar responses are shown by the rhizobacteria, as *R. meliloti* exhibited increased Ca^{2+} requirement for reviving its growth under osmotic stress (Busse and Bottomley 1989). Ca^{2+} helps in cell division, elongation, and stabilization of membrane (Torimitsu et al. 1985). Cells exposed to acidic environment also display Ca^{2+} requirement for maintenance of cytoplasmic pH, as found in *R. meliloti* (O'Hara et al. 1989). The Ca^{2+} also helps in phosphorus mobilization in cells under low pH (Beck and Munns 1984). Ca^{2+} also controls K^+ permeability and activates K^+ uptake through the cytoplasm acidification. Low availability of Ca^{2+} under saline environment has been found to reduce the bacterial colonization on the root hairs of host plant (Zahran and Sprent 1986). Lack of Ca^{2+} under salt or low pH stress disturbs the nodulating capacity of legumes, as Ca^{2+} has been reported to induce nod gene expression (Alva et al. 1990).

Phosphorus (P) is one of the major yield-limiting nutrients affecting the N_2 fixation in legume-rhizobium symbiosis (Pereira and Bliss 1989). The development of legume host plants requires different concentrations of P (First et al. 1987). Likewise, rhizobial strains exhibit differential variation in tolerance under P deficiency (Beck and Munns 1984). Under low pH conditions, the dissolved P is precipitated in the presence of aluminum (Al^{3+}) increasing P deficiency. Alkaline phosphatase activity has been observed in cells under P-limited condition (Smart et al. 1984). P also plays an important role in nodulation and N_2 fixation as nodules act as sinks for P (Hart 1989). Addition of P to soils under low pH has been reported to increase the nodule occupancy of *Trifolium subterraneum* by *R. leguminosarum* bv. *trifolii* (Almendras and Bottomley 1987). Nitrogenase activity has also observed to be increased when supplemented with higher concentration of P (Lynd et al. 1984). P and zinc (Zn^{2+}) have also been seen to interact under salt stress improving the nodulation of legumes (Saxena and Rewari 1991). Increased Zn^{2+} concentration protects plants by reducing the shoot Na^+/K^+ ratio under salt stress.

3.5 Conclusion and Future Prospects

Regarding the N_2-fixing potential under severe environmental conditions, the legume-rhizobium symbiosis is superior to any other N2-fixing systems. Recent studies have revealed the underlying molecular mechanisms of the specificity in the legume-rhizobium symbiosis. The NifA protein plays the key role in regulating the nif genes by controlling several cellular processes in rhizobia. Further research is required to explain the rhizobial infection process in the symbiosis. As very less percentage of microsymbionts of the legume genera have been characterized, more investigation of rhizobial biodiversity is also necessary. Genomic studies on the rhizobial ecology might be essential in revealing the evolutionary history of rhizobia and the control of intracellular infection and nodulation.

References

Ahmed B, Quilt P (1980) Effect of soil moisture stress on yield, nodulation and nitrogenase activity of *Macroptilium atropurpureum* cv. *Sirato* and *Desmodium intortum* cv. *Greenleaf*. Plant Soil 57:187–194

Albrecht SL, Bennett JM, Boote KJ (1994) Relationship of nitrogenase activity to plant water stress in field grown soybeans. Field Crop Res 8:61–71

Almendras AS, Bottomley PJ (1987) Influence of lime and phosphate on nodulation of soil-grown *Trifolium subterraneum* L. by indigenous *Rhizobium trifolii*. Appl Environ Microbiol 53:2090–2097

Alva AK, Assher CJ, Edwards DG (1990) Effect of solution pH, external calcium concentration and aluminum activity on nodulation and early growth of cowpea. Aust J Agric Res 41:359–365

Basu S, Giri RK, Benazir I, Kumar S, Rajwanshi R, Dwivedi SK, Kumar G (2017) Comprehensive physiological analyses and reactive oxygen species profiling in drought tolerant rice genotypes under salinity stress. Physiol Mol Biol Plants 23:837–850

Battaglia M, Ripodas C, Clua J, Baudin M, Aguilar OM, Niebel A, Zanetti ME, Blanco FA (2014) A nuclear factor Y interacting protein of the GRAS family is required for nodule organogenesis, infection thread progression, and lateral root growth. Plant Physiol 164:1430–1442

Baudin M, Laloum T, Lepage A, Ripodas C, Ariel F, Frances L, Crespi M, Gamas P, Blanco FA, Zanetti ME, de Carvalho-Niebel F, Niebel A (2015) A phylogenetically conserved group of nuclear factor-Y transcription factors interact to control nodulation in legumes. Plant Physiol 169:2761–2773

Beck DP, Munns DN (1984) Phosphate nutrition of *Rhizobium* sp. Appl Environ Microbiol 47:278–282

Bordeleau LM, Prevost D (1994) Nodulation and nitrogen fixation in extreme environments. Plant Soil 161:115–124

Breedveld MW, Zevenhuizen LPTM, Zehnder AJB (1991) Osmotically-regulated trehalose accumulation and cyclic *beta*-(1,2)-glucan excreted by *Rhizobium leguminosarum* bv. *trifolii* TA-1. Arch Microbiol 156:501–506

Busse MD, Bottomley PJ (1989) Growth and nodulation responses of *Rhizobium meliloti* to water stress induced by permeating and nonpermeating solutes. Appl Environ Microbiol 55:2431–2436

Cerri MR, Frances L, Laloum T, Auriac MC, Niebel A, Oldroyd GE, Barker DG, Fournier J, de Carvalho-Niebel F (2012) *Medicago truncatula* ERN transcription factors: regulatory interplay with NSP1/NSP2 GRAS factors and expression dynamics throughout rhizobial infection. Plant Physiol 160:2155–2172

Cerri MR, Frances L, Kelner A, Fournier J, Middleton PH, Auriac M-C, Mysore KS, Wen J, Erard M, Barker DG, Oldroyd GE, de Carvalho-Niebel F (2016) The symbiosis-related ERN transcription factors act in concert to coordinate rhizobial host root infection. Plant Physiol 171:1037–1054

Cooper JE (2007) Early interactions between legumes and rhizobia: disclosing complexity in a molecular dialogue. J Appl Microbiol 103:1355–1365

Cordovilla MP, Ligero F, Lluch C (1995a) Influence of host genotypes on growth, symbiotic performance and nitrogen assimilation in Faba bean (*Vicia faba* L.) under salt stress. Plant Soil 172:289–297

Cordovilla MP, Ocana A, Ligero F, Lluch C (1995b) Growth stage response to salinity in symbiosis *Vicia faba-Rhizobium leguminosarum* bv. *viciae*. Plant Physiol 14:105–111

Correa OS, Barneix AJ (1997) Cellular mechanisms of pH tolerance in *Rhizobium loti*. World J Microbiol Biotechnol 13:153–157

Craig GF, Atkins CA, Bell DT (1991) Effect of salinity on growth of *Rhizobium* and their infectivity and effectiveness on two species of *Acacia*. Plant Soil 133:253–262

Delgado MJ, Ligero F, Lluch C (1994) Effects of salt stress on growth and nitrogen fixation by pea, faba-bean, common bean and soybean plants. Soil Biol Biochem 26:371–376

Diedhiou I, Diouf D (2018) Transcription factors network in root endosymbiosis establishment and development. World J Microbiol Biotechnol 34:37

Diedhiou I, Tromas A, Cissoko M, Gray K, Parizot B, Crabos A, Alloisio N, Fournier P, Carro L, Svistoonoff S, Gherbi H, Hocher V, Diouf D, Laplaze L, Champion A (2014) Identification of potential transcriptional regulators of actinorhizal symbioses in *Casuarina glauca* and *Alnus glutinosa*. BMC Plant Biol 14:342

Evans J, Dear B, O'Connor GE (1990) Influence of an acid soil on the herbage yield and nodulation of five annual pasture legumes. Aust J Exp Agric 30:55–60

First AJ, Smith FW, Edwards DG (1987) External phosphorus requirements of five tropical grain legumes grown in flowing-solution culture. Plant Soil 99:75–84

Fischer HM (1994) Genetic regulation of nitrogen fixation in rhizobia. Microbiol Rev 58:352–386

Ford CW (1984) Accumulation of low molecular weight solutes in water stressed tropical legumes. Phytochemistry 23:1007–1015

Foy CD, Lee RH (1987) Different aluminum tolerance of two barley cultures related to organic acids in their roots. J Plant Nutr 10:1089–1101

Frugier F, Poirier S, Satiat-Jeunemaitre B, Kondorosi A, Crespi M (2000) A Kruppel-like zinc finger protein is involved in nitrogen-fixing root nodule organogenesis. Genes Dev 14:475–482

Fujihara S, Yoneyama T (1993) Effects of pH and osmotic stress on cellular polyamine contents in the soybean rhizobia *Rhizobium fredii* p220 and *Bradyrhizobium japonicum* A1017. Appl Environ Microbiol 59:1104–1109

Fujihara S, Yoneyama T (1994) Response of *Rhizobium fredii* P220 to osmotic shock: interrelationships between K^+, Mg^{2+}, glutamate and homospermidine. Microbiology 140:1909–1916

Fujita H, Aoki S, Kawaguchi M (2014) Evolutionary dynamics of nitrogen fixation in the legume-rhizobia symbiosis. PLoS One 9:e93670

Fukutoku Y, Yamada Y (1982) Accumulation of carbohydrates and proline in water stressed soybean (*Glycine max* L.). Soil Sci Plant Nutr 28:147–151

Galibert F, Finan TM, Long SR, Puhler A, Abola P, Ampe F, Barloy-Hubler F, Barnett MJ, Becker A, Boistard P, Bothe G, Boutry M, Bowser L, Buhrmester J, Cadieu E, Capela D, Chain P, Cowie A, Davis RW, Dreano S, Federspiel NA, Fisher RF, Gloux S, Godrie T, Goffeau A, Golding B, Gouzy J, Gurjal M, Hernandez-Lucas I, Hong A, Huizar L, Hyman RW, Jones T, Kahn D, Kahn ML, Kalman S, Keating DH, Kiss E, Komp C, Lelaure V, Masuy D, Palm C, Peck MC, Pohl TM, Portetelle D, Purnelle B, Ramsperger U, Surzycki R, Thebault P, Vandenbol M, Vorholter FJ, Weidner S, Wells DH, Wong K, Yeh KC, Batut J (2001) The composite genome of the legume symbiont *Sinorhizobium meliloti*. Science 293:668–672

Ghittoni NE, Bueno MA (1996) Changes in the cellular content of trehalose in four peanut rhizobia strains cultured under hypersalinity. Symbiosis 20:117–127

Glass ADH (1983) Regulation of ion transport. Annu Rev Plant Physiol 34:311–326

Gouffi K, Pica N, Pichereau V, Blanco C (1999) Disaccharides as a new class of nonaccumulated osmoprotectants for *Sinorhizobium meliloti*. Appl Environ Microbiol 65:1491–1500

Graham PH (1992) Stress tolerance in *Rhizobium* and *Bradyrhizobium* and nodulation under adverse soil conditions. Can J Microbiol 38:475–484

Granqvist E, Sun J, Op den Camp R, Pujic P, Hill L, Normand P, Morris RJ, Downie JA, Geurts R, Oldroyd GED (2015) Bacterial-induced calcium oscillations are common to nitrogen-fixing associations of nodulating legumes and nonlegumes. New Phytol 207:551–558

Hamdi Y (1970) Soil water tension and the movement of rhizobia. Soil Biol Biochem 3:121–126

Hart AL (1989) Nodule phosphorus and nodule activity in white clover. N Z J Agric Res 32:145–149

Hirsch S, Kim J, Munoz A, Heckmann AB, Downie JA, Oldroyd GE (2009) GRAS proteins form a DNA binding complex to induce gene expression during nodulation signaling in *Medicago truncatula*. Plant Cell 21:545–557

Hofferek V, Mendrinna A, Gaude N, Krajinski F, Devers EA (2014) MiR171h restricts root symbioses and shows like its target NSP2 a complex transcriptional regulation in *Medicago truncatula*. BMC Plant Biol 14:199
Hungria M, Franco AA (1993) Effects of high temperature on nodulation and nitrogen fixation by *Phaseolus vulgaris* L. Plant Soil 149:95–102
Hunt PJ, Wollum AG, Matheny TA (1981) Effects of soil water on *Rhizobium japonicum* infection nitrogen accumulation and yield in bragg soybean. Agric J 73:501–505
Jian W, Susheng Y, Jilun L (1993) Studies on the salt tolerance of *Rhizobium meliloti*. Acta Microbiol Sin 33:260–267
Kaneko T, Nakamura Y, Sato S, Minamisawa K, Uchiumi T, Sasamoto S, Watanabe A, Idesawa K, Iriguchi M, Kawashima K, Kohara M, Matsumoto M, Shimpo S, Tsuruoka H, Wada T, Yamada M, Tabata S (2002) Complete genomic sequence of nitrogen-fixing symbiotic bacterium *Bradyrhizobium japonicum* USDA110. DNA Res 9:189–197
Kapuya JA, Barendse GWM, Linskens HF (1985) Water stress tolerance and proline accumulation in *Phaseolus vulgaris*. Acta Bot Neerl 34:295–300
Keck TJ, Wagenet RJ, Campbell WF, Knighton RE (1984) Effects of water and salt stress on growth and acetylene reduction in alfalfa. Soil Sci Soc Am J 48:1310–1316
Kishinevsky BD, Sen D, Weaver RW (1992) Effect of high root temperature on *Bradyrhizobium*-peanut symbiosis. Plant Soil 143:275–282
Kumar G, Purty RS, Sharma MP, Singla-Pareek SL, Pareek A (2009) Physiological responses among *Brassica* species under salinity stress show strong correlation with transcript abundance for SOS pathway-related genes. J Plant Physiol 166:507–520
LaFavre AK, Eaglesham ARJ (1986) The effects of high temperature on soybean nodulation and growth with different strains of bradyrhizobia. Can J Microbiol 32:22–27
Laloum T, Baudin M, Frances L, Lepage A, Billault-Penneteau B, Cerri MR, Ariel F, Jardinaud MF, Gamas P, de Carvalho-Niebel F, Niebel A (2014) Two CCAAT-box-binding transcription factors redundantly regulate early steps of the legume-rhizobia endosymbiosis. Plant J 79:757–768
Laplaze L, Lucas M, Champion A (2015) Rhizobial root hair infection requires auxin signaling. Trends Plant Sci 20:332–334
Laporte P, Lepage A, Fournier J, Catrice O, Moreau S, Jardinaud MF, Mun JH, Larrainzar E, Cook DR, Gamas P, Niebel A (2013) The CCAAT box-binding transcription factor NF-YA1 controls rhizobial infection. J Exp Bot 65:481–494
Lee KB, De Backer P, Aono T, Liu CT, Suzuki S, Suzuki T, Kaneko T, Yamada M, Tabata S, Kupfer DM, Najar FZ, Wiley GB, Roe B, Binnewies TT, Ussery DW, D'Haeze W, Herder JD, Gevers D, Vereecke D, Holsters M, Oyaizu H (2008) The genome of the versatile nitrogen fixer *Azorhizobium caulinodans* ORS571. BMC Genomics 9:271
Lelandais-Briere C, Moreau J, Hartmann C, Crespi M (2016) Noncoding RNAs, emerging regulators in root endosymbioses. Mol Plant Micro Int 29:170–180
Lynd JQ, Hanlon EA, Odell GV Jr (1984) Nodulation and nitrogen fixation by arrow leaf clover: effects of phosphorus and potassium. Soil Biol Biochem 16:589–594
Madsen LH, Tirichine L, Jurkiewicz A, Sullivan JT, Heckmann AB, Bek AS, Ronson CW, James EK, Stougaard J (2010) The molecular network governing nodule organogenesis and infection in the model legume *Lotus japonicus*. Nat Commun 1:10
Mao G, Turner M, Yu O, Subramanian S (2013) miR393 and miR164 influence indeterminate but not determinate nodule development. Plant Signal Behav 8:e26753
Masson-Boivin C, Giraud E, Perret X, Batut J (2009) Establishing nitrogen-fixing symbiosis with legumes: how many rhizobium recipes? Trends Microbiol 17:458–466
Michiels J, Verreth C, Vanderleyden J (1994) Effects of temperature stress on bean nodulating *Rhizobium* strains. Appl Environ Microbiol 60:1206–1212
Mohammad RM, Akhavan-Kharazian M, Campbell WF, Rumbaugh MD (1991) Identification of salt-and drought-tolerant *Rhizobium meliloti* L. strains. Plant Soil 134:271–276
Munns DN (1986) Acid soils tolerance in legumes and rhizobia. Adv Plant Nutr 2:63–91

Nap JP, Bisseling T (1990) Developmental biology of a plant-prokaryote symbiosis: the legume root nodule. Science 250:948–954

Newton WE (2000) In: Pedrosa FO, Hungria M, Yates MG, Newton WE (eds) Nitrogen fixation: from molecules to crop productivity. Kluwer, Dordrecht, pp 3–8

O'Hara GW, Goss TJ, Dilworth MJ, Glenn AR (1989) Maintenance of intracellular pH and acid tolerance in *Rhizobium meliloti*. Appl Environ Microbiol 55:1870–1876

Oldroyd GE, Downie JA (2008) Coordinating nodule morphogenesis with rhizobial infection in legumes. Annu Rev Plant Biol 59:519–546

Orchard VA, Cook FG (1983) Relation between soil respiration and soil moisture. Soil Biol Biochem 15:447–453

Ovchinnikova E, Journet E-P, Chabaud M, Cosson V, Ratet P, Duc G, Fedorova E, Liu W, den Camp RO, Zhukov V, Tikhonovich I, Borisov A, Bisseling T, Limpens E (2011) IPD3 controls the formation of nitrogen-fixing symbiosomes in pea and *Medicago* Spp. Mol Plant-Microbe Interact 24:1333–1344

Pawlowski K, Sprent JI (2007) Comparison between actinorhizal and legume symbiosis. In: Nitrogen-fixing actinorhizal symbioses. Springer, New York, pp 261–288

Pena-Cabriales JJ, Castellanos JZ (1993) Effect of water stress on N_2 fixation and grain yield of *Phaseolus vulgaris* L. Plant Soil 152:151–155

Peoples MB, Ladha JK, Herridge DF (1995) Enhancing legume N_2 fixation through plant and soil management. Plant Soil 174:83–101

Pereira PAA, Bliss FA (1989) Selection of common bean (*Phaseolus vulgaris* L.) for N_2 fixation at different levels of available phosphorus under field and environmentally-controlled conditions. Plant Soil 115:75–82

Perret X, Staehelin C, Broughton WJ (2000) Molecular basis of symbiotic promiscuity. Microbiol Mol Biol Rev 64:180–201

Piha MI, Munnus DN (1987) Sensitivity of the common bean (*Phaseolus vulgaris* L.) symbiosis to high soil temperature. Plant Soil 98:183–194

Pijnenborg JWM, Lie TA, Zehnder ATB (1991) Nodulation of lucerne (*Medicago sativa* L.) in an acid soil: effects of inoculum size and lime pelleting. Plant Soil 131:1–10

Postma J, Van Veen JA, Walter S (1989) Influence of different initial soil moisture contents on the distribution and population dynamics of introduced *Rhizobium leguminosarum* biovar *trifolii*. Soil Biol Biochem 21:437–442

Rai R (1992) Effect of acidity factors on aspects of symbiotic N_2 fixation of *Lens culinaris* in acid soils. J Gen Appl Microbiol 38:391–406

Rai R, Prasad V (1983) Studies on compatibility of nitrogen fixation by high temperature-adapted rhizobium strains and *Vigna radiata* genotype at two levels in calcareous soil. J Agric Sci 101:377–381

Robin C, Shamsun-Noor L, Guckert A (1989) Effect of potassium on the tolerance to PEG-induced water stress of two white clover varieties (*Trifolium repens* L.). Plant Soil 120:143–158

Rubio LM, Ludden PW (2008) Biosynthesis of the iron-molybdenum cofactor of nitrogenase. Annu Rev Microbiol 62:93–111

Sangakkara UR, Hartwig UA, Noesberger J (1996) Soil moisture and potassium affect the performance of symbiotic nitrogen fixation in faba bean and common bean. Plant Soil 184:123–130

Saxena AK, Rewari RB (1991) The influence of phosphate and zinc on growth, nodulation and mineral composition of chickpea (*Cicer arietinum* l.) under salt stress. World J Microbiol Biotechnol 7:202–205

Shoushtari NH, Pepper IL (1985) Mesquite rhizobia isolated from the Sonoran desert: competitiveness and survival in soil. Soil Biol Biochem 17:803–806

Singh S, Katzer K, Lambert J, Cerri M, Parniske M (2014) CYCLOPS, a DNA-binding transcriptional activator, orchestrates symbiotic root nodule development. Cell Host Microbe 15:139–152

Smart JB, Dilwarth MJ, Robson AD (1984) Effect of phosphorus supply on phosphate uptake and alkaline phosphatase activity in rhizobia. Arch Microbiol 140:281–286

Smith LT, Pocard JA, Bernard T, Le Rudulier D (1988) Osmotic control of glycine betaine biosynthesis and degradation in *Rhizobium meliloti*. J Bacteriol 170:3142–3149

Smith LT, Smith GM, Desouza MR, Pocard JM, Le Rudulier D, Madkour MA (1994a) Osmoregulation in *Rhizobium meliloti*: mechanism and control by other environmental signals. J Exp Zool 268:162–165

Smith LT, Allaith AM, Smith GM (1994b) Mechanism of osmotically-regulated N-acetylglutaminylglutamine amide production in *Rhizobium meliloti*. Plant Soil 161:103–108

Soyano T, Kouchi H, Hirota A, Hayashi M (2013) Nodule inception directly targets NF-Y subunit genes to regulate essential processes of root nodule development in *Lotus japonicus*. PLoS Genet 9:e1003352

Susheng Y, Jing Z, Jilun L (1993) The osmoregulation of *Sinorhizobium fredii*. Acta Microbiol Sin 33:86–91

Suzaki T, Ito M, Kawaguchi M (2013) Genetic basis of cytokinin and auxin functions during root nodule development. Front Plant Sci 4:42

Talibort R, Mohamed J, Goouesbet G, Himidi-Kabbab S, Wroblewski H, Blanco C, Bernard T (1994) Osmoregulation in rhizobia: ectoine-induced salt tolerance. J Bacteriol 176:5210–5217

Tang C, Thomson BD (1996) Effects of solution pH and bicarbonate on the growth and nodulation of a range of grain legumes. Plant Soil 186:321–330

Tang C, Barton L, Raphael C (1998) Pasture legume species differ in their capacity to acidify soil. Aust J Agric Res 49:53–58

Tate RL (1995) Soil microbiology (symbiotic nitrogen fixation). Wiley, New York, NY, pp 307–333

Taylor RW, Williams ML, Sistani KR (1991) Nitrogen fixation by soybean-*Bradyrhizobium* combinations under acidity, low P and high Al stresses. Plant Soil 131:293–300

Torimitsu K, Hayashi M, Ohta E, Sakata M (1985) Effect of K^+ and H^+ stress and role of Ca^{2+} in the regulation of intracellular K^+ concentration in mung bean roots. Physiol Plant 63:247–252

Tu JC (1981) Effect of salinity on rhizobium-root hair interaction, nodulation and growth of soybean. Can J Plant Sci 61:231–239

Udvardi M, Poole PS (2013) Transport and metabolism in legume-rhizobia symbioses. Annu Rev Plant Biol 64:781–805

Van Der Watt HVH, Barnard RO, Cronie IJ, Dekker J, Croft GJB, Van Der Watt MM (1991) Amelioration of subsoil acidity by application of a coal-derived calcium fulvate to the soil surface. Nature 350:146–148

Vargas AAT, Graham PH (1988) *Phaseolus vulgaris* cultivar and *Rhizobium* strain variation in acid-pH tolerance and nodulation under acid conditions. Field Crops Res 19:91–101

Vargas AAT, Graham PH (1989) Cultivar and pH effects on competition for nodule sites between isolates of *Rhizobium* in beans. Plant Soil 117:195–200

Venkateswarlu B, Rao AV, Lahiri AN (1983) Effect of water stress on nodulation and nitrogenase activity of guar (*Cyamopsis tetragonoloba* (L.) Taub.). Proc Indian Acad Sci (Plant Sci) 92:297–301

Venkateswarlu B, Maheswari M, Karan NS (1989) Effects of water deficits on N_2 (C_2H_2) fixation in cowpea and groundnut. Plant Soil 114:69–74

Wadisirisuk P, Danso SKA, Hardarson G, Bowen GD (1989) Influence of *Bradyrhizobium japonicum* location and movement on nodulation and nitrogen fixation in soybeans. Appl Environ Microbiol 35:1711–1716

Walsh KB (1995) Physiology of the legume nodule and its response to stress. Soil Biol Biochem 27:637–655

Williams PM, De Mallorca MS (1984) Effect of osmotically induced leaf moisture stress on nodulation and nitrogenase activity of *Glycine max*. Plant Soil 80:267–283

Worrall VS, Roughley RJ (1976) The effect of moisture stress on infection of *Trifolium subterraneum* L. by *Rhizobium trifolii* dang. J Exp Bot 27:1233–1241

Xie F, Murray JD, Kim J, Heckmann AB, Edwards A, Oldroyd GE, Downie JA (2012) Legume pectate lyase required for root infection by rhizobia. Proc Natl Acad Sci 109:633–638

Yan Z, Hossain MS, Arikit S, Valdes-Lopez O, Zhai J, Wang J, Libault M, Ji T, Qiu L, Meyers BC, Stacey G (2015) Identification of microRNAs and their mRNA targets during soybean nodule development: functional analysis of the role of miR393j-3p in soybean nodulation. New Phytol 207:748–759

Young JP, Crossman LC, Johnston AW, Thomson NR, Ghazoui ZF, Hull KH, Wexler M, Curson AR, Todd JD, Poole PS, Mauchline TH, East AK, Quail MA, Churcher C, Arrowsmith C, Cherevach I, Chillingworth T, Clarke K, Cronin A, Davis P, Fraser A, Hance Z, Hauser H, Jagels K, Moule S, Mungall K, Norbertczak H, Rabbinowitsch E, Sanders M, Simmonds M, Whitehead S, Parkhill J (2006) The genome of *Rhizobium leguminosarum* has recognizable core and accessory components. Genome Biol 7:R34

Zahran HH (1991) Conditions for successful *Rhizobium*-legume symbiosis in saline environments. Biol Fertil Soils 12:73–80

Zahran HH (1992) DNA-DNA hybridization of some root-nodule bacteria indigenous in the salt-affected soils of Egypt. Folia Microbiol 37:295–298

Zahran HH (1999) Rhizobium-legume symbiosis and nitrogen fixation under severe conditions and in an arid climate. Microbiol Mol Biol Rev 63:968–989

Zahran HH, Sprent JI (1986) Effects of sodium chloride and polyethylene glycol on root hair infection and nodulation of *Vicia faba* L. plants by *Rhizobium leguminosarum*. Planta 167:303–309

Zahran HH, Rasanen LA, Karsisto M, Lindstrom K (1994) Alteration of lipopolysaccharide and protein profiles in SDS-PAGE of rhizobia by osmotic and heat stress. World J Microbiol Biotechnol 10:100–105

Zahran HH, Mohammad EM, Emam MM, Ismael SS (1997) The chemical composition, structure and ultrastructure of a halotolerant rhizobia isolated from Egypt. In: Proceedings of the 9th microbiology conference, Cairo, Egypt, pp 121–148

Zhu H, Chen T, Zhu M, Fang Q, Kang H, Hong Z, Zhang Z (2008) A novel ARID DNA-binding protein interacts with SymRK and is expressed during early nodule development in *Lotus japonicus*. Plant Physiol 148:337–347

Zou N, Dort PJ, Marcar NE (1995) Interaction of salinity and rhizobial strains on growth and N_2-fixation by *Acacia ampliceps*. Soil Biol Biochem 27:409–413

Chapter 4
A Genome-Wide Investigation on Symbiotic Nitrogen-Fixing Bacteria in Leguminous Plants

Lebin Thomas and Zeeshanur Rahman

Abstract The present article is focused on a wide genome investigation on nodule forming bacteria in 17 different genera. The genera included for the search were *Aminobacter*, *Azorhizobium*, *Bosea*, *Bradyrhizobium*, *Devosia*, *Ensifer/Sinorhizobium*, *Mesorhizobium*, *Methylobacterium*, *Microvirga*, *Neorhizobium*, *Ochrobactrum*, *Pararhizobium*, *Phyllobacterium*, *Rhizobium*, and *Shinella*, belonging to *Alphaproteobacteria*, and *Burkholderia* and *Cupriavidus* from *Betaproteobacteria*. All these genera (possessing full genomes) were scrutinized for the sequences with and without nodule forming capacity in the list of full genome sequences of NCBI. Approximately 1.83% of the total number of bacterial genome sequences were reported with nodule forming capacity. Maximum sequences for nodulation were available for *Burkholderia* followed by *Rhizobium* and *Bradyrhizobium*. Also, a great diversity of nodulation proteins were observed in most of the genera. Different nodulation proteins associated with different genera and their function were documented using genome mining in NCBI, UniProtKB, and literature survey. Different sequences of *Bradyrhizobium*, *Mesorhizobium*, *Rhizobium*, and *Ensifer* exhibited a great variety of nodulation proteins (nod, nol, noe, nfe, and nop). Overall, our investigation presents a molecular understanding about the nodule formation in legume plants and provides better future prospects for various biotechnological approaches to supply nitrogen in legume and nonlegume crops.

4.1 Introduction

Nitrogen is assimilated into plants from atmosphere by a wide diversity of nitrogen-fixing microorganisms. They are prokaryotic and are known as "diazotrophs." Both bacterial and archaebacterial genera can fix nitrogen (Young 1992). For the same, a wide variety of Gram-positive and Gram-negative bacteria and methanogenic

L. Thomas
Department of Botany, Hansraj College, University of Delhi, Delhi, Delhi, India

Z. Rahman (✉)
Department of Botany, Zakir Husain Delhi College, University of Delhi, New Delhi, Delhi, India

A. Varma et al. (eds.), *Plant Microbe Symbiosis*,
https://doi.org/10.1007/978-3-030-36248-5_4

archaea have been identified for reduction of molecular nitrogen into ammonia. This bioreaction is possible in free-living state and symbiotic association of bacteria with plants, corals, and other animals (Franche et al. 2009; Lesser et al. 2018). However, the complex coordination of symbiosis between plant hosts and their nitrogen-fixing bacteria is the major nitrogen-fixing system in biosphere, and the hosts serve as the main reserve of nitrogen for the great value in agriculture (Franche et al. 2009).

The symbiotic nitrogen fixers are divided into two different groups: plant growth-promoting rhizobacteria (PGPRs) and endosymbionts (Mus et al. 2016). PGPRs like *Azospirillum* and *Nostoc* are the associative and rhizospheric colonizers of the tropical grasses (*Oryza*, *Zea*, *Leptochloa*, etc.). They usually inhabit on root surface, root hair, ruptured epidermis, and outer cortex regions in the rhizosphere (James 2000). Nitrogen-fixing endosymbionts involve bacterial interaction with legumes and actinorhizal plants (nodule formation) and cyanobacterial association with bryophytes, cycads, and basal eudicots (Mus et al. 2016).

4.2 Rhizobia and Legume

The process of symbiosis in legume is restricted with the few bacterial lineages of *Proteobacteria* (selected members of only *Alphaproteobacteria* and *Betaproteobacteria*), which are broadly called as "rhizobia." The process of symbiosis involves proliferation of bacteria to the root surface (and occasionally stem) and acquisition of the capacity to develop nodule (Sachs et al. 2018). According to Weir (2016) and other investigation, approximately 13–17 bacterial genera are identified for the legume association. The common genera include *Azorhizobium*, *Bradyrhizobium*, *Devosia*, *Ensifer* (syn. *Sinorhizobium*), *Mesorhizobium*, *Methylobacterium*, *Microvirga*, *Ochrobactrum*, *Phyllobacterium*, *Rhizobium*, and *Shinella*, belonging to order *Rhizobiales* from *Alphaproteobacteria*, and *Burkholderia* and *Cupriavidus* from *Betaproteobacteria*. Less commonly known spp. are *Aminobacter*, *Bosea*, *Neorhizobium*, and *Pararhizobium* from *Rhizobiales*.

The distribution of nodulation in plants is restricted to only one family, i.e., Fabaceae (Leguminosae). However, nodulation within the family is widespread and reported in more than 16,000 species in 650–700 genera (Downie 2014). In the three subfamilies of Fabaceae, nodulation is most and least common in Papilionoideae and Caesalpinioideae, respectively (Sprent 2007). Most common hosts in legumes are *Lupinus* spp., *Pisum sativum*, *Vicia faba*, *Phaseolus vulgaris*, *Trifolium* spp., *Lotus* spp., *Medicago* spp., etc. (Wang et al. 2018; Perret et al. 2000). Other than legumes, *Parasponia*, a member of family Cannabaceae, is also reported to display nodule formation with rhizobia as exception (Op den Camp et al. 2012).

4.3 Nodulation Factors (NFs)

In the legume-symbiotic rhizobia interaction, nodulation factors (NFs) in rhizobia are the key signals to develop the symbiosis in plants. NFs are encoded by accessory genome that varies in composition among different strains within the same species (other than the core genome, which is common in all species and encodes essential functions). The essential genes for NFs and infection of nod signals include *nod/nol/noe*, and the conserved canonical genes for nodulation are *nodABCDIJ* (Remigi et al. 2016). NFs of rhizobia communicate with potential host plants by specific LysM-domain-containing receptor-like kinases (Limpens et al. 2015). Even a slight modification in the NF receptors can bring sufficient change in the expression from delayed nodulation to the inability to elicit infection (Radutoiu et al. 2007). Other than NF regulation, a report by Giraud (2007) suggested that some bradyrhizobia do not possess *nod* genes, but use an alternative pathway for nodulation.

The symbiosis is initiated by the secretion of flavonoids/isoflavonoids from plant roots in soil. Host-specific rhizobia identify these molecules and induce the synthesis of a transcription factor of *nodD* gene, which mediate the activation of various NFs (Via et al. 2016). NFs are lipochitooligosaccharide (LCO) molecules with a lot of chemical variations (Remigi et al. 2016). These molecules offer communication through molecular signaling pathways in very specific and selective way to clearly defined hosts for the infection (Zgadzaj et al. 2015).

LCO has a backbone of four or five N-acetylglucosamine oligosaccharide with adjunction of a fatty acyl chain of varying length at the nonreducing end (Wang et al. 2018). The common *nodABC* operon encodes the core structure of LCO following the steps of chain elongation, deacetylation, and acylation (Perret et al. 2000; Poinsot et al. 2016). Other substituents of NF genes are rewarding in species-specific chemical decorations. For example, *nodEF* leads to the development of polyunsaturated fatty acid on the main residue of NF. *nodIJ* is involved in the transport of NFs. *noeC* is determinant of arabinosylation and *nodZ* and *nolK* for fucosylation, which are the additional sugar attachments of 6-O glycosylation type. *nodH* and *noeE* are characteristics of sulfonation associated with *nodPQ*. Acetylation at both extremities is represented by *nodL*, *nodX*, and *nolL*. N-Methylation and carbamoylation are controlled by *nodS*, *nodU*, and *nolO* and 2-O methylation by *noeI* (Perret et al. 2000). Besides, *nodVW* contributes as another recognition system of two-component regulatory family (other than *nodD*) in response to plant-produced isoflavone signal (Loh et al. 1997). Different nodulation proteins and their function are listed in Table 4.1.

Other than NFs, bacterial cell surface components like extracellular polysaccharides (EPS), lipopolysaccharides (LPS), K polysaccharides, and cyclic glucans are also crucial for infection and nodulation development (Cooper 2007).

Table 4.1 Different nodulation proteins and their functions

Nodulation proteins	Function	References
nodA	N-Acyltransferase activity that links the acyl chain to the NH2-free carbon C-2 of the nonreducing end of the oligosaccharide in biosynthesis of lipochitooligosaccharides (LCO)	Atkinson et al. (1994)
nodB	N-Deacetylase activity removes the N-acetyl moiety from the nonreducing end of the N-acetylglucosamine oligosaccharides	Atkinson et al. (1994)
nodC	Chitin synthase for biosynthesis and initiation of assembly of LCO	Shamseldin (2013); Barny et al. (1996)
nodD	Activates the transcription of other inducible *nod* genes	Perret et al. (2000)
nodE, nodF	Development of polyunsaturated fatty acid on the main residue	Perret et al. (2000)
nodG	Enzymatic activity of an 3-oxoacyl-acyl carrier protein reductase	López-Lara and Geiger (2001)
nodH, nodP nodQ	Responsible for the 6-O-sulfation of the reducing N-acetyl-D-glucosamine of NodRm factors	Roche et al. (1991)
nodI	Lipooligosaccharide transport system ATP-binding protein; are involved in the export of Nod factors	Rudder et al. (2014)
nodJ	The efficiency of secretion of lipochitin oligosaccharides; are involved in the export of Nod factors	Spaink et al. (1995)
nodK, nodY	Specific function in *Bradyrhizobium*	You et al. (2002)
nodL, nodX, nolL	Acetylation at both extremities	Perret et al. (2000)
nodM, nodN	Production of the root hair deformation (Had) factor, where nodM is for d-glucosamine synthetase	Baev et al. (1991)
nodO	Encodes a Ca^{2+}-binding protein	Economou et al. (1990)
nodS	Involved in N-methylation of NodNGR factors	Lewin et al. (1990)
nodT	Efflux transmembrane transporter activity	Surin et al. (1990)
nodU	Carbamoyl transferase involved in 6-O-carbamoylation of NodNGR factors	Lewin et al. (1990), Snoeck et al. (2003)
nodV, nodW	Recognition system of two-component regulatory family in response to plant-produced isoflavone signal	Loh et al. (1997)
nodZ	The transfer of GDP-fucose (fucosylation) to NodNGR factors	Quesada-Vincens et al. (1997)
noeA	The SAM-dependent methyltransferase is involved in alfalfa cultivar-specific nodulation	Du et al. (2005)
noeB	Sulfuric ester hydrolase activity for host-specific nodulation	Schneiker-Bekel et al. (2011)

(continued)

Table 4.1 (continued)

Nodulation proteins	Function	References
noeC, noeH	Arabinosylation of Nod factors	Mergaert et al. (1996), Niehaus and Becker (1998)
noeD	A glucosamine synthase	Lohrke et al. (1998)
noeE	A fucose-specific sulfotransferase which is required for the sulfation of LCOs (nod factors) that acquire the capacity to nodulate	Hanin et al. (1997)
noeI	2-O-Methyltransferase for methylation of the fucose moiety of Nod factors	Madinabeitia et al. (2002)
noeJ	Mannose-1-phosphate guanylyltransferase (GDP) protein	Peralta et al. (2016)
noeK	Intramolecular phosphotransferase activity	Sullivan et al. (2002)
noeL	GDP-D-mannose dehydratase nodulation protein	UniProt
noeO	Dehydrogenase/reductase	UniProt
noeP	No information available	–
noeT	Acetyltransferase for the acetylation of the nodulation factors	Österman et al. (2014)
nolA	DNA-binding activity involved in genotype-specific nodulation	Gillette and Elkan (1996)
nolC	Homologous to heat shock protein	Krishnan and Pueppke (1991)
nolE	Periplasmic protein with no known function	Johnston et al. (2014)
nolF, nolG	Involve in the production of *Medicago*-specific nodulation signal molecule and transmembrane transporter activity	UniProt
nolJ	Involve in efficiency of soybean nodulation and in nodulation delay	UniProt
nolK	Fucosylation of Nod factors; is involved in the synthesis of GDP-fucose	Mergaert et al. (1996)
nolL	Determines nodulation efficiency by mediating the acetylation of the fucosyl residue in the nodulation factor	Corvera et al. (1999)
nolM, nolP, nolS	No information available	
nolO	A carbamoyltransferase for 3 (or 4)-O-carbamoylation of the nonreducing terminus of NodNGR factors	Jabbouri et al. (1998)
nolR	DNA-binding transcription factor activity to downregulate the expression of the activator nodD1 gene, nodD2, nodM, and the common nodABC operon	Cren et al. (1995)
nolB, nolT, nolU, nolV, nolW, nolX	Type III secretion system (TTSS) to deliver effector proteins directly into the cytosol from operon *nolXWBTUV*	Krishnan et al. (2003)
nolY, nolZ	Production of lipooligosaccharide nodulation signals	Dockendorff et al. (1994)

(continued)

Table 4.1 (continued)

Nodulation proteins	Function	References
nopA	An external component (the macromolecular surface appendages) of the TTSS needed for secretion of all other Nops	Deakin et al. (2005)
nopB	Type III effector secreted protein associated with pilus-like surface appendages	Saad et al. (2005), Kimbrel et al. (2013)
nopC	A *Rhizobium*-specific TTSS effector	Jiménez-Guerrero et al. (2015)
nopT	Rhizobial type 3 effector and a functional protease (with autoproteolytic activity) of the YopT-AvrPphB effector family for nodulation to particular host legumes	Dai et al. (2008)
Rj2 and Rfg1	Responsible for host specificity with legumes	Yang et al. (2010)
nap and nos	Involve in nodule regulation	Sánchez et al. (2013)
nfe (nfeA, nfeB, nfeD)	Nodulation efficiency and competitiveness of bacterium with plant root	Sanjuan and Olivares (1989), García-Rodríguez and Toro (2000)
nfeD	Stomatin-like protein (slp)	Green et al. (2009)

4.4 Advances in Nitrogen-Fixing Root Nodule Symbiosis

Many foremost advances have been made in describing plant responses to NF required for bacterial infection and nodule morphogenesis. Also, great efforts have been made in molecular characterization and cloning of several plant genes required for NF signal transduction (Schauser et al. 1999; Endre et al. 2002; Madsen et al. 2003; Lévy et al. 2004; Cerri et al. 2016; Calatrava-Morales et al. 2018; Tsikou et al. 2018). Many investigations have found that the precise bacterial penetration, infection, and development of efficient nitrogen-fixing root nodules are tightly regulated. For instance, the nodulation in roots of *Trifolium repens* involves a specific breakdown of its growing root hair apical cell walls, which is mediated by *CelC2* gene in *Rhizobium leguminosarum* bv. *trifolii* that encodes for an endoglucanse component of cellulase system (Robledo et al. 2018). In many legumes like *Medicago truncatula*, the earliest stages of root nodule development were found to be regulated by transcription factor like ERF required for nodulation (ERN), which on further characterization revealed the presence of a conserved threonine for both DNA binding and transcriptional activity (Cerri et al. 2016). Further, nitrogen-fixing symbiont *Sinorhizobium* (*Ensifer*) *meliloti* is known to utilize interkingdom N-acyl homoserine lactone signals in a population density-dependent mode for regulating highly variable gene expression, needed to establish symbiosis with the legume alfalfa (Calatrava-Morales et al. 2018).

In the wild legume *Lotus japonicus*, the gene *Epr3* is found integrated into symbiosis signal transduction pathways, which encodes for a LysM receptor kinase. This integration is needed for the perception of compatible rhizobial exopolysaccharides at root epidermis, thereby promoting intracellular cortical infection and maintaining bacteria enclosed in host membranes (Kawaharada et al. 2017). In soybean cultivar BARC2, the ineffective nodule formation with *Bradyrhizobium elkanii* USDA61 was found to be partly mediated by the type 3 effector-triggered immunity of the host-specific *Rj4* gene (Faruque et al. 2015). This soybean-*B. elkanii* symbiosis has a low nitrogen-fixation efficiency, but these bacterial strains are highly competitive for nodulation in the cultivars with an *Rj4* allele, which was later found to encode a thaumatin-like protein (Tang et al. 2016). In soybean nodules, characterization of a *GmFWL1* gene (which is specifically expressed in cells of root hairs in response to rhizobia) showed that it encodes for a membrane microdomain-associated protein needed for interacting with various proteins (like remorin, prohibitins, and flotillins) that regulates nodulation, particularly in the early rhizobial infection (Qiao et al. 2017).

Furthermore, in this legume, a microRNA was found to be translocated to roots from shoots for controlling rhizobial infection by a posttranscriptional regulation of the symbiosis suppressor in roots (Tsikou et al. 2018). This was considered to be an inducible autoregulatory process in which the legume host constrains nodule numbers for balancing symbiosis and plant growth. This mRNA functioned as an activator of symbiosis downstream of a histidine kinase-mediated cytokinin perception and root formation genes, when the roots are uninfected.

The phylogenetic sequence analysis of *nodA* gene in *Methylobacterium nodulans* (isolated from *Crotalaria* legumes) revealed its close relatedness to gene *nodA* of *Bradyrhizobium*, which suggested that it was acquired by horizontal gene transfer (Sy et al. 2001). Similarly, the *nodA* gene of *Aminobacter* was highly related to *Mesorhizobium loti* but was differentiated and highly divergent from those of *Mesorhizobium metallidurans* (Maynaud et al. 2012). Rhizobial symbiosis genes that are frequently carried on symbiotic islands or plasmids have different phylogenies (than genome of their host) and are involved in a geographically widespread and non-constrained horizontal transfer among different rhizobial genera. This movement of symbiosis genes allows rhizobia of a particular soil condition to overcome incompatibility with different legumes growing in soil (Andrews et al. 2018).

In nitrogen-fixing rhizobia symbionts (*Rhizobium, Sinorhizobium, Neorhizobium*) of leguminous plants, two types of genome organization including unitary type (chromosomes encoded) and multipartite type (plasmids encoded) have been revealed, in which the latter frequently controls the symbiotic properties in the fast-growing rhizobia. The analysis of the bacterial phylogenetic diversity (including *nodC* gene) modulating *Lupinus micranthus* from different geographical sites identified groups formed by isolates of *Bradyrhizobium*, *Microvirga*, and *Phyllobacterium*. The β-rhizobia *Burkholderia* was recently found to form a nitrogen-fixing symbiosis with several legumes, among which *B. phymatum* and *B. mimosarum* were considered to be highly competitive symbionts for the legumes *Phaseolus vulgaris*, *Macroptilium atropurpureum*, *Vigna unguiculata*, and *Mimosa pudica* (Lardi and Pessi 2018). The less competitive *Cupriavidus* that originated in the neotropics is considered to have acquired its plasmid-borne symbiotic genes from its relative, *Burkholderia* (Remigi et al. 2016). However, it can be a dominant symbiont of *Mimosa* in environments that contain high concentrations of heavy metals like mining areas.

4.5 Genome-Wide Investigation of Nodule Forming Bacteria

A genome-wide investigation was made from the genome list available on NCBI—Genome Information by Organism under the website https://www.ncbi.nlm.nih.gov/genome/browse/#!/overview/. As of November 28, 2018, a total no. of 162,354 genome sequences of bacteria were available on NCBI web portal. In the list, the no. of genome sequences was highest in group *Proteobacteria* (i.e., 49.21%) followed by *Terrabacteria* and FCB group (Fig. 4.1a). The subgroup *Gammaproteobacteria* (73.67%) within *Proteobacteria* dominated for the full genome sequences, though nodule forming bacteria in the legumes were prevalent belonging to *Alphaproteobacteria* and *Betaproteobacteria* with 9.61% (7681 sequences) and 9.79% (7821 sequences), respectively (Fig. 4.1b).

In the present investigation, a total of 17 genera (15 belonging to order *Rhizobiales* and 2 from *Betaproteobacteria*) were scrutinized for nodule forming capacity by genome mining using keywords as nodulation, nodA, nodB, nodC, and nodD in NCBI protein search and UniProtKB. For keyword "nodulation" search, different proteins related to nodulation that appeared in search were covered in our list. The no. of genomes with and without nodule forming capacity for different bacteria was presented using component bar diagram for relative percent comparison (Fig. 4.2). Also, different proteins related to nodule formation for different genera were listed in Table 4.2. A total no. of 3419 genome sequences were studied, out of which 2970 sequences possessed different nodulation proteins. Approximately 1.83% of the total bacterial genome sequences harbored nodule forming capability.

In *Bosea*, out of seven genome sequences, four genomes possessed nodulation proteins. Although nodA-, nodB-, and nodD-like important proteins were absent, other nodulation proteins like nodC, nodE, nodS, nodW, and nolO were reported in different species of *Bosea*. Many investigators have identified *Bosea* as endophytic bacteria, but its ability of nodule formation has not been well investigated (Safronova et al. 2015). The genus *Bradyrhizobium* is very commonly reported to have a nodule forming capacity. Out of 192 genome sequences available, 140 genomes exhibited nodulation proteins. In the search, different nod, nol, and noe were present in their genomes. The essential *nodABC* and other *nodVW* and *nodMN* operon products were dominantly present in many sequences of *Bradyrhizobium*. Other NFs particularly nodD, nodJ, nodS, and nodZ were also present in many species. Two unique nodulation proteins, nodK and nodY, were reported only in the different sequences of *Bradyrhizobium*. For *Ochrobactrum* also, a high proportion of its genome sequences showed the nodule forming properties; however, important proteins like nodA and nodB were absent. Nevertheless, polysaccharide deacetylase, which showed a high nodB homology, was reported in several sequences of *Ochrobactrum*. Other than this, nodN was dominantly present in many sequences. In *Devosia*, almost 50% of the total genomes were identified with nodule forming capacity. nolG was very common in different species of *Devosia*. Other than this, only one other protein nodW was reported in *Devosia*

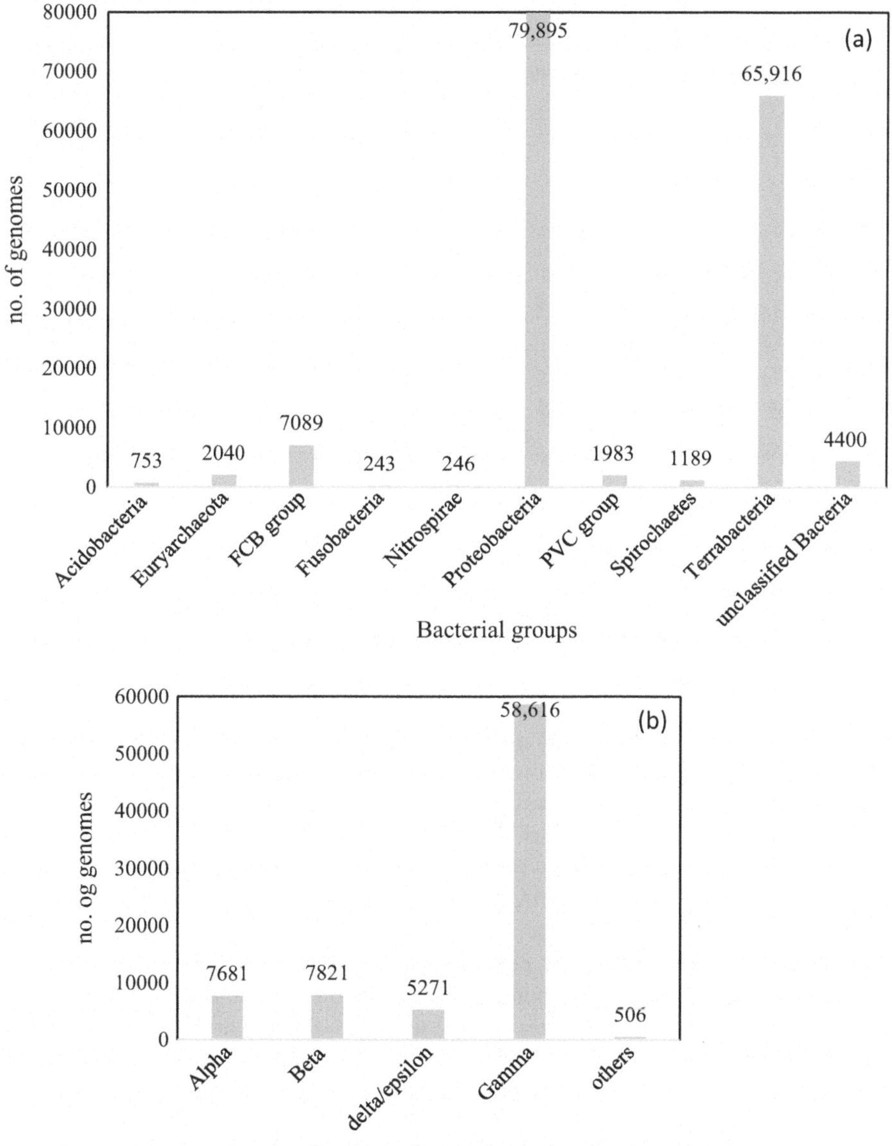

Fig. 4.1 Number of genome sequences available for bacteria in different groups (**a**) and in different subgroups of *Proteobacteria* (**b**) at NCBI

after genome mining; however, other essential nodulation proteins were absent. In another search, nodD was found in the plasmid of one species (*D. neptuniae*) as a partial sequenced protein, but the full genome sequencing of that species was not available. For *Methylobacterium*, nodS was abundantly present in many sequences.

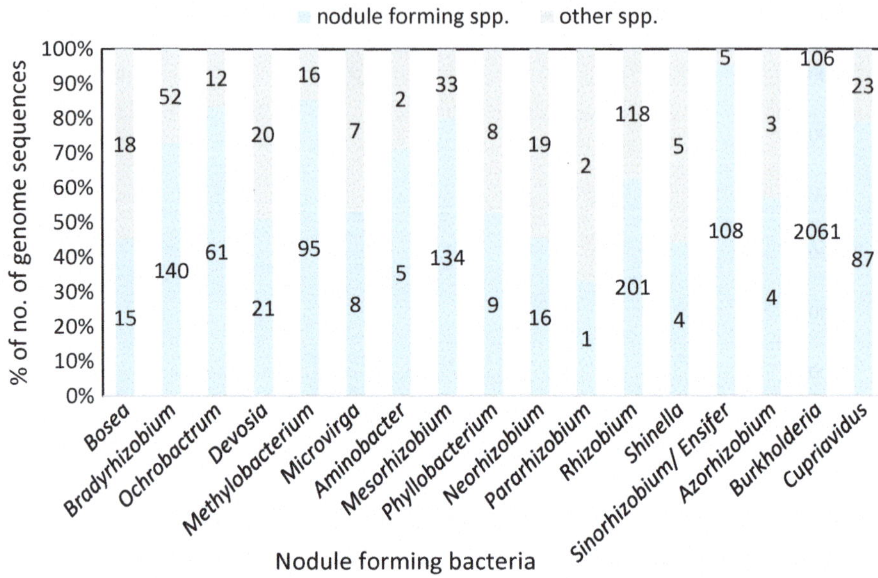

Fig. 4.2 Percentage proportion of number of nodule forming and other bacteria in the genome list of NCBI. Numeric written at the bar represents the number of genome sequences for different bacteria

Other proteins of nod, nol, and nfe types were also reported for nodulation. In *Microvirga*, a total of eight genome sequences were found to possess nodulation proteins (nod and nol types). However, the essential nodB, nodC, and nodD were not reported in mining for this genus. For *Aminobacter*, there were a total of seven species in the genome list, from which four spp. had nodulation proteins like nodA, nodC, nodD, nodF, nodN, and nolA. For *Mesorhizobium*, a high number of genomes (134 sequences) were available for nodulation studies. This bacterium possesses a great variety of different nod, nol, noe, nfe, nop, and rhcL in different sequences. For *Phyllobacterium*, 9 out of 19 genome sequences showed nodulation capacity. Only four different nod types of proteins (nodA, nodC, nodD, and nodN) and nfeD were present in their sequences. In *Neorhizobium*, 16 genome sequences possessed nodulation proteins with many characteristics of nod type and single characteristics ofnol, noe, and nfe types. For *Pararhizobium*, only one out of three genome sequences showed nodule forming capacity containing only four different nodulation proteins. Although *Rhizobium* is very popular for nodulation in legumes, approximately 63% of numbers of the genomes of *Rhizobium* spp. were found with nodule forming properties. Different nodulation proteins like nod, nol, noe, nfe, nop, etc. were present very frequently in their sequences. Another genus, *Shinella*, was less frequent in the genome list for the nodulation. Only four sequences were reported, containing few nod and nol types of proteins. In *Sinorhizobium/Ensifer*, more than 95% of numbers of genome sequences belonged to nodule forming species. Moreover, its different species showed a high diversity of

Table 4.2 Nodule forming bacteria possessing different nodulation proteins

Bacterial groups and subgroups	Genera	nod	nol	noe	Other nodulation proteins
Alphaproteobacteria Rhizobiales					
Bradyrhizobiaceae	*Bosea*	nodC, nodE, nodS, nodW	nolO	–	–
	Bradyrhizobium	nodA, nodB, nodC, nodD, nodI, nodJ, nodK, nodM, nodN, nodS, nodT, nodU, nodV, nodW, nodY, nodZ	nolA, nolB, nolG, nolK, nolL, nolM, nolN, nolO, nolR, nolU, nolV, nolW, nolX, nolY, nolZ	noeA, noeD, noeE, noeI, noeL, nolNO	nfeD, nopA, nopAR, nopB, nopE, nopX, nwsA, nwsB
Brucellaceae	*Ochrobactrum*	nodD, nodG, nodN, nodT, nodV, nodW	nolR	–	–
Hyphomicrobiaceae	*Devosia*	nodW	nolG	–	–
Methylobacteriaceae	*Methylobacterium*	nodA, nodB, nodC, nodD, nodE, nodF, nodH, nodI, nodJ, nodN, nodS, nodW	nolU, nolV, nolW, nolX	–	rhcL
	Microvirga	nodA, nodJ, nodS, nodU, nodZ	nolU, nolV, nolW, nolX	–	–
Phyllobacteriaceae	*Aminobacter*	nodA, nodC, nodD, nodF, nodN	nolA	–	–
	Mesorhizobium	nodA, nodB, nodC, nodD, nodE, nodF, nodG, nodH, nodI, nodJ, nodL,	nolB, nolF, nolL, nolO, nolR, nolU, nolV, nolW, nolX	noeA, noeB, noeC, noeE, noeI, noeK	nfeD, nopA, nopB, nopC, nopT, nopX, rhcL

(continued)

Table 4.2 (continued)

Bacterial groups and subgroups	Genera	nod	nol	noe	Other nodulation proteins
		nodN, nodO, nodS, nodU, nodW, nodX, nodZ			
	Phyllobacterium	nodA, nodC, nodD, nodN	–	–	nfeD
Rhizobiaceae	*Neorhizobium*	nodA, nodB, nodC, nodD, nodE, nodF, nodI, nodJ, nodN, nodO, nodU, nodV, nodW	nolR	noeT	nfeD
	Pararhizobium	nodB, nodN, nodT, nodW	–	–	–
	Rhizobium	nodA, nodB, nodC, nodD, nodE, nodF, nodG, nodH, nodI, nodJ, nodL, nodM, nodN, nodO, nodP, nodQ, nodS, nodT, nodU, nodV, nodW, nodX, nodZ	nolB, nolC, nolE, nolF, nolG, nolJ, nolK, nolL, nolO, nolP, nolR, nolT, nolU, nolV, nolW, nolX	noeA, noeB, noeC, noeE, noeH, noeI, noeJ, noeK, noeO, noeP	nfeD, nopA, nopB, nopL, nopP, nopX, fixJ, NOLO, nfeA, nodUcds, RND
	Shinella	nodB, nodC, nodD,	nolF, nolG, nolR	–	–

(continued)

Table 4.2 (continued)

Bacterial groups and subgroups	Genera	nod	nol	noe	Other nodulation proteins
		nodG, nodN, nodT			
	Sinorhizobium/ Ensifer	nodA, nodB, nodC, nodD, nodE, nodF, nodG, nodH, nodI, nodJ, nolL, nodM, nodN, nodO, nodP, nodQ, nodS, nodU, nodV, nodW, nodX, nodZ	nolB, nolC, nolF, nolG, nolJ, nolK, nolL, nolO, nolR, nolS, nolT, nolU, nolV, nolW, nolX	noeA, noeB, noeE, noeI, noeK, noeJ	nfeA, nfeD, nopA, nopB, nopC, nopL, nopP, nopX
Xanthobacteraceae	*Azorhizobium*	nodA, nodB, nodC, nodD, nodI, nodJ, nodS, nodU, nodW, nodZ	–	noeC, noeO, noeP	–
Betaproteobacteria Burkholderiales					
Burkholderiaceae	*Burkholderia*	nodA, nodB, nodC, nodD, nodJ, nodI, nodL, nodN, nolO, nodS, nodV, nodU	nolA, nolG, nolNO, nolT, nolV, nolX	–	nfeD
	Cupriavidus	nodA, nodB, nodC, nodD, nodH, nodI, nodJ, nodN, nodS, nodT, nodV, nodW	nolG	–	–

nodulation proteins, possessing different nod, nol, noe, nop, and nfe in their genomes. In *Azorhizobium*, out of seven species, four of them were found to possess nodule forming proteins, where nodA and nodB were commonly present in different species, but nodC and nodD were reported in only one species (*A. caulinodans* ORS571). Other than this, few other types of NF and two noe proteins (noeC and noeP) also appeared in the genome mining.

In group *Betaproteobacteria*, *Burkholderia* has been widely sequenced as many as 2167 times for different species, and more than 95% sequences had possessed nodule forming properties. Different types of nod, nol, and nfe proteins were present in their sequences after genome mining. For *Cupriavidus*, a total number of 87 out of 110 genomes were available for nodule forming capacity. Proteins nodA and nodC were very common in their genome sequences.

4.6 Conclusions and Future Prospects

It is concluded that some selected spp. in 17 bacterial genera have the capacity of nodulation in legume plants. Approximately 1.83% of the total number of bacterial genomes can be used for in silico analysis of nodulation. The investigation also enlisted diverse proteins required for the nodule formation in different bacterial genera. A variety of NFs like nod, nol, noe, nop, nfe, nws, and rhcL were studied from different bacteria. This study also provided comparisons among different nodule forming bacteria for possessing different types of NFs. Several bacterial genera were identified containing operons and host-specific nodulation signals and lacking essential nodulation proteins.

The study on the sequences of nodule forming bacterial genomes and their nodulation proteins can bring better understanding on nitrogen supply to legume plants. Particularly, the symbiotic characters can be studied in so much detail. The infection of bacteria in plants can be enhanced by transferring the genes of different NFs of other rhizobia. This research investigation can be useful in speculation in study of evolution of nodulation proteins in bacteria. The geographically widespread non-constrained horizontal transfer of rhizobial symbiosis genes among different rhizobial genera can be studied using different NF-associated genes and their products to analyze a critical ecological link of soil rhizobial bacteria with the floral biodiversity and vegetation community structures. Furthermore, an effort for nitrogen-fixing ability in nonlegume plants can be also projected under biotechnological methodologies for crop improvement as a future prospect.

References

Andrews M, De Meyer S, James E, Stępkowski T, Hodge S, Simon M, Young J (2018) Horizontal transfer of symbiosis genes within and between rhizobial genera: occurrence and importance. Genes 9:321

Atkinson EM, Palcic MM, Hindsgaul O, Long SR (1994) Biosynthesis of *Rhizobium meliloti* lipooligosaccharide nod factors: NodA is required for an N-acyltransferase activity. Proc Natl Acad Sci U S A 91:8418–8422

Baev N, Endre G, Petrovics G, Banfalvi Z, Kondorosi A (1991) Six nodulation genes of *nod* box locus 4 in *Rhizobium meliloti* are involved in nodulation signal production: *nodM* codes for D-glucosamine synthetase. Mol Gen Genet 228:113–124

Barny MA, Schoonejans E, Economou A, Johnston AW, Downie JA (1996) The C-terminal domain of the *Rhizobium leguminosarum* chitin synthase NodC is important for function and determines the orientation of the N-terminal region in the inner membrane. Mol Microbiol 19:443–453

Calatrava-Morales N, McIntosh M, Soto MJ (2018) Regulation mediated by N-acyl homoserine lactone quorum sensing signals in the *Rhizobium*-legume symbiosis. Genes 9:263

Cerri MR, Frances L, Kelner A, Fournier J, Middleton PH, Auriac MC, Mysore KS, Wen J, Erard M, Barker DG, Oldroyd GE (2016) The symbiosis-related ERN transcription factors act in concert to coordinate rhizobial host root infection. Plant Physiol 171(2):1037–1054. https://doi.org/10.1104/pp.16.00230

Cooper J (2007) Early interactions between legumes and rhizobia: disclosing complexity in a molecular dialogue. J Appl Microbiol 103:1355–1365

Corvera A, Promé D, Promé JC, Martínez-Romero E, Romero D (1999) The *nolL* gene from *Rhizobium etli* determines nodulation efficiency by mediating the acetylation of the fucosyl residue in the nodulation factor. Mol Plant-Microbe Interact 12:236–246

Cren M, Kondorosi A, Kondorosi E (1995) NoIR controls expression of the *Rhizobium meliloti* nodulation genes involved in the core nod factor synthesis. Mol Microbiol 15:733–747

Dai WJ, Zeng Y, Xie ZP, Staehelin C (2008) Symbiosis-promoting and deleterious effects of NopT, a novel type 3 effector of *Rhizobium* sp strain NGR234. J Bacteriol 190:5101–5110

Deakin WJ, Marie C, Saad MM, Krishnan HB, Broughton WJ (2005) NopA is associated with cell surface appendages produced by the type III secretion system of *Rhizobium* sp. strain NGR234. Mol Plant-Microbe Interact 18:499–507

Dockendorff TC, Sharma AJ, Stacey G (1994) Identification and characterization of the *nolYZ* genes of *Bradyrhizobium japonicum*. Mol Plant-Microbe Interact 7:173–180

Downie JA (2014) Legume nodulation. Curr Biol 24:R184–R190

Du BH, Jiang JQ, Li XH, Wang L, Yang SS (2005) Cloning, deletion and functional analysis of *noeA* from *Sinorhizobium meliloti* 042BM. Acta Microbiol Sin 45:195–200

Economou A, Hamilton WD, Johnston AW, Downie JA (1990) The rhizobium nodulation gene *nodO* encodes a $Ca2^{(+)}$-binding protein that is exported without N-terminal cleavage and is homologous to haemolysin and related proteins. EMBO J 9:349–354

Endre G, Kerestz A, Kevei Z, Mihacea S, Kalo P, Kiss BG (2002) A receptor kinase gene regulating symbiotic nodule development. Nature 417:962–966

Faruque OM, Miwa H, Yasuda M, Fujii Y, Kaneko T, Sato S, Okazaki S (2015) Identification of *Bradyrhizobium elkanii* genes involved in incompatibility with soybean plants carrying the Rj4 allele. Appl Environ Microbiol 81:6710–6717

Franche C, Lindstrom K, Elmerich C (2009) Nitrogen-fixing bacteria associated with leguminous and non-leguminous plants. Plant Soil 321:35–59

García-Rodríguez FM, Toro N (2000) *Sinorhizobium meliloti nfe* (nodulation formation efficiency) genes exhibit temporal and spatial expression patterns similar to those of genes involved in symbiotic nitrogen fixation. Mol Plant-Microbe Interact 13:583–591

Gillette WK, Elkan GH (1996) *Bradyrhizobium* (*Arachis*) sp. strain NC92 contains two *nodD* genes involved in the repression of *nodA* and a *nolA* gene required for the efficient nodulation of host plants. J Bacteriol 178:2757–2766

Giraud E, Moulin L, Vallenet D, Barbe V, Cytryn E, Avarre JC et al (2007) Legumes symbioses: absence of *Nod* genes in photosynthetic bradyrhizobia. Science 316:1307–1312

Green JB, Lower RP, Young JPW (2009) The NfeD protein family and its conserved gene neighbours throughout prokaryotes: functional implications for stomatin-like proteins. J Mol Evol 69:657

Hanin M, Jabbouri S, Quesada-Vincens D, Freiberg C, Perret X, Promé JC, Fellay R (1997) Sulphation of *Rhizobium* sp NGR234 nod factors is dependent on *noeE*, a new host-specificity gene. Mol Microbiol 24:1119–1129

Jabbouri S, Relić B, Hanin M, Kamalaprija P, Burger U, Promé D, Broughton WJ (1998) *nolO* and *noeI* (HsnIII) of *Rhizobium* sp NGR234 are involved in 3-O-carbamoylation and 2-O-methylation of nod factors. J Biol Chem 273:12047–12055

James EK (2000) Nitrogen fixation in endophytic and associative symbiosis. Field Crops Res 65:197–209

Jiménez-Guerrero I, Pérez-Montaño F, Medina C, Ollero FJ, López-Baena FJ (2015) NopC is a *Rhizobium*-specific type 3 secretion system effector secreted by *Sinorhizobium* (*Ensifer*) *fredii* HH103. PLoS One 10:e0142866

Johnston AWB, Latchford JW, Davis EO, Economou A (2014) Five genes on the *sym* plasmid of *Rhizobium leguminosarum* whose products are located in the bacterial membrane or periplasm. Signal Mol Plants Plant Microbe Interact 36:311–318

Kawaharada Y, Nielsen MW, Kelly S, James EK, Andersen KR, Rasmussen SR, Füchtbauer W, Madsen LH, Heckmann AB, Radutoiu S, Stougaard J (2017) Differential regulation of the Epr3 receptor coordinates membrane-restricted rhizobial colonization of root nodule primordial. Nat Commun 8:14534

Kimbrel JA, Thomas WJ, Jiang Y, Creason AL, Thireault CA, Sachs JL, Chang JH (2013) Mutualistic co-evolution of type III effector genes in *Sinorhizobium fredii* and *Bradyrhizobium japonicum*. PLoS Pathog 9:e1003204

Krishnan HB, Pueppke SG (1991) *nolC*, a *Rhizobium fredii* gene involved in cultivar-specific nodulation of soybean, shares homology with a heat-shock gene. Mol Microbiol 5:737–745

Krishnan HB, Lorio J, Kim WS, Jiang G, Kim KY, DeBoer M, Pueppke SG (2003) Extracellular proteins involved in soybean cultivar-specific nodulation are associated with pilus-like surface appendages and exported by a type III protein secretion system in *Sinorhizobium fredii* USDA257. Mol Plant-Microbe Interact 16:617–625

Lardi M, Pessi G (2018) Functional genomics approaches to studying symbioses between legumes and nitrogen-fixing rhizobia. High-Throughput 7:15

Lesser MP, Morrow KM, Pankey SM, Noonan SH (2018) Diazotroph diversity and nitrogen fixation in the coral *Stylophorapistillata* from the great barrier reef. ISME J 12:813

Lévy J, Bres C, Geurts R, Chalhoub B, Kulikova O, Duc G, Journet EP, Ané JM, Lauber E, Bisseling T et al (2004) A putative Ca^{2+} and calmodulin-dependent protein kinase required for bacterial and fungal symbioses. Science 303:1361–1363

Lewin A, Cervantes E, Wong CH, Broughton WJ (1990) *nodSU*, two new nod genes of the broad host range *Rhizobium* strain NGR234 encode host-specific nodulation of the tropical tree *Leucaena leucocephala*. Mol Plant-Microbe Interact 3:317–326

Limpens E, van Zeijl A, Geurts R (2015) Lipochitooligosaccharides modulate plant host immunity to enable endosymbioses. Annu Rev Phytopathol 53:311–334

Loh J, Garcia M, Stacey G (1997) *NodV* and *NodW*, a second flavonoid recognition system regulating nod gene expression in *Bradyrhizobium japonicum*. J Bacteriol 179:3013–3020

Lohrke SM, Day B, Kolli VK, Hancock R, Yuen JY, De Souza ML et al (1998) The *Bradyrhizobium japonicum noeD* gene: a negatively acting, genotype-specific nodulation gene for soybean. Mol Plant-Microbe Interact 11:476–488

López-Lara IM, Geiger O (2001) The nodulation protein NodG shows the enzymatic activity of an 3-oxoacyl-acyl carrier protein reductase. Mol Plant-Microbe Interact 14:349–357

Madinabeitia N, Bellogín RA, Buendía-Clavería AM, Camacho M, Cubo T, Espuny MR, Soria-Díaz ME (2002) *Sinorhizobium fredii* HH103 has a truncated *nolO* gene due to a-1 frameshift mutation that is conserved among other geographically distant *S fredii* strains. Mol Plant-Microbe Interact 15:150–159

Madsen EB, Madsen LH, Radutoiu S, Olbryt M, Rakwalska M, Szcyglowski K, Sato S, Kaneko T, Tabata S, Sandal N et al (2003) A receptor kinase gene of the *LysM* type is involved in legume perception of rhizobial signals. Nature 425:637–640

Maynaud G, Willems A, Soussou S, Vidal C, Mauré L, Moulin L, Cleyet-Marel JC, Brunel B (2012) Molecular and phenotypic characterization of strains nodulating *Anthyllis vulneraria* in mine tailings, and proposal of *Aminobacter anthyllidis* sp nov, the first definition of *Aminobacter* as legume-nodulating bacteria. Syst Appl Microbiol 35:65–72

Mergaert P, D'Haeze W, Fernández-López M, Geelen D, Goethals K, Promé JC, Holsters M (1996) Fucosylation and arabinosylation of nod factors in *Azorhizobium caulinodans*: involvement of *nolKnodZ* as well as *noeC* and/or downstream genes. Mol Microbiol 21:409–419

Mus F, Crook MB, Garcia K, Costas AG, Geddes BA, Kouri ED et al (2016) Symbiotic nitrogen fixation and challenges to extending it to non-legumes. Appl Environ Microbiol 82 (13):3698–3710. https://doi.org/10.1128/AEM.01055

Niehaus K, Becker A (1998) The role of microbial surface polysaccharides in the *Rhizobium*-legume interaction. In: Biswas BB, Das HK (eds) Sub-cellular biochemistry, plant-microbe interactions, vol 29. Springer Science and Business Media, Boston, MA, pp 73–116

Op den Camp RH, Polone E, Fedorova E, Roelofsen W, Squartini A, Op den Camp HJ et al (2012) Nonlegume *Parasponia andersonii* deploys a broad rhizobium host range strategy resulting in largely variable symbiotic effectiveness. Mol Plant-Microbe Interact 25:954–963

Österman J, Marsh J, Laine PK, Zeng Z, Alatalo E, Sullivan JT, Lindström K (2014) Genome sequencing of two *Neorhizobium galegae* strains reveals a *noeT* gene responsible for the unusual acetylation of the nodulation factors. BMC Genomics 15:500

Peralta H, Aguilar A, Díaz R, Mora Y, Martínez-Batallar G, Salazar E (2016) Genomic studies of nitrogen-fixing rhizobial strains from *Phaseolus vulgaris* seeds and nodules. BMC Genomics 17:711

Perret X, Staehelin C, Broughton WJ (2000) Molecular basis of symbiotic promiscuity. Microbiol Mol Biol Rev 64:180–201

Poinsot V, Crook MB, Erdn S, Maillet F, Bascaules A, Ané JM (2016) New insights into nod factor biosynthesis: analyses of chitooligomers and lipo-chitooligomers of *Rhizobium* sp. IRBG74 mutants. Carbohydr Res 434:83–93

Qiao Z, Brechenmacher L, Smith B, Strout GW, Mangin W, Taylor C, Russell SD, Stacey G, Libault M (2017) The *GmFWL1* (FW2-2-like) nodulation gene encodes a plasma membrane microdomain-associated protein. Plant Cell Environ 40:1442–1455

Quesada-Vincens D, Fellay R, Nasim T, Viprey V, Burger U, Prome JC, Jabbouri S (1997) *Rhizobium* sp. strain NGR234 NodZ protein is a fucosyltransferase. J Bacteriol 179:5087–5093

Radutoiu S, Madsen LH, Madsen EB, Jurkiewicz A, Fukai E, Quistgaard EM, Albrektsen AS, James EK, Thirup S, Stougaard J (2007) LysM domains mediate lipochitin-oligosaccharide recognition and *Nfr* genes extend the symbiotic host range. EMBO J 26:3923–3935

Remigi P, Zhu J, Young JPW, Masson-Boivin C (2016) Symbiosis within symbiosis: evolving nitrogen-fixing legume symbionts. Trends Microbiol 24:63–75

Robledo M, Menéndez E, Jiménez-Zurdo JI, Rivas R, Velázquez E, Martínez-Molina E, Giles Oldroyd Mateos PF (2018) Heterologous expression of rhizobial CelC2 cellulase impairs symbiotic signaling and nodulation in *Medicago truncatula*. Mol Plant-Microbe Interact 31:568–575

Roche P, Debellé F, Maillet F, Lerouge P, Faucher C, Truchet G, Promé JC (1991) Molecular basis of symbiotic host specificity in *Rhizobium meliloti*: *nodH* and *nodPQ* genes encode the sulfation of lipo-oligosaccharide signals. Cell 67:1131–1143

Rudder S, Doohan F, Creevey CJ, Wendt T, Mullins E (2014) Genome sequence of *Ensifer adhaerens* OV14 provides insights into its ability as a novel vector for the genetic transformation of plant genomes. BMC Genomics 15:268

Saad MM, Kobayashi H, Marie C, Brown IR, Mansfield JW, Broughton WJ, Deakin WJ (2005) NopB, a type III secreted protein of *Rhizobium* sp. strain NGR234, is associated with pilus-like surface appendages. J Bacteriol 187:1173–1181

Sachs JL, Quides KW, Wendlandt CE (2018) Legumes versus rhizobia: a model for ongoing conflict in symbiosis. New Phytol 219:1199–1206

Safronova VI, Kuznetsova IG, Sazanova AL, Kimeklis AK, Belimov AA, Andronov EE, Pinaev AG, Chizhevskaya EP, Pukhaev AR, Popov KP, Willems A (2015) *Bosea vaviloviae* sp. nov., a new species of slow-growing rhizobia isolated from nodules of the relict species *Vavilovia Formosa* (Stev.) Fed. Antonie Leeuwenhoek 107:911–920

Sánchez C, Itakura M, Mitsui H, Minamisawa K (2013) Linked expressions of *nap* and *nos* genes in a *Bradyrhizobium japonicum* mutant with increased N_2O reductase activity. Appl Environ Microbiol 79:4178–4180

Sanjuan J, Olivares JOSE (1989) Implication of *nifA* in regulation of genes located on a *Rhizobium meliloti* cryptic plasmid that affect nodulation efficiency. J Bacteriol 171:4154–4161

Schauser L, Rousiss A, Stiller J, Stougaard J (1999) A plant regulator controlling development of symbiotic root nodules. Nature 402:191–195

Schneiker-Bekel S, Wibberg D, Bekel T, Blom J, Linke B, Neuweger H, Pühler A (2011) The complete genome sequence of the dominant *Sinorhizobium meliloti* field isolate SM11 extends the *S meliloti* pan-genome. J Biotechnol 155:20–33

Shamseldin A (2013) The role of different genes involved in symbiotic nitrogen fixation—review. GJBB 8:84–94

Snoeck C, Verreth C, Hernández-Lucas I, Martínez-Romero E, Vanderleyden J (2003) Identification of a third sulfate activation system in *Sinorhizobium* sp strain BR816: the CysDN sulfate activation complex. Appl Environ Microbiol 69:2006–2014

Spaink HP, Wijfjes AH, Lugtenberg BJ (1995) Rhizobium NodI and NodJ proteins play a role in the efficiency of secretion of lipochitin oligosaccharides. J Bacteriol 177:6276–6281

Sprent JI (2007) Evolving ideas of legume evolution and diversity: a taxonomic perspective on the occurrence of nodulation. New Phytol 174:11–25

Sullivan JT, Trzebiatowski JR, Cruickshank RW, Gouzy J, Brown SD, Elliot RM, Weaver JE (2002) Comparative sequence analysis of the symbiosis island of *Mesorhizobium loti* strain R7A. J Bacteriol 184:3086–3095

Surin BP, Watson JM, Hamilton WDO, Economou A, Downie JA (1990) Molecular characterization of the nodulation gene, *nodT*, from two biovars of *Rhizobium leguminosarum*. Mol Microbiol 4:245–252

Sy A, Giraud E, Jourand P, Garcia N, Willems A, De Lajudie P, Prin Y, Neyra M, Gillis M, Boivin-Masson C, Dreyfus B (2001) Methylotrophic *Methylobacterium* bacteria nodulate and fix nitrogen in symbiosis with legumes. J Bacteriol 183:214–220

Tang F, Yang S, Liu J, Zhu H (2016) *Rj4*, a gene controlling nodulation specificity in soybeans, encodes a thaumatin-like protein but not the one previously reported. Plant Physiol 170:26–32

Tsikou D, Yan Z, Holt DB, Abel NB, Reid DE, Madsen LH, Bhasin H, Sexauer M, Stougaard J, Markmann K (2018) Systemic control of legume susceptibility to rhizobial infection by a mobile microRNA. Science 362:233–236

Uniprot. https://www.uniprot.org/

Via VD, Zanetti ME, Blanco F (2016) How legumes recognize rhizobia. Plant Signal Behav 11: e1120396

Wang Q, Liu J, Zhu H (2018) Genetic and molecular mechanisms underlying symbiotic specificity in legume-rhizobium interactions. Front Plant Sci 9:313

Weir BS (2016) The current taxonomy of rhizobia. NZ Rhizobia website. https://www.rhizobia.co.nz/taxonomy/rhizobia Last updated: X Jan, 2016

Yang S, Tang F, Gao M, Krishnan HB, Zhu H (2010) *R* gene-controlled host specificity in the legume–rhizobia symbiosis. Proc Natl Acad Sci U S A 107(43):18735–18740. https://doi.org/10.1073/pnas.1011957107

You Z, Marutani M, Borthakur D (2002) Diversity among *Bradyrhizobium* isolates nodulating yardlong bean and sunnhemp in Guam. J Appl Microbiol 93:577–584

Young P (1992) Phylogenetic classification of nitrogen-fixing organisms. In: Stacey G, Burris RH, Evans HJ (eds) Biological nitrogen fixation. Chapman and Hall Inc, New York, pp 43–86

Zgadzaj R, James EK, Kelly S, Kawaharada Y, de Jonge N, Jensen DB et al (2015) A legume genetic framework controls infection of nodules by symbiotic and endophytic bacteria. PLoS Genet 11:e1005280

Chapter 5
Symbiotic Signaling: Insights from Arbuscular Mycorrhizal Symbiosis

Rinku Dhanker, Suman Chaudhary, Anju Kumari, Rakesh Kumar, and Sneh Goyal

Abstract Arbuscular mycorrhiza is an evolutionary symbiotic association between roots of terrestrial plants and fungi of phylum Glomeromycota. The development of this association resulted from the exchange of signaling molecules between the two partners, which leads to reciprocal benefits. Different stages of symbiosis are regulated by various plant hormones, different genes and miRNAs. While plant-derived strigolactone hormones stimulate the fungal hyphal branching and its metabolism, fungi releases lipochitooligosaccharides (Myc-Lcos) which elicit pre-symbiotic responses in the host root. These signaling molecules develop a molecular dialogue between AM fungi and plant roots, which generates a cascade of co-evolutionary events leading to the preparation of both the partners for successive root colonization.

5.1 Introduction

Mycorrhizal associations are defined as mutual symbiotic relationships between plants and fungi confined in root and/or -like structures where energy transfers mainly from plants to the fungi while other inorganic nutrients and water move from fungi to plants. This interaction should not be confused with rhizobial interactions, which are mutualistic relationships with bacteria responsible for nitrogen fixation (Miyasaka et al. 2003; IJdo et al. 2011). Mycorrhizal associations take place between almost all families of land plant roots and some specific fungi (monophyletic phylum, *Glomeromycota* and the order Glomales). Based on molecular analysis, the SSrRNA sequences the phylum *Glomeromycota* was found to be in relation with Ascomycota and *Basidiomycota* (Morton and Benny 1990; Schussler et al. 2001; Hibbett et al. 2007; Prasad et al. 2017).

R. Dhanker · S. Chaudhary · R. Kumar (✉) · S. Goyal
Department of Microbiology, CCS Haryana Agricultural University, Hisar, Haryana, India
e-mail: sehrawatrk@hau.ernet.in

A. Kumari
Center of Food Science and Technology, CCS Haryana Agricultural University, Hisar, Haryana, India

A. Varma et al. (eds.), *Plant Microbe Symbiosis*,
https://doi.org/10.1007/978-3-030-36248-5_5

Arbuscular mycorrhiza (AM) is probably the utmost prevalent symbiosis in land (Fitter 2005) formed by 70–90% of plant species, including liverworts, both gametophytic and sporophytic club mosses as well as horsetails, gymnosperms, hornworts and angiosperms (Smith and Read 2008; Goltapeh et al. 2008). In terms of geographical coverage also, AM is one of the most extensive symbiosis on earth (Newman and Reddell 1987).

The word mycorrhiza originates from two Greek words, *mycos* and *rhiza*, where mycos means fungus while rhiza means root; therefore, mycorrhiza literally means "fungus root" (Ianson 2015), and the word arbuscular mycorrhiza is derived from the Latin word *arbusculum* which means little tree. In fact, the fungi often form little tree-like structures, arbuscules, in plant root cells with the help of numerous fine branched fungal hyphae. Although the fungal hyphae forming arbuscules in mycorrhizal root cortex is intra-cellular, both associates stay away due to their plasma membranes and are called as symbiosomes. For mycorrhizal fungi, this symbiosis is obligatory in nature to complete the asexual stage of their life cycle (Remy et al. 1994).

The origin of arbuscular mycorrhizal symbiosis traces back more than 400 million years (Remy et al. 1994), and fossil records of 460 million years old were found to have mycorrhizal fungi very similar to the fungi found in today's plants. Robert Hartig in 1840 first described the fine roots of mycorrhiza into a pine but could not recognize them as a separate identity. In 1847, S. Reissek recognized and explained fungal cells associated with orchids. The German pathologist A.B. Frank in 1885 identified and observed fungus-root structures and explained increased growth of plants having mycorrhizal associations (Frank 1885).

5.1.1 Types of Mycorrhiza

On the basis of phylogenetic place of fungal partners and structures formed by symbiosis, many types of mycorrhiza have been defined such as ectomycorrhiza, ericoid, orchid and arbuscular mycorrhiza (Smith and Read 1997; Hause and Fester 2005). Different types of mycorrhiza may be described as follows (Fig. 5.1).

5.1.1.1 Ectomycorrhizae (ECM)

The root of ectomycorrhizae lacks root hairs and is covered by a covering or sheath of fungal hyphae, which is almost similar to the host tissue in appearance. This outer coating is called as the pseudoparenchymatous sheath. The mycelia of fungus spread into the root cortical cells to form a complex network called as "Hartig net" and outside into the soil around the roots. The fungal hyphae also form a mantle on the root surface. They increase the surface area and hence afford better nutrient uptake from the surrounding soil. Hyphae do not penetrate into cells, but contact with roots is very close and metabolites are transferred in both directions. ECM makes associations with so many woody plants ranging from shrubs to forest trees. Examples of

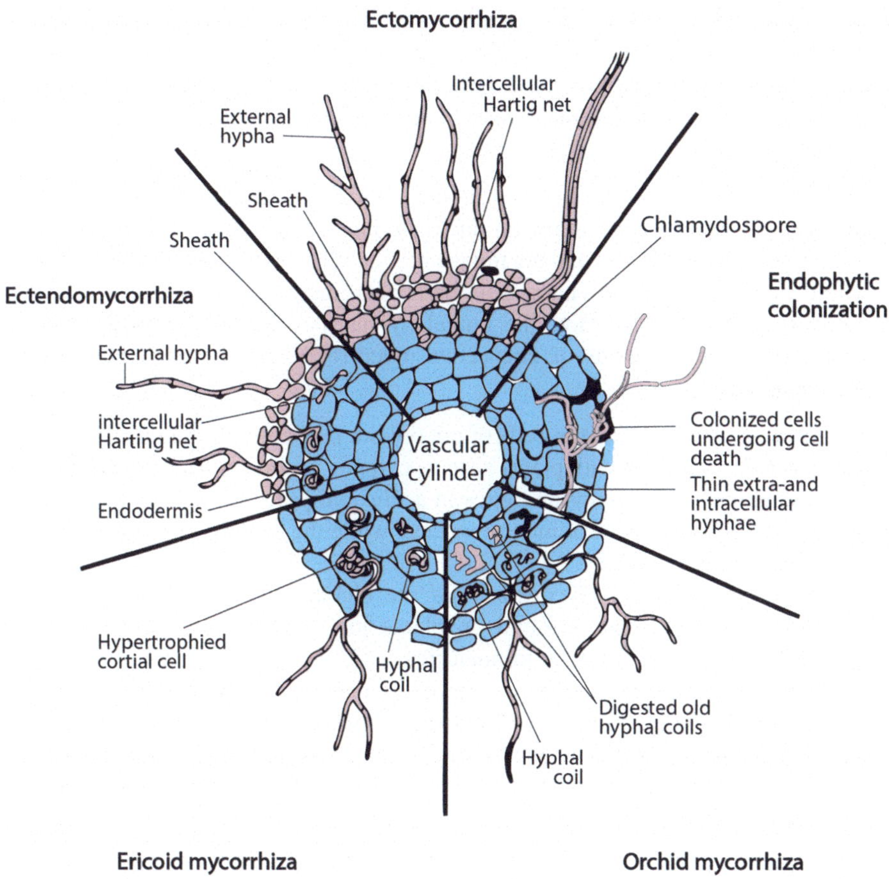

Fig. 5.1 Different types of mycorrhizal associations

host plants of ECM belong to the families Fagaceae, Pinaceae, Myrtaceae and Betulaceae and a few others but no grasses and fungi predominantly from Basidiomycota and Ascomycota (Smith and Read 1997). This type of fungus is not having the property of cellulolysis or lignolysis and so depends upon carbohydrates formed by their host plants.

5.1.1.2 Endomycorrhizae

In this type of association, the fungus penetrates intra-cellularly into the cortical cells of host roots along with outside extension into the surrounding soil. Firstly, the fungus grows in between cortical cells, and after that penetrates into the host cell wall and spreads within the cells. Endomycorrhizae is common in species of herbaceous angiosperms, flowering plants, annual and perennial crops and many of the gymnosperm genera (Cazares and Smith 1996). It forms arbuscules which help the fungus to penetrate into the plant root cell. Here the fungus structure is formed entirely within the host root. Endomycorrhizae can be further divided into three major groups: orchidaceous, ericoid and arbuscular endomycorrhiza, and two other minor groups.

1. *Orchidaceous mycorrhiza*: In this type, the fungus propagates in the plant cells by infolding the cell membrane and producing hyphal coils in the cell. The plant host of this type is orchidaceae and the fungus is generally found from basidiomycetes. These supply carbon and vitamins to the developing embryo.
2. *Ericaceous mycorrhiza*: Here the fungal hyphae penetrate the cortical cells of roots. Three major forms of ericaceous mycorrhiza are ericoid, arbutoid and monotropoid.

   (a) *Ericoid*: The inner cortex cells of ericoid are packed with fungal hyphae and generally associate with plants such as calluna, rhododendron and vaccinium having fine root systems and grow in acidic and peaty soils. The plant host of this fungus is ericales or monotropaceae while the fungus type is basidiomycetes and ascomycetes.
   (b) *Arbutoid*: Arbutoid mycorrhiza has the characteristics of both ecto- and endomycorrhiza and is found on plants like Arbutus and Arctostaphylos while fungi involved is basidiomycetes.
   (c) *Monotropoid*: Fungi colonize chlorophyllous plants in Monotropaceae, producing Hartig net or mantle. Monotropoid mycorrhiza is found in the subfamily Monotropoideae of the Ericaceae. These have heterotrophic or mixotrophic mode of nutrition, and so retrieve their energy source from the fungus partner. Thus, it is a non-mutualistic, parasitic type of symbiosis.

5.1.1.3 Arbuscular Mycorrhiza

The main difference among the arbuscular mycorrhiza and ectomycorrhizal interaction is that the AM symbiosis does not create a protective layer around the root as the ECM does. As an alternative, its hyphae penetrate the plant cells producing highly branched arbuscules within root cortical cells. Other structures formed by AM fungi are vesical, auxiliary cells and asexual spores. This association is made by Zygomycetes fungi involving six genera namely *Glomus*, *Gigospora*, *Acaulospora*, *Enterophospora*, *Sclerocystic* and *Scutellospora*.

Arbuscules are complex haustoria with different branches formed within a root cortex cell. Arbuscules initiate to form nearly 2 days after entering the root. They enter inside each cell of the root cortex but always remain outside of their cytoplasm due to the inside-folding property of plasma membrane. These are considered as the main location for exchange of nutrients between partners, the fungus and the host plant. The formation of arbuscules takes place after hyphal growth toward outside from the entry point. These are short-lived and thus begin to breakdown after a few days, but the hyphal structure and vesicles remain in host roots for months or years.

Vesicles develop to collect the products to be stored in most of the AM symbiosis. Vesicle formation initiates just after the first arbuscule and continues to form even after the arbuscules senesce. These are swellings of hyphae present in the cortex of the root containing cytoplasm and lipids. Vesicles may be either intra- or intercellular. They may form dense walls in old roots and also work as propagules. While auxiliary cells generally develop in the soil and may be coiled or knobby structures, their functions are still unknown.

Spores may develop as swellings on ≥ 1 fungal hypha present in the roots or soil. Spores generally have cytoplasm, lipids and nuclei. These structures usually form dense layers of walls and may work as propagules. They may be grouped together to form sporocarps, which may comprise specialized hyphae and are enclosed in an outward layer called peridium. Spores actually develop when nutrients are depleted from the root system and the mycorrhizal associations are senescing. Spores mainly act as propagules, resting stages and storage structures. They may also develop particular germination assemblies or hyphae by emerging through the hyphae or may grow directly from walls.

In the last few years, advancement has been made toward identifying the mycorrhizal signals and mechanisms of signaling pathways, which leads to the outlining of a refined model on signaling of early and late AM associations (Bucher et al. 2009). This has led to better understanding of mechanisms at the molecular level involving symbiosis. Interaction between both partners of AM is complemented by cellular morphological changes in these partners, hyphal growth and sequential root colonization process by the fungus (Parniske 2008; Varma et al. 2017).

5.2 Arbuscular Mycorrhizal Development

5.2.1 Asymbiotic Stage

The fungi involved in AM symbiosis are obligate biotrophs. The large resting multinucleate spores of AM fungi can spontaneously grow in the soil and can persist for years, but they are unable to complete their life cycle during this asymbiotic phase, which is only limited to germination of spores and production of mycelium in a very limited amount (Akiyama et al. 2005; Reddy et al. 2008). The AM fungal spores may have thousands of nuclei per spore depending on the species. The hyphae are always coenocytic after its germination. It has been proven that during the hyphal growth several nuclei moved into the growing hyphae and some of them undergo mitosis during nuclear division. AM fungi need host plants for further hyphal growth (Garg and Chandel 2010). If the signals from the roots of host plant are absent, their growth is arrested, and there is apical vacuolization, septation and retraction of cytoplasm including nuclei. In asymbiotic growth the fungus is dependent mainly on its glycogen and triacylglycerides reserves since it is insufficient in uptake of hexoses (Requena et al. 2007). It has been hypothesized by Requena et al. (2000) that in starvation conditions the cell cycle is being controlled by a gene GmTOR 2 in

Glomus mosseae, a homolog to the gene TOR2 from *Saccharomyces cerevisiae*, which is the gene for cell cycle checkpoint.

Requena et al. (2000) characterized a two-domain structured GmGIN1 gene from *G. mosseae*, which is assumed to have self-splicing activity. During the asymbiotic stage, GmGIN1 is highly expressed and during symbiosis it is completely silenced. It is assumed that this gene is probably involved in the spore germination signaling before the symbiosis process.

5.2.2 Mutual Recognition of Symbiotic Partners

5.2.2.1 Strigolactones

There are many studies which have exemplified the process of pre-symbiotic chemical cross-talk between host plant roots and AM fungi. Many researchers have recognized various signal molecules produced by plants, which stimulate the changes in morphological, physiological and mitochondrial activity in the multinucleate fungal spores, the hyphal growth and its branching (Buee et al. 2000; Bucher et al. 2009). These branching factors are assumed to act as the first possible signal in the AM symbiosis process. In *Lotus japonicum* the plant signal molecule was identified as carotenoid-derived phytohormones known as strigolactones. They were first illustrated 50 years ago as seed germination molecules in parasitic plant species *Orobanche* spp. and *Striga* spp. (Bucher et al. 2009; Xie et al. 2010). Along with strigolactones, 2-hydroxydodecanoic acid and 2-hydroxytetradecanoic acid are two other hyphal branching inducing factors identified in roots exudates of carrot, which stimulate the hyphal growth of *Gigaspora gigantea* (Nagahashi and Douds 2011; Gutjahr et al. 2012). An extensive range of dicots and monocot plants produce these signal molecules (Akiyama et al. 2005). Probably strigolactones are produced in the cortex of roots under phosphate- and nitrogen-limiting conditions and from hypodermal passage cells as they are being released into the rhizosphere by PDR1, an ATP-binding cassette transporter (Sasse et al. 2015; Kobae et al. 2017). The cell membrane of AM fungi appears to have highly sensitive and strong receptors for SLs, indicating that minute concentrations of as low as 10 nM of SLs can induce hyphal proliferation (Gutjahr et al. 2012). Matusova et al. (2005) illustrated that SLs production is regulated by the phytohormone abscisic acid, which is derived from the carotenoid biosynthetic pathway.

The naturally occurring SLs contain a tricyclic ABC lactone ring that connects via an enol ether bridged to a methylbutenolide (Fig. 5.2a). It is synthesized during the methylerythritol phosphate pathway upon the cleavage of C-40 carotenoids (Matusova et al. 2005; Bucher et al. 2009). In PDR1 mutants of Petunia, root exudates induced weak hyphal branching and colonization of AM fungi was reduced and delayed, leading to a slow rate of initiation of hyphopodium. In rice mutant D17 and D10, colonization is less efficient; however, fungal hyphal morphology remains the same. Exogenous application of synthetic model SLGR24 induced high rate of

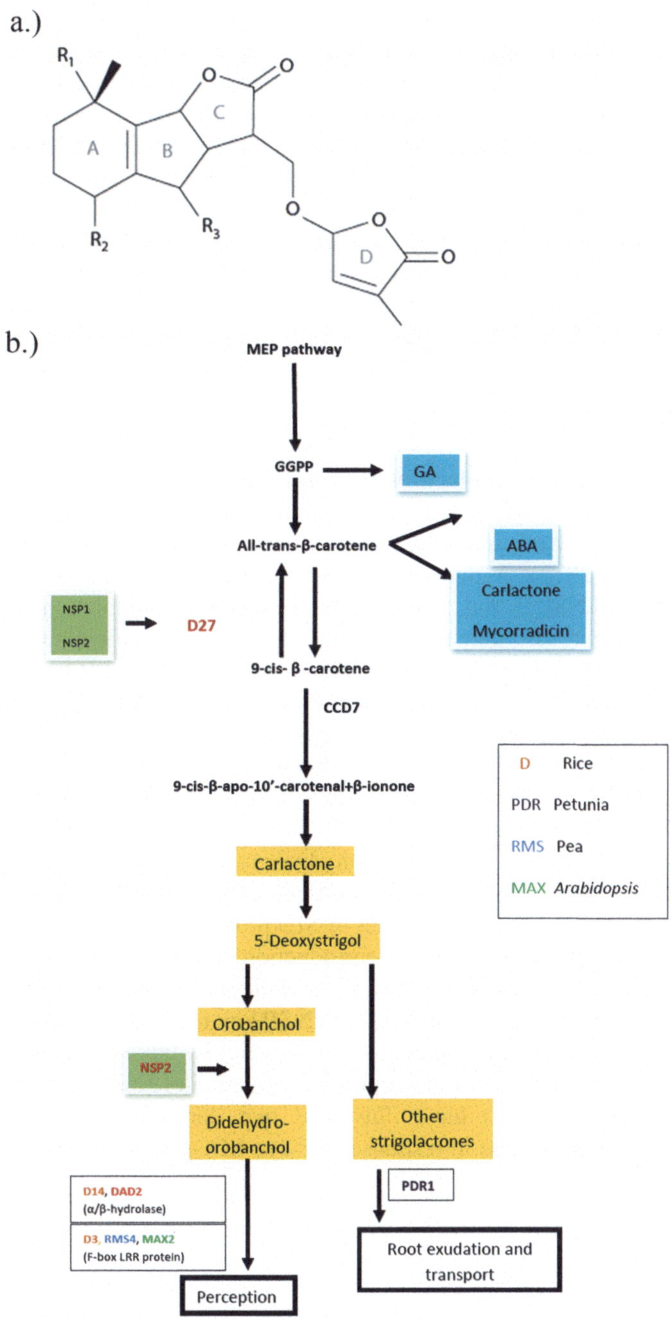

Fig. 5.2 Biosynthesis of strigolactones (SLs) and further derivatives of MEP pathway with their association in arbuscular mycorrhiza symbiosis. (**a**) Molecular structure of strigolactones. In the complexes isolated till date, R is a keto, hydroxyl or acetyl group. (**b**) The pathway of biosynthesis of SLs and its regulation by the transcription factors NSP1 and NSP2. NSP1 and NSP2 (green) promote the expression of the gene DWARF27 (D27) (Liu et al. 2011). Mutants of nsp2 accumulate

fungal respiration, nuclear division, hyphal ATP content and other hyphal proliferation activities by influencing the mitochondrial activity in SL-deficient mutants, thus indicating that SLs induce successful development of colonization in the host plant (Kretzschmar et al. 2012; Gutjahr et al. 2012). In the development of AM fungi, infection units which are composed of internal mycelium are formed. Kobae et al. (2017) illustrated that in SL-deficient mutants subsequent growth of infection units (IUs) was not attenuated whereas hyphopodium formation was severely suppressed. However, at a later mycorrhization stage the growth of colonized regions were suppressed. SL biosynthesizing genes have a major role in the efficient formation of hyphopodium, which leads to the entry of hyphae into the root cortex, that substantially influences the whole mycorrhization process (Kobae et al. 2016, 2017).

5.2.2.2 Regulation of Strigolactone Biosynthesis

The biosynthesis and exudation of SLs are promoted under low-phosphate conditions (Yoneyama et al. 2012). A pair of GRAS-type transcription factors, nodulation signaling pathway NSP 1 and 2, which are essential for nodulation in legume–*Rhizobium* symbiosis, also play a vital role in the regulation of synthesis of SLs in plant roots under phosphate-limiting conditions (Delaux et al. 2013; Takeda et al. 2013). In both *Medicago truncatula* and rice double mutants of NSP1 and NSP2, biosynthesis of SLs and AM colonization gets decreased. These two GRAS-type transcription factors are essential for the regulation of β-carotene isomerase–encoding gene D27 (DWARF27) (Liu et al. 2011). Beta-carotene isomerase catalyzes the first committed step of SL biosynthesis (Fig. 5.2b). In *M. truncatula* two different types of SLs are produced, orobanchol and didehydo-orobanchol NSP1 and NSP2 both different regulates the synthesis of SL's (Alder et al. 2012; Maillet et al. 2011) (Fig. 5.3). There is no significant amount of SLs produced in the NSP1 mutant, but in case of the NSP2 mutant there is more production of orobanchol than those found in wild-type plants. This shows that in *M. truncatula*, orobanchol biosynthesis can be controlled by low expression levels of D27 in NSP2 mutants. Didehydro-orobanchol is subsequently derived from its precursor orobanchol by three enzymatic activities which are assumed to be under transcriptional regulation of NSP2 (Liu et al. 2011). Lauressergues et al. (2012) have illustrated that during root colonization by *Rhizophagus irregularies* and in responses to Myc-LCOs, a micro RNA targets the

---

Fig. 5.2 (continued) orobanchol; however, the misregulated gene liable for its accumulation is not known. PDR1 is essential for the transport of SL inside plant tissues and for root exudation (Kretzschmar et al. 2012). There are evidences for plant SLs' perception through a complex between the F-box leucine-rich repeat (LRR) E3 ligase D3/RMS4/MAX2 and the α/β-fold hydrolase D14/DAD2 (Hamiaux et al. 2012). The exact mechanism for SLs perception by AM fungi is not known. Products of diverging pathways of biosynthesis (blue) also play a vital role in AM (Herrera-Medina et al. 2007; Flo et al. 2008)

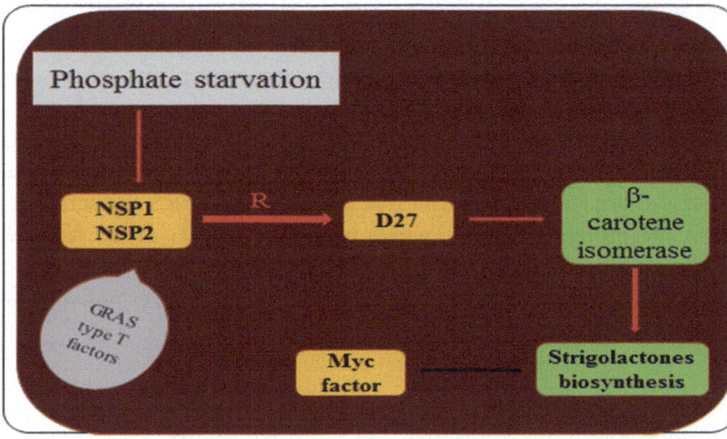

Fig. 5.3 Regulation of SLs synthesis by a pair of GRAS-type transcription factors, NSP1 and NSP2, under low-phosphate conditions, which regulates synthesis of β-carotene isomerase encoding gene D27 (DWARF27)

NSP2, which resulted in up-regulated expression of miR171h in the elongation zone of the root, which usually remains uncolonized. Over-expression of miR171h in roots can reduce fungal colonization. However in plants expressing miR171-resistant NSP2, much increased fungal colonization extending into the elongation zone of the roots can be seen. miR171h regulation of NSP2 is probably conserved in mycotrophic plants. This suggests that Myc-LCOs trigger a regulatory mechanism that inhibits the colonization by mediating negative regulation of NSP2 by miRNA (Gutjahr et al. 2012).

Recent studies by Guillotin et al. (2016) have shown that the expression of D27, NSP1 and MAX1 is regulated by an auxin signaling-related gene S1-IAA27, which is also necessary for the process of mycorrhization. S1-IAA27 expression is triggered before any physical contact between the roots and the fungus, which suggest that SLs biosynthesis in roots is induced by some diffusible fungal signals. In *M. truncatula* after the application of Myc-lipo-chitooligosaccharide, accumulation of D27 slightly increased, which suggests that SL biosynthesis is upregulated by a mix of fungal Myc-LCOs and COs (Hohnjec et al. 2015; Sun et al. 2015). It has been reported that to perceive SLs' signals the F-box ubiquitin E3 ligase complex, protein D3/MAX2/RMS4, interacts with the soluble α/β fold hydrolase family receptor V protein D14/DAD2/HTD2 (Zhao et al. 2014). Evidences have been collected that once SL molecules bind to D14/DAD2/HTD2 the Ser-His-Asp catalytic property of this α/β hydrolase receptor cleaves SL molecules resulting in its conformational changes by free hydroxylated D ring. Also in rice and pea for AM colonization, the F-box protein D3/MAX2/RMS4 is required in the early phase of symbiotic association and SL-biosynthetic pathway. The role of D27 and D14 genes have been suggested by some early reports (Yoshida et al. 2012). In earlier studies only a few transcription factors gene were identified as regulated by mycorrhiza. In *L. japonicum*, according to the plant transcription factors database there are 46 GRAS family members. In recent studies by Xue et al. (2015) RAD1, GRAS transcription factors have been identified, which is basically conserved in AM-competent plants. Mutants of allele rad1 showed less number of arbuscules, which further degenerated very fast. In further studies, two closest homologs of RAM1 were identified, which enhance cutin biosynthesis to up-regulate hyphopodia formation by regulating glycerol-3-phosphate acyl transferase activity. In rice silencing of GARS transcription factor, Della Interacting Protein 1 (DIP1), which interacts with rice DELLA, RAM1 and SLR1 resulted in subsequent low AM fungal colonization. These results give strong evidences about the existence of several proteins involved in the establishment of AM symbiosis in different plant species (Gobbato et al. 2012).

5.2.2.3 Fungal Signaling Molecules and Plant Receptors

For a successful symbiosis between AM fungi and its host, it is imperative to have concerted co-operation and suitable control of cellular responses of plant and its gene expression. Between the AM fungus and the host plant, a molecular dialog is developed which prepares both associated partners for successive root colonization (Balzergue et al. 2013; Mohanta and Bae 2014). In response to SLs secreted by the host plant root, the fungal hyphae derive some diffusible and soluble compounds known as "Myc factors" that are analogous to Nod factors of rhizobial symbiosis recognized by plant roots (Fig. 5.4) (Bucher 2010; Gutjahr and Parniske 2013; Delaux et al. 2013). These fungal signaling molecules induce structural changes and transcriptional activation of mycorrhizal-responsive genes such as ENOD11.

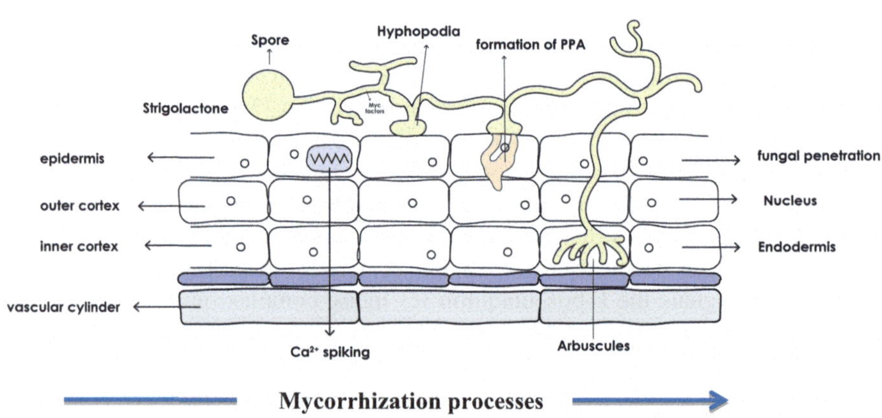

Fig. 5.4 Schematic view of the genes involved in mycorrhization process. Plant host roots–derived SL induces hyphal branching in AM fungi. At the same time, Ca^{2+} spiking is induced by the Myc factors secreted by AM fungi. Hyphopodia are formed at the tips of branched fungal hyphae. Just below the hyphopodia, the pre-penetration apparatus (PPA) grows further to guide hyphal penetration. Arbucules develop at the inner cortex of plant tissue and act as the site for nutrient exchange between the AM fungi and host plants

Kosuta et al. (2003) illustrated that a symbiosis-specific ENOD11-promoter GUS (β-glucornidase) gene is activated in the plant roots of *M. truncatula* when there is no contact between the fungus and the plant (Parniske 2004, 2008). Furthermore, AMF-released signal molecules generate a cascade of reactions, which are perceived by cellular membranes and finally carried to the nucleus; such activities include a rapid and transient Ca^{+} spiking in rhizodermal cells and also lateral root formation in *M. truncatula* (Chabaud et al. 2011; Maillet et al. 2011; Mukherjee and Ane 2010). Signaling induced by these signal molecules depends on DMI1, DMI2 and DMI3, the symbiosis-related genes, which are common in both AMF and rhizobial symbiosis with plants (Balestrini and Lanfranco 2006). Myc factors have been shown to consist of (NAG) chitooligosaccharide, lipochitooligosaccharides and active molecules that include tetra- or pentachitooligosaccharides which are analogous to Nod factors in rhizobia; both of them have Lys M domain in their extracellular domains. In rice and *A. thaliana*, two different putative plant-receptor-like molecules for fungal chitin derivatives perception have been identified (Parniske 2008; Walker et al. 2000). In *Parasponia andersonii* a single Lys M receptor kinase has been recognized to perceive signals from both rhizobia and AM fungi. In Lys M receptor kinase mutants, loss of nodulation and AM formation have been observed (Op den Camp et al. 2011; Gutjahr et al. 2012).

In the case of *Arabidopsis* CERK1 is a chitin receptor, which is an analogue to Nod factor receptor (NFR1). The fact that in AM-competent plants large gene families encode Lys-M-Type receptor kinases indicates that various different Myc factors are generally perceived by plants. This would demonstrate the extensive spectrum of host in AM (Ercolin and Reinhardt 2011).

5.2.3 Formation of Appresorium/Hyphopodium

Morphologically, symbiosis initiation and its development are marked by the formation of the hyphopodium/appressorium. This is the site, where the fungal and the plants cells come in direct contact with each other. The appresorium is differentiated as a locus for the penetration of fungal hyphae into the host root (Garg and Chandel 2010). Despite the fact that segments of root exudates are efficient for the development of fungal hyphae and its branching, cannot evoke the appressoria development initially, which were initiated only after intact contact with the host root. The advancement of appressoria can be thought to be the consequence of fruitful pre-symbiotic acknowledgement events when the plants and fungal partners are focused on interaction.

Nagahashi and Douds (1997) had illustrated in his study that in vitro cell walls purified from the roots of the carrot plant, which is the host for fungi *Gigaspora margarita*, were marked by development of appressorium by *G. margarita*, but the appressorium was not formed on the cell walls of non-host plant roots of sugar beet. It has been also assumed that the signals for the formation of the appressorium are carried only by the epidermal cell walls because the fungal calls could not initiate the

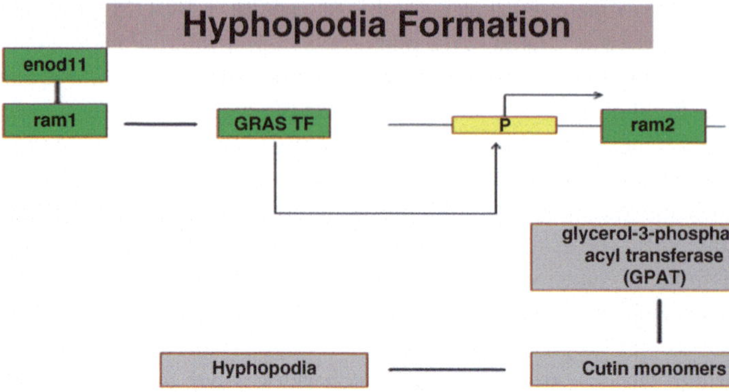

Fig. 5.5 Two *M. truncatula* genes required for arbuscular mycorrhiza, (ram) 1 and 2, are found to be responsible for hyphopodium formation. *RAM1* encodes a GRAS transcription factor that is required for *RAM2* expression and binds to its promoter. *RAM2* encodes a glycerol-3-phosphate acyl transferase (GPAT) that is involved in the production of cutin monomers

development of the appressorium on vascular or cortical call walls of the host plant. The vast hyphal branching could not be elicited by the isolated plant cell wall segment as in intact roots; so it can be affirmed from the studies that the signal for the extensive hyphal branching is either excreted by the roots or it is loosely attached to the roots. Apparently the isolated cell walls contain a mixture of some proteins and polysaccharides like polygalacturonones and cellulases. In various other plant fungal collaborations sugar particles go about as signs and are likely contenders for eliciting the formation of the appressorium. Appressoria are different from hyphae morphologically by being smooth and having oblonged hyphal tips attached to the host cell wall surface by unknown means. Hyphopodia development is diminished in ram 1 and ram 2 *M. truncatula* mutants. A GRAS transcription factor needed for the expression of RAM2 and attaching to its promoter was found to be encoded by RAM1 (Gobbato et al. 2012). A glycerol-3-phosphate acyl transferase is encoded by RAM2, which produces cutin monomers (Fig. 5.5). In ram 2 mutants, development of hyphopodia could be regained with the application of some cutin precursor compound like 1,16-hexadecanediol, C16-monomers or 16-hydroxy-hexadecanoic acid. It is still not clear whether these compounds acted as signals diffusing to elicit fungal differentiation. The appressorium development was also eradicated on ram2 *Medicago* mutants by the pathogen *Phytophthora palmivora* (Wang et al. 2012; Gutjahr and Parniske 2013). It has been assumed that among pathogenic oomycetes, symbiotic fungi and pathogenic fungi, which colonizes the plant roots. Cutin perception is probably the conserved feature. Cutin monomers are even created by liverwort thalli, which likely have a place with developmentally the most antiquated plant organs colonized by AMF. The impression of cutin might in this way speak to an archaic mechanism of the parasitic plant surface acknowledgement (Parniske 2008).

5.2.4 PPA Formation

When hyphopodia are formed, plant cells acknowledge the intra-radical hyphae in an extremely dynamic way. There is a crucial role played by the plant cell. A key revelation was the finding of a finger-shaped or tunnel-like structure known as PPA that the plant cells frame in the foresight of parasitic infection. The PPA development is preceded by epidermal cell walls with a differentiated order of cellular reorganization events, which includes migration of the nucleus just beneath the appressorium, which later directs its direction of growth through the cell, pushing itself forward of the developing PPA (Fig. 5.4). This process leaves behind an accumulation of actin microfilaments, microtubules and ER cisternae, forming an empty tube inside the PPA that associates the main core with the site of appressorial contact (Parniske 2008; Reddy et al. 2008). The PPA characterizes direction through the cell that correctly presages the way of the penetrating fungal hypha. Only after the construction of this novel compartment the hyphae of the fungi can penetrate the plant cell. Endoplasmic reticulum film that lines the passage is preferably situated for the amalgamation of the perifungal layer. During the development of PPA and before its formation, the gene ENOD11 has been assumed to be activated in the epidermal cell walls. Yet the signal molecules that elicit the development of PPA are not known. The displacement of the nucleus can be induced purely by a mechanical incitement of the plant cell even with a needle. This may be the underlying trigger amid AM as this reaction is autonomous of the basic plant SYM genes DMI3 and DMI2 (Parniske 2008). However, to instigate the arrangement of the PPA, extra compound signs are presumably expected to give specificity. The basically related "pre-infection thread" of legumes, which are formed in bacterial infection by rhizobia, presumably developed from the PPA (Fournier et al. 2008). Strikingly, at any rate at the cytological level, PIT and PPA arrangements show up to be completely reversible as seen in cells in which PIT and PPA arrangements were not trailed by microbial intrusion (Sieberer et al. 2012). Related with this redifferentiation, nuclei of the plant cell experience endoreduplication in any event amid arbuscule formation (Bainard et al. 2011; Berta et al. 2000). This together with the perception of limited plant cell wall decomposition or debilitation is predictable with the possibility that the infection thread and, furthermore, PPA development co-selected prior functionalities from the cell division program (Brewin 1998; Parniske 2000). On the off chance that this were valid then during the PPA formation genes related to cell division program would be anticipated to be expressed.

5.2.4.1 Plant Genes Required for PPA Formation

The common seven symbiosis signaling genes, which are involved in root nodule symbiosis, act up-stream of PPA formation in AM symbiosis (Gutjahr et al. 2012). Proteins encoded by these genes are somehow associated in signal transduction, leading to the formation of PPA and intra-cellular accommodation structures for bacteria in the host cell. In common symbiosis gene mutants, the fungal infection was aborted in the outer cell layers, which is at least in the case of Castor-2, associated with cell death of the infected plant cell and the invading fungus. In CCaMK and Symrk (dmi2-2 and dmi3-1) mutants of *M. truncatula*, the development of PPA was aborted; this analysis shows that for eliciting PPA formation the common Sym genes DMI-2 and DMI-3 are necessary (Genre et al. 2005). Signal transduction studies have revealed that in common Sym gene mutants, most of the AM-induced genes are silenced. PPA-consisting cells in *L. japonicus* are manifested by the marker gene SbtM1 expression, particularly, which is expressed during AM (Kistner et al. 2005; Takeda et al. 2012). Imperatively, an universally expressed, deregulation variant of CCaMK, comprising exclusively of a kinase domain model to a nuclear restricted signal, could instigate the SbtM1 promoter-driven signal peptide expression. Cells with evident promoter activity, which showed cortical cell re-differentiation implicative of PPA development, indicate that AM-connected cellular re-differentiation can be induced by CCaMK alone (Takeda et al. 2012). This indicates that the common symbiotic pathways have a major role in enactment of the intra-cellular modification program. The cells were not consistently dispersed along the root, which were experiencing this re-differentiation; however, in the nearby cortex cells they happened to be patches that looked like the course of action of early colonization units (Gutjahr et al. 2012). High reoccurrence calcium spiking has been observed by Sieberer et al. (2012) in the nuclei of cells consisting of PPA just before the fungal infection, which indicates that CCaMK gets activated before the fungal invasion just after the formation of PPA. In reaction to Nod factor, an examination of calcium spiking in *L. japonicum* affirmed that mutants of Symrk, pollux, castor, nup133 and nup85 are damaged for Ca^{+} spiking while CYCLOPS and CCaMK act down-stream (Miwa et al. 2006; Harris et al. 2003). Infection threads are not formed by mutants with flawed common SYM genes, and nodule organogenesis is also not started except for Cyclops mutants (Szczyglowski et al. 1998; Catoira et al. 2000). This indicates that the incitation of calcium splicing in and around the nucleus and its decoding are the result of common SYM gene products, and they are also responsible for inducing gene expression of early symbiosis. There are very fine deviations in the AM phenotypes of mutants of SYM genes in the cortical cell layers, epidermis and AM-forming cells (Parniske 2004). For instance in arbuscule formation, CCaMK and CYCLOPS are clearly required, but in Symrk mutants arbuscules can normally develop. This is demonstrative of considerable versatility and, most likely, functionality of specific cell types in the network of signaling which is characterized by basic SYM proteins (Kosuta et al. 2003; Siciliano et al. 2007).

5.2.5 The Common SYM Pathway

With the emergence of a large number of mutants, the tools for the analysis of molecular mechanisms and the genetic dissection for endosymbiotic interactions have been best developed. Root nodule and tAM symbiosis are the two mutualistic

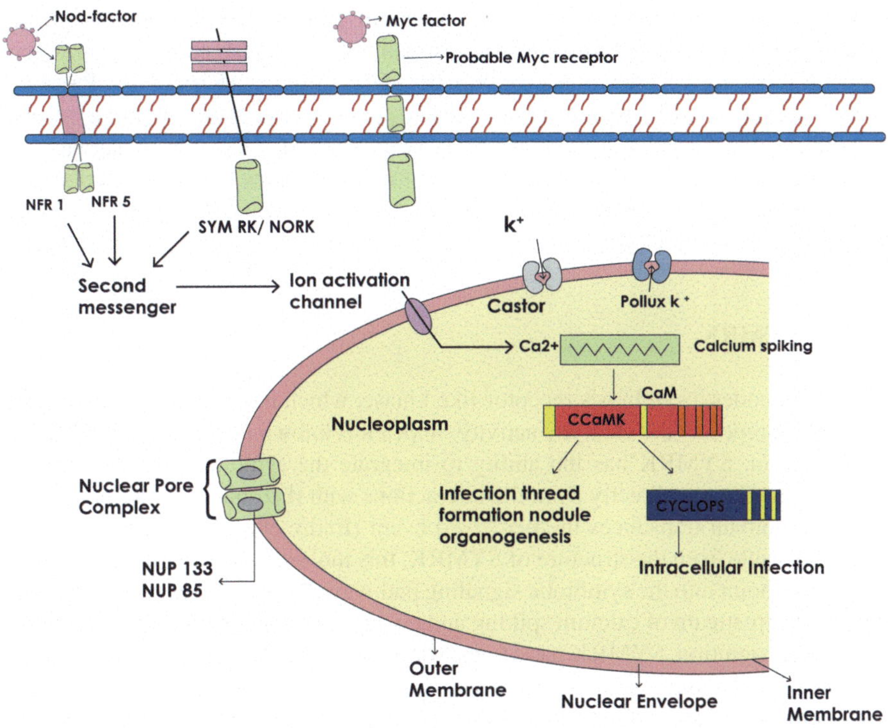

Fig. 5.6 Figure representing the signaling pathway involved in AM symbiosis. The receptor molecules with perception of Nod/Myc signals pass the signal down-stream that activates the secondary messengers present in the cytoplasm of the cell. The second messengers, in turn, activate POLLUX or DMI1 and CASTOR-like molecules. With the activation of POLLUX or DMI1 and CASTOR molecules, the K^+ ion gets released from the nucleoplasm, which makes the cytoplasm hyperpolarized. With this, the perinuclear Ca^{2+} ion enters into the nucleoplasm resulting in Ca^{2+} spiking. Then the Ca^{2+} ions bind to the CcaMK that perceives the signal and passes it to the CYCLOPOS gene, which is responsible for the regulation process of mycorrhization

interaction systems in legumes (Nakagawa and Imaizumi-Anraku 2015). Fifty percent of nod mutants were also found non-mycorrhization phenotypes, i.e., myc^-, when tested for AM interactions (Oldroyd 2013). This demonstrated a typical regulatory pathway for the two mutualistic interaction systems. Consequently the detected genes are collectively known as common SYM genes and the respective mutants are known as common SYM mutants (Kouchi et al. 2010; Nakagawa and Imaizumi-Anraku 2015). The fractional overlay between the molecular mechanisms for both the fungal and bacterial endosymbioses is defined by SYM genes, and they characterized a typical SYM pathway (Fig. 5.6). Since the RNS is much younger than the AM interaction, it appears that the RNS has evolved many functions from the evolutionary AM symbiosis. Therefore, from both an evolutionary and mechanistic viewpoint, the common SYM genes are at the center of interest. However, the basic contrasts among AM and RNS, for example, cytology and life structures of the separate structures, nature of the microsymbiont and the supplements included, recommended that up-stream and down-stream of the SYM pathway particular segments must have been evolved for both the endosymbioses (Kistner and Parniske 2002; Nakagawa and Imaizumi-Anraku 2015). Some presumptive signaling events, including a nuclear calcium splicing at its core, suggest a cascade of signals are encoded by common SYM genes. This common SYM pathway is characterized by at least seven genes which define different signaling steps (Reddy et al. 2008; Oldroyd and Downie 2006).

5.2.5.1 SYMRK

SYMRK encodes a symbiosis receptor-like kinase, which has a kinase domain with leucine-rich repeats with catalytic activity. In pea it is known as NORK and DMI2 in *M. truncatula*. SYMRK has the ability to integrate the symbiotic signals that are released directly or indirectly from the interactions with rhizobia and AM fungi, and its kinase domain transduces the perceived event (Endre et al. 2002; Stracke et al. 2002). Inferable from the structure of SYMRK, this molecule is normally depicted as the passage point into the symbiotic signaling pathway (Parniske 2008). This receptor kinase acts up-stream of calcium spiking and down-stream of the microbial signaling molecule recognition. SYMRK has a kinase domain on the plasma membrane with an apoplastic site having leucine-rich repeats, a central transmembrane domain and on the cytoplasmic side a C-terminal kinase domain (Stracke et al. 2002). On the perception of the signaling molecule, the SYMRK protein encounters a proteolytic cleavage followed by its intra-cellular malectin-like domain liberation and C-terminal decomposition. In different plant varieties, according to the domain structure, the length and function of the SYMRK proteins are of three types. In monocots it is of the shortest type, the intermediate type is in dicots (non-leguminous) and legumes have the longest type which is enough to establish AM symbiosis, and the only nodules which fix N_2 have the longest type. Hence for legumes to gain the potential for root nodule formation, SYMRK evolution seems to be a necessary event (Markmann et al. 2008). It is thus convincing that several extracellular SYMRK domains can bind to distinct extracellular ligands during root nodule formation and AM symbiosis. Recognition specificity is not marked by SYMRK and apparently it does not bind to the Nod factor directly. Two nodulation-specific Nod factor receptor–like kinase, NFR1/NFP and NFR1/LYK3, which contain Lys M motifs, get activated upon Nod factor perception. Genes involved in the first markable common step in signal transduction pathways for both AM and rhizobial symbiosis were found in *M. truncatula* and *L. japonicus*. Receptors for the perception of AM fungal signals are somewhat similar to Nod factors since both are derivatives of LCOs. This assumption is supported by the fact that a single homolog copy of NFR5/NFP is involved in AM and RN symbiosis of *Parasponia* (Op den Camp et al. 2011). SYMRK is assumed to transmit and integrate the fungal and bacterial signals, which involve NFR1 and NFR5. But it is still not clear that whether SYMRK transmits the signals via making heterocomplexes with NFR or indirectly through secondary signals released from the NFR1/NFR5 complex. Among different plant species, distinct homologs of NFR1/LYK3 and NFR5/NFP are conserved. The nearest homologs of NFR1/LYK3 in rice are known as OsCERK1 that perceives chitin or peptidoglycan molecules and induces plant immunity responses (Gobbato 2015). RLKs network approach has been recently developed from various studies, which not only links both AM and RNs symbioses, but also has been widened to the responses of pathogens. It has been illustrated from recent studies that both AM colonization and AM marker gene activation were hindered by mutation in LYK3/NFR1 but not in NFP/NFR5; this was likely through a defect in initiation of Ca^+ spiking in response to AM fungi or AM LCOs (Zhang et al. 2015; Gobbato 2015).

5.2.5.2 CASTOR and POLLUX

CASTOR and POLLUX are two nuclear membrane proteins which encode potassium-permeable channels, predicted to have a role in calcium channel regulation and in calcium release from the nuclear envelope. In pea, the protein is known as SYMS and DMI1 in *M. truncatula* (Charpentier et al. 2008; Imaizumi-Anraku et al. 2005). Overall the domain structures of CASTOR and POLLUX are similar and they have high sequence similarity. Being nuclear membrane–localized proteins, CASTOR and POLLUX are steady with their prospective roles as counter ion channels, which recoup for the charge imbalance that is generated during the calcium spiking (Parniske 2008; Imaizumi-Anraku et al. 2005).

5.2.5.3 Nucleoporins

Two genes of *L. japonicus*, homologous to the nucleoporins NUP85 and NUP133, are supposed to be essential for the induction of symbiotic signal transduction, which is temperature dependent. NUP85 and NUP133 may constitute a specific nuclear pore sub-complex (Reddy et al. 2008). In yeast and humans both NUP85 and NUP133 are associated with the same nuclear pore sub-complex NUP107-160, and there is no direct connection between the proteins and the substrates of the recognized import and export pathways (Alber et al. 2007). But whether these proteins are involved in transport of proteins more than 75 kDa toward the inner nuclear envelope remains to be elucidated (Lusk et al. 2007). Since the proteins NUP85 and NUP133 of *L. japonicus* operate up-stream of calcium oscillations, the plant adaptation of vertebrate NUP107-160 sub-complex may be associated with CASTOR and POLLUX transportation to the inner nuclear envelope (Parniske 2008). According to the studies of Venkateshwaran et al. (2012), *M. truncatula* DMI1 is a homologue of POLLUX of *L. japonicus*, and in the absence of CASTOR and POLLUX, it can induce the process of symbiosis alone.

5.2.5.4 CCaMK

The gene CCaMK of *L. japonicus* and DMI3 of *M. truncatula* encode a calcium- and calmodulin-dependent protein kinase, which constitutes three main domains including a kinase, calmodulin-binding and EF-hand domains. This makes an exciting feasibility that CCaMK is a master candidate to decode and integrate the nuclear calcium-spiking signal, which leads to the phosphorylation event (Gobbato 2015). This shows that calcium oscillations are necessary components of signaling events, which lead to AM colonization. However, it is not seen in interactions of AM. Strikingly, over-expression and de-regulation of CCaMK-mutant proteins, having point mutations in the autophosphorylation residues Thr-271/Thr-265 or C-terminal CCaMK/DMI3 truncations, both trigger spontaneous nodule formation, which suggests that CCaMK deregulation is sufficient enough to initiate the entire process of empty nodule formation and C-terminal has a negative self-regulatory activity (Gleason et al. 2006; Tirichine et al. 2006). It has been discovered from the functional analysis of CCaMK that CaMBD and EF-hand domains are essential for AM symbiosis since CaMBD and EF-hand domain mutants can accommodate AM fungi. However, for the infection of rhizobial bacteria, complete CCaMK is required. This makes a possibility that a more complicated regulation of CCaMK by CaMBD/EF hand domains is required for RN symbiosis. It has been reported by Genre et al. (2009) that CCaMK/DMI3 plays another important function in protecting the cells from death, during physical contact, which enhances its biological importance outside of the common symbiotic pathway.

5.2.5.5 CYCLOPS

The *L. japonicus* gene CYCLOPS (IPD3 in *Medicago*) encodes a nuclear protein with a coiled-coil domain and a nuclear localization signal (Yano et al. 2008). This protein product combined with CCaMK/DMI3 acts as a substrate for its kinase activity in the nuclei of plants and yeasts. Mutants of CYCLOPS diminish RN symbiosis and also cause defects in arbuscule development during fungal symbiosis (Horvath et al. 2011). CYCLOPS mutants exhibit normal nodule formation during the RN symbiosis but lead to impair the interaction of infection thread formation; this indicates that CYCLOPS has a crucial role in the infection-specific division of symbiotic signaling pathway. It has been shown by phosphomimetic and phosphoablative mutagenesis that for the process of symbiosis, phosphorylation of S50 and S154 by CCaMK is necessary (Singh et al. 2014; Gobbato 2015).

5.2.6 *Arbuscule Development*

In AM fungi, for the exchange of various nutrients with the plants, there is a special intra-cellular structure known as the arbuscule. It comprises of highly extended hyphae with fine tips that result in increased surface area for efficient nutrient transfer (Dickson and Kolesik 1999). Arbuscule development is correlated with intense changes in sub-cellular structures of plant cells, and their functions are almost filled up by a vacuole (Gutjahr et al. 2012). There are two morphological patterns (Paris and Arum) of arbuscules depending upon the type of species of the plant and associated AM fungi. The Paris-type colonization is characterized by intercalary fungal coils, which entail fungal hyphae from cell to cell having very less or no inter-cellular space. However, in the arum type an inter-cellular highly branched hyphal tree–like structure, known as terminal arbuscule, is formed in the cortical cells (Bonfante and Genre 2008). Most of the plant species consist of the intermediate form of these two types, basically known as "arum–Paris-type continuum" (Dickson et al. 2007). Once the host plant cell is colonized by arbuscules, its structure undergoes several remarkable changes. During the formation of an arbuscule, the cortex cell of the root responds with distinct transcriptional and cellular adaptations. The nucleus of the cell becomes large and migrates towards the center, which is enveloped by fine hyphal branches of arbusculate coil (Ayling et al. 2001). Further the central large vacuole is divided into several small vacuoles. Despite the arbuscules are developed intra-cellularly they are separated from host plant cell cytoplasm by a plant-derived periarbuscular membrane (PAM), which is in continuum with the plasmalemma and surrounds the branches of the arbuscule (Fig. 5.7). The cytoskeleton of the plant cell develops a network all over the arbuscule branches on the cytoplasmic side of the PAM (Gutjahr et al. 2012). Eventually, the development process of the arbuscule terminates with the formation of a symbiotic interphase between the fungal membrane, PAM and the interspatial matrix between them. The plant cells containing arbuscules have distinct machinery for the active transport of nutrients. Despite direct contact with the plasma

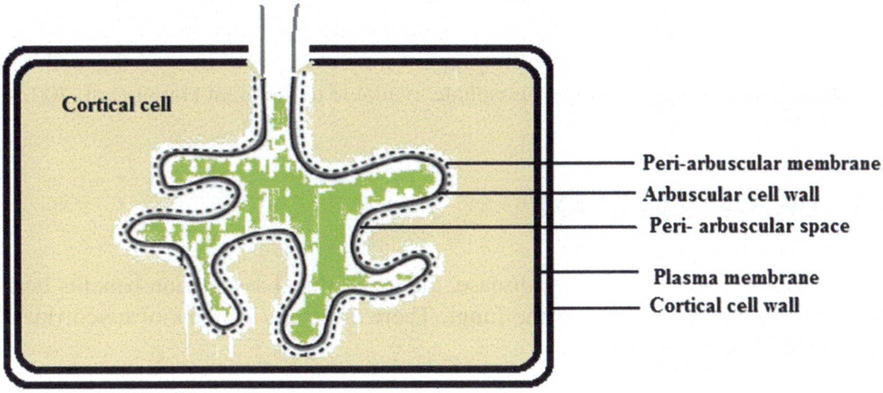

Fig. 5.7 The arbuscule: nutrient exchange site. The symbiotic structure that fills almost the complete cell volume. These structures are enclosed by a plant-derived periarbuscular membrane (PAM) that is continuous with the plasma membrane of the plant cell and separates the plant cytoplasm from the fungus. The apoplastic space between the plant-derived PAM and fungal plasma membrane is known as periarbuscular space (PAS)

membrane, a distinct protein resides in the PAM: the arbuscular branches have PT4 (phosphate transporter 4) whereas the arbuscular trunks and the peripheral plasma membrane contain BCP1 (Blue Copper Binding Protein 1) (Javot et al. 2007). The phosphate transported by the arbuscule is assumed to take up by the PT4 (Harrison et al. 2002). An electrochemical gradient is generated through the H^+-ATPases present in the fungal membrane and the PAM of the arbuscule, which is required for active nutrient transport (Gianinazzi-Pearson et al. 2000). The arbuscules have a short life span of about 8.5 days; subsequently, a single cell of the host plant is assumed to be competent for various rounds of consecutive invasion of fungal hyphae (Parniske 2008). It has been found that the development of the arbuscule is aborted once it reaches its maximum size after which it gets separated from the cytoplasm by septation; consequently, they get collapsed and eventually disappear. This growth phase of the arbuscule is costly in terms of fungal and plant resources. The arbuscules with PT4 mutants of PAM degrade in the premature phase; this suggests that the life span of the arbuscules is affected by their capability of transporting phosphate and several other nutrients. This is how the plants degrade the inefficient arbuscules and maintain the efficient ones (Javot et al. 2007). This process also allows the plant species to differentiate between the efficient and non-efficient fungal species, along with the potential of removing "good" fungal species, which is connected to the poor source of phosphate. This allows the different fungal species to compete for the formation of arbuscules. Over time the distribution of nutrients would change in the soil and the "non-providers" would be replaced by well-connected hyphae (Parniske 2008). Hence, the short life span of arbuscules allows constant rewiring and renewal of the network of hyphae and permits to make successful connections with the most efficient providers. Lysophosphatidylcholine (LPC) is likely to play an important role in the process of arbuscule development. It is the compound derived from phospholipid metabolism and ascribed to act as a aignaling molecule, which activates the genes of phosphate transporters, including PT3, a potato gene. Compounds consisting of phosphate such as LPC act as autonomous molecular measures of the cell for determining the concentration of phosphate available to the plant (Javot et al. 2007).

5.3 Diverse Roles of AM

Except the fungus, which causes disease, the mycorrhizal association benefits both the partners, the host plant and the fungi. There are many benefits of mycorrhizal symbiosis to plants.

5.3.1 Improved Absorption of Water and Nutrients

Though plants uptake water and nutrients via their fine root hairs, the mycorrhizal association enhanced the amount of water and nutrients (Allen 1991; Govindarajulu et al. 2005; Karandashov and Bucher 2005; Finlay 2008) by providing fungal hyphae, which acts in a similar way. The fungal hyphae have three main benefits over the plant's root hairs: (1) the hyphae can reach more distance in the soil and cover more surface area than the root hairs; the hyphal network of the AM fungi may be in excess of 100 m of fungal hyphae/cubic cm of soil (Miller et al. 1995), (2) hyphae are more attracted toward nutrients in comparison to root hairs, and (3) they are more fine than root hairs, so can easily penetrate into the soil space in which the root hairs cannot. These hyphae improve the amount of soil, which can be accessed by plants up to 100,000 times. Absorption of trace elements like copper, boron, zinc and molybdenum is also enhanced by Arbuscular mycorrhizae. Thus, it will be accurate to say that "the mycorrhizae, not roots, are the principal organs of nutrient uptake by terrestrial plants" (Smith and Read 2008).

5.3.2 Improved Phosphorus Uptake

The presence of numerous fine hyphae enhances the growth of bacteria, which can extract phosphorous (P) from organic matter components. The P released by the bacteria is thus absorbed by hyphae and transferred to the plant. The AM also increase the plant's uptake of potassium (K) copper, iron, nickel, sulfur and zinc apart from P.

5.3.3 Improvement in Nitrogen-Fixing Capacity of Nodules in Legumes

The improved absorption of phosphorous, a peculiar characteristic of a mycorrhizal interactions, helps legumes like peas to fix nitrogen. Phosphorous promotes the colonization of *Rhizobium* within the roots of plants. These bacteria utilize nitrogen from the air and convert it into inorganic forms, which can be utilized by plants. AM and ECM mostly have well-developed additional radical mycelial stages, which are able to overcome nutrient depletion zones around the roots of plants. The mycelia penetrate microsites in soil, thus enlarging the surface area of the root system.

5.3.4 Enhanced Plant Growth Hormone Production

Mycorrhizal association mostly improved the levels of hormones like cytokinins and gibberellins, which are responsible for seed germination, cell division, stem elongation and other functions of plants.

5.3.5 Suppression of Root Disease

Mycorrhizae protect host plants from disease by both means, physically and chemically. The fungi inhibit the disease-causing organisms by producing antibiotics and improving plant nutrition, which increases plant strength. Healthy plants can efficiently tolerate or resist pathogens such as *Rhizoctonia*, *Fusarium*, *Phythium* and *Phytophthora* (root rots) and *Verticillum* (stem infecting). The protective covering or mantle formed by the ECM physically protects the plant root from diseases.

5.3.6 Improvement of Soil's Physical Characteristics

Compounds excreted by fungal hyphae like glomalin (a carbohydrate/protein molecule) act like glue and keep the soil particles stick together. These soil aggregates are also resistant to breakdown by water and also improve the soil's physical characteristics like movement of water and air in the soil.

5.3.7 Harsh Conditions Tolerance

Fungi are generally more tolerable to high soil temperatures, acidity, elemental toxicity and also, in some cases (ECM), provide a shield to the root from these conditions.

5.3.8 More Survival of Seedlings

Mycorrhizal association promotes the survival of new seedlings and also out-planted container. Survival of inoculated plants is found upto five times more than the uninoculated plants. The improved survival is due to combination effects of mycorrhizal benefits, faster growth, capacity to overtop weeds, defence from pathogens, and better drought tolerance (Liu et al. 2007).

5.3.9 Protection Against Heavy Metals

Fungal hyphae block the uptake of heavy metals like cadmium, zinc and manganese from soil having excessive levels of these metals. This protection enhances the plant's capability to reestablish and stabilize the soils of mines that may have high quantity of heavy metals.

5.3.10 Protection Against Pathogens

Mycorrhizal association promotes beneficial bacteria that may be responsible for protection of plants against root pathogens. Also, the root colonization with AM fungi leads the tolerance of plants toward abiotic stress. This dual protection may be due to upgraded plant fitness or some specific unknown defense reactions persuaded by AM fungi (Liu et al. 2007; Marschner 2012).

5.3.11 Tool for Studying Molecular Mechanisms and Improving Productivity

AM can be inoculated with crops like rice to study their mechanisms although present rice cultivars are sown under anaerobic conditions, 70% of total inorganic phosphate (Pi) is uptaken through AM symbiosis only. Rice highly depends on the mycorrhizal pathway for Pi uptake under both aerobic and anaerobic field conditions. Though the flooding conditions in rice have a negative effect on fungal growth, the fungi that had already entered into roots previously remain viable during anaerobic conditions. So basically, the functioning capacity of AM symbiosis is not harmed by flooding (Vallino et al. 2014). Consequently, AM symbiosis may be a potential target for breeding to enhance productivity.

5.4 Conclusion/Future Perspectives

An array of various experimental evidences from mutants of various crops and other model organisms has revealed the basic signaling events and the molecules involved with their functions in developing the symbiotic association and also the genetics involved in the regulation of different stages of symbiosis. The important symbiotic genes necessary for the association are conserved in both the associated partners. Detailed study of plant genetics and proteomics will always be an important tool for further identification of the genes required for AM development and function. To reveal the significance of AM symbiosis in sustainable agriculture it is necessary to analyze its molecular mechanism deeply. It is also crucial to investigate the responsiveness of AM symbiosis within important agricultural crop plants. The aim should be to develop a unique combination of crop and fungi that would reduce the necessity of application of chemical fertilizers, which are depleting the fertility of soil day by day.

Acknowledgment We sincerely thank Mr. Manu Phogat for preparing the figures used in this chapter.

References

Akiyama K, Matsuzaki K, Hayashi H (2005) Plant sesquiterpenes induce hyphal branching in arbuscular mycorrhizal fungi. Nature 435:824–827

Alber F, Dokudovskaya S, Veenhoff LM, Zhang W, Kipper J, Devos D, Suprapto A, Karni-Schmidt O, Williams R, Chait BT, Sali A, Rout MP (2007) The molecular architecture of the nuclear pore complex. Nature 450:695–701

Alder A, Jamil M, Marzorati M, Bruno M, Vermathen M, Bigler P, Ghisla S, Bouwmeester H, Beyer P, Al-Babili S (2012) The path from β-carotene to carlactone, a strigolactone-like plant hormone. Science 335:1348–1351

Allen MF (1991) The ecology of mycorrhizae. Cambridge University Press, New York

Ayling SM, Smith SE, Smith FA (2001) Colonisation by arbuscular mycorrhizal fungi changes the relationship between phosphorus uptake and membrane potential in leek (Allium porrum) seedlings. Aust J Plant Physiol 28:391–399

Bainard LD, Klironomos JN, Gordon AM (2011) Arbuscular mycorrhizal fungi in tree-based intercropping systems: a review of their abundance and diversity. Pedobiologia 54:57–61

Balestrini R, Lanfranco L (2006) Fungal and plant gene expression in arbuscular mycorrhizal symbiosis. Mycorrhiza 16:509–524

Balzergue C, Chabaud M, Barker DG, Bécard G, Rochange SF (2013) High phosphate reduces host ability to develop arbuscular mycorrhizal symbiosis without affecting root calcium spiking responses to the fungus. Front Plant Sci 4:426

Berta G, Fusconi A, Sampo S, Lingua G, Perticone S, Repetto O (2000) Polyploidy in tomato roots as affected by arbuscular mycorrhizal colonization. Plant Soil 226:37–44

Bonfante P, Genre A (2008) Plants and arbuscular mycorrhizal fungi: an evolutionary-developmental perspective. Trends Plant Sci 13:492–498

Brewin N (1998) Tissue and cell invasion by Rhizobium: the structure and development of infection threads and symbiosomes. In: Spaink H, Kondorosi A, Hooykaas P (eds) The Rhizobiaceae. Kluwer Academic Publishers, Dordrecht, pp 417–429

Bucher M (2010) A novel lipid signal in the arbuscular mycorrhizal symbiosis within eyesight? New Phytol 185:593–595

Bucher M, Wegmuller S, Drissner D (2009) Chasing the structures of small molecules in arbuscular mycorrhizal signaling. Curr Opin Plant Biol 12:500–507

Buee M, Rossignol M, Jauneau A, Ranjeva R, Becard G (2000) The pre-symbiotic growth of arbuscular mycorrhizal fungi is induced by a branching factor partially purified from plant root exudates. Mol Plant-Microbe Interact 13:693–698

Catoira R, Galera C, De Billy F, Penmetsa RV, Journet EP, Maillet F, Rosenberg C, Cook D, Gough C, Denarie J (2000) Four genes of Medicago truncatula controlling components of a Nod factor transduction pathway. Plant Cell 12:1647–1666

Cazares E, Smith JE (1996) Occurrence of vesicular–arbuscular mycorrhizae in Pseudotsuga menziesii and Tsuga heterophylla seedlings grown in Oregon Coast Range soils. Mycorrhiza 6:65–67

Chabaud M, Genre A, Sieberer BJ, Faccio A, Fournier J, Novero M, Barker DG, Bonfante P (2011) Arbuscular mycorrhizal hyphopodia and germinated spore exudates trigger Ca2+ spiking in the legume and nonlegume root epidermis. New Phytol 189:347–355

Charpentier M, Bredemeier R, Wanner G, Takeda N, Schleiff E, Parniske M (2008) *Lotus japonicus* CASTOR and POLLUX are ion channels essential for perinuclear calcium spiking in legume root endosymbiosis. Plant Cell 20:3467–3479

Delaux P-M, Bécard G, Combier J-P (2013) NSP1 is a component of the Myc signaling pathway. New Phytol 199:59–65

Dickson S, Kolesik P (1999) Visualisation of mycorrhizal fungal structures and quantification of their surface area and volume using laser scanning confocal microscopy. Mycorrhiza 9:205–213

Dickson S, Smith FA, Smith SE (2007) Structural difference in arbuscular mycorrhizal symbioses: more than 100 years after Gallaud, where next? Mycorrhiza 17(5):375–393

Endre G, Kereszt A, Kevei Z, Mihacea S, Kalo P, Kiss GB (2002) A receptor kinase gene regulating symbiotic nodule development. Nature 417:962–966

Ercolin F, Reinhardt D (2011) Successful joint ventures of plants: arbuscular mycorrhiza and beyond. Trends Plant Sci 16:356–362

Finlay RD (2008) Ecological aspects of mycorrhizal symbiosis: with special emphasis on the functional diversity of interactions involving the extraradical mycelium. J Exp Bot 59:1115–1126

Fitter AH (2005) Darkness visible: reflections on underground ecology. J Ecol 93:231–243

Flo DS, Hause B, Lange PR, Küster H, Strack D, Walter MH (2008) Knock-down of the MEP pathway isogene 1-deoxy-D-xylulose 5-phosphate synthase 2 inhibits formation of arbuscular mycorrhiza-induced apocarotenoids, and abolishes normal expression of mycorrhiza-specific plant marker genes. Plant J 56:86–100

Fournier J, Timmers ACJ, Sieberer BJ, Jauneau A, Chabaud M, Barker DG (2008) Mechanism of infection thread elongation in root hairs of Medicago truncatula and dynamic interplay with associated rhizobial colonization. Plant Physiol 148:1985–1995

Frank B (1885) Ueber die auf Wurzelsymbiose beruhende Ernährung gewisser Bäume durch unterirdische Pilze. Ber Dtsch Bot Ges 3:128–145

Garg N, Chandel S (2010) Arbuscular mycorrhizal networks: process and functions. A review. Agron Sustain Dev 30:581–599

Genre A, Chabaud M, Timmers T, Bonfante P, Barker DG (2005) Arbuscular mycorrhizal fungi elicit a novel intracellular apparatus in *Medicago truncatula* root epidermal cells before infection. Plant Cell 17:3489–3549

Genre A, Ortu G, Bertoldo C, Martino E, Bonfante P (2009) Biotic and abiotic stimulation of root epidermal cells reveals common and specific responses to arbuscular mycorrhizal fungi. Plant Physiol 149:1424–1434

Gianinazzi-Pearson V, Arnould C, Oufattole M, Arango M, Gianinazzi S (2000) Differential activation of H+-ATPase genes by an arbuscular mycorrhizal fungus in root cells of transgenic tobacco. Planta 211:609–613

Gleason C, Chaudhuri S, Yang T, Muñoz A, Poovaiah BW, Oldroyd GE (2006) Nodulation independent of rhizobia induced by a calcium-activated kinase lacking auto inhibition. Nature 441:1149–1152

Gobbato E (2015) Recent developments in arbuscular mycorrhizal signaling. Curr Opin Plant Biol 26:1–7

Gobbato E, Marsh J, Vernié T, Wang E, Maillet F et al (2012) A GRAS-type transcription factor with a specific function in mycorrhizal signaling. Curr Biol 22:2236–2241

Goltapeh EM, Danesh YR, Prasad R, Varma A (2008) Mycorrhizal fungi: what we know and what should we know? In: Varma A (ed) Mycorrhiza, 3rd edn. Springer-Verlag, Berlin, pp 3–27

Govindarajulu M, Pfeffer PE, Jin H, Abubaker J, Douds DD, Allen JW, Bücking H, Lammers PJ, Shachar-Hill Y (2005) Nitrogen transfer in the arbuscular mycorrhizal symbiosis. Nature 435 (7043):819–823

Guillotin B, Etemadi M, Audran C, Bouzayen M, Becard G, Combier JP (2016) Sl-IAA27 regulates strigolactone biosynthesis and mycorrhization in tomato (var. MicroTom). New Phytol 213:1124–1132

Gutjahr C, Parniske M (2013) Cell and developmental biology of arbuscular mycorrhiza symbiosis. Annu Rev Cell Dev Biol 29:593–617

Gutjahr C, Radovanovic D, Geoffroy J, Zhang Q, Siegler H, Chiapello M, Casieri L, An K, An G, Guiderdoni E, Kumar CS, Sundaresan V, Harrison MJ, Paszkowski U (2012) The half-size ABC transporters STR1 and STR2 are indispensable for mycorrhizal arbuscule formation in rice. Plant J 69:906–920

Hamiaux C, Drummond RSM, Janssen BJ, Ledger SE, Cooney JM et al (2012) DAD2 is an α/β hydrolase likely to be involved in the perception of the plant branching hormone, strigolactone. Curr Biol 22:2032–2036

Harris JM, Wais R, Long SR (2003) Rhizobium induced calcium spiking in *Lotus japonicus*. Mol Plant-Microbe Interact 16:335–341

Harrison MJ, Dewbre GR, Liu J (2002) A phosphate transporter from Medicago truncatula involved in the acquisition of phosphate released by arbuscular mycorrhizal fungi. Plant Cell 14:2413–2429

Hause B, Fester T (2005) Molecular and cell biology of arbuscular mycorrhizal symbiosis. Planta 221:184–196

Herrera-Medina M, Steinkellner S, Vierheilig H, Ocampo Bote J, Garcia Garrido J (2007) Absicic acid determines arbuscule development and functionality in the tomato arbuscular mycorrhiza. New Phytol 175:554–564

Hibbett DS, Binder M, Bischoff JF, Blackwell M, Cannon PF, Eriksson OE, Huhndorf S, James T, Zhang N et al (2007) A higher-level phylogenetic classification of the fungi. Mycol Res 111:509–547

Hohnjec N, Czaja-Hasse LF, Hogekamp C, Kuster H (2015) Preannouncement of symbiotic guests: transcriptional reprogramming by mycorrhizal lipochitooligosaccharides shows a strict co-dependency on the GRAS transcription factors NSP1 and RAM1. BMC Genomics 16:994

Horvath B, Yeun LH, Domonkos A, Halasz G, Gobbato E, Ayaydin F, Miro K, Hirsch S, Sun J, Tadege M (2011) Medicago truncatula IPD3 is a member of the common symbiotic signaling pathway required for rhizobial and mycorrhizal symbioses. Mol Plant-Microbe Interact 24:1345–1358

Ianson D (2015) Mycorrhizae in the Alaska Landscape by, Mycorrhizast and Jeff Smeenk, Extension Horticulture Specialist. University of Alaska Fairbanks Cooperative Extension Service in cooperation with the United States Department of Agriculture Heidi Rader, Tribes Extension Educator 1-877-520-5211: 1–7

IJdo M, Cranenbrouck S, Declerck S (2011) Methods for large-scale production of AM fungi: past, present, and future. Mycorrhiza 21:1–16

Imaizumi-Anraku H, Takeda N, Charpentier M, Perry J, Miwa H, Umehara Y, Kouchi H, Murakami Y, Mulder L, Vickers K, Pike J, Downie JA, Wang T, Sato S, Asamizu E, Tabata S, Yoshikawa M, Murooka Y, Wu G-J, Kawaguchi M, Kawasaki S, Parniske M, Hayashi M (2005) Plastid proteins crucial for symbiotic fungal and bacterial entry into plant roots. Nature 433:527–531

Javot H, Pumplin N, Harrison M (2007) Phosphate in the arbuscular mycorrhizal symbiosis: transport properties and regulatory roles. Plant Cell Environ 30:310–322

Karandashov V, Bucher M (2005) Symbiotic phosphate transport in arbuscular mycorrhizas. Trends Plant Sci 10(1):22–29

Kistner C, Parniske M (2002) Evolution of signal transduction in intracellular symbiosis. Trends Plant Sci 7:511–518

Kistner C, Winzer T, Pitzschke A, Mulder L, Sato S et al (2005) Seven *Lotus japonicus* genes required for transcriptional reprogramming of the root during fungal and bacterial symbiosis. Plant Cell 17:2217–2229

Kobae Y, Ohmori Y, Saito C, Yano K, Ohtomo R, Fujiwara T (2016) Phosphate treatment strongly inhibits new arbuscule development but not the maintenance of arbuscule in mycorrhizal rice roots. Plant Physiol 171:566–579

Kobae Y, Kameoka H, Sugimura Y, Saito K, Ohtomo R, Fujiwara T, Kyozuka J (2017) Strigolactone biosynthesis genes of rice are required for the punctual entry of arbuscular mycorrhizal fungi into the roots. Plant Cell Physiol 59:544–553

Kosuta S, Chabaud M, Lougnon G, Gough C, Denarie J, Barker DG, Becard G (2003) A diffusible factor from arbuscular mycorrhizal fungi induces symbiosis-specific MtENOD11 expression in roots of Medicago truncatula. Plant Physiol 131:952–962

Kouchi H, Imaizumi-Anraku H, Hayashi M, Hakoyama T, Nakagawa T, Umehara Y, Suganuma N, Kawaguchi M (2010) How many peas in a pod? Legume genes responsible for mutualistic symbioses underground. Plant Cell Physiol 51:1381–1397

Kretzschmar T, Kohlen W, Sasse J, Borghi L, Schlegel M et al (2012) A petunia ABC protein controls strigolactone-dependent symbiotic signalling and branching. Nature 483:341–344

Lauressergues D, Delaux P-M, Formey D, Lelandais-Brière C, Fort S et al (2012) ThemicroRNA miR171h modulates arbuscular mycorrhizal colonization of *Medicago truncatula* by targeting NSP2. Plant J 72:512–522

Liu J, Maldonado-Mendoza I, Lopez-Meyer M, Cheung F, Town CD, Harrison MJ (2007) Arbuscular mycorrhizal symbiosis is accompanied by local and systemic alterations in gene expression and an increase in disease resistance in the shoots. Plant J 50:529–544

Liu W, Kohlen W, Lillo A, Op den Camp R, Ivanov S, Hartog M, Limpens E, Jamil M, Smaczniak C, Kaufmann K, Yang WC, Hooiveld GJ, Charnikhova T, Bouwmeester HJ, Bisseling T, Geurts R (2011) Strigolactone biosynthesis in *Medicago truncatula* and rice requires the symbiotic GRAS-type transcription factors NSP1 and NSP2. Plant Cell 23:3853–3865

Lusk CP, Blobel G, King MC (2007) Highway to the inner nuclear membrane: rules for the road. Nat Rev Mol Cell Biol 8:414–420

Maillet F, Poinsot V, Andre O, Puech-Pages V, Haouy A, Gueunier M, Cromer L, Giraudet D, Formey D, Niebel A, Martinez EA, Driguez H, Becard G, Denarie J (2011) Fungal lipochitooligosaccharide symbiotic signals in arbuscular mycorrhiza. Nature 469:58–63

Markmann K, Giczey G, Parniske M (2008) Functional adaptation of a plant receptor-kinase paved the way for the evolution of intracellular root symbioses with bacteria. PLoS Biol 6(3):e68

Marschner P (2012) Marschner's mineral nutrition of higher plants, 3rd edn. https://doi.org/10.1016/C2009-0-63043-9

Matusova R, Rani K, Verstappen FWA, Franssen MCR, Beale MH, Bouwmeester HJ (2005) The strigolactone germination stimulants of the plant-parasitic *Striga* and *Orobanche* spp. are derived from the carotenoid pathway. Plant Physiol 139:920–934

Miller RM, Reinhardt DR, Jastrow JD (1995) External hyphal production of vesicular-arbuscular mycorrhizal fungi in pasture and tallgrass prairie communities. Oecologia 103:17–23

Miwa H, Sun J, Oldroyd GE, Downie JA (2006) Analysis of Nod-factor-induced calcium signaling in root hairs of symbiotically defective mutants of *Lotus japonicus*. Mol Plant-Microbe Interact 19:914–923

Miyasaka SC, Habte M, Friday JB, Johnson EV (2003) Manual on arbuscular mycorrhizal fungus production and inoculation techniques. Soil Crop Manag 5:1–4

Mohanta TK, Bae H (2014) Functional genomics and signaling events in mycorrhizal symbiosis. J Plant Interact 10:21–40

Morton JB, Benny GL (1990) Revised classification of arbuscular mycorrhizal fungi (zygomycetes): a new order, glomales, two new suborders, glomineae and gigasporineae, and two new families, acaulorporaceae and gigasporaceae with an amendation of glomaceae. Mycotaron 37:471–491

Mukherjee A, Ane J-M (2010) Germinating spore exudates from arbuscular mycorrhizal fungi: molecular and developmental responses in plants and their regulation by ethylene. Mol Plant-Microbe Interact 24:260–270

Nagahashi G, Douds DD (1997) Appressorium formation by AM fungi on isolated cell walls of carrot roots. New Phytol 136:299–304

Nagahashi G, Douds DD Jr (2011) The effects of hydroxy fatty acids on the hyphal branching of germinated spores of AM fungi. Fungal Biol 115:351–358

Nakagawa T, Imaizumi-Anraku H (2015) Rice arbuscular mycorrhiza as a tool to study the molecular mechanisms of fungal symbiosis and a potential target to increase productivity. Rice 8:32

Newman EI, Reddell P (1987) The distribution of mycorrhizas among families of vascular plants. New Phytol 106:745–751

Oldroyd GE (2013) Speak, friend, and enter: signaling systems that promote beneficial symbiotic associations in plants. Nat Rev Microbiol 11:252–263

Oldroyd GE, Downie JA (2006) Nuclear calcium changes at the core of symbiosis signalling. Curr Opin Plant Biol 9:351–357

Op den Camp R, Streng A, De Mita S, Cao Q, Polone E, Liu W, Ammiraju JS, Kudrna D, Wing R, Untergasser A, Bisseling T, Geurts R (2011) LysM-type mycorrhizal receptor recruited for rhizobium symbiosis in nonlegume Parasponia. Science 331:909–912

Parniske M (2000) Intracellular accommodation of microbes by plants: a common developmental program for symbiosis and disease? Curr Opin Plant Biol 3:320–328

Parniske M (2004) Molecular genetics of the arbuscular mycorrhizal symbiosis. Curr Opin Plant Biol 7:414–421

Parniske M (2008) Arbuscular mycorrhiza: the mother of plant root endosymbioses. Nat Rev Microbiol 6:763–775

Prasad R, Bhola D, Akdi K, Cruz C, Sairam KVSS, Tuteja N, Varma A (2017) Introduction to mycorrhiza: historical development. In: Varma A, Prasad R, Tuteja N (eds) Mycorrhiza. Springer, Cham, pp 1–7

Reddy SDMR, Svistoonoff S, Breuillin F, Wegmüller S, Bucher M, Reinhardt D (2008) Development and function of the arbuscular mycorrhizal symbiosis in Petunia. In: Gerats T, Strommer J (eds) Petunia: evolutionary, developmental and physiological genetics. Springer-Verlag, Cham, pp 131–156

Remy W, Taylor TN, Hass H, Kerp H (1994) Four hundred-million-year old vesicular arbuscular mycorrhizae. Proc Natl Acad Sci USA 91:11841–11843

Requena N, Mann P, Franken P (2000) A homologue of the cell cycle check point TOR2 from Saccharomyces cerevisiae exists in the arbuscular mycorrrhizal fungus Glomus mosseae. Protoplasma 212:89–98

Requena N, Serrano E, Ocon A, Breuninger M (2007) Plant signals and fungal perception during arbuscular mycorrhizal establishment. Phytochemistry 68:33–40

Sasse J, Simon S, Gubeli C, Liu GW, Cheng X, Friml J, Bouwmeester H, Martinoia E, Borghi L (2015) Asymmetric localizations of the ABC transporter PaPDR1 trace paths of directional strigolactone transport. Curr Biol 25:647–655

Schussler A, Schwarzott D, Walker C (2001) A new fungal phylum, the Glomeromycota: phylogeny and evolution. Mycol Res 105:1413–1421

Siciliano V, Genre A, Balestrini R, Cappellazzo G, deWit PJGM, Bonfante P (2007) Transcriptome analysis of arbuscular mycorrhizal roots during development of the prepenetration apparatus. Plant Physiol 144:1455–1466

Sieberer BJ, Chabaud M, Fournier J, Timmers ACJ, Barker DG (2012) A switch in Ca2+ spiking signature is concomitant with endosymbiotic microbe entry into cortical root cells of Medicago truncatula. Plant J 69:822–830

Singh S, Katzer K, Lambert J, Cerri M, Parniske M (2014) CYCLOPS, a DNA-binding transcriptional activator, orchestrates symbiotic root nodule development. Cell Host Microbe 15:139–152

Smith SE, Read DJ (1997) Mycorrhizal symbiosis, 2nd edn. Academic, San Diego, CA

Smith SE, Read D (2008) Mycorrhizal symbiosis, 3rd edn. Academic, New York, NY

Stracke S, Kistner C, Yoshida S, Mulder L, Sato S, Kaneko T, Tabata S, Sandal N, Stougaard J, Szczyglowski K, Parniske M (2002) A plant receptor-like kinase required for both bacterial and fungal symbiosis. Nature 417:959–962

Sun J, Miller JB, Granqvist E, Wiley-Kalil A, Gobbato E, Maillet F, Cottaz S, Samain E, Venkateshwaran M, Fort S, Morris RJ, Ane JM, Denarie J, Oldroyd GE (2015) Activation of symbiosis signaling by arbuscular mycorrhizal fungi in legumes and rice. Plant Cell 27:823–838

Szczyglowski K, Shaw RS, Wopereis J, Copeland S, Hamburger D, Kasiborski B, Dazzo FB, de Bruijn FJ (1998) Nodule organogenesis and symbiotic mutants of the model legume Lotus japonicus. Mol Plant-Microbe Interact 11:684–697

Takeda N, Maekawa T, Hayashi M (2012) Nuclear-localized and deregulated calcium- and calmodulin-dependent protein kinase activates rhizobial and mycorrhizal responses in *Lotus japonicus*. Plant Cell 24:810–822

Takeda N, Tsuzuki S, Suzaki T, Parniske M, Kawaguchi M (2013) CERBERUS and NSP1 of *Lotus japonicus* are common symbiosis genes that modulate arbuscular mycorrhiza development. Plant Cell Physiol 54:1711–1723

Tirichine L, Imaizumi-Anraku H, Yoshida S, Murakami Y, Madsen LH, Miwa H, Nakagawa T, Sandal N, Albrektsen AS, Kawaguchi M, Downie A, Sato S, Tabata S, Kouchi H, Parniske M, Kawasaki S, Stougaard J (2006) Deregulation of a Ca2+/calmodulin dependent kinase leads to spontaneous nodule development. Nature 441:1153–1156

Vallino M, Fiorilli V, Bonfante P (2014) Rice flooding negatively impacts root branching and arbuscular mycorrhizal colonization, but not fungal viability. Plant Cell Environ 37(3):557–572

Varma A, Prasad R, Tuteja N (2017) Mycorrhiza: eco-physiology, secondary metabolites, nanomaterials. Springer, Cham. ISBN 978-3-319-57849-1. http://www.springer.com/us/book/9783319578484

Venkateshwaran M, Cosme A, Han L, Banba M, Satyshur KA, Schleiff E, Parniske M, Imaizumi-Anraku H, Ané J-M (2012) The recent evolution of a symbiotic ion channel in the legume family altered ion conductance and improved functionality in calcium signaling. Plant Cell 24:2528–2545

Walker S, Viprey V, Downie J (2000) Dissection of nodulation signalling using pea mutants defective for calcium spiking induced by Nod factors and chitin oligomers. Proc Natl Acad Sci USA 97:13413–13418

Wang E, Schornack S, Marsh JF, Gobbato E, Schwessinger B (2012) A common signaling process that promotes mycorrhizal and oomycete colonization of plants. Curr Biol 22:2242–2246

Xie X, Yoneyama K, Yoneyama K (2010) The strigolactone story. Annu Rev Phytopathol 48:93–117

Xue L, Cui H, Buer B, Vijayakumar V, Delaux PM, Junkermann S, Bucher M (2015) Network of GRAS transcription factors involved in the control of arbuscule development in Lotus japonicus. Plant Physiol 167:854–871

Yano K, Yoshida S, Müller J, Singh S, Banba M, Vickers K, Markmann K, White C, Schuller B, Sato S, Asamizu E, Tabata S, Murooka Y, Perry J, Wang TL, Kawaguchi M, Imaizumi-Anraku H, Hayashi M, Parniske M (2008) CYCLOPS, a mediator of symbiotic intracellular accommodation. Proc Natl Acad Sci USA 105:20540–20545

Yoneyama K, Xie X, Kim H, Kisugi T, Nomura T et al (2012) How do nitrogen and phosphorus deficiencies affect strigolactone production and exudation? Planta 235:1197–1207

Yoshida S, Kameoka H, Tempo M, Akiyama K, Umehara M, Yamaguchi S, Hayashi H, Kyozuka J, Shirasu K (2012) The D3 F-box protein is a key component in host strigolactone responses essential for arbuscular mycorrhizal symbiosis. New Phytol 196:1208–1216

Zhang X, Dong W, Sun J, Feng F, Deng Y, He Z, Oldroyd GE, Wang E (2015) The receptor kinase CERK1 has dual functions in symbiosis and immunity signalling. Plant J 81:258–267

Zhao J, Wang T, Wang M, Liu Y, Yuan S, Gao Y, Yin L, Sun W, Peng L, Zhang W, Wan J, Li X (2014) DWARF3 participates in an SCF complex and associates with DWARF14 to suppress rice shoot branching. Plant Cell Physiol 55:1096–1109

Chapter 6
Contribution of Beneficial Fungi for Maintaining Sustainable Plant Growth and Soil Fertility

Rakesh Suchitra, Kaushik Rajaram, Nagarathinam Arunkumar, and D. Siva Sundara Kumar

Abstract Beneficial fungi like mycorrhiza act as a natural bio-fertilizer for more than 80% of the plant species. Mycorrhizae facilitates water and nutrient uptake and in protecting host plants from pathogens and abiotic stresses in exchange of photosynthetic products. It also improves nitrogen fixation, heavy metal tolerance, nodulation, leghemoglobin content, and polyamine contents in plants. Mycorrhizae (ecto/endo) serves as a safe, effective, and environmentally friendly alternative to conventional methods for maintaining sustainable plant growth and soil fertility.

6.1 Introduction

Rhizosphere represents a significant proportion of fungal microflora and exerts an enormous influence on the growth of the plants. The beneficial association between plant roots and mycorrhizae increases the surface area of the plant root, and hence, plant absorbs more nutrients and water. There are two main types of mycorrhizae, i.e., ectomycorrhizae and endomycorrhizae. If the mycorrhizal fungal growth is on the outer surface of the root, then they are ectotrophic mycorrhizae, and if growth is inside the root, then they are called endotrophic mycorrhizae. Mycorrhizal plants can withstand low nutrient soils, exhibit higher growth rate, and are more disease resistant and also have the capacity to absorb more nutrients from the soil. The nutrient uptake by non-mycorrhizal plants extends to a few millimeters beyond the root zone, whereas ectomycorrhizal plants can absorb nutrients from up to 20 cm and endomycorrhizal plants up to 8 cm from the root zone.

Ectomycorrhizae are commonly found on roots of woody angiosperms and gymnosperms. The ectomycorrhizal fungal hyphae colonize the roots of trees and woody plants, with restricted growth largely between the epidermal cells and often

R. Suchitra · K. Rajaram · N. Arunkumar · D. S. S. Kumar (✉)
Department of Microbiology, School of Life Sciences, Central University of Tamil Nadu, Thiruvarur, India
e-mail: d.sivasundarakumar@cutn.ac.in

A. Varma et al. (eds.), *Plant Microbe Symbiosis*,
https://doi.org/10.1007/978-3-030-36248-5_6

the root cortex as well, but the fungus hardly penetrates into the root cells. Mycorrhizal infection in the roots produce a compact mass in the epidermal layers of the root called as the "Hartig net." Also, in ectomycorrhizal plants, the growth of plant root hairs is suppressed, and mycorrhizal hyphae take up the function of root hairs for water and nutrient absorption from the soil. Mycorrhizal fungal hyphae help in sequestering phosphorus, nitrogen, zinc, calcium, sulfur, iron, and various other mineral elements for plant growth from soil.

Endomycorrhizae is commonly found in symbiotic association with herbaceous plants, vascular plants, and bryophytes. Plants provide organic acids, carbohydrates, and carbon compounds to the *endomycorrhizae*, while the mycorrhizal fungi facilitate improved nutrient uptake for plant growth and development. Endomycorrhizae contains two organelles, viz., vesicles and arbuscules. Vesicles of arbuscular mycorrhizae (AM) usually contain lipids, glycogen, proteins, and other compounds, while arbuscules are intracellular highly branched treelike structures. Hence, endomycorrhizae are also known as VAM (vesicular arbuscular mycorrhizae).

Mycorrhizae are considered as natural bio-fertilizers as they are present in more than 80% of the plants where they facilitate nutrient and water absorption and provide protection against pathogens to the host plant (Berruti et al. 2016; Prasad et al. 2017). Mycorrhizae are also involved in protecting the plants from various abiotic stresses (Miransari 2010; Evelin et al. 2009). Mycorrhizae performs various other functions like secondary metabolite production, improves nitrogen fixation, and enhances photosynthetic rate and resistance against abiotic and biotic stress conditions (Shinde et al. 2013; Goltapeh et al. 2008). Zhang et al. (2010) has also reported the role of endomycorrhizae in heavy metal tolerance, drought, salinity, and pathogen resistance.

6.2 Role of Mycorrhizae in Maintaining Soil Fertility

Several researchers have studied the role of mycorrhizae in improving soil quality and sustainable agriculture (Harrier and Watson 2003; Barea et al. 2005). Mycorrhizae are involved in heavy metal tolerance in plants and, hence, can be further utilized for enhancing plant growth on soils contaminated with heavy metals (Vahedi 2013; Gaur and Adholeya 2004). Increased *M. sativa* growth in AM inoculated heavy metal-contaminated soils as reported by Chen et al. (2007). Similar results were further shown by Liang et al. (2009) and Zhang et al. (2010). In heavy metal-contaminated soils, mycorrhizae promote plant growth in two ways, i.e., either by decreasing the toxic metals uptake or by enhancing growth metal uptake by the plants.

Endomycorrhizae produce a special glycoprotein called glomalin, which will bind with heavy metals and, hence, affect the metal uptake of plants in heavy metal-contaminated soils (Bedini et al. 2009). A greenhouse study was conducted to analyze the effect of endomycorrhizae on lead uptake in maize plant (Zhang et al. 2010). Different concentrations of lead were compared with control to evaluate the performance of mycorrhizal plants. Mycorrhizal plants reported higher plant height, biomass, an antioxidant enzyme, superoxide dismutase, etc. as compared with control. Andrade et al. (2009) has reported that under lead-contaminated soils, the reason for efficient mycorrhizal activity might be due to the amino acid composition, presence of antioxidant enzymes, and lipid peroxidation. Similar results were reported by Asif and Bhabatosh (2013) and Vahedi (2013).

Extrametrical hyphae of ectomycorrhizal fungi facilitate heavy metal sequestration and hence reduce their toxicity to the plants. Joner et al. (2000) have reported that hyphae for *Glomus mosseae* absorb heavy metals more efficiently than other non-mycorrhizal fungi. Another study has shown the transport of Cd from the soil to the vesicles of AM fungi within the roots, due to which Cd is immobilized, thereby restricting its transfer to other plant tissues (Joner and Leyval 1997). Similar results were showed by Tonin et al. (2001) for clover plant.

Mycorrhizae help in preserving the soil structure by fascinating soil aggregate formation (Ryan and Graham 2002). They maintain the soil structure by its external hyphae, which holds the soil particles together to form soil clots and hence prevents soil erosion (Miller and Jastrow 2000).

In addition, a mycorrhizal protein, i.e., glomalin, is involved in soil aggregate stabilization as it has longer resistance time in soil compared to hyphae (Langley and Hungate 2003; Staddon et al. 2003; Sharma et al. 2017). The crop management practices usually enhance soil stability, hence considered as factors aimed at in maintaining extraradical hyphae and glomalin.

6.3 Role of Mycorrhizae in Stress Tolerance to Plants

Many researchers have clearly mentioned the benefits of mycorrhizae in improving plant growth and development under adverse environmental conditions (Shinde et al. 2013). Many reports have suggested improved phosphorus absorption under salinity and water deficit conditions, which can be considered as a key mechanism in plants for promoting stress tolerance (Colla et al. 2008). Many reports have shown that mycorrhizae have the ability to enhance soil enzyme activities like phosphatase (Mar Vazquez et al. 2000). Mycorrhizae increase proline content in plants by influencing physiological processes, which act as osmoregulator under various stress situations (Ashraf and Foolad 2007). Mechanisms by which mycorrhizae make the plant tolerant to stress conditions involve improvement in plant nutrition, modification in physiological and enzymatic reactions, variation in ion uptakes like K and Na, and also modification in root morphology for facilitating maximum water and nutrient uptake (Zhang et al. 2011; Zolfaghari et al. 2013). Mycorrhizal fungi are also involved in various functions of plants like improving water use efficiency of plants, stomatal conductance and exchange rate of carbon dioxide (Birhane et al. 2012). Mycorrhizae improve nitrogen availability to the plant under drought conditions. Mycorrhizae absorb more water and nutrients, even from water-deficient areas due to the presence of extraradical hyphae, which can move beyond the nutrient depletion zone in soil (Khalvati et al. 2005; Berta et al. 2005). Jahromi et al. (2008)

have reported that under drought conditions, mycorrhizae affect plant growth positively by enhancing the abscisic acid concentration of plants. Recent studies have reported high abscisic acid percentage in mycorrhizal infected plants. It is also reported that mycorrhizae increase salinity tolerance in plants by enhancing the water uptake (Jahromi et al. 2008). Mycorrhizae also enhance concentration of soluble sugar and electrolyte in plants by hydrolysis of starch to sugars as reported by Porcel and Ruiz-Lozano (2004) and Al-Garni (2006). Gamalero et al. (2010) reported that mycorrhizae affect the expression of many antioxidant enzymes in order to withstand stress conditions. Some of the activities which are increased due to the expression of these enzymes under salinity stress involve improved nodulation, nitrogenase activity, leghemoglobin content, and polyamine contents (Yaseen et al. 2012).

Mycorrhizae enhance plant growth under salinity stress by improving the K^+/Na^+ ratio in plants (Zhang et al. 2011). Mycorrhizae maintain water potential gradient for better absorption of water from the soil (Porcel and Ruiz-Lozano 2004).

Hence, mycorrhizae affect plant development and growth under adverse environmental conditions by various mechanisms like production of hormones, nutrient regulation, production of antioxidant enzyme, and regulation of physiological processes. The effectiveness of all the above mentioned mechanisms depend on the extent of mycorrhizal colonization with the host plant.

6.4 Role of Mycorrhizae in Plant Disease Control

Biological pathogen control is an important method for improving crop yield. Mycorrhizae can be used as a biological control for plant pathogens (Varma et al. 2017). These mycorrhizae are obligate biotrophs, which utilize the photosynthates from the host plant for their growth. Mycorrhizal specificity for crop disease control is an important factor for mitigating any non-target effects on beneficial group of microorganisms. Mycorrhizal root colonization facilitates plant disease control, as aggressive mycorrhizal root colonization prevents the invasion of phytopathogenic fungi (Xavier and Boyetchko 2003). Many researchers have reported AM fungi mediated root rot reduction in plants (Slezack et al. 2000). *Phytophthora* spp. which are involved in many of the plant's diseases can also be controlled by mycorrhizal infestation to plant (Norman and Hooker 2000). The various ways mycorrhizae mediate disease control include host nutritional effect (involves tolerance to the pathogens, improved plant nutrition, quantitative and qualitative alterations in pathogen population), competition, biochemical and physiological alterations in host like systemic induced resistance, phytoalexin production, and antibiosis (Norman and Hooker 2000).

6.5 Role of Mycorrhiza in Plant Nutrition

Mycorrhizae affect nutrient uptake of plants even in low nutrient conditions and follow a symbiotic pathway for phosphorus ions (Bucher 2007; Smith and Smith 2011). There is an exchange of nutrients and carbon compounds between fungi and plants (Cavagnaro 2008). Smith et al. (2011) have reported that plants can obtain 100% of phosphorus through mycorrhizal fungi and around 4–20% of plant carbon is transferred for mycorrhizal growth (Cavagnaro 2008). Most of the terrestrial plants are able to interact with mycorrhizae naturally and have evolved via co-multiplication over 450 million years (Smith and Read 2008). Mycorrhizal interaction with plants helps in acquiring phosphorus from soil and its mobilization to cortical cells of the root. Mycorrhizae reduce the depletion of inorganic phosphate in the rhizosphere. Hence, mycorrhizae reduce the impact of Pi depletion in the rhizosphere. Mycorrhizal plants help in absorbing P in two different pathways, i.e., via Pi transporter and phosphorous access from the various regions and volumes of soil (Smith et al. 2011). Direct nutrient uptake by root epidermis involves root hairs, which absorb phosphorus from the soil solution near the rhizosphere. Genes encoding affinity for Pi mobilizers is present in root apex and hairs (Gordon-Weeks et al. 2003) and suppressed in more mature regions, and their expression often declines with increased phosphorus supply and by mycorrhizal colonization (Javot et al. 2007).

6.6 Role of Mycorrhizae in Nitrogen Fixation

Mycorrhizae exhibit high potential to improve nitrogen fixation in leguminous crops. An enhanced nutrient absorption and the synergistic effect of other rhizospheric microbes are instrumental in improving nitrogen fixation under adverse environmental conditions. Many studies have clearly reported the efficacy of AM fungi along with diazotrophic bacteria in improving the yield of leguminous crops (Lesueur and Sarr 2008). Since mycorrhizal fungi naturally exist under adverse conditions like a highly saline environment, their association could be very useful in improving growth and vigor of plants under stress conditions (Kumar et al. 2010). Garg and Chandel (2011) studied the beneficial interaction of pigeon pea with *G. mosseae* and observed improved plant dry mass and nitrogen-fixing potential of nodules under salt stress. The inoculation of mycorrhizae with *Rhizobium* could be effective for enhancing nitrogen fixation under adverse environmental situations (Franzini et al. 2009). Rabie and Almadini (2005) showed that mycorrhizae protect mung bean plants from the deleterious effects of salts. Shokri and Maadi (2009) reported an improved root length, nutrient uptake, and total biomass under salinity stress conditions of *Trifolium alexandrinum*.

Mycorrhizal application along with plant growth-promoting rhizobacteria enhances the nitrification in majority of the plants. Bisht et al. (2009) reported positive response of both *Rhizobium leguminosarum* and *P. fluorescens* with mycorrhizae. Under adverse abiotic conditions, there are limits in nitrogen fixation in plants, although such abiotic stresses can be overcome by inoculating mycorrhizae in association with the PGPR group of microorganisms.

6.7 Conclusion

Mycorrhizae play a significant role in improving crop productivity by using existing resources, enhancing endurance to development of chemical pesticide resistance, facilitating pollution and risk-free disease control, and conforming to sustainable agricultural practices. In the future, mycorrhizosphere management must become a viable and eco-friendly solution not only for plant disease control but also for maintaining overall plant growth and soil fertility.

References

Al-Garni SMS (2006) Increasing NaCl-salt tolerance of a halophytic plant *Phragmites australis* by mycorrhizal symbiosis. Am Eur J Agric Environ Sci 1:19–26

Andrade SAL, Gratao PL, Schiavinato MA, Silveira APD, Azevedo RA, Mazzafera P (2009) Zn uptake, physiological response and stress attenuation in mycorrhizal jack bean growing in soil with increasing Zn concentrations. Chemosphere 75:1363–1370

Ashraf M, Foolad MR (2007) Roles of glycine betaine and proline in improving plant abiotic stress resistance. Environ Exp Bot 59:206–216

Asif M, Bhabatosh M (2013) Effects of inoculation with stress-adapted arbuscular mycorrhizal fungus *Glomus deserticola* on growth of *Solanum melogena* L. and *Sorghum sudanese* Staph. seedlings under salinity and heavy metal stress conditions. Arch Agron Soil Sci 59:173–173

Barea J, Pozo M, Azcon R, Aguilar C (2005) Microbial co-operation in the rhizosphere. J Exp Bot 56:1761–1778

Bedini S, Pellegrino E, Avio L, Pellegrini S, Bazzoffi P, Argese E (2009) Changes in soil aggregation and glomalin-related soil protein content as affected by the arbuscular mycorrhizal fungal species *Glomus mosseae* and *Glomus intraradices*. Soil Biol Biochem 41:1491–1496

Berruti A, Lumini E, Balestrini R, Blanciotto V (2016) Arbuscular mycorrhizal fungi as natural biofertilizers: let's Benefit from past successes. Front Microbiol 6:1559

Berta G, Sampo S, Gamalero E, Massa N, Lemanceau P (2005) Suppression of *Rhizobium* root-rot of tomato by *Glomus mosseae* BEG12 and *Pseudomonas fluorescens* A6RI is associated with their effect on the pathogen growth and on the root morphogenesis. Eur J Plant Pathol 111:279–288

Birhane E, Streck FJ, Fetene M, Bongers F, Kuyper TW (2012) Arbuscular mycorrhizal fungi enhance photosynthesis, water use efficiency, and growth of frankincense seedlings under pulsed water availability conditions. Oecologia 169:895–904

Bisht R, Chaturvedi S, Srivastava R, Sharma AK, Johri BN (2009) Effect of arbuscular mycorrhizal fungi, *Pseudomonas fluorescens* and *Rhizobium leguminosarum* on the growth and nutrient status of dalbergia sissoo Roxb. Trop Ecol 2:231–242

Bucher M (2007) Functional biology of plant phosphate uptake at root and mycorrhizal interfaces. New Phytol 173:11–26

Cavagnaro TR (2008) The role of arbuscular mycorrhizas in improving plant zinc nutrition under low soil zinc concentrations: a review. Plant Soil 304:315–325

Chen BD, Zhu TG, Duan J, Xiao XY, Smith SE (2007) Effects of the arbuscular mycorrhizal fungus *Glomus mosseae* on growth and metal uptake by four plant species in copper mine tailings. Environ Pollut 147:374–380

Colla G, Rouphael Y, Cardarelli M, Tullio M, Rivera CM, Rea E (2008) Alleviation of salt stress by arbuscular mycorrhizal in zucchini plants grown at low and high phosphorus concentration. Biol Fertil Soil 44:501–509

Evelin H, Kapoor R, Giri B (2009) Arbuscular mycorrhizal fungi in alleviation of salt stress: a review. Ann Bot 104:1263–1280

Franzini VI, Azcon R, Mendes FL, Aroca R (2009) Interactions between *Glomus* species and *Rhizobium* affect the nutritional physiology of drought-stressed legume hosts. J Plant Physiol 167:614–619

Gamalero E, Berta G, Massa N, Glick BR, Lingua G (2010) Interactions between Pseudomonas putida UW4 and Gigaspora rosea BEG9 and their consequences for the growth of cucumber under salt-stress conditions. J Appl Microbiol 108(1):236–245

Garg N, Chandel S (2011) Effect of mycorrhizal inoculation on growth, nitrogen fixation, and nutrient uptake in *Cicer arietinum* (L) under salt stress. Turk J Agric For 35:205–214

Gaur A, Adholeya A (2004) Prospects of arbuscular mycorrhizal fungi in phytoremediation of heavy metal contaminated soils. Curr Sci 86:528–534

Goltapeh EM, Danesh YR, Prasad R, Varma A (2008) Mycorrhizal fungi: what we know and what should we know? In: Varma A (ed) Mycorrhiza, 3rd edn. Springer-Verlag, Berlin, pp 3–27

Gordon-Weeks R, Tong YP, Davies TGE, Leggewie G (2003) Restricted spatial expression of a high-affinity phosphate transporter in potato roots. J Cell Sci 116:3135–3144

Harrier LA, Watson CA (2003) The role of arbuscular mycorrhizal fungi in sustainable cropping systems. Adv Agron 79:185–122

Jahromi F, Aroca R, Porcel R, Ruiz-Lozano JM (2008) Influence of salinity on the in vitro development of *Glomus intraradices* and on the in vivo physiological and molecular responses of mycorrhizal lettuce plants. Microb Ecol 55:45–53

Javot H, Penmetsa RV, Terzaghi N, Cook DR, Harrison MJ (2007) A. *Medicago truncatula* phosphate transporter indispensable for the arbuscular mycorrhizal symbiosis. Proc Natl Acad Sci USA 104:1720–1725

Joner EJ, Leyval C (1997) Uptake of 109Cd by roots and hyphae of *Glomus mosseae/Trifolium subterraneum* mycorrhiza from soil amended with high and low concentration of cadmium. New Phytol 135:353–360

Joner EJ, Van Aarle IM, Vosatka M (2000) Phosphatase activity of extraradical arbuscular mycorrhizal hyphae: a review. Plant Soil 226:199–210

Khalvati MA, Hu Y, Mozafar A, Schmidhalter U (2005) Quantification of water uptake by arbuscular mycorrhizal hyphae and its significance for leaf growth, water relations, and gas exchange of barley subjected to drought stress. Plant Boil 7:706–712

Kumar A, Sharma S, Mishra S (2010) Influence of arbuscular mycorrhizal (AM) fungi and salinity on seedling growth, solute accumulation and mycorrhizal dependency of *Jatropha curcas* L. J Plant Growth Regul 29:297–306

Langley JA, Hungate BA (2003) Mycorrhizal controls on belowground litter quality. Ecology 84:2256

Lesueur D, Sarr A (2008) Effects of single and dual inoculation with selected microsymbionts (rhizobia and arbuscular mycorrhizal fungi) on field growth and nitrogen fixation of *Calliandra calothyrsus* Meissn. Agrofor Syst 73:37–45

Liang CC, Li T, Xiao YP, Liu MJ, Zhang HB, Zhao ZW (2009) Effects of inoculation with arbuscular mycorrhizal fungi on maize grown in multi-metal contaminated soils. Int J Phytoremediation 11:692–703

Mar Vazquez M, Cesar S, Azcon R, Barea JM (2000) Interactions between arbuscular mycorrhizal fungi and other microbial inoculants (*Azospirillum, Pseudomonas, Trichoderma*) and their effects on microbial population and enzyme activities in the rhizosphere of maize plants. Appl Soil Ecol 15:261–272

Miller RM, Jastrow JD (2000) Mycorrhizal fungi influence soil structure. In: Kapulnik Y, Douds DD Jr (eds) Arbuscular mycorrhizas: physiology and function. Kluwer Academic, Dordrecht, pp 3–18

Miransari M (2010) Contribution of arbuscular mycorrhizal symbiosis to plant growth under different types of soil stress. Plant Biol 12:563–569

Norman JR, Hooker JE (2000) Sporulation of *Phytophthora fragariae* shows greater stimulation by exudates of non-mycorrhizal than by mycorrhizal strawberry roots. Mycol Res 104:1069–1073

Porcel R, Ruiz-Lozano JM (2004) Arbuscular mycorrhizal influence on leaf water potential, solute accumulation and oxidative stress in soybean plants subjected to drought stress. J Exp Bot 55:1743–1750

Prasad R, Bhola D, Akdi K, Cruz C, Sairam KVSS, Tuteja N, Varma A (2017) Introduction to mycorrhiza: historical development. In: Varma A, Prasad R, Tuteja N (eds) Mycorrhiza. Springer, Cham, pp 1–7

Rabie GH, Almadini AM (2005) Role of bioinoculants in development of salt tolerance of *Vicia faba* plants under salinity stress. Afr J Biotechnol 4:210–222

Ryan MH, Graham JM (2002) Is there a role for arbuscular mycorrhizal fungi in production agriculture? Plant Soil 244:263–271

Sharma S, Prasad R, Varma A, Sharma AK (2017) Glycoprotein associated with *Funneliformis coronatum*, *Gigaspora margarita* and *Acaulospora scrobiculata* suppress the plant pathogens in vitro. Asian J Plant Pathol 11(4):199–202. https://doi.org/10.3923/ajppaj.2017

Shinde SK, Shinde BP, Patale SW (2013) The alleviation of salt stress by the activity of AM fungi in growth and productivity of onion (*Allium cepa* I.) plant. Int J Life Sci Pharma Res 3:11–15

Shokri S, Maadi B (2009) Effect of arbuscular mycorrhizal fungus on the mineral nutrition and yield of *Trifolium alexandrinum* plants under salinity stress. J Agron 8:79–83

Slezack S, Dumas-Gaudot E, Paynot M, Gianinazzi S (2000) Is a fully established arbuscular mycorrhizal symbiosis required for bioprotection of *Pisum sativum* roots against *Aphanomyces euteiches*. Mol Plant-Microbe Interact 13:238–241

Smith SE, Read DJ (2008) Mineral nutrition, toxic element accumulation and water relations of arbuscular mycorrhizal plants. In: Smith SE, Read DJ (eds) Mycorrhizal symbiosis. Academic, London, pp 145–148

Smith SE, Smith FA (2011) Roles of arbuscular mycorrhizas in plant nutrition and growth: new paradigms from cellular to ecosystem scales. Annu Rev Plant Biol 62:227–250

Smith SE, Jakobsen I, Gronlund M, Smith FA (2011) Roles of arbuscular mycorrhizas in plant phosphorus nutrition: interactions between pathways of phosphorus uptake in arbuscular mycorrhizal roots have important implications for understanding and manipulating plant phosphorus acquisition. Plant Physiol 156:1050–1057

Staddon PL, Ramsey CB, Ostle N, Ineson P, Filler AH (2003) Rapid turnover of hyphae of mycorrhizal fungi determined by AMS microanalysis of ^{14}C. Science 300:1138–1140

Tonin C, Vandenkoornhuyse P, Joner EJ, Straczek J, Leyval C (2001) Assessment of arbuscular mycorrhizal fungi diversity in the rhizosphere of *Viola calaminaria* and effect of these fungi on heavy metal uptake by clover. Mycorrhiza 10:161–168

Vahedi A (2013) The investigation of arbuscular mycorrhizal fungal effect on growth and nutrients in *Trifolium pratense* in the multi metals contaminated soil. Int J Biol 5:71–78

Varma A, Prasad R, Tuteja N (2017) Mycorrhiza: nutrient uptake, biocontrol, ecorestoration. Springer, Cham. ISBN 978-3-319-68867-1. http://www.springer.com/us/book/9783319688664

Xavier LJC, Boyetchko SM (2003) Arbuscular mycorrhizal fungi in plant disease control. In: Arora D, Bridge P, Bhatnagar D (eds) Fungal biotechnology in agricultural, food and environmental applications. Marcel Dekker, New York, pp 182–194

Yaseen T, Burni T, Hussain F (2012) Effect of arbuscular mycorrhizal inoculation on nutrient uptake, growth and productivity of chickpea (*Cicer arietinum*) varieties. Int J Agron Plant Prod 3:334–345

Zhang HH, Tang M, Chen H, Zheng C, Niu Z (2010) Effect of inoculation with AM fungi on lead uptake, translocation and stress alleviation of *Zea mays* L. seedlings planting in soil with increasing lead concentrations. Eur J Soil Biol 46:306–311

Zhang YF, Wang P, Yang YF, Bi Q, Tian SY, Shi XC (2011) Arbuscular mycorrhizal fungi improve reestablishment of *Leymus chinensis* in bare saline-alkaline soil: implication on vegetation restoration of extremely degraded land. J Arid Environ 75:773–778

Zolfaghari M, Nazeri V, Sefidkon F, Rejali F (2013) Effect of arbuscular mycorrhizal fungi on plant growth and essential oil content and composition of *Ocimum basilicum* L. Iran J Plant Physiol 3:643–650

Chapter 7
Biofertilizers Toward Sustainable Agricultural Development

G. Chandramohan Reddy, R. K. Goyal, Shriniketan Puranik, Vijaykumar Waghmar, K. V. Vikram, and K. S. Sruthy

Abstract Modern intensive farming technologies enhance crop production but over time are associated with more problems, causing environment pollution that is hazardous to human health, and, ultimately poor production of crops and the threat to food security of the growing world population. Therefore, sustainable agricultural production is the major challenge to encounter the huge demand of food grain production by the emerging population in an environmentally safe and cost-effective manner. Biofertilizers are one of the key sources in sustainable agriculture production and organic farming to meet consumer preferences and quality crop production. Biofertilizers are live microorganisms that enhance the supply of adequate nutrients to the crop plants through nitrogen fixation, phosphorus and potassium solubilization, and production of plant growth hormones, ensuring optimum growth and development of crops, and ultimately facilitate higher crop production and productivity. Biofertilizers, being essential components of sustainable farming, are vital in maintaining long-term soil fertility and the sustainability of crop production. The modern agricultural production system needs the widespread use of biofertilizers and potent sources for inclusive and sustainable development of agriculture without damaging the ecosystem.

7.1 Introduction

The world population is now 7.7 billion. India alone contains 1.36 billion people, which number is swelling day by day, placing pressure on the agricultural production system, and on our natural resources, which are required for food production for

G. C. Reddy (✉) · R. K. Goyal
Department of Horticulture, CCS Haryana Agricultural University, Hisar, Haryana, India

S. Puranik · V. Waghmar
Department of Microbiology, Indian Agricultural Research Institute, New Delhi, India

K. V. Vikram · K. S. Sruthy
Department of Agricultural Microbiology, University of Agricultural Sciences, GKVK, Bangalore, Karnataka, India

A. Varma et al. (eds.), *Plant Microbe Symbiosis*,
https://doi.org/10.1007/978-3-030-36248-5_7

this huge population with limited land. According to the 15th Census of India in 2011, the observed population decadal growth was 17.64%, of which about 68.84% is composed of rural populations mainly dependent on agriculture. This growing human population imparts stresses on the agriculture sector to meet the demand of food security, which forces the farming sector to follow modern intensive cultivation methods with heavy usage of synthetic fertilizers and pesticides for increased crop productivity (Santos et al. 2012). Such prolonged usage of synthetic fertilizers causes significant deterioration of plant roots, increasing the susceptibility of plants to disease, insect pests, and abiotic stress such as soil acidification, salinity changes (Chun-Li et al. 2014), and eutrophication of the groundwater and other water bodies (Youssef and Eissa 2014). Nitrogen fertilizers such as nitrates leach to groundwater, polluting the water resources and causing blue baby syndrome (acquired methemoglobinemia), which affects future generations (Knobeloch et al. 2000).

In this regard, eco-friendly and sustainable crop production with optimal natural resources usage approaches are now gaining popularity in meeting the demands for food security. The use of biofertilizers in modern crop production systems thus has a major role in achieving sustainable agriculture production (Giri et al. 2019). The universal market for biofertilizers is likely to exceed a market worth of USD 10.5 billion by 2020. European and Latin American countries are the principal consumers of biofertilizers because of their strict protocols on the usage of inorganic fertilizers; ultimately chemical fertilizers are to be replaced by biofertilizers in agricultural production (Raja 2013).

Biofertilizers are denoted as "substances which contain living microorganisms that colonize the rhizosphere or the interior of the plants and promote the growth of plants by increasing the availability of nutrients to the host crops, when applied to soils, seeds, plant surfaces" (Mazid et al. 2011; Malusa et al. 2012). Biofertilizers are preparations of organic origin containing cells of microorganisms that may be N-fixers, P- or Zn-solubilizers and absorbers, mobilizers, S-oxidizers or organic matter decomposers, and mainly act in conversion of unavailable nutrients to available forms through their routine metabolic activities (Vessey 2003). The crop plant utilizes only 10–40% of applied nutrients; the rest, 60–90%, is not available or is lost in the various forms by immobilization, leaching, runoff, or volatilization. In these conditions, biofertilizers may help provide slow, steady, and continuous release of nutrients by their metabolism and consequently form a significant component in the integrated nutrient management (INM) system to achieve sustainable agricultural production and productivity (Adesemoye and Kloepper 2009).

7.2 Role of Biofertilizers in Agriculture

Biofertilizers are eco-friendly and cost-effective. Continuous use of these fertilizers boosts soil fertility. Inclusion of biofertilizers in crop production system significantly enhances crop production by various mechanisms such as nitrogen fixation, P-solubilization, P-mobilization, K-solubilization, micronutrient solubilization, organic matter enhancement, and excretion of growth hormones, and also reduces the harmful influences of inorganic fertilizers on crops and soil productivity (Jeyabal and Kupuswamy 2001; Mahdi et al. 2012; Singh et al. 2011). Certain genera of microorganisms are capable of nitrogen fixing, as shown in the following subsections.

7.2.1 Azotobacter

Azotobacter is an aerobic, Gram-negative, heterotrophic, free-living nitrogen-fixing bacteria, with the capability of autonomous biological nitrogen fixation (BNF) (Martyniuk and Martyniuk 2003). *Azotobacter* species are normally present in neutral and alkaline soils frequently occurring in fertile soils. The genus *Azotobacter* comprises several species: *A. chroococcum*, *A. vinelandii*, *A. beijerinckii*, *A. paspali*, *A. armeniacus*, *A. nigricans*, and *A. salinestri* (Gothandapani et al. 2017), having a motile and mesophilic nature with ability for fixing on average 20 kg N ha^{-1} per year (Rawia et al. 2009). The application of *Azotobacter* as biofertilizer has a key role in plant metabolism such as synthesis of antibiotics, secretion of plant growth hormones (Pandey and Kumar 1989), vitamins, and coloring pigments (Jimenez et al. 2011), and antifungal activity (Sudhir et al. 1983). Dudeja et al. (1981) reported the maximum percent increase in yield of crops resulting from application of *Azotobacter* over synthetic fertilizers in different crops (Table 7.1).

The maximum fruit size of strawberry plants was 37.62 × 28.01 mm with the application of 25% nitrogen through farmyard manure (FYM) augmented with *Azotobacter*, which was on par with the plant with percent nitrogen in the form of urea in combination with *Azotobacter* (Iqbal et al. 2009).

Table 7.1 Effect of *Azotobacter* on crop yields

Crop	Increase in yield over chemical fertilizers (%)
Wheat	8–10
Rice	5
Maize	15–20
Sorghum	15–20
Other	13
Potato	16
Carrot	40
Cauliflower	2–24
Tomato	7–27
Cotton	9–24

Source: Dudeja et al. (1981)

7.2.2 Azospirillum

Azospirillum are Gram-negative, aerobic, non-nodule-forming, nitrogen-fixing bacteria belonging to the family *Spirillaceae*, and are associated with symbiosis chiefly with roots of C_4 cycle crops having a dicarboxylic pathway of photosynthesis because they nurture and fix nitrogen on the organic salts of malic and aspartic acid (Mishra and Dash 2014). *Azospirillum* significantly affects root development and exudation (Trabelsi and Mhamdi 2013) and can fix 20–40 kg nitrogen per ha^{-1} under aerobic conditions. Important species of *Azospirillum* are *A. lipoferum*, *A. brasilense*, *A. amazonense*, *A. halopreferans*, and *A. trakense*, mainly used for production of field crops. Steenhoudt and Vanderleyden (2000) observed that inoculation of *A. brasilense* recorded maximum vegetative growth and yield in maize crop.

Generally, *Azospirillum* spp. are used as seed treatment just before sowing in different crops. Apart from BNF, *Azospirillum* also has the ability to produce growth-promoting phytohormones and phosphate solubilization (Puente et al. 2004), antifungal activity (Bashan and de-Bashan 2010), and tolerance of crops against biotic and abiotic stress such as soil salinity or acidity (Creus et al. 1997). Subraya et al. (2017) noted that application of NPK at 80:40:40 kg ha^{-1} along with *Azosprillum* and phosphate-solubilizing bacteria (PSB) showed maximum vegetative growth and yield per plant in strawberry.

7.2.3 Rhizobium

The rhizobium is a symbiotic association colonizing roots that fixes atmospheric nitrogen in legume crops and utilizes photosynthates as the energy source from plants, fixing nitrogen from the atmosphere for their host plants. Rhizobacteria are the most efficient biofertilizer for leguminous crops for higher quantity of nitrogen fixation and inoculation, with *Rhizobium* enhancing nutrient uptake and photosynthetic rate, ultimately enhancing the production and productivity of different crops (Jehangir et al. 2017). Cross-inoculation has an important role in the host specificity of *Rhizobium* and host plant compatibility (Table 7.2) (Ponmurugan and Gopi 2006).

Table 7.2 Major inoculation groups with inoculant and host plants

Cross-inoculation group	*Rhizobium* species	Host legume
Pea group	*R. leguminosarum*	Pea, sweet pea
Alfalfa group	*R. meliloti*	Sweet clover
Clover group	*R. trifoli*	Clover/berseem
Bean group	*R. phaseoli*	All beans
Soybean group	*Bradyrhizobium japonicum*	Lupins
Cowpea group	*Rhizobium* sp.	Cowpea, arhar, urd, moong, groundnut

Source: Ponmurugan and Gopi (2006)

Khaitov et al. (2016) noted that *Rhizobium* inoculation enhances N-fixation, increases N-nutrition, and increases yields in chickpea cultivation on salinated soils in Uzbekistan. Adeyeye et al. (2017) recorded that inoculation of soybean seed with *Rhizobium* and compost (4 t^{-1}) significantly enhanced the grain yield (35%) compared to noninoculated plots.

7.2.4 *Cyanobacteria*

Cyanobacteria, also known as blue-green algae (BGA), are prokaryotic and autotrophic, fix atmospheric nitrogen, and are associated with fungi, liverworts, ferns, and flowering plants (RoyChowdhury et al. 2014). *Nostoc*, *Anabaena*, *Cylindrospermum*, *Gloeotrichia*, *Tolypothrix*, *Aulosira*, and *Aphanothece* are examples of BGA used in rice cultivation. Blue-green algae extensively used in rice cultivation enhance yields (15–38%) by producing growth hormones such as auxins and gibberellins, and fix about 20–30 kg N ha^{-1} under wetland systems (Mishra et al. 2013). Cyanobacterial inoculation has reported significant results in terms of yields and quality of produce in various crops such as cotton, sugarcane, oats, barley, tomato, radish, maize, and chilli (Thajuddin and Subramanian 2005).

7.2.5 Azolla

Azolla is a free-floating, symbiotic, aquatic fern extensively used in rice cultivation; they decompose easily in the soil, providing organic manure, contain a higher nitrogen percentage, and provide all macro- and micronutrients to the rice crop. Generally, *Azolla* is considered as an organic nitrogen fertilizer in rice crops as an alternative to synthetic nitrogen fertilizer because of its sustainable supplementation of nitrogen (fixes 30–50 kg N ha^{-1}) to rice crops, and reduces weed growth and increases soil fertility by adding organic manure (Yao et al. 2018). Sundaravarathan and Kannaiyan (2002) stated that application of *Azolla microphylla* at 15 t ha^{-1} increased yield by 29.2% in a rice crop. In India, *Azolla pinnata* is commercially used for rice cultivation more than other species (Mazid and Khan 2015).

7.2.6 Gluconacetobacter diazotrophicus

Gluconacetobacter diazotrophicus is a non-spore-forming, non-nodule-producing, nitrogen-fixing bacterium that belongs to the family *Acetobacteraceae*, initially found in monocot plants and then subsequently noticed in various crops.

Gluconacetobacter diazotrophicus is neither crop nor plant specific and is naturally found in wild or unrelated plant species, requiring no modifications or complexes such as nodules. Inoculation of *G. diazotrophicus* to the host plants enhances plant growth and development through stimulating plant growth hormones, and solubilizing the P and Zn elements; also acting as an antibiotic, it helps the plants to tolerate some soil-borne diseases (Intorne et al. 2009; Logeshwarn et al. 2011; Eskin et al. 2014).

7.3 P-Solubilization

Phosphorus is the element of major importance after nitrogen in the nourishment of crop plants, with a key role in plant metabolic processes including photosynthesis, transfer of energy, signal transduction, and nitrogen fixation in legume plants. Although phosphorus is plentiful in soils in both organic and inorganic forms, it is not available to the crops because it is mostly present in the insoluble mineral form. In these situations, biofertilizers with phosphorus-solubilizing microorganisms such as some bacteria, *Pseudomonas putida* and *Bacillus megaterium*, and fungi, such as *Aspergillus* and *Penicillium*, solubilize the insoluble phosphorus sources in the soil to make it available for growth and development of crop plants. These biofertilizers fulfill the maximum 20–25% phosphorus requirement of the crop plants during the crop period and reduce the costs of phosphate fertilizers (Chang and Yang 2009).

Amid the P-solubilizing microbial population in soil, bacteria (PSB) constitute 1–50%, and fungi (PSF) constitute 0.1–0.5% in P-solubilization potential (Chen et al. 2006). Recently, actinomycetes (*Actinomyces*, *Streptomyces*) have been used for P-solubilization, and these are gaining great attention because they can survive better in drought and extreme heat conditions, with more production of plant hormones and antibiotics. Hamdali et al. (2008) noted that nearly 20% of actinomycetes have the capacity to solubilize P in soils.

7.3.1 Bacteria

The phosphorus-solubilizing bacteria (PSB), also known as phosphobacteria, are Gram-positive, rod-shaped bacteria found in soils and the plant rhizosphere, able to solubilize mineral phosphorus in soils, to facilitate root uptake and enhance biological nitrogen fixation by nitrogen-fixing microorganisms (Lach et al. 1990; Mohammadi and Sohrabi 2012). Application of P-solubilizers to the soil may release organic and inorganic acids, which enhances the phosphatase enzymes that facilitate mineralization of organic P compounds in soils (Stevenson 1986), and also have solubilizing potential of other elements such as Zn, K, Fe, and Mn (Amalraj et al. 2012). Various strains of bacteria, such as *Pseudomonas*, *Bacillus*, *Rhizobium*,

Agrobacterium, *Acetobacter*, *Micrococcus*, and *Erwinia* have greater ability to solubilize P sources than do other bacteria (Diriba et al. 2013).

7.3.2 Fungi

Some fungi, such as *Aspergillus* and *Penicillium*, have a great capacity to solubilize P into a form available to crop plants by producing some organic acids that can solubilize insoluble P in the soils (Vassilev et al. 2007; Oliveira et al. 2009). These fungi are generally found in arable soils and can enhance plant growth by 15–20% when applied to the crops (Kucey and Paul 1982; Gunes et al. 2009). Kapri and Tewari (2010) observed that *Aspergillus niger*-treated chickpea plants showed maximum increase of dry biomass by 22–33% compared with noninoculated control plants. Similarly, Ram et al. (2015) studied the potential use of phosphorus-solubilizing fungi (PSF) (*Penicillium bilaiae*) as bioinoculants with 50% of recommended P-fertilizer doses; results revealed that higher wheat grain yield was recorded in PSF-inoculated plots than in controls.

7.4 P-Mobilizers

Some of the microorganisms (arbuscular mycorrhizae fungi) increase the P-uptake by mobilizing the rich-P environment to a low-P environment instead of solubilizing P; these microorganisms are known as P-mobilizers. Generally, these biofertilizers increase P use efficiency by increasing its mobility from a higher P source to a lower P source in soil, reduce P fixation, and facilitate more P being available to the crop plants (Ghorbanian et al. 2012). Ewa et al. (2013) noted that inoculation of *Pseudomonas luteola* significantly increased the total shoot length in apple grown in pots and observed higher P concentration in the soil.

7.4.1 Mycorrhiza

Mycorrhizae are a group of fungi belonging to phylum Glomeromycota, derived from the Greek words 'mukes,' meaning fungus, and 'rhiza,' meaning roots: they are associated symbiotically with host plant roots in the rhizosphere (Prasad et al. 2017). Mycorrhiza significantly enhance the uptake of P by plant root systems through production of extensive external hyphae and mycelium in the rhizosphere that allow plants to draw more water and nutrients (Bolan 1991; Jakobsen et al. 1992; Varma et al. 2017). Yao et al. (2001) observed that inoculation of arbuscular mycorrhiza fungus (AMF) significantly enhanced P-availability in the soil by mobilizing and solubilizing phosphates through production of some organic acids.

Generally, mycorrhizae are of two types: ectomycorrhizae and endomycorrhizae, further divided into different groups such as ectomycorrhizae (EM), ectendomycorrhizae, arbuscular (AM), monotropoid, arbutoid, ericoid, and orchid, but ectomycorrhizae (EM) and arbuscular mycorrhizae (AM) are the more prominent types. Mycorrhizal inoculation not only enhances the nutrient uptake with a key role in bioremediation, but provides a defense mechanism to the host plant against disease-causing pathogens in different crops. Berdeni et al. (2018), reported that mycorrhizal fungi inoculation in apple tress significantly enhanced resistance to the fungal pathogen *Neonectria ditissima*.

7.5 K-Solubilizing Bacteria

Potassium (K) is one of the main macroelements needed for plant growth and productivity throughout the crop period. Generally, larger amounts of K than other macronutrients are present in soils, but most K is in a form not available for effective plant uptake: only 1–2% of this element is available to the plants (Sparks and Huang 1985). Some of the microorganisms such as bacteria, fungi, and actinomycetes are known for solubilization of K present in soil; among microorganisms, bacteria mostly solubilize K by production of organic and inorganic acids, complexolysis, acidolysis, chelation, and exchange reactions in the soil and rhizosphere of the host plant (Archana et al. 2013; Meena et al. 2015).

Saha et al. (2016) observed that inoculation of *Bacillus licheniformis* and *Pseudomonas azotoformans* in rice fields showed higher K-solubilizing capability than other isolated rhizobacteria in rice crops. Similarly, Prajapati and Modi (2016) stated that K and chlorophyll content significantly improved through inoculation of *Enterobacter hormaechei* in a cucumber crop. In tobacco, Subhashini (2015) recorded a higher potassium content in the leaf of plants inoculated with *Frateuria aurantia* than in the control. Bacteria such as *Bacillus mucilaginosus*, *B. edaphicus*, *B. circulans*, *Pseudomonas*, *Burkholderia*, *Acidithiobacillus ferrooxidans*, and *Paenibacillus* spp. also have the ability of K-solubilization in different crops, and using these K-solubilizing bacteria effectively in a crop production system reduces the cost of cultivation and enhances high-quality production.

7.6 Plant Growth-Promoting Rhizobacteria (PGPR)

The bacteria that colonize roots of plants or the rhizosphere in the soil, stimulating plant growth, are collectively known as plant growth-promoting rhizobacteria (PGPR). Plant growth-promoting rhizobacteria may improve plant growth and productivity by producing plant growth regulators (auxin, gibberellins, cytokinins, etc.), by solubilizing and mineralizing organic and inorganic phosphate or other nutrients, fixing atmospheric nitrogen, suppressing plant disease, facilitating the uptake of nutrients, and preventing toxic effects on the soil as produced by synthetic fertilizers (Glick et al. 2007; Prasad et al. 2015). Among the different genera of rhizobacteria, *Bacillus* spp. and *Pseudomonas fluorescens* are mainly considered as PGPRs (Podile and Kishore 2006).

PGPRs are used as biocontrol agents in crop production because of their antagonism to soil-borne pathogens and indirect stimulation of plant growth through activation of siderophores and antibiotics. Rhizobacteria encourage resistance to biotic stress through the salicylic acid-dependent SAR pathway, or require jasmonic acid and ethylene perception from the plant for induced systemic resistance (ISR); *Pseudomonas* and *Bacillus* especially are major influences on ISR. Resistance-inducing and antagonistic rhizobacteria are a key in framing new inoculants with combinations of different biofertilizers, leading to a more efficient use as biocontrol agents in organic farming systems (Oliveira et al. 2009; Beneduzi et al. 2012).

7.7 Zinc Solubilizers

Zinc is an important micronutrient for plant growth and development: it is involved in carbohydrate metabolism, in auxin metabolism, and acts as a significant antioxidant (Alloway 2004). Generally, zinc availability to the plants in soil is dependent on various factors, but most of the zinc is in the form of insoluble complexes, leading to zinc deficiency in crops.

Some of the microorganisms have the capability to solubilize zinc by producing organic acids and converting the insoluble zinc sulfide, zinc oxide, and zinc carbonate into available Zn^{+} through reducing the soil pH and breaking down the complexity to increase crop growth and yield and soil fertility. Mahdi et al. (2012) observed that biofertilizer application (*Bacillus* sp.) enhances zinc availability more than control treatments. Several PGPR, such as *Pseudomonas*, *Rhizobium*, *Bacillus*, and *Azospirillum*, are reported as significantly capable of zinc solubilization (Deepak et al. 2013; Hussain et al. 2015; Naz et al. 2016).

7.8 Innovation Approaches of Biofertilizers for Sustainable Agriculture Production

Sustainable agricultural production denotes a method of agricultural production in which natural resource usage aims to meet the demands of food security with consideration for both environmental and human health and of economic development. Biofertilizers are one of the ways to achieve sustainable agricultural production without hampering environmental and human health. The following approaches are used for reviving sustainable agriculture production by adopting biofertilizer technology:

- Identification of specific strains for nitrogen fixing, and for P, K, Zn, and iron solubilizing and mobilization, to suit diverse climatic situations, soils, and crops.
- Adoption of modern technologies for strain development, such as biotechnological approaches.
- Development of improved protocols for exchanging strains among nations and evaluation techniques to obtain good strains and evade natural mutants.
- Development of suitable alternative formulations, such as liquid inoculants, or powdered or solid formulations, for all bioinoculants for better utilization and storage.
- Engaging microbiology and pathology scientists in manufacturing units to monitor the production process and quality evaluation and labeling of products.
- Adequate construction of cold storage in production units for long-term storage of valuable biostrains without harmful results.
- Engagement of technical and field training schedules on production and application methods to the producers and farmers, and interpretation of practical guidance and projects to manufacturers for better production and utilization.
- Creating awareness of biofertilizer advantages in agriculture production through electronic and print media, newspapers and extension bulletins, leaflets, brochures, etc.

7.9 Conclusion and Future Prospects

Modern-day intensive cropping systems need higher inputs and energy to obtain higher yields, and prolonged use of these chemical inputs causes land degradation, poor soil, and low crop productivity. Hence, sustainable agricultural production methods are the only way to both meet food demand and conserve ecosystems for future generations. To challenge this problem, biofertilizers may be essential components of sustainable agricultural farming. Biofertilizers are the greatest option for farmers to increase production and productivity per unit area and time in sustainable farming for an era of wealth and safer environments. Complete dependence on chemical fertilizers and pesticides not only pollutes the environment but also increases the costs of cultivation, ultimately leading to crises among the farmers. Hence, systematic use of biofertilizers in production and plant protection systems has a key role in sustainable agricultural production and the economic development of farmers, and also contributes to a sustainable ecosystem and the holistic well-being of the nation.

References

Adesemoye AO, Kloepper JW (2009) Plant–microbes interactions in enhanced fertilizer use efficiency. Appl Microbiol Biotechnol 1:1–12

Adeyeye AS, Togun AO, Olaniyan AB, Akanbi WB (2017) Effect of fertilizer and *Rhizobium* inoculation on growth and yield of soya bean variety (*Glycine max* L. Merrill). Adv Crop Sci Tech 5:255

Alloway BJ (2004) Zinc in soil and crop nutrition. International Zinc Association, Belgium

Amalraj EDL, Maiyappan S, John Peter A (2012) In vivo and in vitro studies of *Bacillus megaterium* var. *phosphaticum* on nutrient mobilization, antagonism and plant growth promoting traits. J Ecobiotechnol 1:35–42

Archana D, Nandish M, Savalagi V, Alagawadi A (2013) Characterization of potassium solubilizing bacteria (KSB) from rhizosphere soil. Bioinfolet Q J Life Sci 10:248–257

Bashan Y, de-Bashan L (2010) How the plant growth-promoting bacterium *Azospirillum* promotes plant growth. A critical assessment. Adv Agron 108:77–136

Beneduzi A, Ambrosini A, Passaglia LMP (2012) Plant growth-promoting rhizobacteria (PGPR): their potential as antagonists and biocontrol agents. Genet Mol Biol 35(4):1044–1051

Berdeni D, Cotton TEA, Daniell TJ, Bidartondo MI, Cameron DD, Evans KL (2018) The effects of arbuscular mycorrhizal fungal colonisation on nutrient status, growth, productivity, and canker resistance of apple (*Malus pumila*). Front Microbiol 9:1461

Bolan NS (1991) A critical review on the role of mycorrhizal fungi in the uptake of phosphorus by plants. Plant Soil 134:189–207

Chang CH, Yang SS (2009) Thermo-tolerant phosphate-solubilizing microbes for multi-functional biofertilizer preparation. Bioresour Technol 100:1648–1658

Chen YP, Rekha PD, Arunshen AB, Lai WA, Young CC (2006) Phosphate solubilizing bacteria from subtropical soil and their tricalcium phosphate solubilizing abilities. Appl Soil Ecol 34:33–41

Chun-Li W, Shiuan-Yuh C, Chiu-Chung Y (2014) Present situation and future perspective of bio-fertilizer for environmentally friendly agriculture. Annu Rep 2014:1–5

Creus C, Sueldo R, Barassi C (1997) Shoot growth and water status in *Azospirillum* inoculated wheat seedlings grown under osmotic and salt stresses. Plant Physiol Biochem 35:939–944

Deepak J, Geeta N, Sachin V, Anita S (2013) Enhancement of wheat growth and Zn content in grains by zinc solubilizing bacteria. Int J Agric Environ Biotechnol 6:363–370

Diriba M, Fassil A, Elisabet B, Granhall UF (2013) Phosphate solubilising rhizobacteria associated with *Coffea arabica* L. in natural coffee forests of southwestern Ethiopia. J Saudi Soc Agric Sci 12:73–84

Dudeja SS, Khurana AL, Kundu BS (1981) Effect of rhizobium and phosphorus- micro-organisms on yield and nutrient uptake in chickpea. Curr Sci 50:503–505

Eskin N, Vessey K, Tian L (2014) Research progress and perspectives of nitrogen fixing bacterium, *Gluconacetobacter diazotrophicus*, in monocot plants. Int J Agron 2014:1–13. https://doi.org/10.1155/2014/208383

Ewa K, Ewa O, Piotr S, Anna S, Jolanta JS (2013) Effect of *Pseudomonas luteola* on mobilization of phosphorus and growth of young apple trees (Ligol): pot experiment. Sci Hortic 164:270–276

Ghorbanian D, Harutyunyan S, Mazaheri D, Rasoli V, Mohebl A (2012) Influence of arbuscular mycorrhizal fungi and different levels of phosphorus on the growth of corn in water stress conditions. Afr J Agric Res 7(16):2575–2580

Giri B, Prasad R, Wu Q-S, Varma A (2019) Biofertilizers for sustainable agriculture and environment. Springer, Cham. ISBN 978-3-030-18932-7. https://www.springer.com/gp/book/9783030189327

Glick BR, Cheng Z, Czarny J, Duan J (2007) Promotion of plant growth by ACC deaminase-producing soil bacteria. Eur J Plant Pathol 119:329–339

Gothandapani S, Soundarapandian S, Jasdeep CP (2017) *Azotobacter chroococcum*: utilization and potential use for agricultural crop production: an overview. Int J Adv Res Biol Sci 4:35–42

Gunes A, Ataoglu N, Turan M, Esitken A, Ketterings QM (2009) Effects of phosphate-solubilizing microorganisms on strawberry yield and nutrient concentrations. J Plant Nutr Soil Sci 172:385–392

Hamdali H, Bouizgarne B, HafidI M, Lebrihi A, Virolle MJ, Ouhdouch Y (2008) Screening for rock phosphate solubilizing actinomycetes from Moroccan phosphate mines. Appl Soil Ecol 38:12–19

Hussain A, Arshad M, Zahir ZA, Asghar M (2015) Prospects of zinc solubilizing bacteria for enhancing growth of maize. Pak J Agric Sci 52:915–922

Intorne AC, Oliveira MVV, Lima ML, Silva DJF, Olivares FL, De Souza FA (2009) Identification and characterization of *Gluconacetobacter diazotrophicus* mutants defective in the solubilization of phosphorus and zinc. Arch Microbiol 191(5):477–483

Iqbal U, Wali VK, Kher R, Jamawal M (2009) Effect of FYM, urea and *Azotobacter* on growth, yield and quality of strawberry cv. Chandler. Notulae Botanicae Horti Agrobotanici Cluj-Napoca 37(1):139–143

Jakobsen I, Abbott LK, Robson AD (1992) External hyphae of vesicular-arbuscular mycorrhizal fungi associated with *Trifolium subterraneum* L. I. Spread of hyphae and phosphorus inflow into roots. New Phytol 120:371–380

Jehangir IA, Mir MA, Bhat MA, Ahangar MA (2017) Biofertilizers an approach to sustainability in agriculture: a review. Int J Pure Appl Biosci 5:327–334

Jeyabal A, Kupuswamy G (2001) Recycling of organic wastes for the production of vermicompost and its response in rice legume cropping system and soil fertility. Eur J Agron 15:153–170

Jimenez DJ, Jose SM, Maria MM (2011) Characterization of free nitrogen fixing bacteria of the genus *Azotobacter* in organic vegetable-grown Colombian soils. Braz J Microbiol 42 (3):846–858

Kapri A, Tewari L (2010) Phosphate solubilization potential and phosphatase activity of rhizospheric *Trichoderma* spp. Braz J Microbiol 41:787–795

Khaitov B, Kurbonov A, Abdiev A, Adilov M (2016) Effect of chickpea in association with *Rhizobium* to crop productivity and soil fertility. Eur J Soil Sci 5(2):105–112

Knobeloch L, Salna B, Hogan A, Postle J, Anderson H (2000) Blue babies and nitrate-contaminated well water. Environ Health Perspect 108:675–678

Kucey RMN, Paul FA (1982) Carbon flux, photosynthesis and nitrogen fixation in mycorrhizal and nodulated faba beans (*Vicia faba* L.). Soil Biol Biochem 14:407–412

Lach D, Sharma VK, Vary PS (1990) Isolation and characterization of a unique division mutant of *Bacillus megaterium*. J Gen Microbiol 3:545–553

Logeshwarn P, Thangaraju M, Rajasundari K (2011) Antagonistic potential of *Gluconacetobacter diazotrophicus* against *Fusarium oxysporum* in sweet potato (*Ipomea batatus*). Arch Phytopathol Plant Protect 44:216–223

Mahdi SS, Talat MA, Hussain Dar M, Hamid A, Ahmad L (2012) Soil phosphorus fixation chemistry and role of phosphate solubilizing bacteria in enhancing its efficiency for sustainable cropping: a review. J Pure Appl Microbiol 6(4):1–7

Malusa E, Sas-Paszt L, Ciesielska J (2012) Technologies for beneficial micro-organisms inoculation used as biofertilizers. Sci World J 2012:491206. https://doi.org/10.1100/2012/491206

Martyniuk S, Martyniuk M (2003) Occurrence of Azotobacter Spp. in some polish soils. J Environ Stud 12:371–374

Mazid M, Khan TA (2015) Future of bio-fertilizers in Indian agriculture: an overview. Int J Agric Food Res 3(3):10–23

Mazid M, Khan TA, Mohammad F (2011) Potential of NO and H_2O_2 as signaling molecules in tolerance to abiotic stress in plants. J Ind Res Technol 1:56–68

Meena VS, Maurya BR, Verma JP, Aeron A, Kumar A, Kim K, Ajpai VK (2015) Potassium solubilizing rhizobacteria (KSR): isolation, identification, and K-release dynamics from waste mica. Ecol Eng 81:340–347

Mishra P, Dash D (2014) Rejuvenation of biofertilizer for sustainable agriculture and economic development. J Sustain Dev 11:41–61

Mishra DJ, Rajivir S, Mishrra UK, Kumar SS (2013) Role of biofertilizers in organic agriculture: a review. Res J Recent Sci 2:39–41

Mohammadi K, Sohrabi Y (2012) Bacterial biofertilizers for sustainable crop production: a review. ARPN J Agric Biol Sci 7(5):307–316

Naz I, Ahmad H, Khokhar SN, Khan K, Shah AH (2016) Impact of zinc solubilizing bacteria on zinc contents of wheat. Am Eurasian J Agric Environ Sci 16:449–454

Oliveira CA, Alvesb VMC, Marreib IE, Gomesb EA, Scottia MR, Carneiro NP (2009) Phosphate solubilizing microorganisms isolated from rhizosphere of maize cultivated in an oxisol of the Brazilian Cerrado Biome. Soil Biol Biochem 41:1782–1787

Pandey A, Kumar SJ (1989) Soil beneficial bacterial and their role in plant growth promotion. Sci Indian Res 48:134–144

Podile AR, Kishore GK (2006) Plant growth-promoting *Rhizobacteria*. In: Gnanamanickam SS (ed) Plant-associated bacteria. Springer, Dordrecht, pp 195–230

Ponmurugan P, Gopi C (2006) Distribution pattern and screening of phosphate solubilizing bacteria isolated from different food and forage crops. J Agron 5:600–604

Prajapati K, Modi H (2016) Growth promoting effect of potassium solubilizing *Enterobacter hormaechei* (KSB-8) on cucumber (*Cucumis sativus*) under hydroponic conditions. Int J Adv Res Biol Sci 3:168–173

Prasad R, Kumar M, Varma A (2015) Role of PGPR in soil fertility and plant health. In: Egamberdieva D, Shrivastava S, Varma A (eds) Plant Growth-Promoting Rhizobacteria (PGPR) and medicinal plants. Springer, Cham, pp 247–260

Prasad R, Bhola D, Akdi K, Cruz C, Sairam KVSS, Tuteja N, Varma A (2017) Introduction to mycorrhiza: historical development. In: Varma A, Prasad R, Tuteja N (eds) Mycorrhiza. Springer, Cham, pp 1–7

Puente M, Li C, Bashan Y (2004) Microbial populations and activities in the rhizoplane of rock-weathering desert plants. II. Growth promotion of cactus seedlings. Plant Biol 6:643–650

Raja N (2013) Biopesticides and biofertilizers: ecofriendly sources for sustainable agriculture. J Biofertil Biopestic 4:112

Ram H, Malik SS, Dhaliwal SS, Kumar B, Singh Y (2015) Growth and productivity of wheat affected by phosphorus-solubilizing fungi and phosphorus levels. Plant Soil Environ 61:122–126

Rawia EA, Nemat MA, Hamouda HA (2009) Evaluate effectiveness of bio and mineral fertilization on the growth parameters and marketable cut flowers of *Matthiola incana* L. Am Eur J Environ Sci 5:509–518

RoyChowdhury DE, Paul MA, Banerjee SK (2014) A review on the effects of biofertilizers and biopesticides on rice and tea cultivation and productivity. Int J Sci Eng Technol 2:96–106

Saha M, Maurya BR, Meena VS, Bahadur I, Kumar A (2016) Identification and characterization of potassium solubilizing bacteria (KSB) from Indo-Gangetic plains of India. Biocatal Agric Biotechnol 7:202–209

Santos VB, Araujo SF, Leite LF (2012) Soil microbial biomass and organic matter fractions during transition from conventional to organic farming systems. Geoderma 170:227–231

Singh JS, Pandey VC, Singh DP (2011) Efficient soil microorganisms: a new dimension for sustainable agriculture and environmental development. Agric Ecosyst Environ 140(3):339–353

Sparks DL, Huang PM (1985) Physical chemistry of soil potassium. In: Munson RD (ed) Potassium in agriculture. American Society of Agronomy, Crop Science Society of America, and Soil Science Society of America, Madison, WI, pp 201–276

Steenhoudt O, Vanderleyden J (2000) *Azospirillum*, a free living nitrogen-fixing bacterium closely associated with grasses: genetic, biochemical and ecological aspects. FEMS Microbiol Rev 24:487–506

Stevenson FJ (1986) Cycles of soil: carbon, nitrogen, phosphorus, sulfur, micronutrients. Wiley, New York, pp 231–284

Subhashini DV (2015) Growth promotion and increased potassium uptake of tobacco by potassium-mobilizing bacterium *Frateuria aurantia* grown at different potassium levels in vertisol. Commun Soil Sci Plant Anal 46(2):210–220

Subraya BK, Madaiah D, Dinesh Kumar M (2017) Effect of integrated nutrient management on growth and physiological parameters of strawberry (*Fragaria* × *ananassa* Duch) under naturally-ventilated polyhouse. Int J Farm Sci 7(3):72–75

Sudhir U, Meshram A, Jager G (1983) Antagonism of *Azotobacter chroococcum* isolates to *Rhizoctonia solani*. Eur J Plant Pathol 89:91–197

Sundaravarathan S, Kannaiyan S (2002) Influence of *Azolla* and *Sesbania rostrata* application on changes in microbial population and enzymes in rice soils. In: Kannaiyan S (ed) Biotechnology of biofertilizers. Alpha Science International, Pangbourne, pp 251–225

Thajuddin N, Subramanian G (2005) Cyanobacterial biodiversity and potential applications in biotechnology. Curr Sci 89:47–57

Trabelsi D, Mhamdi R (2013) Microbial inoculants and their impact in microbial soil microbial communities: a review. Biomed Res Int 2013:11

Varma A, Prasad R, Tuteja N (2017) Mycorrhiza: nutrient uptake, biocontrol, ecorestoration. Springer, Cham. ISBN 978-3-319-68867-1. http://www.springer.com/us/book/9783319688664

Vassilev N, Vassileva M, Bravo V, Fernandez M, Nikolaev I (2007) Simultaneous phytase production and rock phosphate solubilization by *Aspergillus niger* grown on dry olive wastes. Ind Crop Prod 2007:332–336

Vessey JK (2003) Plant growth promoting rhizobacteria as biofertilizers. Plant Soil 255:571–586

Yao Q, Xiaolin L, Gu F, Peter C (2001) Mobilization of sparingly soluble inorganic phosphates by the external mycelium of an arbuscular mycorrhizal fungus. Plant Soil 230:279–285

Yao YB, Zhanga B, Yuhua T, Miao Z, Ke ZB, Bowen Z, Meng ZB, Bin Y (2018) Azolla biofertilizer for improving low nitrogen use efficiency in an intensive rice cropping system. Field Crops Res 216:158–164

Youssef MMA, Eissa MFM (2014) Biofertilizers and their role in management of plant parasitic nematodes: a review. Biotechnol Pharm Res 5(1):1–6

Chapter 8
Plant Microbiome: Trends and Prospects for Sustainable Agriculture

Arjun Singh, Murugan Kumar, Shaloo Verma, Prassan Choudhary, and Hillol Chakdar

Abstract The plant microbiome or microbial assemblage present in plants is known to have evolved along with the plants. With the help of high-throughput community analyses methods, next-generation sequencing techniques, etc., the black box of plant microbiome has been revealed to a significant extent. A great deal of microbial diversity exists in the plant which is known to be influenced by the genotype, soil properties, environmental factors, etc., and even in some cases they are organ or tissue specific. Despite their structural variation, they contribute significantly in the plant growth and development. Plant-associated microflora are known to contribute in nutrient mobilization, tolerance to biotic and abiotic stresses and even in many physiological functions of plants. Nowadays, study of plant microbiome has claimed much attention as engineering the microbiome can be a sustainable future option to tackle many of the issues pertaining to crop production and protection.

8.1 Introduction

Prokaryotic microorganisms are the ancestors of the present-day eukaryotes. Due to their enormous metabolic diversity and genomic plasticity, microorganisms are known to be present in any conceivable ecological niche. Symbiotic or mutualistic relationships of microorganisms with eukaryotes are well known throughout the history of evolution. The most well-known example is the endosymbiosis of aerobic and photosynthetic bacteria in progenitor cells leading to evolution of eukaryotic cells containing mitochondria and chloroplast. During the evolution, plants and microorganisms have evolved together making a complex, intricate relationship. This relationship helps both the partners in multiple ways like nutrition, immunity, stress tolerance, etc. Microorganisms are distributed within and on every plant parts like fruit, flower, leaf, stem, root, etc. The entire microbial ssemblage of a plant is known as plant microbiome, while the microbial communities are sometime specific

A. Singh · M. Kumar · S. Verma · P. Choudhary · H. Chakdar (✉)
ICAR-National Bureau of Agriculturally Important Microorganisms (NBAIM), Kushmaur, Maunath Bhanjan, Uttar Pradesh, India

A. Varma et al. (eds.), *Plant Microbe Symbiosis*,
https://doi.org/10.1007/978-3-030-36248-5_8

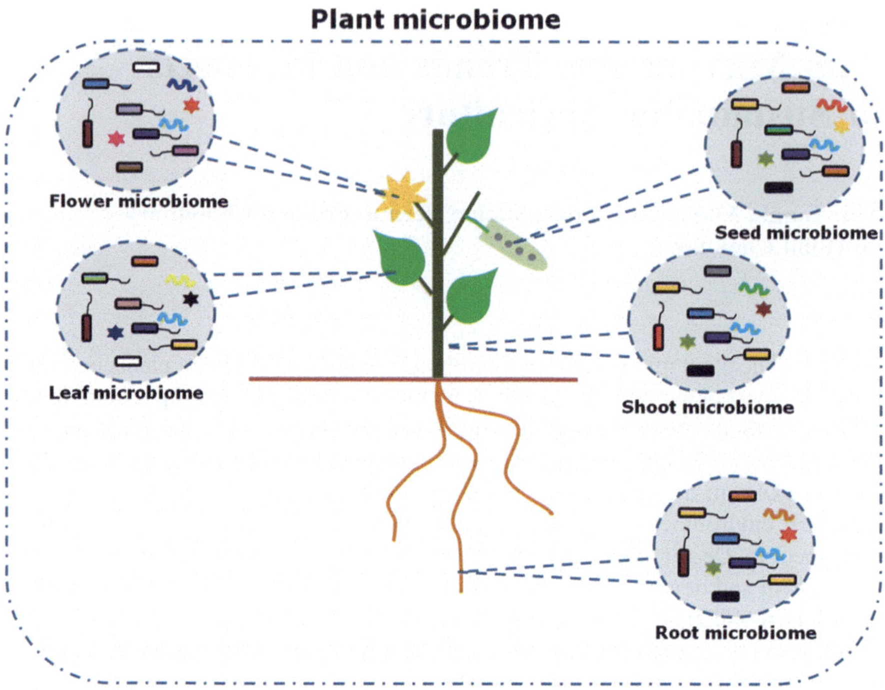

Fig. 8.1 Diagrammatic representation of variation of microbiota in different plant parts constituting the plant microbiome

to plant parts also (Fig. 8.1). For example, Wagner et al. (2016) reported that the microbiome of phyllosphere was dominated by members of *Sphingomonadaceae* and *Cytophagaceae*, while *Bradyrhizobiaceae* and *Nocardiaceae* prevailed in the root microbiome of a perennial wild mustard (Wagner et al. 2016). The differences of microbiome composition within the plant parts are associated with differences in the functions they impart. Microbiome structure and function also differ in different plant genotypes or their growth conditions.

During the course of domestication of crop plants, their microbiome has changed significantly leading to an adverse impact on the association. The change in microbiome has also led to altered functioning of the microbiome. A healthy plant microbiome is always required for optimum phenotypic expression of any plant. However, a native plant microbiome is always influenced by agronomic and nutrient management practices. Understanding the interaction of microbiome with plants and other environmental factors is very much required. Thanks to the available high-throughput sequencing technologies which have unravelled the structure and functions of microbiomes of a number of crop plants along with their interaction with environment and cultivation practices. Such information are highly useful to devise strategies to manipulate the microbiome for higher nutrient acquisition and better tolerance to biotic and abiotic stresses leading to higher crop productivity (Prasad et al. 2018).

8.2 Plant Microbiome Concept

Plants and microorganisms have co-evolved and formed a strong association to fulfil each other's needs and demands. The association has grown to such an extent that microorganisms are behaving as extended biological entity. Plant microbiome has developed specialized strategies to counter many difficulties like nutrient acquisition and abiotic and biotic stress tolerance. Co-evolution of the plant-associated microbes has induced changes in its genome to accumulate more genes responsible for plant colonization, mutualism, defence, carbohydrate metabolism, etc. Many of the classical and omics techniques were able to produce many facts and figures for the development of plant microbiome concept. In the following sections, the concept of plant microbiome will be elaborated.

8.2.1 Evolution of Microbial Interaction: The Hologenome Concept

The origin and evolution of prokaryotic interactions dates back to more than 3500 million years ago. As the time passed by, eukaryotic multicellular life forms originated, and also the interactions got advanced from microbe-microbe to microbe-eukaryotic interactions. The teaming up of the microbes with the multicellular eukaryotes and in particular with plants has helped both the members in gaining multiple fitness and survival traits. Some of the benefits imparted to their host plants includes better nutrient acquisition, development of intrinsic resistance against various abiotic and biotic stresses, etc. These arrangements between the plant and the microbes are not unilateral but are more of mutualistic or symbiotic interactions. The plant host provides the microbes with metabolites and conditions to adopt (West et al. 2002; Jones and Dangl 2006). With the advent of culture-independent high-throughput techniques like SYBR gold staining and community metagenomics, host plants were found to be associated with diverse group of microorganisms which were dynamic in nature and varied with the existing environmental cues (Fig. 8.2). Further this kind of long-term mutualistic associations had changed the course of evolution by increasing the plant fitness trait and also acting as a level of selection.

Rosenberg et al. (2007) published a landmark paper which stated that coral's probiotic determines its survivability under immense pressure of biotic stresses (Rosenberg et al. 2007). The observed phenomenon was explained like this: during the event of *Vibrio shiloi* infection on corals, the symbiotic microbiota or the holobiont (host and its symbionts) changes to accommodate microflora which can cope up with the prevailing condition (Reshef et al. 2006). The series of action taken up by corals to accommodate stress-adoptive microflora resulted in reduced disease incidence. Hence the microbe-induced resistance in corals were much rapid as compared to the resistance acquired by genetic mutations and natural selections of coral host. As a matter of fact, the corals don't have any kind of innate immunity

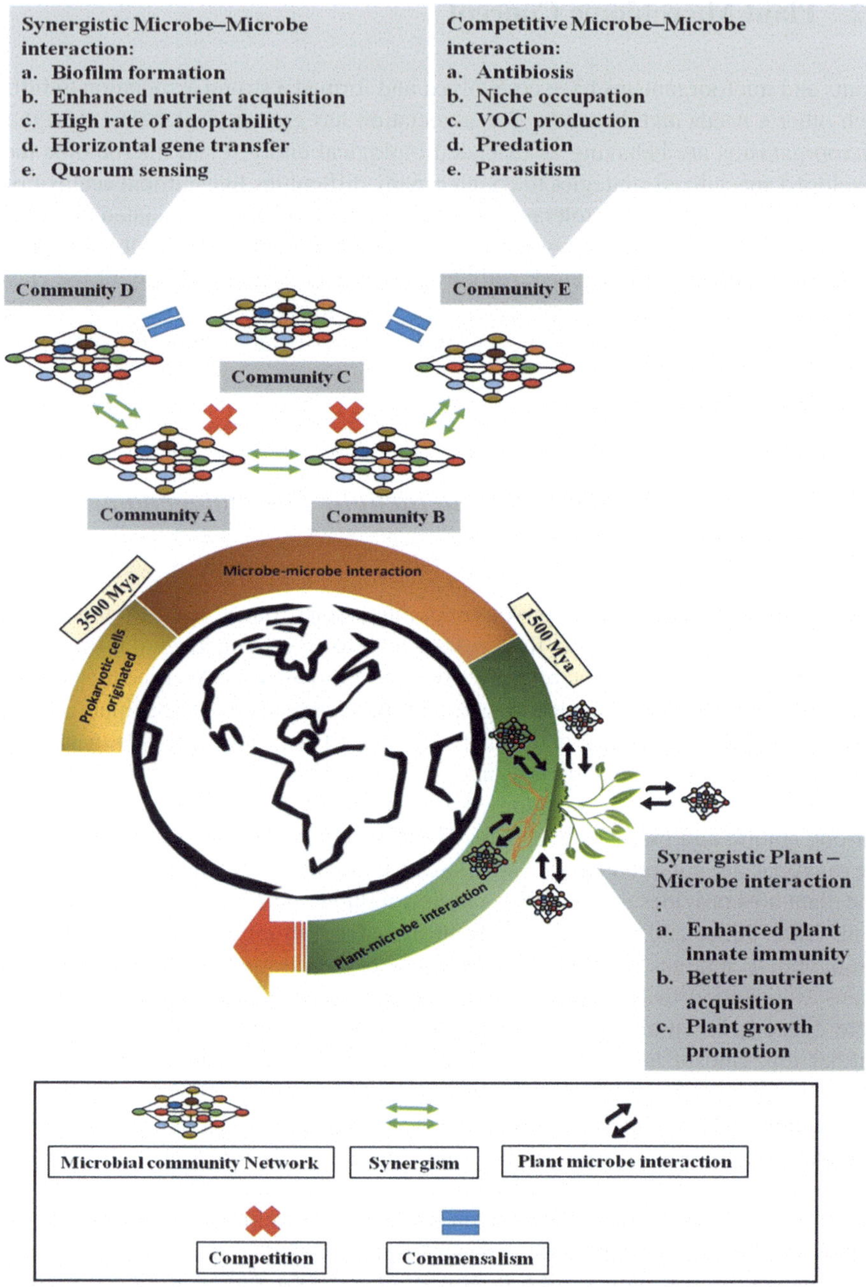

Fig. 8.2 Evolution of microbe-microbe and plant-microbe interaction. The timeline also shows the evolution of various kinds of outcomes of interactions in the form of synergism, competition and commensalism

system; the microbial recruitment carried out by the corals acted as the super shield and protected the corals from existing threats. Taking that facts and figures into account, hologenome concept of evolution was deduced which states that holobiont (plant host plus its microbiome) with its hologenome (host genome and sum total of genomic contributions made by microbes inhabiting the host) behaves like a biological feature and therefore determines the course of selection during evolution.

There are several examples where fitness of the holobionts is being determined by the microorganisms, some of the important ones include respiration and ATP production (mitochondria), photosynthesis (chloroplast), protection against abiotic and biotic stresses, development (root nodule), etc. (Rosenberg and Zilber-Rosenberg 2016). But still the present concept of hologenome evolution was not able to explain co-existence of diverse group of microbial consortia which inhabits the host body due to many of the complex interactions and different ways by which the microbiome is getting inherited by the host system (Etesami 2018).

8.2.2 Composition of Plant-Associated Microflora

Over the time, bacteria, archaea, fungi and protists have colonized above- and belowground parts of the plants. Numerous interactions of the plant with its microflora have extended it with the better survival traits under stress conditions. Looking at the terrestrial plant architecture, the belowground parts of the plants, i.e. roots and the nodules, recruit most of its microbiome from the soil. Several soil intrinsic factors such as its physical properties; temperature; chemical properties including its nutrient status, EC, pH; etc. determine the composition of soil microbiota (Schreiter et al. 2014).

The next region of the root is the rhizosphere which acts as the highly transient active zone of microbial successions and plant microbiome assemblage (Singh et al. 2019). Root exudation pattern of the plant is determined by its genotype which has a major influence on shaping the rhizospheric microbial communities (Ofek-Lalzar et al. 2014). Rhizoplane or the root tissue region acts as the dwelling place of the microorganisms attached to the root surface which forms the root epiphytic communities. The ectomycorrhizal communities are the well-established root epiphytic microbial communities known for their soil nutrition mobilizing activities (Sangabriel-Conde et al. 2014). Sometimes the microbial communities penetrate the root surface and enters the sub-epidermal layer and colonizes it, and these established communities become part of the plant endophytic community (Santoyo et al. 2016). The diversity of the plant endophytic community is less as compared to rhizospheric and epiphytic microbial communities (Gaiero et al. 2013). The endophytic microbial recruitment is highly specialized because only the selected few gets the chance to dwell inside the plant body (Vandenkoornhuyse et al. 2015).

It has been reported that aboveground parts of the plant mostly leaves and floral parts also act as the dwelling grounds of the plant-associated microflora (Vandenkoornhuyse et al. 2015). As compared to the plant roots, leaves experience

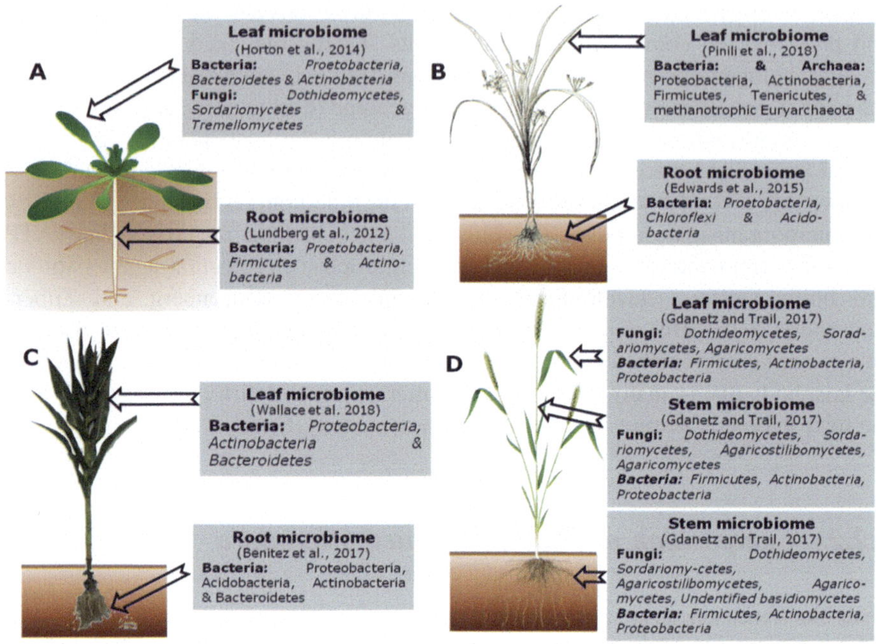

Fig. 8.3 Composition of microbiome of few well studied and important plant species (**a**) arabidopsis, (**b**) rice, (**c**) maize, (**d**) wheat

much harsher environmental stress like radiation, high osmotic pressure, temperature fluctuation, etc. But still it harbours huge microbial population; it is estimated that approximately 10^{26} cells of the epiphytic microorganisms are present on leaves (Vorholt 2012). Functional diversity of the microorganisms revealed that majority was harbouring pigments which preferentially absorb green spectrum of light; hence plant photosynthetic light reactions were not affected (Atamna-Ismaeel et al. 2012). Also they were surviving on the methanol excreted out from the leaf surface; thus there were no direct dependencies on the carbon compounds being utilized by the plant (Vorholt 2012).

Majority of the studies in the area of plant microbiome composition were done on *Arabidopsis thaliana*, *Hordeum vulgare*, *Zea mays*, *Triticum aestivum*, *Oryza sativa*, and few more crop and tree plants (Engelbrektson et al. 2012; Horton et al. 2014; Edwards et al. 2015; Benitez et al. 2017; Gdanetz and Trail 2017; De Leon et al. 2018; Wallace et al. 2018) (Fig. 8.3). The microbial flora of the plants is not random; rather it depends on the multitude of the factors like plant architecture, its genotype, root structure, soil type and prevailing environmental conditions.

8.2.3 *Impact of Anthropogenic Interventions on Plant Microbiome Composition*

Humans have domesticated and selectively cultivated particular plant species which can cater their food and fuel demands. In order to attain maximal yield potential out of the crops, selective sets of conditions have been imposed since the beginning of their domestication which has influenced changes in plant architecture and growth, for instance, selective breeding of high-yielding varieties. Due to this, the domesticated plants are having altered plant-associated microflora as compared to their wild relatives (Pérez-Jaramillo et al. 2016). There was a level of variation in the mycorrhizal dependency among the land races and hybrid maize varieties. It was found that 60% of P intake in the local land races of maize was influenced by their mycorrhizal symbionts in contrast to that of hybrid maize varieties developed under fertilizer-intensive agricultural practices leading to very low dependency on mycorrhizal P uptake (Sangabriel-Conde et al. 2014). Similar observations were earlier recorded in the case of wheat (Graham and Abbott 2000). The wild relatives of barley, *Cardamine*, lettuce and common bean, showed significant abundance of the *Bacteroidetes*, whereas the improved varieties of these plants were showing significant abundance of *Actinobacteria* and *Proteobacteria* (Pérez-Jaramillo et al. 2016). Studies showed that under the influence of the anthropogenic pressures, there was significant loss of microbial communities and beneficial functions.

8.2.4 *Taking the Plant Microbiome Association to the Next Level: Genomic Level*

Plant and its associated microbiota had always co-evolved due to existent natural selection. Further bacterial microbiota of the plant microbiome has always constantly evolved their genes mainly responsible for plant growth promotional activity, carbohydrate metabolism, signal transductions, etc (Singh et al. 2019). Two groups of gene categories had been found in the plant symbiotic bacterial microbiota, viz. plant-associated and root-associated genes (Melnyk and Haney 2017). The comparative genomics was performed on 4000 plant-associated bacterial species. Results revealed that the genome size of the plant-associated bacteria were larger as compared to the non-associative bacterial species. Further there was high copy number of genes responsible for bacterial adherence, carbohydrate metabolism, motility, cell division, etc. (Levy et al. 2018). These genes were not only linked to their original genomes, but their sequence homologues was also found in other plant-associated bacteria and also the eukaryotic cells, indicating horizontal gene transfer events. Such high-throughput genomic studies were able to throw light on the underlying molecular mechanisms of plant-microbe interactions and high rate of adoption strategies.

8.3 Plant Associative Microbiota Leases Each Other Services for Survival

It was established by Hardoim et al. (2008) that associative fungal and bacterial flora is needed for sustaining plant growth, but the existence of these relationship is not unidirectional (Hardoim et al. 2008), although there were not many evidences available showing impact of the host on the survivability and development of the symbiotic microflora. The major fact supporting this view is: if the host hasn't had extended its favourable condition for its inherited microflora, the plant microbiome would had devised the strategies to break the relationship and created the condition for its growth and survival (Mushegian and Ebert 2016). In the legume-rhizobia interaction, the leguminous plant serves as the host for rhizobia in which the *Rhizobium* transforms itself to grow under anaerobic condition of the root nodules to favour energy-intensive nitrogen fixation for the host. The relationship seems to be only favouring the plant host in a way that it is getting ample amount of biologically fixed nitrogen to fulfil its nutritional demand, but that is not true. The story is the other way around: during the process of biological nitrogen fixation in the root nodules, a huge amount of flavonoids are secreted to attract more rhizobia populations and increase its density in the root nodules thus favouring the survivability of the kin species. Also the host provides the rhizobia with ample amount of metabolites and shelter for its growth and survival (West et al. 2002). Plant microbiota also plays important role in acquisition of orthophosphate from the clay complex phosphates. The plants have in-built mechanism to sense phosphate-starved condition known as phosphate starvation response (PSR); the stimuli is an outcome of plant-microbe interaction. The genetic network of PSR triggers the change in root microbiota under various levels of orthophosphate in soil (Finkel et al. 2017). Further in the same study, it was also demonstrated that plant immune system, nutrient acquisition and microbiome assemblages are linked.

The bipartite plant-microbe relationship acts as a protection shield against invasions from non-member microbial community for occupancy of the host niche; in this case the existing plant microflora increases its assemblages and doubles its metabolic rate so that the host niche is fully occupied and creates a competitive condition for the invading foreign microflora for nutrient resources and habitat (Gou et al. 2015). Although it appears that the plant is protected against the pathogenic invasion, actually the inherited plant microflora are trying to protect their niche from getting occupied by invading microflora.

Biofilms are the result of exopolysaccharide secretion from the population of single or multispecies microbial communities co-exiting with each other in a single niche. Formation of biofilms provides selective advantage to the adhering bacterial communities by protecting it against antimicrobial compounds and abiotic and biotic stresses and acts as exchange space for horizontal gene transfer (Danhorn and Fuqua 2007). Recently, it has been demonstrated that during the invasion of *Hyaloperonospora arabidopsidis* to *Arabidopsis*, the plant's innate immune system worked in tandem to recruit specifically three bacterial genera, viz., *Xanthomonas*,

Microbacterium, and *Stenotrophomonas* sp. These bacterial species behaved synergistically and assembled a biofilm which created a defence wall against the existing invasion from the *Hyaloperonospora* (Berendsen et al. 2018).

Most of the discussion was focused on the interplay of the plant microbiome on nutrient acquisition and development of innate immunity against biotic stresses. There are also some basic plant functions which get influenced due to the presence of microorganism. For instance, no germination of seeds would have been observed in the orchids if there was no colonization of *Rhizoctonia* (Jacquemyn et al. 2015). It was also observed that later on the *Rhizoctonia* was killed by the orchids. In a recent study conducted on *Boechera stricta*, it was revealed that there was a direct impact of the microbiome on the flowering time (Wagner et al. 2014). In another study supporting this view, induction of early and late flowering was successfully demonstrated in *Brassica rapa* by transplanting early and late flowering soil microbiome of *Arabidopsis* (Panke-Buisse et al. 2015).

Apart from the plants, bacteria and fungus also possessed phytohormone production capabilities. The phytohormone production pathways got evolved independently and have independent phylogenetic lineage. Natural selection has allowed for the selective assemblage of the phytohormone-producing group of microorganism (Ozioko et al. 2015). Presence of these microorganisms in its holobiont has helped in improving the plant's ability to sense any kind of upcoming abiotic and biotic stresses. Also it acted as the signal response and communication channel between the host and the microorganism. Plant senescence hormone ethylene is produced whenever the plant faces any kind of abiotic stress conditions such as drought, flood, heat, cold, etc. Glick (2014) advocated that 1-aminocyclopropane-1-carboxylate (ACC) deaminase an enzyme which converts ethylene precursor ACC to ammonia and α-ketobutyrate can be used to alleviate ethylene-induced plant growth inhibition under abiotic stress condition (Glick 2014). The presence of this enzyme in the plant-associated microorganisms has been well documented and has been exploited to induce better plant survival under prevailing stress conditions (Bouffaud et al. 2018). Volatilomes or volatile organic compound metabolome of the plant-associated microbiota has been mined which showed the presence of diverse group of microbial volatiles such as 1-undecene, dimethyl disulphide, dimethyltrisulfide, s-methyl methanethiosulfonate, HCN, etc. (Bailly and Weisskopf 2017). The microbial volatile organic compounds have been demonstrated to have antimicrobial activities, hence imparting the plant with extra immunity against the pathogenic attacks; they are also acting as the signalling system to mediate plant-microbe interactions.

8.4 Approaches to Study Plant Microbiome

The necessity to study plant microbiome arises out of the fact that it is fundamentally an important factor for plant health and productivity. Understanding plant microbiome can help generate strategies to reduce the incidence of diseases, manage abiotic stresses and increase yield, thereby having an impact in environment by reducing the use of chemical inputs in agriculture. Therefore the study of plant microbiome received enormous attention in the recent years (García-Fraile et al. 2015). Traditionally microbiome studies were carried out by isolating and culturing the microorganisms using different nutrient combination and physiological conditions depending on the type of organisms targeted. Although this technique has the advantage of bringing the target organisms into culture, culture-dependant studies grossly underestimate the diversity of microorganisms in any environment, and the full picture on the role of microbiome in plant sustenance can never be achieved (Turner et al. 2013). The advent of new technologies with reduced cost allows scientists to use high-throughput "omics" approaches to understand and manipulate plant microbiome. The different omics approaches to study plant microbiome directly are metagenomics (to study genes directly from the environmental sample), metatranscriptomics (to study transcripts like mRNA and rRNA directly from environmental samples), metaproteomics (study of environmental proteins) and metabolomics (study of metabolites directly from the environment) (López-Mondéjar 2017). Each of these approaches will be elaborately discussed in this section.

8.4.1 *Metagenomics in Plant Microbiome Studies*

Metagenomics is the direct study of genetic material of microbial communities present in environmental samples, by utilizing high-throughput sequencing tools. Thanks to technological advancement in sequencing genetic material, and increased computational power, it is now possible to study microbial community in every conceivable environment circumventing the need to isolate and culture microorganisms. Due to reduced sequencing costs, an astonishing number of plant microbiome have been unravelled in the recent past employing metagenomics studies (Kobayashi et al. 2015; Tsurumaru et al. 2015; Sarria-Guzmán et al. 2016; Gdanetz and Trail 2017; Moronta-Barrios et al. 2018). Two main types of sequencing are commonly employed in plant microbiome studies: (1) sequencing the amplicons of universal genetic markers like 16S rRNA gene for bacteria and archaea and internal transcribed spacer (ITS) region for eukaryotic microorganisms and (2) shotgun sequencing of the entire metagenome of a particular sample (Sergaki et al. 2018). The most frequently used practice to identify community composition of plant microbiome is amplicon sequencing of universal markers followed by comparison with reference database (López-Mondéjar 2017). The benefit of employing amplicon sequencing in microbiome studies is that a large number of samples can be studied at a comparatively low cost and shorter time. Amplicon sequencing has been used to study rhizo-microbiome of rice (Edwards et al. 2015; Shi et al. 2018), wheat (Mavrodi et al. 2018) and maize (Gomes et al. 2018) and phyllosphere microbiome of common bean, soybean, canola (Copeland et al. 2015), indoor ornamentals (Ortega et al. 2016) and maize (Wallace et al. 2018). Amplicon sequencing for microbiome studies can lead to primer bias, and in order to get a more representative picture of microbiome associated with plants, the entire metagenome can be sequenced. Sequencing whole metagenome for microbiome studies has been used to reveal endosphere microbiome of rice (Sessitsch et al. 2012), rhizoplane and rhizosphere microbiome of wheat and cucumber (Ofek-Lalzar et al. 2014), structural and functional microbial diversity of wild and domesticated barley (Bulgarelli et al. 2015) and phyllosphere microbiome of grapes (Salvetti et al. 2016). Sequencing whole metagenome can give a clear idea of not just the composition but also the functionalities of the microbiome. But these advantages do not come without challenges like assembling the sequence reads into a high-quality metagenome assembly; assigning taxonomy and function to assembled reads, especially for the samples with high degree of heterogeneity; and getting rid of the plant DNA and filtering out the DNA of microbiome (Levy et al. 2018).

8.4.2 Metatranscriptomics in Plant Microbiome Studies

Plant microbiome includes not just the microorganisms living in association with plant, rather it includes the whole set of genes and the interaction among all the associated microorganisms and interaction between microorganisms and plants. Such interactions cannot be revealed by concluding the studies with knowing just what is present, but it can be determined by the levels of gene expression and activities of gene products (López-Mondéjar 2017). Metatranscriptomic studies for plant microbiome became important due to the fact that functional capability and diversity of microbiome is more important than mere structural diversity. Metatranscriptome analysis of leaf samples of soybean in Illinois led to the identification of 22 new mycovirus (Marzano and Domier 2016). This approach has also been used to study wheat rhizosphere microbiome in *Rhizoctonia solani* suppressive and non-suppressive soils. Non-suppressive soils showed greater expression of genes involved in antibiotic production and detoxifying reactive oxygen species and superoxide radicals, while suppressive soils showed greater expression of cold shock proteins (Hayden et al. 2018). In another work where effects on glyphosate treatment on rhizosphere microbiome of corn and soybean was studied, it was shown that glyphosate application upregulates genes involved in protein metabolism and respiration while downregulates genes involved in carbohydrate metabolism (Newman et al. 2016). Analysis of metatranscriptome data of 20 wheat rhizosphere samples with special reference to degradation of pollutants revealed 21 different pathways for aromatic compound degradation and six different pathways for xenobiotics degradation (Singh et al. 2018). Like in the case of metagenomics studies, metatranscriptomic studies also come with many difficulties like plant transcripts generally outnumber the microbial transcripts, and even within the microbial transcripts, gene products of housekeeping genes mask the mRNA. Achieving required quantity of microbial mRNA transcripts for differential and sequencing studies therefore become a challenging task (Levy et al. 2018). To overcome these problems in one of the studies on metatranscriptome, a technique was used to separate bacterial cells from plant cells after grinding the leaf samples in a RNA stabilization buffer, and then RNA isolation was carried out (Nobori et al. 2018). Although metatranscriptomic studies through sequencing possess immense potential in microbiome studies, lack of high-quality reference genomes limits the usage of this technique. Techniques alternative to sequencing-based metatranscriptomics are evolving like hybridization-based NanoString technology, and with the improvement in enrichment and detection of microbial transcripts, metatranscriptomic approaches can further enhance our understanding of plant microbiome (Levy et al. 2018).

8.4.3 Metaproteomics in Plant Microbiome Studies

Metaproteomics is defined as a large-scale characterization of the entire protein complement of environmental microbiota at a given point in time, and this term was proposed by Wilmes and Bond (2004). Metaproteomics studies are generally based on LC-tandem mass spectrometry which involves a series of steps like sample collection from the environment, protein extraction, fractionation employing liquid chromatography, mass spectrometry and then finally comparison with a proteome database. Proteomic approaches possess the potential to provide a more detailed and accurate information on the active pathways in an environment as compared to metagenomics and metatranscriptomics approaches (Levy et al. 2018). Initially this approach was used to study the proteome of laboratory-scale activated sludge, employing 2D gel electrophoresis followed by quadrupole-TOF mass spectrometry and was able to identify only three proteins (Wilmes and Bond 2004). Due to the advancement in technology, a decade later this number increased to more than 7000 proteins being identified from a soil microbiome in an active layer of permafrost (Hultman et al. 2015). Metaproteomics approach has been used to identify microorganisms responsible for methane oxidation and nitrogen fixation under field condition by studying the proteome of root-associated bacteria. In this study, type II methanotrophs were identified as responsible for methane oxidation and nitrogen fixation employing a combination of metaproteomics approach and catalysed reporter deposition-fluorescence in situ hybridization (CARD-FISH) (Bao et al. 2014). Liquid chromatography-high-resolution mass spectrometry-based metaproteomics approach was used to study microbiome of rhizosphere of plants thriving in heavy metal-contaminated serpentine soil. More than 800 proteins involved in transport of metals and nutrients and proteins involved in response to stimulus were identified employing this approach (Mattarozzi et al. 2017). Metaproteomics has also been used to study phyllosphere microbiome of trees in Brazilian Atlantic Forest, and it was found that in spite of the variation in the bacterial communities within different trees, their metaproteomes are functionally redundant when it comes to traits responsible for survival on the leaf surface (Lambais et al. 2017). Just like metatranscriptomics, metaproteomic studies also face lots of constraints like low concentration of high-quality proteins for processing and further studies, presence of large quantities of host proteins as compared to microbial proteins and lack of extensive reference database to identify novel proteins that are often identified in metaproteomic studies (Levy et al. 2018). Owing to these facts, the use of metaproteomics for studies on plant microbiome is limited. It is expected that with the advent of metaproteogenomics (where metaproteomes are identified based on metagenome database) more and more usage of proteomics approaches will be used to study plant microbiome.

8.4.4 Metabolomics in Plant Microbiome Studies

The term "metabolomics" was coined by Oliver Fiehn and is defined as a comprehensive and quantitative analysis of all metabolites (small molecules) in a biological system (Fiehn 2001). Metabolomics studies can be used to uncover the changes in specific metabolite levels in response to a specific treatment like PGPR, pathogen or any other. Alternatively, the effect of metabolome (like root exudates, saps from plants, leaf oozes) on the microbial community can also be studied by employing a combination of approaches. Metabolomics has been successfully used to profile root exudates of many plant species like *Zea mays*, *Solanum lycopersicum*, *Medicago truncatula*, *Beta vulgaris* and *Arabidopsis thaliana* (van Dam and Bouwmeester 2016). Application of soil microorganisms in the root zone have been found to alter the leaf metabolome of *Arabidopsis thaliana*, which suggest a link between below- and aboveground plant parts (Badri et al. 2013). Metabolomics study recently established that the bacteria with substrate preference to particular exudates are successionally enriched when the respective exudate concentration is higher in the rhizosphere (Zhalnina et al. 2018). Metabolomics studies are often hindered with challenges like high cost of analysis, lack of availability of equipments and technical expertise and an inadequate reference database. Just as the case of proteomics, a spur in the technology is expected before metabolomics studies are used in a frequent manner to uncover plant microbiome and plant-microbe interactions.

8.5 Microbiome and Crop Production

Plant microbiomes which include rhizobacteria, epiphytes and endophytes are novel resources for sustainable agricultural productivity (Singh and Trivedi 2017). As mentioned earlier, microbiome is an integral part of plants and offers a number of functions improving the fitness. Microbes help plants by enhancing growth, yield and adaptation to adverse stress conditions including plant diseases (Kumar and Verma 2018). Some of the important plant beneficial functions are discussed below.

8.5.1 Phytohormone Production

Increased productions of phytohormones like IAA, gibberellin (GA) and cytokinin by microbiomes of host plant have been extensively studied. *Pseudomonas*, *Rhizobium*, *Azospirillum*, *Azotobacter*, *Pantoea*, *Acetobacter*, *Herbaspirillum*, *Burkholderia*, *Bacillus* and *Streptomyces* are some of the well-known IAA-producing endophytic bacteria (Rashid et al. 2012; Duca et al. 2014). IAA plays an important role in cell division and elongation; initiation of root systems; development of leaves, flowers and fruits; as well as senescence (Duca et al. 2014). *Firmicutes*, *Actinobacteria*, *Alphaproteobacteria* and *Gammaproteobacteria* were detected from the stem of ginseng (*Panax ginseng*) with *Firmicutes* being the most prominent IAA producers (Vendan et al. 2010). *Acetobacter diazotrophicus* and *Herbaspirillum seropedicae* have been reported to produce gibberellins and IAA in graminaceae species and help in seed germination, stem elongation, etc. (Bömke and Tudzynski 2009). Fungi have also been found to produce GA. *Aspergillus fumigatus* and *Scolecobasidium tshawytschae*, isolated from drought and salt-stressed cultivars of soybean, were reported to produce high level of GA in treated plants resulting in increased plant growth and chlorophyll content (Hamayun et al. 2009a, b). *Phoma glomerata* LWL2 and *Penicillium* sp. LWL3 provided protection to cucumber plants under salinity and drought stress (Waqas et al. 2012). Under salinity stress, cucumber plants treated with *Phoma glomerata* LWL2 and *Penicillium* sp. LWL3 upregulated SA levels, altered JA levels and downregulated ABA levels (Waqas et al. 2012). Cytokinin promotes cell division in plant roots, shoots and growing buds.

8.5.2 Nutrient Acquisition in Plants

Micro- or macro-nutrient are mostly present in insoluble form in the soil. Plant has to employ different strategies for uptake of nutrients from the soil. The associated plant microbiomes help in the uptake of essential micronutrients by solubilizing or mineralizing it and making it available in bioavailable form by acidification (Chen et al. 2014; Jog et al. 2014; Oteino et al. 2015), excretion of proton, production of siderophore molecules and increase in hydrolytic enzymes such as phosphatase or phytase (Li et al. 2016). Arbuscular mycorrhizal fungi (AMF), rhizospheric and endophytic fungi as well as bacteria benefit plant by nutrient acquisition from soil by solubilization of P, Zn, Fe, Ca, K, and S (Behie et al. 2013; Gaiero et al. 2013). *Gluconacetobacter*, *Azospirillum*, *Azotobacter*, *Bacillus*, *Burkholderia*, *Herbaspirillum*, *Pseudomonas*, *Achromobacter*, *Klebsiella*, *Chryseobacterium*, *Pantoea*, *Citrobacter* and *Streptomyces* isolated from different crops such as rice, wheat, maize and legumes improve plant growth through stimulation of root development and uptake of micronutrients (Sharma et al. 2013; Rascovan et al. 2016; Suman et al. 2016; Yadav et al. 2016, 2017a, b; Gaba et al. 2017). Siderophores produced by endophytes help plant to uptake iron from soil which cannot enter inside the plant cell through the transporters because of its unavailable form present in the soil. *Acetobacter diazotrophicus*, *Herbaspirillum* sp. and *Azoarcus* sp. are root endophytes reported to fix nitrogen. *Actinobacteria*, *Gammaproteobacteria* and *Bacillus* are three major diazotrophic endophytic communities identified as atmospheric N_2 fixer in rice (Ji et al. 2014; Sengupta et al. 2017). Rhizobia belonging to *Alphaproteobacteria*, *Betaproteobacteria* subphylum (predominantly *Burkholderiales*) form root nodules on leguminous plant and convert atmospheric N_2 into plant-available NH_3 in return for carbon compounds released by the plant (Gyaneshwar et al. 2011; Oldroyd et al. 2011). Root microbiomes of *Eucalyptus* plant containing *Actinobacteria* (*Kineococcus*, *Microbacterium*, *Nocardia*, and *Rhodococcus*), *Proteobacteria* (*Rhizobium*, *Mesorhizobium*, *Burkholderia* and *Methylobacterium*) and *Firmicutes* (*Bacillus*, *Paenibacillus*, and *Brevibacillus*) were found to be involved in nitrogen fixation (da Silva Fonseca et al. 2018).

8.5.3 Alleviation of Abiotic Stress

Under water stress, the microorganisms present on and within the plant can interfere with the plant physiological functions and can reduce the stress. The abundance of the member of *Streptomycetaceae* in root increased significantly under drought condition (Fitzpatrick et al. 2018). This increase was correlated with improved drought tolerance of the studied angiosperms due to better colonization traits and plant hormone production by the actinobacterial group. Rolli et al. (2015) also showed that inoculation of selected member of grape root-associated microbiome could improve the plants' adaptation to drought stress via drought-induced promotion mechanism (Rolli et al. 2015). Fonseca-García et al. (2016) showed that microbial symbionts of cacti microbiome can be vertically inherited and contribute for the drought tolerance of cacti growing in arid and semi-arid ecosystems (Fonseca-García et al. 2016). Under high salinity, plant growth decreases drastically due to a number of physiological and metabolic changes like low water uptake, reduced photosynthesis, increased ion toxicity, etc. In the plants growing under saline condition, the microbiome many a times is specifically adapted for halotolerance. Al-Mailem et al. (2010) reported a large number of halotolerant bacteria and archaebacteria associated with *Halocnemum strobilaceum*—a halophyte growing in a hypersaline costal region of Arabian Gulf (Al-Mailem et al. 2010). Many of such halotolerant microorganism could improve plant fitness under salinity stress through better nutrient acquisition, phytohormone production, exopolysaccharides production, ion homeostasis, etc. Yuan et al. (2016) reported that superior halotolerance of seepweed was strongly associated with a specialised halotolerant belowground microbiome (Yuan et al. 2016). They found that the genome of rhizospheric and root endospheric bacteria were enriched in genes associated with salinity stress acclimatization, nutrient solubilization as well as competitive root colonization. *Piriformospora indica* a well-known fungal root endophyte induced salt tolerance in barley (Baltruschat et al. 2008) and drought tolerance in Chinese cabbage plants (Sun et al. 2010) by increasing the level of antioxidant (Prasad et al. 2013). Bacterial endophyte *Burkholderia phytofirmans* strain PsJN alleviated drought tolerance levels in maize (Naveed et al. 2014b) and wheat plants (Naveed et al. 2014a). Halotolerant ACC deaminase producing bacteria associated with halophytes have also been implicated for salt stress alleviation. Jha et al. (2012) reported a large number of novel diazotrophic, halotolerant, ACC deaminase producing bacteria from *Salicornia brachiata* (Jha et al. 2012). Re-inoculation of such bacteria improved fitness and growth of *Salicornia brachiata* under salt stress. ACC is the immediate precursor of ethylene, but under stress condition, ACC deaminase is activated by pyridoxal 5′-phosphate and cleaves ACC in α-ketobutyrate and ammonia and thereby decreases the level of ethylene which can hamper the plant growth (Kour et al. 2017; Verma et al. 2017; Yadav and Saxena 2018). ACC deaminase-containing *Burkholderia phytofirmans* PsJN-treated switchgrass has shown enhanced biomass production, root growth, tillering and greater early-season plant growth vigour (Lowman et al. 2015). ACC deaminase-producing halotolerant *Brachybacterium paraconglomeratum* isolated from the surface-sterilized roots of *Chlorophytum borivilianum* could promote plant growth by lowering the oxidative and osmotic damages caused by salinity (Barnawal et al. 2016).

8.5.4 Biocontrol

The pathogen, insect or pests attack plants and inhibit their growth, yield and development. Plant-associated microflora are known to produce a number of metabolites which can protect plants from pest and pathogen attack. Phyllospheric microbes isolated from different plant species showed dominance of *Firmicutes* producing VOCs which helped to protect crop plant from a variety of bacterial and fungal pathogens (Ortega et al. 2016). Plant microbiome provides protection from the invading plant pathogens, herbivores and insects via antibiosis or via induced systemic resistance (ISR). Flagella, antibiotics, *N*-acyl homoserine lactones, salicylic acid, jasmonic acid, siderophores, volatiles (e.g., acetoin) and lipopolysaccharides produced by bacterial endophytes are reported to induce ISR in plants (van Loon et al. 2008; Bordiec et al. 2011). Fungal endophytes belonging to *Ascomycota*, *Basidiomycota*, *Glomeromycota* and *Zygomycota* groups produced growth inhibitory compounds such as alkaloids, steroids, terpenoids, peptides, polyketones, flavonoids, quinols, phenols and chlorinated compound and protected plant from pathogens, insects and herbivores (Higginbotham et al. 2013). Actinomycetes have been well studied for its production for antimicrobial activity against plant pathogenic microbial strain. *Streptomyces* spp. have been reported to produce a number of antimicrobial compounds like munumbicins (Castillo et al. 2002), kakadumycins (Castillo et al. 2003), coronamycin (Ezra et al. 2004), indolo-sesquiterpene antimicrobial compounds (Ding et al. 2011). Siderophore have been reported to induce ISR in plant and contribute in biological control, e.g. endophytic *Methylobacterium* strains involved in suppression of *Xylella fastidiosa* (causing citrus variegated chlorosis in *Citrus* trees) by production of siderophore (Araújo et al. 2008). Mendes et al. (2011) reported that rhizospheric members of *Proteobacteria*, *Firmicutes* and *Actinobacteria* were associated with suppression of *Rhizoctonia solani* in sugar beet and *Gammaproteobacteria* were able to suppress the disease through non-ribosomal peptide synthesis (NRPS) (Mendes et al. 2011). Klein et al. (2013) reported increased abundance of bacteria like *Bacillus*, *Paenibacillus*, *Rhizobium* and *Streptomyces* in root microbiome of cucumber during its growth in a suppressive soil (Klein et al. 2013). Jack and Nelson (2018) showed that a seed recruited microbiome could produce zoosporlytic compounds to suppress the infection by *Pythium aphanidermatum* (Jack and Nelson 2018). Badri et al. (2013) reported that alteration in leaf microbiome of *Arabidopsis thaliana* due to inoculation of different soil microbiome protected it from caterpillar *Trichoplusia ni* (Badri et al. 2013).

8.6 Future Prospects

With the changing climate, crop production throughout the world is seriously impeded by a number of biotic and abiotic stresses. Management of such biotic and abiotic stresses has always been a problem as there is no universal, low-cost, eco-friendly strategy available. Breeding-tolerant/breeding-resistant varieties are time-consuming approaches. Microbes as individual or in a community can solve many of these problems sustainably. Due to their immense metabolic diversity and intricate interaction with soil and plants, they hold considerable potential to manage many of the problems associated with crop production and protection. Use of microorganisms in agriculture is age old, but the usage of single or a consortium of few microbes is mostly practiced. However, in any given ecological niche, a huge number of microbes contribute in any ecological function, and only a very limited portion of them can be captured by the available microbiological techniques. The whole lot of microbial assemblage in soil or in plant has now been unveiled to us through next-generation sequencing techniques. Such microbial assemblage which play an important role in plant survival and fitness can be more effective than single or consortial inoculants. A large number of recent microbiome studies have established that plants thriving under unmanaged soil with harsh environmental conditions sustain extreme stresses with the help of its microbiomes. Hence, transplantation of such microbiomes in cultivation of crop plants can holistically manage many of the biotic and abiotic stresses. Engineering the existing crop microbiome through cultural management or application of plant probiotics can also be another option for targeted enrichment of beneficial microbes in the microbiome. Despite all these potentialities which if realized can completely transform the present agricultural scenario, the research on microbiome is still in its infancy. As the microbiome is influenced by a number of environmental factors, soil characteristics, plant genotypes, etc., its study and understanding is complex. More attention should be paid to study and understand the microbiomes of major crop plants along with their wild relatives so that appropriate strategies can be devised for better management of biotic and abiotic stresses.

Acknowledgements The authors acknowledge the infrastructural facility provided by ICAR-NBAIM, Mau, under the projects entitled "Deciphering molecular mechanism for eliciting drought tolerance in model plant by drought stress alleviating bacteria" and "Identification of key biological indicators of pesticide contamination at agricultural fields of Indo-Gangetic Plains of Eastern Uttar Pradesh" and "Development of bacterial/archaeal indicators for soil health as influenced by management practices".

References

Al-Mailem DM, Sorkhoh NA, Marafie M et al (2010) Oil phytoremediation potential of hypersaline coasts of the Arabian Gulf using rhizosphere technology. Bioresour Technol 101(15):5786–5792

Araújo WL, Lacava PT, Andreote FD, Azevedo JL (2008) Interaction between endophytes and plant host: biotechnological aspects. Plant-Microbe Interact 1:1–21

Atamna-Ismaeel N, Finkel OM, Glaser F et al (2012) Microbial rhodopsins on leaf surfaces of terrestrial plants. Environ Microbiol 14:140–146

Badri DV, Zolla G, Bakker MG et al (2013) Potential impact of soil microbiomes on the leaf metabolome and on herbivore feeding behavior. New Phytol 198(1):264–273

Bailly A, Weisskopf L (2017) Mining the volatilomes of plant-associated microbiota for new biocontrol solutions. Front Microbiol 8:1638

Baltruschat H, Fodor J, Harrach BD et al (2008) Salt tolerance of barley induced by the root endophyte *Piriformospora indica* is associated with a strong increase in antioxidants. New Phytol 180(2):501–510

Bao Z, Okubo T, Kubota K et al (2014) Metaproteomic identification of diazotrophic methanotrophs and their localization in root tissues of field-grown rice plants. Appl Environ Microbiol 80(16):5043–5052

Barnawal D, Bharti N, Tripathi A et al (2016) ACC-deaminase-producing endophyte *Brachybacterium paraconglomeratum* strain SMR20 ameliorates *Chlorophytum* salinity stress via altering phytohormone generation. J Plant Growth Regul 35:553–564

Behie SW, Padilla-Guerrero IE, Bidochka MJ (2013) Nutrient transfer to plants by phylogenetically diverse fungi suggests convergent evolutionary strategies in rhizospheric symbionts. Commun Integr Biol 6(1):e22321

Benitez MS, Osborne SL, Lehman RM (2017) Previous crop and rotation history effects on maize seedling health and associated rhizosphere microbiome. Sci Rep 7(1):15709

Berendsen RL, Vismans G, Yu K et al (2018) Disease-induced assemblage of a plant-beneficial bacterial consortium. ISME J 12:1496–1507

Bömke C, Tudzynski B (2009) Diversity, regulation, and evolution of the gibberellin biosynthetic pathway in fungi compared to plants and bacteria. Phytochemistry 70(15–16):1876–1893

Bordiec S, Paquis S, Lacroix H et al (2011) Comparative analysis of defence responses induced by the endophytic plant growth-promoting rhizobacterium *Burkholderia phytofirmans* strain PsJN and the non-host bacterium *Pseudomonas syringae* pv. pisi in grapevine cell suspensions. J Exp Bot 62(2):595–603

Bouffaud ML, Renoud S, Dubost A et al (2018) 1-Aminocyclopropane-1-carboxylate deaminase producers associated to maize and other Poaceae species. Microbiome 6(1):114

Bulgarelli D, Garrido-Oter R, Münch PC et al (2015) Structure and function of the bacterial root microbiota in wild and domesticated barley. Cell Host Microbe 17(3):392–403

Castillo UF, Strobel GA, Ford EJ et al (2002) Munumbicins, wide-spectrum antibiotics produced by *Streptomyces* NRRL 30562, endophytic on *Kennedia nigricans*. Microbiology 148 (Pt 9):2675–2685

Castillo U, Harper JK, Strobel GA et al (2003) Kakadumycins, novel antibiotics from *Streptomyces* sp. NRRL 30566, an endophyte of *Grevillea pteridifolia*. FEMS Microbiol Lett 224(2):183–190

Chen Y, Fan JB, Du L et al (2014) The application of phosphate solubilizing endophyte *Pantoea dispersa* triggers the microbial community in red acidic soil. Appl Soil Ecol 84:235–244
Copeland JK, Yuan L, Layeghifard M et al (2015) Seasonal community succession of the phyllosphere microbiome. Mol Plant-Microbe Interact 28(3):274–285
da Silva Fonseca E, Peixoto RS, Rosado AS et al (2018) The microbiome of Eucalyptus roots under different management conditions and its potential for biological nitrogen fixation. Microb Ecol 75:183–191
Danhorn T, Fuqua C (2007) Biofilm formation by plant-associated bacteria. Annu Rev Microbiol 61:401–422
De Leon MP, Montecillo AD, Pinili DS et al (2018) Bacterial diversity of bat guano from Cabalyorisa Cave, Mabini, Pangasinan, Philippines: a first report on the metagenome of Philippine bat guano. PLoS One 13:e0200095
Ding L, Maier A, Fiebig HH et al (2011) A family of multicyclic indolosesquiterpenes from a bacterial endophyte. Org Biomol Chem 9(11):4029–4031
Duca D, Lorv J, Patten CL et al (2014) Indole-3-acetic acid in plant-microbe interactions. Antonie van Leeuwenhoek 106(1):85–125
Edwards J, Johnson C, Santos-Medellín C et al (2015) Structure, variation, and assembly of the root-associated microbiomes of rice. Proc Natl Acad Sci USA 112(8):E911–E920
Engelbrektson A, Kunin V, Engelbrektson A et al (2012) Defining the core *Arabidopsis thaliana* root microbiome. Nature 488(7409):86–90
Etesami H (2018) Can interaction between silicon and plant growth promoting rhizobacteria benefit in alleviating abiotic and biotic stresses in crop plants? Agric Ecosyst Environ 253:98–112
Ezra D, Castillo UF, Strobel GA et al (2004) Coronamycins, peptide antibiotics produced by a verticillate Streptomyces sp. (MSU-2110) endophytic on *Monstera* sp. Microbiology 150:785–793
Fiehn O (2001) Combining genomics, metabolome analysis, and biochemical modelling to understand metabolic networks. Comp Funct Genomics 2(3):155–168
Finkel OM, Castrillo G, Herrera Paredes S et al (2017) Understanding and exploiting plant beneficial microbes. Curr Opin Plant Biol 38:155–163
Fitzpatrick CR, Copeland J, Wang PW et al (2018) Assembly and ecological function of the root microbiome across angiosperm plant species. Proc Natl Acad Sci USA 115(6):E1157–E1165
Fonseca-García C, Coleman-Derr D, Garrido E et al (2016) The cacti microbiome: interplay between habitat-filtering and host-specificity. Front Microbiol 7:150
Gaba S, Singh RN, Abrol S et al (2017) Draft genome sequence of *Halolamina pelagica* CDK2 isolated from natural salterns from Rann of Kutch, Gujarat, India. Genome Announc 5(6): e01593–e01516
Gaiero JR, McCall CA, Thompson KA et al (2013) Inside the root microbiome: bacterial root endophytes and plant growth promotion. Am J Bot 100(9):1738–1750
García-Fraile P, Menéndez E, Rivas R (2015) Role of bacterial biofertilizers in agriculture and forestry. AIMS Bioeng 2(3):183–205
Gdanetz K, Trail F (2017) The wheat microbiome under four management strategies, and potential for endophytes in disease protection. Phytobiomes 1:158–168
Glick BR (2014) Bacteria with ACC deaminase can promote plant growth and help to feed the world. Microbiol Res 169(1):30–39
Gomes EA, Lana UGP, Quensen JF et al (2018) Root-associated microbiome of maize genotypes with contrasting phosphorus use efficiency. Phytobiomes 2:129–137
Gou W, Tian L, Ruan Z et al (2015) Accumulation of choline and glycinebetaine and drought stress tolerance induced in maize (*Zea mays*) by three plant growth promoting rhizobacteria (PGPR) strains. Pakistan J Bot 47(2):581–586
Graham JH, Abbott LK (2000) Wheat responses to aggressive and non-aggressive arbuscular mycorrhizal fungi. Plant Soil 220:207–218
Gyaneshwar P, Hirsch AM, Moulin L et al (2011) Legume-nodulating betaproteobacteria: diversity, host range, and future prospects. Mol Plant-Microbe Interact 24(11):1276–1288

Hamayun M, Khan SA, Khan MA et al (2009a) Gibberellin production by pure cultures of a new strain of *Aspergillus fumigatus*. World J Microbiol Biotechnol 25(10):1785–1792

Hamayun M, Khan SA, Kim HY et al (2009b) Gibberellin production and plant growth enhancement by newly isolated strain of *Scolecobasidium tshawytschae*. J Microbiol Biotechnol 19 (6):560–565

Hardoim PR, van Overbeek LS, van Elsas JD (2008) Properties of bacterial endophytes and their proposed role in plant growth. Trends Microbiol 16:463–471

Hayden HL, Savin KW, Wadeson J et al (2018) Comparative metatranscriptomics of wheat rhizosphere microbiomes in disease suppressive and non-suppressive soils for *Rhizoctonia solani* AG8. Front Microbiol 9:859

Higginbotham SJ, Arnold AE, Ibañez A et al (2013) Bioactivity of fungal endophytes as a function of endophyte taxonomy and the taxonomy and distribution of their host plants. PLoS One 8(9): e73192

Horton MW, Bodenhausen N, Beilsmith K et al (2014) Genome-wide association study of Arabidopsis thaliana leaf microbial community. Nat Commun 5:5320

Hultman J, Waldrop MP, Mackelprang R et al (2015) Multi-omics of permafrost, active layer and thermokarst bog soil microbiomes. Nature 521(7551):208–212

Jack ALH, Nelson EB (2018) A seed-recruited microbiome protects developing seedlings from disease by altering homing responses of *Pythium aphanidermatum* zoospores. Plant Soil 422:209–222

Jacquemyn H, Brys R, Waud M et al (2015) Mycorrhizal networks and coexistence in species-rich orchid communities. New Phytol 206:1127–1134

Jha B, Gontia I, Hartmann A (2012) The roots of the halophyte *Salicornia brachiata* are a source of new halotolerant diazotrophic bacteria with plant growth-promoting potential. Plant Soil 356 (1–2):1–13

Ji SH, Gururani MA, Chun SC (2014) Isolation and characterization of plant growth promoting endophytic diazotrophic bacteria from Korean rice cultivars. Microbiol Res 169(1):83–98

Jog R, Pandya M, Nareshkumar G, Rajkumar S (2014) Mechanism of phosphate solubilization and antifungal activity of Streptomyces spp. isolated from wheat roots and rhizosphere and their application in improving plant growth. Microbiology (United Kingdom) 160(Pt 4):778–788

Jones JDG, Dangl JL (2006) The plant immune system. Nature 444(7117):323–329

Klein E, Ofek M, Katan J et al (2013) Soil suppressiveness to fusarium disease: shifts in root microbiome associated with reduction of pathogen root colonization. Phytopathology 103 (1):23–33

Kobayashi A, Kobayashi YO, Someya N, Ikeda S (2015) Community analysis of root- and tuber-associated bacteria in field-grown potato plants harboring different resistance levels against common scab. Microbes Environ 30(4):301–309

Kour D, Rana KL, Verma P et al (2017) Drought tolerant phosphorus solubilizing microbes: diversity and biotechnological applications for crops growing under rainfed conditions. In: Proceeding of national conference on advances in food science and technology, p 166–167

Kumar A, Verma JP (2018) Does plant—microbe interaction confer stress tolerance in plants: a review? Microbiol Res 207:41–52

Lambais MR, Barrera SE, Santos EC et al (2017) Phyllosphere metaproteomes of trees from the Brazilian Atlantic Forest show high levels of functional redundancy. Microb Ecol 73 (1):123–134

Levy A, Conway JM, Dangl JL, Woyke T (2018) Elucidating bacterial gene functions in the plant microbiome. Cell Host Microbe 24:475–485

Li G, Kronzucker HJ, Shi W (2016) The response of the root apex in plant adaptation to iron heterogeneity in soil. Front Plant Sci 7:344

López-Mondéjar L (2017) Una aproximación a los espacios sagrados en el conjunto ibérico de Lorca (Murcia) entre los periodos ibérico y romano: problemas y perspectivas de trabajo

Lowman JS, Lava-Chavez A, Kim-Dura S et al (2015) Switchgrass field performance on two soils as affected by bacterization of seedlings with *Burkholderia phytofirmans* strain PsJN. Bioenergy Res 8:440–449
Marzano SYL, Domier LL (2016) Reprint of "Novel mycoviruses discovered from metatranscriptomics survey of soybean phyllosphere phytobiomes". Virus Res 213:332–342
Mattarozzi M, Manfredi M, Montanini B et al (2017) A metaproteomic approach dissecting major bacterial functions in the rhizosphere of plants living in serpentine soil. Anal Bioanal Chem 409 (9):2327–2339
Mavrodi DV, Mavrodi OV, Elbourne LDH et al (2018) Long-term irrigation affects the dynamics and activity of the wheat rhizosphere microbiome. Front Plant Sci 9:345
Melnyk RA, Haney CH (2017) Plant pathology: plasmid-powered evolutionary transitions. elife 6: e33383
Mendes R, Kruijt M, De Bruijn I et al (2011) Deciphering the rhizosphere microbiome for disease-suppressive bacteria. Science 332(6033):1097–1100
Moronta-Barrios F, Gionechetti F, Pallavicini A et al (2018) Bacterial microbiota of rice roots: 16S-based taxonomic profiling of endophytic and rhizospheric diversity, endophytes isolation and simplified endophytic community. Microorganisms 6(1):E14
Mushegian AA, Ebert D (2016) Rethinking "mutualism" in diverse host-symbiont communities. BioEssays 38:100–108
Naveed M, Hussain MB, Zahir ZA et al (2014a) Drought stress amelioration in wheat through inoculation with *Burkholderia phytofirmans* strain PsJN. Plant Growth Regul 73:121–131
Naveed M, Mitter B, Reichenauer TG et al (2014b) Increased drought stress resilience of maize through endophytic colonization by *Burkholderia phytofirmans* PsJN and Enterobacter sp. FD17. Environ Exp Bot 97:30–39
Newman MM, Lorenz N, Hoilett N et al (2016) Changes in rhizosphere bacterial gene expression following glyphosate treatment. Sci Total Environ 553:32–41
Nobori T, Velásquez AC, Wu J et al (2018) Transcriptome landscape of a bacterial pathogen under plant immunity. Proc Natl Acad Sci USA 115(13):E3055–E3064
Ofek-Lalzar M, Sela N, Goldman-Voronov M et al (2014) Niche and host-associated functional signatures of the root surface microbiome. Nat Commun 5:4950
Oldroyd GED, Murray JD, Poole PS, Downie JA (2011) The rules of engagement in the legume-rhizobial symbiosis. Annu Rev Genet 45:119–144
Ortega RA, Mahnert A, Berg C et al (2016) The plant is crucial: specific composition and function of the phyllosphere microbiome of indoor ornamentals. FEMS Microbiol Ecol 92(12):fiw173
Oteino N, Lally RD, Kiwanuka S et al (2015) Plant growth promotion induced by phosphate solubilizing endophytic *Pseudomonas* isolates. Front Microbiol 6:745
Ozioko FU, Chiejina NV, Ogbonna JC (2015) Effect of some phytohormones on growth characteristics of *Chlorella sorokiniana* IAM-C212 under photoautotrophic conditions. Afr J Biotechnol 14:2367–2376
Panke-Buisse K, Poole AC, Goodrich JK et al (2015) Selection on soil microbiomes reveals reproducible impacts on plant function. ISME J 9(4):980–989
Pérez-Jaramillo JE, Mendes R, Raaijmakers JM (2016) Impact of plant domestication on rhizosphere microbiome assembly and functions. Plant Mol Biol 90:635–644
Prasad R, Gill SS, Tuteja N (2018) Crop improvement through microbial biotechnology. Elsevier, Amsterdam. ISBN 9780444639882. https://www.elsevier.com/books/crop-improvement-through-microbialbiotechnology/prasad/978-0-444-63987-5
Prasad R, Kamal S, Sharma PK, Oelmueller R, Varma A (2013) Root endophyte *Piriformospora indica* DSM 11827 alters plant morphology, enhances biomass and antioxidant activity of medicinal plant *Bacopa monnieri*. J Basic Microbiol 53(12):1016–1024
Rascovan N, Carbonetto B, Perrig D et al (2016) Integrated analysis of root microbiomes of soybean and wheat from agricultural fields. Sci Rep 6:28084
Rashid S, Charles TC, Glick BR (2012) Isolation and characterization of new plant growth-promoting bacterial endophytes. Appl Soil Ecol 61:217–224

Reshef L, Koren O, Loya Y et al (2006) The coral probiotic hypothesis. Environ Microbiol 8:2068–2073
Rolli E, Marasco R, Vigani G et al (2015) Improved plant resistance to drought is promoted by the root-associated microbiome as a water stress-dependent trait. Environ Microbiol 17(2):316–331
Rosenberg E, Zilber-Rosenberg I (2016) Microbes drive evolution of animals and plants: the hologenome concept. MBio 7:e01395–e01315
Rosenberg E, Koren O, Reshef L et al (2007) The role of microorganisms in coral health, disease and evolution. Nat Rev Microbiol 5:355
Salvetti E, Campanaro S, Campedelli I et al (2016) Whole-metagenome-sequencing-based community profiles of *Vitis vinifera* L. cv. Corvina berries withered in two post-harvest conditions. Front Microbiol 7:937
Sangabriel-Conde W, Negrete-Yankelevich S, Maldonado-Mendoza IE, Trejo-Aguilar D (2014) Native maize landraces from Los Tuxtlas, Mexico show varying mycorrhizal dependency for P uptake. Biol Fertil Soils 50:405–414
Santoyo G, Moreno-Hagelsieb G, del Carmen Orozco-Mosqueda M, Glick BR (2016) Plant growth-promoting bacterial endophytes. Microbiol Res 183:92–99
Sarria-Guzmán Y, Chávez-Romero Y, Gómez-Acata S et al (2016) Bacterial communities associated with different *Anthurium andraeanum* L. plant tissues. Microbes Environ 31(3):321–328
Schreiter S, Ding G-C, Heuer H et al (2014) Effect of the soil type on the microbiome in the rhizosphere of field-grown lettuce. Front Microbiol 5:144
Sengupta S, Ganguli S, Singh PK (2017) Metagenome analysis of the root endophytic microbial community of Indian rice (*O. sativa* L.). Genomics Data 12:41–43
Sergaki C, Lagunas B, Lidbury I et al (2018) Challenges and approaches in microbiome research: from fundamental to applied. Front Plant Sci 9:1205
Sessitsch A, Hardoim P, Döring J et al (2012) Functional characteristics of an endophyte community colonizing rice roots as revealed by metagenomic analysis. Mol Plant-Microbe Interact 25 (1):28–36
Sharma SB, Sayyed RZ, Trivedi MH, Gobi TA (2013) Phosphate solubilizing microbes: sustainable approach for managing phosphorus deficiency in agricultural soils. Springerplus 2:587
Shi S, Tian L, Nasir F et al (2018) Impact of domestication on the evolution of rhizomicrobiome of rice in response to the presence of *Magnaporthe oryzae*. Plant Physiol Biochem 132:156–165
Singh BK, Trivedi P (2017) Microbiome and the future for food and nutrient security. Microb Biotechnol 10(1):50–53
Singh RP, Mishra S, Jha P et al (2018) Effect of inoculation of zinc-resistant bacterium *Enterobacter ludwigii* CDP-14 on growth, biochemical parameters and zinc uptake in wheat (*Triticum aestivum L.*) plant. Ecol Eng 116:163–173
Singh D, Raina TK, Kumar A, Singh J, Prasad R (2019) Plant microbiome: a reservoir of novel genes and metabolites. Plant Gene 18:100177. https://doi.org/10.1016/j.plgene.2019.100177
Suman A, Nath Yadav A, Verma P (2016) Endophytic microbes in crops: diversity and beneficial impact for sustainable agriculture. In: Singh DP, Abhilash PC, Ratna P (eds) Microbial inoculants in sustainable agricultural productivity, research perspectives, vol 1. Springer, New Delhi
Sun C, Johnson JM, Cai D et al (2010) *Piriformospora indica* confers drought tolerance in Chinese cabbage leaves by stimulating antioxidant enzymes, the expression of drought-related genes and the plastid-localized CAS protein. J Plant Physiol 167(12):1009–1017
Tsurumaru H, Okubo T, Okazaki K et al (2015) Metagenomic analysis of the bacterial community associated with the taproot of sugar beet. Microbes Environ 30(1):63–69
Turner TR, Ramakrishnan K, Walshaw J et al (2013) Comparative metatranscriptomics reveals kingdom level changes in the rhizosphere microbiome of plants. ISME J 7:2248
van Dam NM, Bouwmeester HJ (2016) Metabolomics in the rhizosphere: tapping into belowground chemical communication. Trends Plant Sci 21(3):256–265

van Loon LC, Bakker PAHM, van der Heijdt WHW et al (2008) Early responses of tobacco suspension cells to rhizobacterial elicitors of induced systemic resistance. Mol Plant-Microbe Interact 21:1609–1621

Vandenkoornhuyse P, Quaiser A, Duhamel M, Le Van A, Dufrense A (2015) The importance of the microbiome of the plant holobiont. New Phytol 206:1196–1206

Vendan RT, Yu YJ, Lee SH, Rhee YH (2010) Diversity of endophytic bacteria in ginseng and their potential for plant growth promotion. J Microbiol 48(5):559–565

Verma P, Yadav AN, Kumar V et al (2017) Beneficial plant-microbes interactions: biodiversity of microbes from diverse extreme environments and its impact for crop improvement. In: Singh D, Singh H, Prabha R (eds) Plant-Microbe interactions in agro-ecological perspectives. Springer, Singapore. https://doi.org/10.1007/978-981-10-6593-4_22

Vorholt JA (2012) Microbial life in the phyllosphere. Nat Rev Microbiol 10(12):828–840

Wagner MR, Lundberg DS, Coleman-Derr D et al (2014) Natural soil microbes alter flowering phenology and the intensity of selection on flowering time in a wild *Arabidopsis* relative. Ecol Lett 17(6):717–726

Wagner MR, Lundberg DS, Del Rio TG et al (2016) Host genotype and age shape the leaf and root microbiomes of a wild perennial plant. Nat Commun 7:12151

Wallace JG, Kremling KA, Buckler ES (2018) Quantitative genetic analysis of the maize leaf microbiome. bioRxiv. https://doi.org/10.1101/268532

Waqas M, Khan AL, Kamran M et al (2012) Endophytic fungi produce gibberellins and indoleacetic acid and promotes host-plant growth during stress. Molecules 17(9):10754–10773

West SA, Kiers ET, Simms EL, Denison RF (2002) Sanctions and mutualism stability: why do rhizobia fix nitrogen? Proc R Soc B Biol Sci 269(1492):685–694

Wilmes P, Bond PL (2004) The application of two-dimensional polyacrylamide gel electrophoresis and downstream analyses to a mixed community of prokaryotic microorganisms. Environ Microbiol 6(9):911–920

Yadav AN, Saxena AK (2018) Biodiversity and biotechnological applications of halophilic microbes for sustainable agriculture. J Appl Biol Biotechnol 6:48–55

Yadav AN, Rana KL, Kumar V, Dhaliwal HS (2016) Phosphorus solubilizing endophytic microbes: potential application for sustainable agriculture. EU Voice 2:21–22

Yadav AN, Verma P, Kour D et al (2017a) Plant microbiomes and its beneficial multifunctional plant growth promoting attributes. Int J Environ Sci Nat Resour 3:1–8

Yadav AN, Verma P, Sachan SG, Saxena AK (2017b) Biodiversity and biotechnological applications of psychrotrophic microbes isolated from Indian Himalayan regions. EC Microbiol Eco 1:48–54

Yuan Z, Druzhinina IS, Labbé J et al (2016) Specialized microbiome of a halophyte and its role in helping non-host plants to withstand salinity. Sci Rep 6:32467

Zhalnina K, Louie KB, Hao Z et al (2018) Dynamic root exudate chemistry and microbial substrate preferences drive patterns in rhizosphere microbial community assembly. Nat Microbiol 3 (4):470–480

Chapter 9
Plants and Microbes: Bioresources for Sustainable Development and Biocontrol

Prachi Bhargava, Neeraj Gupta, Rajesh Kumar, and Siddharth Vats

Abstract This chapter considers the application of bioresources obtained from plants and microbes and biomass obtained from their interactions, their types, their categories, and how these can be used as bioresources for sustainable development. Biomass wastes of plants and microbial origin are underutilized. Globally, 10–50 $\times$ 10^9 tons of lignocellulose biomass is generated from forest, agricultural, and fruit and vegetable processing wastes. This biomass provides an opportunity to be used as a potential biosource for the generation of valuable products by recycling and conservation of cellulose, hemicellulose, and lignin. Similarly, use of microbial biomass for food, biofuels, bioremediation biofertilizers, biocomposites, material to remove heavy metals from the wastewater, and in biocontrol are some key areas where further research is needed. This chapter has also dedicated some sections to the use of plant- and microbes-based biosources for biocontrol. This chapter touches on all these areas, and suggests a few areas where bioresources obtained from plants, microbes, and the biomass obtained from their interactions can be a potential source for sustainable development and biocontrol.

9.1 Introduction

The need for sustainable development is greater than ever. Industries such as agriculture, food, health, medicines, fuel, and the environment are lead producers of bio-wastes, and these wastes are underutilized (Vats and Negi 2013; Vats et al. 2012, 2013a, b; Negi and Vats 2013). These bio-wastes can be used as valuable bioresources for the generation of value-added products (Maurya et al. 2013, 2014; Vats et al. 2014). Bioresources obtained from plants and microbes can be used for sustainable development and biocontrol, directly and indirectly. Use of plants and

P. Bhargava · N. Gupta · S. Vats (✉)
Institute of Biosciences and Technology, Shri Ramswaroop Memorial University, Lucknow, Uttar Pradesh, India

R. Kumar
Univesity Institute of Engineering and Technology, Kurukshetra University, Kurukshetra, Haryana, India

A. Varma et al. (eds.), *Plant Microbe Symbiosis*,
https://doi.org/10.1007/978-3-030-36248-5_9

microbes interactions for sustainable utilization of bioresources is creating a new approach for bioremediation. Biomass obtained from microbes and plants can find uses in bioremediation of heavy metals from wastewater (Ahluwalia and Goyal 2007; Goel et al. 2017; Tandon and Vats 2016; Ojha et al. 2013). Compost produced by the application of microbes present in cow dung on plant biomass can reduce dependence on chemical-based fertilizers (Fan et al. 2006). Microbes are diverse and are cosmopolitan in nature. Use of microbes as biofertilizers, where organic fertilizers prepared from plant biomass contain lignocellulolytic microbes, offers easy availability of primary nutrients (Bhargava et al. 2017).

9.2 Biocontrol for Sustainable Development

Microbes and plants interact for nitrogen fixation and exchange of nutrients. Microbes provide nutrients to plants by decomposing the organic matter. Plants are also used as bioherbicides. In plant-based biofertilizers, plant residues enriched with microbes form a symbiotic relationship with plants, increasing the availability of the micro- and macronutrients to the plants and crops by colonizing the root zone or rhizosphere (Gupta et al. 2018; Bhargava et al. 2017). Interaction of plants and microbes has reduced dependence on chemical-based fertilizers and promoted the concept of organic farming. Plant diseases such as root knot disease in soybean are controlled by application of biofungicide and BINA fertilizers. Application of potassium solubilizers, nitrogen fixers, phosphorus solubilizers, and the combination of fungi and bacteria in consortia improves the health of the plants and increases the availability and acceptability of nutrients to the plants. *Azotobacter*, mycorrhizae, phosphate-solubilizing bacteria, *Azospirillum*, and *Rhizobium* are microbes used in eco-friendly agricultural practices (Bhargava et al. 2017). Biopesticides from microbes and plants are also a bioresource finding application in sustainable agriculture and biocontrol. Plants such as *Azadirachta* and *Chrysanthemum*, with such microbes as *Bacillus thuringiensis*, *Trichoderma*, and nuclear polyhedrosis virus, are used for biocontrol. Microbe-based biopesticides (biocontrol agents) work against phytopathogens by producing biomolecules such as hydrolytic enzymes and antibiotics, controlling the activity of phytopathogens by production of siderophores and HCN (Chandler et al. 2008; Sharma et al. 2014).

9.3 Bioresources

Any resource of biological origin is called a bioresource. These life-generated materials are non-fossil resources and have no limitations for use by humans for food production, substantial products, or energy transport. The direct value of bioresources includes use in agriculture, medicine, and the formation of by-products. Indirect application of bioresources provides ecosystem services: nutrient cycling, pollination, dispersal.

India is known for its rich plant biodiversity, and many of these plants have medicinal value. There are different climatic conditions in the different regions of India, so different types of medicinal plants are found. Many parts of these plants are useful for treatment of certain chronic diseases. In present conditions, as growth of the human population increases more and more, much new disease is invited. So, the healthcare sector is also growing rapidly, and the next generation of antibiotics is coming to the market. However, such synthetic drugs result in poor human health. Now, people are interested in bioresources that because of their natural properties do not cause side effects as do other medicines.

Types of Bioresources

(a) Primary bioresources
(b) Secondary bioresources
(c) Tertiary bioresources
(d) Quaternary bioresources

(a) Primary Bioresources
Primary bioresources are generated in forests, agriculture, or aquaculture, enabling the generation of food, energy, and value-added products. Examples are grains, fish, potato, wood, algae, and bamboo, generated for a specific purpose.

(b) Secondary Bioresources
Secondary bioresources are generated by the processing of primary resources. These type of resources contain a low content of impurities.

(c) Tertiary Bioresources
Tertiary bioresources are also separated by processing from virgin materials, but they have a lower value than secondary bioresources.

(d) Quaternary Bioresources
Quaternary bioresources are those after a product is used. On the basis of timeframe of their generation after start of utilization, they are distinguished into short-, mid-, and long-term categories (Yilmaz et al. 2018).

9.3.1 Need for Sustainable Development

The concept of sustainable development in all walks of life may be controversial, but it captures a set of concerns about our living strategies that are the result of the coevolution of natural and socioeconomic systems (Altieri 1995). Up to now the pattern of agriculture and its allied fields has been that of intensive farming. The successes of the past patterns of agriculture can be summarized as follows:

- Food supply in abundance to the developed world
- Availability of fresh fruits and vegetables all year

- Cost-effectiveness and accessibility of food
- Non-cereal crops such as coffee, tea, cocoa, and spices easily available around the world
- Mechanization produces high labor efficiency
- Use of chemicals has improved yields.
- Problems related to the production mitigated by easy availability of agricultural inputs

The failures of these past strategies have outnumbered the successes. The excessive use of all the natural resources has caused many problems that can be summarized as follows:

- Soil loss-related issues
- Concern for food safety, use of chemicals, antibiotic resistance, food poisoning outbreaks, toxins in food
- Water, land, and air pollution
- Habitat and ecosystem loss
- Water contamination and depletion
- Chronic diseases linked to agricultural chemicals
- Reliability on nonrenewable energy such as fossil fuels
- World issues such as global warming
- Damage to farmland fertility and development
- Ugly and unmanageable countryside

To counteract these challenges, sustainable development is the only possible solution. It is the organizing principle that helps to meet all the goals and needs for human development while sustaining the natural systems for future generations. Sustainability has three roots: the environment, the economy, and society.

Economically sustainable implies that the development should provide secure and cost-effective living for farm families, provide a secure and safe living to all the workers in the food system, and provide access to healthy food for all. Environmental sustainability is achieved when such development helps to preserve the quality of soil, water, and air, and works with and is modeled on natural systems (Vats and Miglani 2011). Development is socially sustainable when it is good for families, supports our communities, and is fair to all involved.

9.3.2 *Plants as Bioresources*

Plants provide food, shelter, and a source of energy to human beings. Food produced from the plants finds direct use for animal and human consumption and the survival of life on earth. Material other than food products generated from plants is lignocellulosic waste (Sharma et al. 2018). This waste that is generated by the plants can be the source of various valuable products. Biomass of lignocellulosic origin is the most abundantly available biomass on the Earth and is used for animal feedstock, proteins

Table 9.1 Continent-wise production of lignocellulose biomass

		Waste generated (millions of tons)			
Sample no.	Continent/country	Plantation crop	Pulse crop	Cereal crop	Oilseed crop
1	South America	147	37	153	10
2	Asia	174	51	1135	61
3	Europe	1	10	550	8
4	Australia	12	1	35	2
5	Central America	21	49	500	84
6	Canada	NA	2	60	1
7	India	18	16	240	14
8	USA	15	44	440	19
9	Africa	34	9	165	11
10	World	548	166	2946	142

for food, paper, and pulp industries, fuel, and the production of chemicals. Globally, $10–50 \times 10^9$ tons of lignocellulose biomass is generated from wastes from forest and agricultural uses and fruit and vegetable processing. This biomass can be used for the generation of value-enhanced products by recycling and conservation of cellulose, hemicellulose, and lignin. In the year 2050, global annual fuel production will be reduced to 5 billion barrels (Vats and Negi 2013; Vats et al. 2013a; b). Plant-generated biomass has uses such as microbial-based delignification for making papers and cellulosic polymers (Maurya et al. 2013), biogas production by anaerobic fermentation, solid-state fermentation for edible mushroom-growing media, biosorbents for heavy metals (Ahluwalia and Goyal 2007), biofuels by pretreating and fermentation with the help of yeasts, and production of phytochemicals for applications in healthcare (Vats and Negi 2013; Vats et al. 2013a; b; Vats and Bhargava 2017; Jain et al. 2011).

9.3.2.1 Source of Lignocellulosic Biomass Wastes

Total dry biomass produced by plants on Earth is 155 billion tons/year, two thirds by plants on land and one third by plants in the sea. The continent-wide production of lignocellulosic biomass is given in Table 9.1. Of total biomass, terrestrial biomass is comparatively easy to access compared to the plant biomass of the oceans. Of total biomass, 65% is generated by forests, and land under cultivation throughout the globe contributes 15%; 1.25% of biomass is consumed by human beings for food and 9% of this biomass is lost during processing (Saini et al. 2015; Kuhad and Singh 1993; Vats 2017).

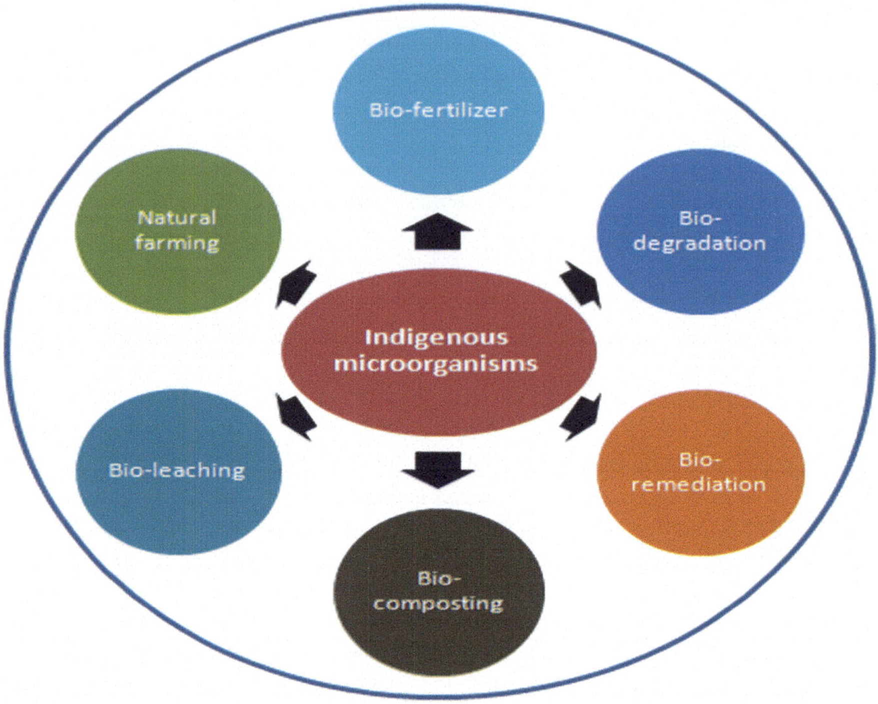

Fig. 9.1 Microbes as bioresource

9.3.2.2 Microbes as a Bioresource

The microbial world is the largest unknown reservoir of benefits that can be explored for the betterment of mankind, consisting of bacteria, yeast, fungi, archaea, algae, and protozoa. To bring advancement and sustainability into the agricultural system of today and to enhance the quality of agricultural products and merchandise, microbes have beneficial final application in all spheres. Native and natural microbes can be used with responsibility to gain benefits at social, monetary, and environmental levels; this is inherently promising and determines a magnificent evolution of research from traditional technologies to modern techniques to provide a system able to protect our environment and new methods of environmental monitoring (Gupta et al. 2018). Figure 9.1 represents the various roles microbes execute for maintaining the sustainability of the ecosystem. They can be seen as a magic wand with the power of biodegradation, biofertilization, bioleaching, biocomposting, and bioremediation that can solve many problems as well as maintaining environmental sustainability.

9.4 Plants and Microbial Interactions

The interactions between plants and microbes occur constantly in every possible way we can imagine. All the parts and organs of a plant do interact and communicate with different microbes for a specific phase of its life cycle. In many ways, the microbes are beneficial for the plants, and in other ways microbes are so harmful that they can be deadly for plants. The plants serve as shelter for the microbes that colonize the roots, shoot surface areas, or areas adjacent to the plant surfaces. Besides providing shelter, plants are a good source of nutrition while they are alive. Many of the plants release compounds that attract and feed associated microbes when they die. The related microbes in association in response secrete compounds that favor the growth of plants (plant growth-promoting factors, PGPR), aid plant resistance toward biotic and abiotic stresses, and defend the plant against more malignant microbes. PGPR also help in bioremediation of soil contamination with heavy metals (phytoremediation) and production of bioethanol (Vats and Mishra 2016). They also induce tolerance toward drought and high salinity in plants. Arbuscular mycorrhizal fungus (AMF) have been proved to have a positive role in the growth of different varieties of cotton. We therefore need to explore the various possibilities and our knowledge of the functioning of the systems to tackle the challenges of the future and ensure the positive growth of plants under extreme adverse conditions.

Just as yin and yang move together, not all microbes are a boon to plant health. Some microorganisms are notoriously damaging to the plant and the related environment also. In the category of plant pathogenic microbes, fungi are the most threatening. Some are specific to a host, but switching of hosts is very common as the basis of emerging fungal diseases, so that the molecular mechanisms that determine host specificity is a hot research topic in plant pathology. Biotrophic plant pathogenic fungi live in very close association to the plant and feed from the tissues of the living plant. These fungi also subvert the defense systems of the plant by secreting effector proteins that interact with plant proteins to the advantage of the pathogen. *Fusarium* is known to cause necrotic spots leading to death in various crops. Harmful microbes and the diseases these cause are summarized in Table 9.2.

Apart from roots, the seeds are also surrounded with microbes in endophytic or ectophytic form. The microbes associated with seeds have enormous influence on seed germination and the ecology, health, and productivity of plants.

9.5 Beneficial Interaction of Plants and Microbes for Bioresource Production

Plants and microbes are major sources of biomass, and similarly their interaction also produces various biomass directly and indirectly. All plant production systems can be capable of feeding the global population if optimized for high yield under the new challenges coming from climate change (Vats et al. 2013a). Demands for

Table 9.2 Disease caused by microorganisms

Sample no.	Name of microorganism	Classification	Disease caused	References
1	*Acidovorax avenae*	Bacteria	Red stripe, top and spindle rot in sugarcane	Mehnaz (2011)
2	*Herbaspirillum rubrisubalbicans*	Bacteria	Mottled stripe in sugarcane	
3	*Pectobacterium chrysanthemi*	Bacteria	Bacterial mottle in sugarcane	
4	*Pseudomonas syringae* pv. *syringae*	Bacteria	Red streak in sugarcane	
5	*P. syringae*	Bacteria	Bacterial apical necrosis in mango	Arrebola et al. (2015)
6	*Pseudomonas fuscovaginae* (Pfv)	Bacteria	Sheath brown rot disease in rice, grain discoloration and sterility	Patel et al. (2014)
7	*Xanthomonas campestris* pv. *campestris*	Bacteria	Black rot disease, *Brassica* plants	Boulanger et al. (2014)
8	*Clavibacter michiganensis*	Bacteria	Goss's bacterial wilt and blight, maize	Baek et al. (2018)
9	*Pseudomonas corrugata*	Bacteria	Pith necrosis, tomato	Catara (2007)
10	*Dickeya solani*	Bacteria	Blackleg and slow wilt, potato	Van Der Wolf et al. (2014)
11	*Alternaria alternata*	Fungus	Grape bunch rot, grapes	Lorenzini and Zapparoli (2014)
12	*Stemphylium* spp.	Fungus	Leaf blight, garlic	Gálvez et al. (2016)
13	*Fusarium graminearum*	Fungus	*Fusarium* head blight, cereals	Sella et al. (2014)
14	*Rhizoctonia cerealis*	Fungus	Sharp eyespot, wheat	Hamada et al. (2011)
15	*Puccinia triticina*	Fungus	Leaf rust, wheat	Kiran et al. (2016)
16	*Pythium iwayamae*	Fungus	Stalk rots, maize	Song et al. (2015)
17	*Magnaporthe grisea*	Fungus	Blast disease, rice	Choi et al. (2013)
18	*Thielaviopsis basicola*	Fungus	Black root rot, cotton	Niu et al. (2008)
19	*Botrytis cinerea*	Fungus	Grey mold, apple, oilseed	Plesken et al. (2015)
20	*Colletotrichum coccodes*	Fungus	Black dot of potato, potato	Nitzan et al. (2002)
21	*Cercospora beticola*	Fungus	Leaf spot disease, sugarbeet	Weiland and Koch (2014)

(continued)

Table 9.2 (continued)

Sample no.	Name of microorganism	Classification	Disease caused	References
22	Carrot red leaf virus Carrot mottle virus	Internal necrosis	Carrot	
23	Wheat yellow mosaic virus	Wheat yellow mosaic	Wheat	Suzuki et al. (2015)
24	Ugandan cassava brown streak virus Cassava brown streak virus	Cassava brown streak disease	Cassava	Patil et al. (2015)
25	Zucchini yellow mosaic virus	Severe yellow mosaic disease	Jatropha	Srivastava et al. (1995)
26	Begomovirus	Tomato leaf curl disease	Tomato	Li et al. (2004)
27	Maize fine streak virus	Chlorotic streaks	*Zea mays*	Redinbaugh et al. (2002)
28	Sugarcane yellow leaf virus	Intense yellowing of the midrib followed by issue necrosis	Sugarcane	El-Sayed et al. (2015)
29	*Hemileia vastatrix*	Coffee leaf rust	*Coffee arabica*	Talhinhas et al. (2014)
30	Turnip curly top virus	Curly top symptoms	Turnip and sugar beet	Razavinejad et al. (2013)

bioresources for feedstocks, construction, animal and human feed, biofuels, and bioenergy are continuously increasing as the population increases. Chemically based agricultural practices are causing negative effects on the ecosystem by the biomagnification of toxic chemicals in consumer systems. Low input of chemicals and highly economically efficient agricultural practices are required for sustainable high-yield agriculture (Carvalho 2006; Vats and Bhargava 2017). Plant–microbe interactions can be important in increasing biomass production. Many crops other than food crops, such as crops for bioenergy and biofuel production, are not domesticated: this is an opportunity lost. To maximize this opportunity, plant–microbe interactions (beneficial, negative, or neutral) can be exploited. Of all such interactions, only a few of them are understood and exploited but other interactions are untapped. Using microbe–plant interactions as an alternative to chemicals can have greater benefits. Use of microbes for biopesticides, biofertilizers, bioherbicides, bioinsecticides, etc. may seem unsuitable for food crops but can make bioenergy crop production sustainable (Farrar et al. 2014).

Plant–microbe interactions can be used for the production of biomass as in mushroom cultivation. Lignocellulosic biomass from plants provides a suitable substrate for mushroom cultivation. Edible fungal fruiting bodies such as *Pleurotus*,

Agaricus, *Lentinus*, and *Volvariella* are grown on agricultural lignocellulosic biomass for waste composting.

Fungal fruiting can be a good source for protein, better than the lignocellulosic-based animal feeds, such as *Pleurotus ostreatus*. *Agaricus bisporus* is the mushroom variety most cultivated throughout the world and is grown on composted straw. Bed logs, wood, and tree stumps can be used for its growth but this is not economical.

Energy saved is energy produced. Use of plant–microbe interactions in biological pulping and bleaching has reduced the overall cost of the process as much as 20%. Biolignolytic microbes such as fungi (white rot fungi) find applications in saving energy in the thermomechanical pulping process. White rot fungi have produced high-quality products with less energy input. When chemically based oxidizing systems, such as the chlorine-based oxidizing system, are incorporated into the biobleaching process, the consumption of chlorine is reduced (Abdullah et al. 2017).

Increasing the digestibility of fodder is another area in which the interaction of microbes with lignocellulosic biomass can produce high-quality fodder. Solid-state fermentation with the help of white rot fungi such as *Coriolus hirsutus*, *Pleurotus* spp., *Cyathus* sp., *Dichomitus squalens*, and *Clirysosporium* makes the substrate more digestible (Rouches et al. 2016).

9.6 What is Biocontrol?

Biocontrol is a strategy that relies upon such methods as parasitism, predation, and herbivory to control pests such as insects, mites, weeds, or pathogens, with active control by human management of this mechanism. This method is used to reduce populations of an invasive species by using its natural enemies. Thus, it is also called biological suppression, that is, reducing the population of a target species to an acceptable level.

9.6.1 Need of Biocontrol

Since time immemorial, agriculture has been the basic method of fulfilling the hunger demands of the world. Agriculture is concerned with crops, livestock, and the related fields. Plants throughout their life cycle encounter many pathological attacks: many they resist, and to many they succumb. Plants are sprayed with pesticides, fungicides, etc. to escape the invasion of disease. However, effective these methods may be, they pose a threat to the environment. Here, biological disease control, which has proved an attractive alternative strategy to counteract plant diseases in a sustainable way, comes into the picture. It is important to understand its mechanism through the interactions between pathogens and antagonists to help us to explore, screen, and construct new and more effective biocontrol agents and manipulate the pedosphere environment. The important factors to be kept in mind when designing a biocontrol system include a suitable antagonist capable of maintaining itself on or in the host plant and the environment of the plant–microbe interaction. The environment governs the choice of antagonist in terms of its optimal temperature, humidity, etc. Besides reducing the disease frequency and percentage of bio- and environmental hazards, biocontrol agents have also been noted to trigger plant growth. An organism that suppresses a pathogen is called a biocontrol agent (BCA). Based on the substrate they attack, biocontrol agents are called biopesticides, biofungicides, bioherbicides, or bioweedicides. To effectively reduce disease development and enhance productivity in different crop systems, integrated pest management should be used with a focus on biocontrol.

Plants and weeds compete for the growth requisites such as water, minerals, and light. To win these battles, plants produce phytochemicals that may inhibit the growth of other plants, or they utilize help from the microbes by providing them with some food or shelter. Microbes directly or indirectly help the host plant meet its growth needs. In allelopathic interactions among plants/plants or plants/microbes, each harms or benefits the other. The same interaction can be used for biocontrol by herbicides (Putnam et al. 1983). Bioherbicides are another category of biocontrol agents prepared from microbes that are added to soil, seed, or roots of the crops or plants. Bioherbicides are prepared from protozoa, fungi, bacteria, and even viruses. Bioherbicides repress the growth of weeds by regulating root and shoot growth. Plant-based herbicides release important valuable oils. Conventional strategies use natural anti-weed species that spread and control the target weed (Frantzen and Müller-Schärer 2006). Onen et al. (2002) used oil extracts from *Artemisia vulgaris*, *Mentha spicata* subsp. *spicata*, *Thymbra spicata* subsp. *spicata*, *Salvia officinalis*, and *Ocimum basilicum* against eight weeds from various families: *Amaranthus*, *Chenopodium album*, *Cardaria draba*, *Echinochloa crus-galli*, *Rumex crispus*, *Reseda lutea*, *Trifolium pratense*, and *Agrostemma githago*. Table 9.3 provides details about the biocontrol agents used.

9.6.2 Major Factors for the Success of Biocontrol

The various factors for successful biocontrol can be biotic, abiotic, and procedural. Among the biotic factors for the plant community, the density of the plant host and its succession rate are factors affecting the efficacy of the biocontrol agents. The type of interaction between the host and the agent can be affected by its nature, whether predation, competition, or parasitism. The agent used has its own characteristics such as its genetic diversity, physiological parameters, emigration nature, and its synchronization with the other species and host. Abiotic factors that affect the efficacy of the biocontrol are the temperature at which is used and the precipitation at the site of action. Site of action, soil type, texture of the soil, slope, shade and sunlight, latitude, season, and day length also affect the efficacy of the biocontrol. Other abiotic disturbances such as fire and floods may also affect the action of biocontrols. Among factors related to biocontrol action are the methods of application,

Table 9.3 Various plants used as biocontrol agents against weeds

Sample no.	Plants	Application	References
1	*Portulaca oleracea*	Inhibit growth of weeds	Bhargava et al. (2017)
2	*Thuja occidentalis*	Inhibit growth of weeds, produces compounds of medicinal value also like neurotoxic composite	Oster et al. (1990) and Cregg and Schutzki (2009)
3	*Eucalyptus* spp.	Biocontrol agent, essential oil produced from eucalyptus act as natural pesticides; pharmaceutically valuable products, repellant, industrial applications also	Batish et al. (2008)
4	*Chamaecyparis lawsoniana*	Inhibit growth of weeds	Bhargava et al. (2017)
5	*Acroptilon repens*	Stimulate growth of plants and inhibit growth of weeds; phytotoxic poly acetylenes obtained from this plant	Quintana et al. (2008)
6	*Amaranthus retroflexus*	Inhibit growth of weeds	Bhargava et al. (2017)
7	*Rosmarinus officinalis*	Volatile compounds act against weed, its oil and blend of other leaf leachate, leaf litter as bio-herbicide	Chen et al. (2013)

procedural methods such as selection of site, site of collection of biocontrol and its colony, how it will be released, time of release, lifespan, shelf life, management of the site after the release, the training and experience of the worker, and follow-up. All the factors are important aspects that are needed to be taken into consideration to have a high success rate with the application of biocontrol.

9.6.3 Role of Cyanobacteria for Biotechnology: Environment and Sustainability

Cyanobacteria are oxygenic phototrophic prokaryotes that are very commonly present in almost all habitats on the Earth. These bacteria may be found in suspended or benthic form when present as single cells or as single trichome or bundles when present in filamentous form. Usually they perform oxygenic photosynthesis, but some species also have the ability to perform anoxygenic photosynthesis using sulfide as an electron donor. Some of the species can also fix atmospheric nitrogen and serve as biofertilizers. The cyanobacteria are now also known to produce certain metabolites that include antibacterial (Jaki et al. 2000), antiviral (Patterson et al. 1994), antiplasmodial (Papendorf et al. 1998), algicide (Abed et al. 2009), anticancer (Gerwick et al. 1994), antifungal, and immunosuppressive agents (Abed et al. 2009). Newly screened cyanobacteria have been shown to accumulate polyhydroxyl alkanoates (PHA), thus opening another avenue of biodegradable plastics to explore.

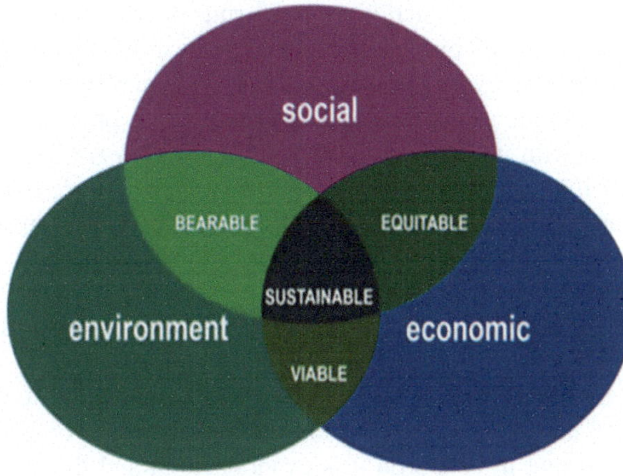

Fig. 9.2 Sustainable development for social, economic, and environment protection

They have been demonstrated to clean up oil-contaminated sediments and wastewaters (Abed et al. 2009). Broadly, the benefits can be categorized as suggested by the following (Fig. 9.2).

9.6.3.1 Cyanobacterial Bioactive Compounds

Cyanobacteria (approximately 19 strains) have been explored for production of more than 20 bioactive compounds that belong to various groups such as alkaloids, indoles, polyketides, fatty acids, amides, and lipopeptides (Bhadury and Wright 2004; Dahms et al. 2006). Structurally, the compounds are lipoproteins and metabolically perform antiviral, antibacterial, anti-algal, antifungal, and anti-protozoan activities. *Phormidium* sp. has antimicrobial ability toward yeasts, fungi, and Gram-negative and Gram-positive bacterial strains. *Lyngbya majuscula* produces nitrogen-containing compounds, such as polyketides, lipopeptides, and cyclic peptides, and can be used as a PGPR. The biological activity of these compounds includes protein kinase C activators, promoters of certain oncogenes, inhibitors of microtubulin assembly, and inhibitors of certain sodium channels.

9.6.3.2 Cyanobacterial Bioplastics

Polyhydroxyl alkanoates (PHAs) are lipidic biomaterials resulting from accumulation of biomass produced by microbes on consumption of carbon sources in abundance (Anderson and Dawes 1990). A PHA is crystalline in nature and has properties similar to thermoplastic petrochemical-based plastic and comparable to that of polypropylene (Doi 1990). It has been the most sought-after research topic for many decades as a potential alternative for nonbiodegradable plastic because it is biodegradable and biocompatible. On degradation, it can break into carbon dioxide and water by the action of occurring microorganisms, which has applications in biomedical and biopharmaceutical fields. *Spirulina platensis*, a cyanobacteria, accumulates PHA with minimal nutrient media with acetate showing phototrophic and/or mixotrophic growth. Cyanobacteria use CO_2 present in the atmosphere and produce PHA, with nitrogen-limiting conditions. *Chlorogloea fritschii*, *Spirulina platensis*, *Spirulinia maxima*, *Oscillatoria limosa*, *Trichodesmium thiebautii*, *Synechocystis platensis*, and *Nostoc muscorum* are common PHA-producing cyanobacteria.

9.6.3.3 Cyanobacterial Consortia for Bioremediation Purposes

Cyanobacteria are able to act on oil-rich compounds and oil components and oxidize them into as surfactants and herbicides (complex organic compounds). Studies have found that in consortia it is cyanobacteria and the associated aerobic organotropic bacteria that cause the degradation of the compounds (Radwan et al. 2002). Consortia consisting of aerobic organotrophs and cyanobacteria help in the degradation of petroleum and other organic compounds. Consortia prepared from cyanobacteria and organotrophs aid wastewater treatment.

(a) Cyanobacteria as Alternative Energy Sources
Dutta et al. (2005) suggested the use of cyanobacteria for the production of hydrogen gas.

(b) Cyanobacteria as a Healthy Source of Food
Cyanobacterial strains of *Nostoc*, *Anabaena*, and *Spirulina* are very commonly used as a source for single-cell protein (SCP) in many countries including Chile, Mexico, Peru, and Philippines. SCP finds its applications as a food supplement. It is rich in nutrients and has good digestibility. SCP is rich in proteins, riboflavin, thiamine, carotene, and minerals and is one of the richest sources of vitamin B_{12}. SCP can be grown on large-scale outdoor ponds, and in bioreactors of a sophisticated type, and the value-added product is marketed in the form of flakes, powder, tablets, and capsules. Cyanobacteria growing in marine conditions (*Phormidium valderianum*) find application in feeding of fishes in the aquaculture industry in India.

9.7 Integrated Resource Management

Natural resource management is an important method for sustainable development. It can minimize the excessive use of natural resources, provide an effective and strict governing system for smooth development and administration, conduct awareness programs, providing formal as well as informal education to illiterate people, promote integration of community involvement, and enhance technical knowledge. Integrated resource management (IRM) involves decision making and planning to minimize the use of resources and their optimization for sustainable benefits for the long term. An IRM program has been implemented by the USDA (Natural

Resources Conservation as the policy through the CA Department of Fish and Game Strategic Vision and California Department of Conservation Watershed Program).

9.8 Applications of Plants and Microbially Derived Biomass in Sustainable Development

There are many applications for biomass generated from plants and microbes in agriculture, such as enhancing sustainability, wastewater treatment, recycling of various organic materials, and production of fuel.

9.8.1 Agriculture Application

Agriculture is the cultivation of crops and the breeding of animals and plants. Agriculture provides foods, fiber, and medicines from medicinal plants, and many other products that are required to sustain life. About 10,000 years ago, humans developed the ability to cultivate their own foods. Farming improved the yield of food plants and allowed people to have food more available. As the human population increases, causing a crisis of productive land, there is a need for sustainable agriculture. Sustainable agriculture is a combined management practice with a key role to enhance and sustain soil fertility and quality (Prasad et al. 2014, 2017). Organic farming is a better choice to increase the sustainability of soil and decrease the applications of chemical fertilizers and pesticides. Organic farming is a fast-growing field worldwide and is growing at a 20% growth rate annually. Worldwide, 24 million hectares or more of land is under organic farming cultivation (Ademir Sérgio et al. 2010). In contrast to conventional farming, organic farming has potential advantages in improving food quality and safety and also building up a large and active soil microbial biomass pool, which is important to enhance soil fertility to result in good production yield. Large amounts of plant and crop biomass residues are generated through agricultural activity. This biomass decreases soil erosion and recovers nutrients back into the soil. The connection between biomass and agriculture is a very close one. Both industries together can form a profitable cycle if more work is done to improve the processes of converting raw biomass into usable fuels.

9.8.2 Composting

Compost is an organic material that is mixed with soil to boost soil fertility and help plants to grow. Soluble phosphorus is an important element required for crop production. Because of the poor solubility of P salts, fertilizers composed of soluble phosphorus of high concentrations increase plant P availability, but fertilizer P reacts with aluminum and iron in low-pH soils and with calcium in high-pH soils (Bertrand et al. 2003; Khan and Joergensen 2009). Fertilizer with high concentrations of soluble P also causes many adverse effects on the water ecosystem such as eutrophication (Song et al. 2007). Organic phosphorus that is obtained from manure, plant residues, and compost is more useful in overcoming this problem. Organic farming is very useful for sustaining soil health and decreasing the cost of farming, especially for small farmers. Microbial-rich compost is used for the commercial production of the fruiting bodies of *Agaricus bisporus*, commonly known as button mushrooms (Vos et al. 2017). The organic compost prepared by efficient microorganisms is used to increase the flower number and pigment content of calendula and marigold flowers (Sharma et al. 2017). Moreover, this compost has also enhanced soil biological health by improving soil enzyme activities that participate in geochemical cycling. The compost derived from decomposed fruit wastes by commercial formulation of effective microorganisms (EM-1) is very effective for the plant growth of *Vigna mungo* (Karthick Raja and Arvind Bharani 2012).

9.8.3 Wastewater Treatment

Metal processing industries, especially the heavy metal-based industry, release heavy metals that adversely affect the environment. Older methods such as chemical precipitation, electrochemical treatment, ion exchange, chemical oxidation, filtration, reverse osmosis, evaporation, and other membrane-based strategies for wastewater treatment are not economical if the metal concentration is in the range of 100 mg/l and the sludge released has a high content of toxic heavy metals. The persistent nature and nonbiodegradable nature of heavy metals poses a threat to the health of the ecosystem. Many of the industries generate waste in the form of water. Heavy metal-associated activities such as mining, metallurgical processes, and mineral processing produce toxic wastewater. To overcome all these problems, use of biological material or biomass has appeared to be more eco-friendly and cost-effective. Biomaterials are materials of plant, animal, and microbial origin. Dead biomass of plant origin that is metabolically inactive has a unique chemical composition wherein metal ions or their complexes, which are toxic to water bodies, can be sequestered. Plant materials have advantages over microbe-based heavy metal removal and sequestering as it is time-consuming to find an appropriate microbe, and also microbes are metal specific. Use of biomaterial for sequestration depends upon many factors such as the origin of the biomaterial (plant, microbe, or animal), and whether the microbial biomass that is to be used is an industrial by-product (safety is a large concern because of the pathogenic nature of microbes), availability of organisms, cost of the production and rearing of the organisms (cultivation or propagation), post-absorption activity, and use of the biomaterials. It will be beneficial to use plant-based biomass that is easily and readily available, although many types of nonliving biomass have been used for metal adsorption such as crab shells, yeast biomass, molds, sea wood, and bacteria (Tiemann et al. 1999; Volesky and Holan 1995). Metals in water causes toxicity and make the water non-habitable and also non-drinkable for humans and animals. Major sources of heavy metals in water are wastewater from industries related to tannery, textiles, leather, dyes and pigments, metallurgical work, galvanizing, paints, electroplating, and other small and heavy industries working on metals and its processing. Heavy metal ions released from these industries causes health issues for animals and humans (Kaur et al. 2010). Most ecotoxic heavy metals such as cadmium, mercury, lead, and chromium are responsible for biophysical distress in both humans and animals. Heavy metals lead to bioaccumulation and transfer throughout the food chain, so organisms at higher trophic levels are at greater risk. Empty mines and mine tailings, and waste effluent (liquid/solid) coming from non-ferrous (Zn, Cu, Cr(III), Ni, Cr, Hg, Pb, V) industries, ferrous industries (Fe), aluminum and aircraft industries (Al), galvanization and alloys (Zn), brass production, copper wire industries, bulb industries (Cu), and cadmium (Cd) all affect the health of the ecosystem and its inhabitants. The half-life of cadmium is 10–30 years, so it is one of the major metals to become accumulated and cause health issues in skin, kidney, and bones (Goel et al. 2017). Table 9.4 gives details of the microbe-based biosorbents for metal removal.

Chromium metal ions are nephrotoxic. Conventionally, heavy metals have been removed by coagulation, adsorption, electrodialysis, ion-exchange cementation, electrowinning, etc., of biomass from the plants and microbial biomass (inactive), allowing the binding of the heavy metals even at very low levels of concentration. Biomass from microbes (fungi, bacteria, algae) acts as a biosorber because of its negative charge, acting as an ion exchanger. Microbial and plant-based biosorbers hold the advantages of being cheaper and growth independent, and are not affected by the toxicity of the heavy metals (no requirement of media for rearing microbes, no wastage of unused media). Biomass from microbial cells is independent as only living cells are influenced by the heavy metals. Biomass from the industries (fermentation/bioprocess) is cost free for disposal issues. These unused wastes can be used to remove the heavy metals. This biomass has no physicochemical limitations (pH, temperature, concentration), acts very rapidly, and has high metal uptake in a wide range of working conditions, even under sepsis. This process is rapid for passive metal sequestration, and its efficiency is dependent upon the affinity, specificity, and capacity of the biosorbents; their nature affects the chemical and physical conditions in the given solution. However, the major limitations are to perform desorption of already bound metals before the next use, and the potential of microbial cells is very limited so the metal valence state cannot be altered.

Understanding the mechanisms for metal adsorption by biomass generated by microbes and plants is important for their better use in removal of toxic metals from industrial wastewaters. In the case of nonliving matter, there is no active metabolism, so binding of the metal ions takes place by adsorption (chemical, physical, ionic). Metal chelation occurs on the fungal cell wall by various ligands such as hydroxyl, amine, carboxyl, sulfhydryl, and phosphate. Toxic metal ions bind to the negatively charged surface of the cells of metabolically inactive fungal biomass.

Table 9.4 Microbial-based biosorbent used for metal removal

Bio-sorbent	Microbe type	Metals	Adsorption capacity
Mucor miehei	Fungal biomass	Cr	–
Aspergillus niger	Fungal biomass	Cu	5 (mg g^{-1})
A. niger	Fungal biomass	Th	30 (mg g^{-1})
A. niger	Fungal biomass	U	29 (mg g^{-1})
A. niger	Fungal biomass	Zn	–
A. niger	Fungal biomass	Pb	30 (mg g^{-1})
A. niger	Fungal biomass	Fe, Cr	–
A. niger	Fungal biomass	Co	95 (mg g^{-1})
A. niger	Fungal biomass	Au	200 (mg g^{-1})
Penicillium chrysogenum	Fungal biomass	Cu, Cd, Pb.	9, 11, 116 (mg g^{-1})
P. chrysogenum	Fungal biomass	Th	142 (mg g^{-1})
P. chrysogenum	Fungal biomass	U	70 (mg g^{-1})
P. chrysogenum	Fungal biomass	Zn	6.5 (mg g^{-1})
P. chrysogenum	Fungal biomass	Cd	56 (mg g^{-1})
P. chrysogenum	Fungal biomass	Zn, Cd, Pb, Cu	–
Candida utilis	Fungal biomass	Pb, Cd, Cu	–
Cladosporium resinae	Fungal biomass	Pb	–
Cladosporium resinae	Fungal biomass	Cu	6 (mg g^{-1})
Mucor rouxii	Fungal biomass	Ni, Pb, Zn Cd	5.24, 17, 4.89, 6.94 (mg g^{-1})
Aspergillus foetidus	Fungal biomass	Cr (VI)	2 (mg g^{-1})
Rhizopus nigricans	Fungal biomass	Pb, Cr	47 (mg g^{-1})
R. nigricans	Fungal biomass	Pb, Ni, Cd	166, 5, 19 (mg g^{-1})
R. nigricans	Fungal biomass	Zn	14 (mg g^{-1})
R. oligosporus	Fungal biomass	Cr	126 (mg g^{-1})
R. oligosporus	Fungal biomass	Cd	17.09 (mg g^{-1})
R. arrhizus	Fungal biomass	Cu, Pb, Ni, Cd, Zn	9.5, 56, 18, 27, 14 (mg g^{-1})
R. arrhizus	Fungal biomass	Cd	30 (mg g^{-1})
R. arrhizus	Fungal biomass	Cu	10 (mg g^{-1})
R. arrhizus	Fungal biomass	Co	2.9 (mg g^{-1})
R. arrhizus	Fungal biomass	Cr	11 (mg g^{-1})
R. arrhizus	Fungal biomass	Co	–
R. arrhizus	Fungal biomass	Co	–
Rhodotorula glutinis	Fungal biomass	Pb	73.5 (mg g^{-1})
Penicillium italicum	Fungal biomass	Th	–
P. italicum	Fungal biomass	Cu	–
Saccharomyces cerevisiae	Fungal biomass	Cu	10 (mg g^{-1})
S. cerevisiae	Fungal biomass	Cd	–
S. cerevisiae	Fungal biomass	Co	5.8 (mg g^{-1})
S. cerevisiae	Fungal biomass	Cr	–
S. cerevisiae	Fungal biomass	Th	11.9 (mg g^{-1})
S. cerevisiae	Fungal biomass	U	55–140 (mg g^{-1})

(continued)

Table 9.4 (continued)

Bio-sorbent	Microbe type	Metals	Adsorption capacity
Ascophyllum nodosum	Algal biomass	Cd	215 (mg g^{-1})
A. nodosum	Algal biomass	Ni, Pb	30, 270–360 (mg g^{-1})
A. nodosum	Algal biomass	Co	100 (mg g^{-1})
Chlorella vulgaris	Algal biomass	Cd	111 (mg g^{-1})
C. vulgaris	Algal biomass	Cu	43 (mg g^{-1})
C. vulgaris	Algal biomass	Zn	133 (mg g^{-1})
C. vulgaris	Algal biomass	U	3.95 (mg g^{-1})
C. vulgaris	Algal biomass	Cr	–
Lyngbya taylori	Algal biomass	Pb, Cd, Cu, Zn	–
Laminaria japonica	Algal biomass	Cd	–
Sargassum spp.	Algal biomass	Cd, Cu, Zn	157, 77, 118 (mg g^{-1})
Sargassum spp.	Algal biomass	Cd	120 (mg g^{-1})
Streptomyces noursei	Bacterial biomass	Ni, Cd, Cu	0.8, 3.4, 9 (mg g^{-1})
S. longwoodensis	Bacterial biomass	U, Pb	440, 100 (mg g^{-1})
S. rimosus	Bacterial biomass	Zn	30 (mg g^{-1})
Pseudomonas fluorescens	Bacterial biomass	U, Th	6, 15 (mg g^{-1})
Thiobacillus ferrooxidans	Bacterial biomass	Zn	82 (mg g^{-1})
Bacillus megaterium	Bacterial biomass	Cr (VI)	30.7 (mg g^{-1})
Zoogloea ramigera	Bacterial biomass	Cr (VI)	2 (mg g^{-1})
Ocimum basilicum	Bacterial biomass	Cr	–

Table adapted from Ahluwalia and Goyal (2007)

9.9 Conclusion

The use of bioresources obtained from plants or microbes or generated from their interaction provides an opportunity for sustainable development. In the use of plant biomass for heavy metal accumulation and extraction, adsorption removes the heavy metals. Wastewater generated from industries can be pretreated before discharge to the environment, reducing the seepage of heavy metals to the water table and the soil. Use of biomass generated by microbes and plants provides an alternative for production of biofuels: utilization of this unexploited biomass has a great advantage. There is a need for an interdisciplinary approach for the proper utilization of this biomass. Eco-friendly approaches are needed for the revival of chemical-free agriculture industries. Biocontrol of pests, insects, and diseases of the plants need further study for an integrated agricultural practice, making agriculture self-reliant and organic food products inexpensive.

References

Abdullah JJ, El-Imam AA, Greetham D, Du C, Tucker GA (2017) The application of fungi for bioleaching of municipal solid wastes for the production of environmental acceptable compost production. J Environ Sci Public Health 1:167–194

Abed RM, Dobretsov S, Sudesh K (2009) Applications of cyanobacteria in biotechnology. J Appl Microbiol 106:1–12

Araújo Ademir Sérgio Ferreira de, Melo Wanderley José de (2010) Soil microbial biomass in organic farming system. Ciência Rural. Santa Maria 40(11):2419–2426

Ahluwalia SS, Goyal D (2007) Microbial and plant derived biomass for removal of heavy metals from wastewater. Bioresour Technol 98:2243–2257

Altieri MA (1995) Agroecology: The science of sustainable agriculture, 2nd edn. CRC Press, Boulder

Anderson AJ, Dawes EA (1990) Occurrence, metabolism, metabolic role, and industrial uses of bacterial polyhydroxyalkanoates. Microbiol Rev 54:450–472

Arrebola E, Carrión VJ, Gutiérrez-Barranquero JA, Pérez-García A, Rodríguez-Palenzuela P, Cazorla FM, de Vicente A (2015) Cellulose production in *Pseudomonas syringae* pv. *syringae*: a compromise between epiphytic and pathogenic lifestyles. FEMS Microbiol Ecol 91:fiv071

Baek KY, Lee HH, Son GJ, Lee PA, Roy N, Seo YS, Lee SW (2018) Specific and sensitive primers developed by comparative genomics to detect bacterial pathogens in grains. Plant Pathol J 34:104

Batish DR, Singh HP, Kohli RK, Kaur S (2008) Eucalyptus essential oil as a natural pesticide. For Ecol Manag 256:2166–2174

Bertrand I, Holloway RE, Armstrong RD, McLaughlin MJ (2003) Chemical characteristics of phosphorus in alkaline soils from Southern Australia. Aust J Soil Res 41:61–76

Bhadury P, Wright PC (2004) Exploitation of marine algae: biogenic compounds for potential antifouling applications. Planta (Berl) 219:561–578

Bhargava P, Singh AK, Goel R (2017) Microbes: bioresource in agriculture and environmental sustainability. In: Plant–microbe interactions in agro-ecological perspectives. Springer, Singapore, pp 361–376

Boulanger A, Zischek C, Lautier M, Jamet S, Rival P, Carrère S, Arlat M, Lauber E (2014) The plant pathogen *Xanthomonas campestris* pv. *campestris* exploits *N*-acetylglucosamine during infection. MBio 5:e01527–e01514

Carvalho FP (2006) Agriculture, pesticides, food security and food safety. Environ Sci Pol 9:685–692

Catara V (2007) *Pseudomonas corrugata*: plant pathogen and/or biological resource? Mol Plant Pathol 8:233–244

Chandler D, Davidson G, Grant WP, Greaves J, Tatchell GM (2008) Microbial biopesticides for integrated crop management: an assessment of environmental and regulatory sustainability. Trends Food Sci Technol 19:275–283

Chen F, Peng S, Chen B, Ni G, Liao H (2013) Allelopathic potential and volatile compounds of *Rosmarinus officinalis* L. against weeds. Allelopathy J 32:57

Choi J, Park SY, Kim BR, Roh JH, Oh IS, Han SS, Lee YH (2013) Comparative analysis of pathogenicity and phylogenetic relationship in *Magnaporthe grisea* species complex. PLoS One 8:e57196

Cregg BM, Schutzki R (2009) Weed control and organic mulches affect physiology and growth of landscape shrubs. Hort Sci 44:1419–1424

Dahms HU, Ying X, Pfeiffer C (2006) Antifouling potential of cyanobacteria: a mini-review. Biofouling 22:317–327

Doi Y (1990) Microbial polyesters. VCH, New York, p 108

Dutta D, De D, Chaudhuri S, Bhattacharya SK (2005) Hydrogen production by cyanobacteria. Microb Cell Factories 4:36

El-Sayed AI, Komor E, Boulila M, Viswanathan R, Odero DC (2015) Biology and management of sugarcane yellow leaf virus: an historical overview. Arch Virol 160:2921–2934
Fan YT, Zhang YH, Zhang SF, Hou HW, Ren BZ (2006) Efficient conversion of wheat straw wastes into biohydrogen gas by cow dung compost. Bioresour Technol 97:500–505
Farrar K, Bryant D, Cope-Selby N (2014) Understanding and engineering beneficial plant–microbe interactions: plant growth promotion in energy crops. Plant Biotechnol J 12:1193–1206
Frantzen J, Müller-Schärer H (2006) Modeling the impact of a biocontrol agent, *Puccinia lagenophorae*, on interactions between a crop, *Daucus carota*, and a weed, *Senecio vulgaris*. Biol Control 37:301–306
Gálvez L, Gil-Serna J, García M, Iglesias C, Palmero D (2016 *Stemphylium* leaf blight of garlic (*Allium sativum*) in Spain: taxonomy and in vitro fungicide response. Plant Pathol J 32:388
Gerwick WH, Proteau PJ, Nagle DG, Hamel E, Blokhin A, Slate DL (1994) Structure of curacin A, a novel antimitotic, antiproliferative and brine shrimp toxic natural product from the marine cyanobacterium *Lyngbya majuscula*. J Org Chem 59:1243–1245
Goel R, Bhargava P, Gupta N, Vats S (2017) Health issues and heavy metals. Austin J Environ Toxicol 3(1):1018
Gupta N, Vats S, Bhargava P (2018) Sustainable agriculture: role of metagenomics and metabolomics in exploring the soil microbiota. In: In silico approach for sustainable agriculture. Springer, Singapore, pp 183–199
Hamada MS, Yin Y, Chen H, Ma Z (2011) The escalating threat of *Rhizoctonia cerealis*, the causal agent of sharp eyespot in wheat. Pest Manag Sci 67:1411–1419
Jain P, Miglani K, Vats S (2011) Aptamers: potential applications in diagnostics and therapeutics. Everymans Sci XLV(6):361
Jaki B, Orjala J, Heilmann J, Linden A, Vogler B, Sticher O (2000) Novel extracellular diterpenoids with biological activity from the cyanobacterium *Nostoc commune*. J Nat Prod 63:339–343
Karthick Raja Namasivayam S, Arvind Bharani RS (2012) Effect of compost derived from decomposed fruit wastes by effective microorganism (EM) technology on plant growth parameters of *Vigna mungo*. J Bioremediat Biodegrad 3:167
Kaur A, Vats S, Rekhi S, Bhardwaj A, Goel J, Tanwar RS, Gaur KK (2010) Physico-chemical analysis of the industrial effluents and their impact on the soil microflora. Procedia Environ Sci 2:595–599
Khan KS, Joergensen RG (2009) Changes in microbial biomass and P fractions in biogenic household waste compost amended with inorganic P fertilizers. Bioresour Technol 100:303–309
Kiran K, Rawal HC, Dubey H, Jaswal R, Devanna BN, Gupta DK, Bhardwaj SC, Prasad P, Pal D, Chhuneja P, Balasubramanian P (2016) Draft genome of the wheat rust pathogen (*Puccinia triticina*) unravels genome-wide structural variations during evolution. Genome Biol Evol 8:2702–2721
Kuhad RC, Singh A (1993) Lignocellulose biotechnology: current and future prospects. Crit Rev Biotechnol 13:151–172
Li ZH, Zhou XP, Zhang X, Xie Y (2004) Molecular characterization of tomato-infecting begomoviruses in Yunnan, China. Arch Virol 149:1721–1732
Lorenzini M, Zapparoli G (2014) Characterization and pathogenicity of *Alternaria* spp. strains associated with grape bunch rot during post-harvest withering. Int J Food Microbiol 186:1–5
Maurya DP, Vats S, Rai S, Negi S (2013) Optimization of enzymatic saccharification of microwave pretreated sugarcane tops through response surface methodology for biofuel. Indian J Exp Biol 51:992
Maurya DP, Singh D, Vats S (2014) Cellulase production and utilization. In Jian A (ed) Category: chemical technology. LAP LAMBERT, Academic.
Mehnaz S (2011) Plant growth-promoting bacteria associated with sugarcane. In: Maheshwari DK (ed) Bacteria in agrobiology: crop ecosystems. Springer, Heidelberg, pp 165–187

Negi S, Vats S (2013) Pine forest litter based bio-refinery for biofuels and value-added phytochemicals. In: Singh RS, Pandey A, Larroche C (eds) Advances in industrial biotechnology. IK International Publishing, Delhi, pp 98–116

Nitzan N, Hazanovsky M, Tal M, Tsror L (2002) Vegetative compatibility groups in *Colletotrichum coccodes*, the causal agent of black dot on potato. Phytopathology 92:827–832

Niu C, Lister HE, Nguyen B, Wheeler TA, Wright RJ (2008) Resistance to *Thielaviopsis basicola* in the cultivated A genome cotton. Theor Appl Genet 117(8):1313–1323

Ojha AK, Forster S, Kumar S, Vats S, Negi S, Fischer I (2013) Synthesis of well-dispersed silver nanorods of different aspect ratios and their antimicrobial properties against gram positive and negative bacterial strains. J Nanobiotechnol 11:42

Onen H, Ozer Z, Telci I (2002) Bioherbicidal effects of some plant essential oils on different weed species. Zeitschrift für Pflanzenkrankheiten und Pflanzenschutz-Sonderheft 18:597–606

Oster U, Spraul M, Rüdiger W (1990) Natural inhibitors of germination and growth. V. Possible allelopathic effects of compounds from *Thuja occidentalis*. Zeitschrift für Naturforschung C 45:835–844

Papendorf O, König GM, Wright AD (1998) Hierridin B and 2,4-dimethoxy-6-heptadecyl-phenol, secondary metabolites from the cyanobacterium *Phormidium ectocarpi* with antiplasmodial activity. Phytochemistry 49:2383–2386

Patel HK, Matiuzzo M, Bertani I, de Paul Bigirimana V, Ash GJ, Höfte M, Venturi V (2014) Identification of virulence associated loci in the emerging broad host range plant pathogen *Pseudomonas fuscovaginae*. BMC Microbiol 14:274

Patil BL, Legg JP, Kanju E, Fauquet CM (2015) Cassava brown streak disease: a threat to food security in Africa. J Gen Virol 96:956–968

Patterson GM, Larsen LK, Moore RE (1994) Bioactive natural products from blue-green algae. J Appl Phycol 6:151–157

Plesken C, Weber RW, Rupp S, Leroch M, Hahn M (2015) *Botrytis pseudocinerea* is a significant pathogen of several crop plants but susceptible to displacement by fungicide-resistant *B. cinerea* strains. Appl Environ Microbiol 81(20):7048–7056

Prasad R, Kumar V, Prasad KS (2014) Nanotechnology in sustainable agriculture: present concerns and future aspects. Afr J Biotechnol 13(6):705–713

Prasad R, Bhattacharyya A, Nguyen QD (2017) Nanotechnology in sustainable agriculture: recent developments, challenges and perspectives. Front Microbiol 8:1014. https://doi.org/10.3389/fmicb.2017.01014

Putnam AR, Defrank J, Barnes JP (1983) Exploitation of allelopathy for weed control in annual and perennial cropping systems. J Chem Ecol 9:1001–1010

Quintana N, Weir TL, Du J, Broeckling CD, Rieder JP, Stermitz FR, Vivanco JM (2008) Phytotoxic polyacetylenes from roots of Russian knapweed (*Acroptilon repens* (L.) DC.). Phytochemistry 69:2572–2578

Radwan SS, Al-Hasan RH, Salamah S, Al-Dabbous S (2002) Bioremediation of oily sea water by bacteria immobilized in biofilms coating macroalgae. Int Biodeterior Biodegr 50:55–59

Razavinejad S, Heydarnejad J, Kamali M, Massumi H, Kraberger S, Varsani A (2013) Genetic diversity and host range studies of turnip curly top virus. Virus Genes 46:345–353

Redinbaugh MG, Seifers DL, Meulia T, Abt JJ, Anderson RJ, Styer WE, Gordon DT (2002) Maize fine streak virus, a new leafhopper-transmitted rhabdovirus. Phytopathology 92:1167–1174

Rouches E, Herpoël-Gimbert I, Steyer JP, Carrere H (2016) Improvement of anaerobic degradation by white-rot fungi pretreatment of lignocellulosic biomass: a review. Renew Sust Energ Rev 59:179–198

Saini JK, Saini R, Tewari L (2015) Lignocellulosic agriculture wastes as biomass feedstocks for second-generation bioethanol production: concepts and recent developments. 3 Biotech 5:337–353

Sella L, Gazzetti K, Castiglioni C, Schäfer W, Favaron F (2014) *Fusarium graminearum* possesses virulence factors common to *Fusarium* head blight of wheat and seedling rot of soybean but differing in their impact on disease severity. Phytopathology 104:1201–1207

Sharma KM, Kumar R, Vats S, Gupta A (2014) Production, partial purification and characterization of alkaline protease from *Bacillus aryabhattai* K3. Int J Adv Pharm Biol Chem 3:290–298
Sharma A, Saha TN, Arora A, Shah R, Nain L (2017) Efficient microorganism compost benefits plant growth and improves soil health in calendula and marigold. Hortic Plant J 3:67–72
Sharma D, Javed S, Arshilekha, Saxena P, Babbar P, Shukla D, Srivastava P, Vats S (2018) Food additives and their effects: a mini review. Int J Curr Res 10:69999–70002
Song C, Han S, Tang C (2007) Changes in phosphorus fractions, sorption and release in Udic Mollisols under different ecosystems. Biol Fertil Soil 44:37–47
Song FJ, Xiao MG, Duan CX, Li HJ, Zhu ZD, Liu BT, Sun SL, Wu XF, Wang XM (2015) Two genes conferring resistance to *Pythium* stalk rot in maize inbred line Qi319. Mol Gen Genomics 290:1543–1549
Srivastava KM, Hallan V, Raizada RK, Chandra G, Singh BP, Sane PV (1995) Molecular cloning of Indian tomato leaf curl vims genome following a simple method of concentrating the supercoiled replicative form of viral DNA. J Virol Methods 51:297–304
Suzuki T, Murai MN, Hayashi T, Nasuda S, Yoshimura Y, Komatsuda T (2015) Resistance to wheat yellow mosaic virus in Madsen wheat is controlled by two major complementary QTLs. Theor Appl Genet 128:1569–1578
Talhinhas P, Azinheira HG, Vieira B, Loureiro A, Tavares S, Batista D, Da Silva C (2014) Overview of the functional virulent genome of the coffee leaf rust pathogen *Hemileia vastatrix* with an emphasis on early stages of infection. Front Plant Sci 5:88
Tandon S, Vats S (2016) Microbial biosynthesis of cadmium sulfide (Cds) nanoparticles and their characterization. Eur J Pharm Med Res 3:545–550
Tiemann KJ, Gardea-Torresdey JL, Gamez G, Dokken K, Sias S, Renner MW, Furenlid LR (1999) Use of X-ray absorption spectroscopy and esterification to investigate Cr (III) and Ni (II) ligands in alfalfa biomass. Environ Sci Technol 33:150–154
Van Der Wolf JM, Nijhuis EH, Kowalewska MJ, Saddler GS, Parkinson N, Elphinstone JG, Pritchard L, Toth IK, Lojkowska E, Potrykus M, Waleron M (2014) *Dickeya solani* sp. nov., a pectinolytic plant-pathogenic bacterium isolated from potato (*Solanum tuberosum*). Int J Syst Evol Microbiol 64:768–774
Vats S (2017) Methods for extractions of value-added nutraceuticals from lignocellulosic wastes and their health application. In: Grumezescu A, Holban A-M (eds) Ingredients extraction by physicochemical methods in food. Elsevier, San Diego, pp 1–64
Vats S, Bhargava P (2017) Alternate energy: fuel for "Modi's India" and "smart cities.". Int J Curr Res 9:49090–49097
Vats S, Miglani K (2011) Synergistic antimicrobial effect of cow urine and *Azadirachta indica* on infectious microbes. Int J Pharm Sci Res 2:1781
Vats S, Mishra A (2016) Soil agro-ecological management by vermicompost: a potential organic nutrient source for the state of Uttar Pradesh. Eur J Pharm Med Res 3:604–609
Vats S, Negi S (2013) Use of artificial neural network (ANN) for the development of bioprocess using *Pinus roxburghii* fallen foliages for the release of polyphenols and reducing sugars. Bioresour Technol 140:392–398
Vats S, Kumar R, Negi S (2012) Natural food that meet antibiotics resistance challenge: in vitro synergistic antimicrobial activity of *Azadirachta indica*, *Terminalia chebula*, *Piper nigrum* and photoactivated cow urine. Asian J Pharm Biol Res 2(2)):122–126
Vats S, Maurya DP, Jain A, Mall V, Negi S (2013a) Mathematical model-based optimization of physico-enzymatic hydrolysis of *Pinus roxburghii* needles for the production of reducing sugars. Indian J Exp Biol 5:944–953
Vats S, Maurya DP, Agarwal A, Shamoon M, Negi S (2013b) Development of a microbial consortium for the production of blend of enzymes for the hydrolysis of agricultural wastes into sugars. J Sci Ind Res 72:585–790
Vats S, Kumar R, Maurya DP (2014) Alkaline amylase from multi resistant microbes and its applications. In Alexei E (ed) Category: Microbiology. LAP LAMBERT, Academic
Volesky B, Holan ZR (1995) Biosorption of heavy metals. Biotechnol Prog 11:235–250

Vos AM, Heijboer A, Boschker HT, Bonnet B, Lugones LG, Wösten HA (2017) Microbial biomass in compost during colonization of *Agaricus bisporus*. AMB Express 7:12

Weiland J, Koch G (2014) Sugarbeet leaf spot disease (*Cercospora beticola* Sacc.). Mol Plant Pathol 5:157–166

Yilmaz N, Atmanli A, Vigil FM (2018) Quaternary blends of diesel, biodiesel, higher alcohols and vegetable oil in a compression ignition engine. Fuel 212:462–469

Chapter 10
Plant-Microbiome Interactions in Hydrocarbon-Contaminated Soils

Ana Carolina Agnello, Irma Susana Morelli, and María Teresa Del Panno

Abstract The use of green remediation technologies (i.e., phytoremediation, bioremediation, mycoremediation) for the restoration of hydrocarbon-contaminated sites is one of the keys for sustainable development. These technologies rely on the joint action of biotic components of the ecosystem, namely, plants, bacteria, and fungi. Despite the fact that previous studies showed that the clean-up of hydrocarbons could be achieved individually by plants or microorganisms, present investigations suggest that the interaction of plants with their surrounding microbiome determines the outcomes of green remediation technologies. This book chapter reviews the state of the art to explain the two-way relationship established between plants and their associated microbiome in hydrocarbon-polluted soils. Special focus is put on stressing the results obtained in recent studies that employ omics approaches.

10.1 Introduction

Petroleum hydrocarbons (HCs) are a large family of heterogeneous organic compounds that are found in crude oil, its derived materials (e.g., diesel, gasoline, kerosene), and waste by-products, which have in common C and H atoms as their main chemical constituents. As a function of the chemical structure, four HC fractions can be separated from crude oil: saturates, aromatics, resins, and asphaltenes (SARA)

A. C. Agnello (✉) · M. T. Del Panno
Centro de Investigación y Desarrollo en Fermentaciones Industriales (CINDEFI), Consejo Nacional de Investigaciones Científicas y Técnicas (CONICET), Universidad Nacional de La Plata (UNLP), Buenos Aires, Argentina
e-mail: aagnello@biotec.quimica.unlp.edu.ar; tere@biol.unlp.edu.ar

I. S. Morelli
Centro de Investigación y Desarrollo en Fermentaciones Industriales (CINDEFI), Consejo Nacional de Investigaciones Científicas y Técnicas (CONICET), Universidad Nacional de La Plata (UNLP), Buenos Aires, Argentina

Comisión de Investigaciones Científicas de la Provincia de Buenos Aires (CIC-PBA), Buenos Aires, Argentina
e-mail: guri@biol.unlp.edu.ar

A. Varma et al. (eds.), *Plant Microbe Symbiosis*,
https://doi.org/10.1007/978-3-030-36248-5_10

(Aske et al. 2001). Saturated HCs are aliphatics that contain C and H joined together in saturated straight (i.e., n-alkanes), branched (i.e., paraffins), or cyclic (i.e., cycloalkanes) chains. Aromatic HCs are formed by one or more benzene rings: when two or more rings are fused, they originate polycyclic aromatic hydrocarbons (PAHs) (Abdel-Shafy and Mansour 2016). On the other hand, the fractions of resins (i.e., pyridines, quinolines, carbazoles, sulfoxides, and amides) and asphaltenes (i.e., phenols, fatty acids, ketones, esters, and porphyrins) do not have a definite structure like saturates and aromatics but a very complex constitution with the addition of heteroatoms like N, S, and O (Mullins 2008). Due to the extensive worldwide use of petroleum products, contamination with HCs is not uncommon either during the exploration, as a consequence of the generation of refinery waste by-products, or as a result of accidental spills throughout transportation and storage processes (Krahforst and Healey 2017). Indeed, petroleum-HC contamination is of great concern as it poses severe toxic effects for the whole ecosystem affecting both environmental compartments (e.g., soil, air, water bodies) and human health (Ahmed and Fakhruddin 2018; Tormoehlen et al. 2014).

Considering that environmental protection is one of the pillars of sustainable development, improving green remediation technologies appears to be a suitable approach for the restoration of HC-contaminated sites (USEPA 2008). In this sense, technologies such as bioremediation and phytoremediation are passive energy remediation systems driven by little or no external energy, which maximize remediation sustainability. Bioremediation takes advantage of heterotrophic microorganisms, which obtain the energy by the oxidation of electron donors in their environment (Abatenh et al. 2017; Singh et al. 2017). This feature can be exploited to achieve the complete mineralization of xenobiotics to non-toxic end products such as CO_2 and H_2O (Das and Chandran 2011; Varjani 2017). Indeed, bioremediation is not limited to bacteria but extended to fungi, which contribute to pollutant removal through specific mechanisms in a distinct type of bioremediation, i.e., mycoremediation (Morelli et al. 2013; Prasad 2017, 2018). On the contrary, as phototrophic organisms, green plants use the energy from light to convert CO_2 and H_2O into carbohydrates, and thus do not rely on HC catabolism as a source of carbon and energy to sustain their metabolic functions. Still, plants may play an important role for the remediation of HCs through phytodegradation, which can take place both inside the plant and/or within the rhizosphere zone (Newman and Reynolds 2004). Plants can either take an active part (i.e., adsorption, accumulation, degradation, and/or volatilization) in cleaning-up organic pollutants or play a secondary role sustaining rhizosphere microbial communities responsible for pollutant removal throughout rhizoremediation (Correa-García et al. 2018).

The current trend toward developing reliable and predictable green remediation technologies focuses on the application of integrative omics tools to explore and harness the microbiome of polluted soils (Bell et al. 2014b; Quiza et al. 2015). Indeed, sustaining soil microbiome diversity certainly plays a major role in HC biodegradation. For instance, Bell et al. (2014a) demonstrated that the highly diverse initial soil microbiome of a polluted soil degraded more crude oil than a more specialized but less diverse bacterial assemblage selected on crude oil media. Furthermore, the diversity of the soil microbiome is subject to the selective pressure that exerts the coexistence of HCs and plants. For example, Tardif et al. (2016) observed that increasing contamination levels of petroleum HCs were related to large shifts in the microbiome composition of bulk soil, favoring HC degraders and microorganisms associated with plant health. Besides, these shifts were moderated in the plant surroundings (i.e., rhizosphere, root, and stem tissues), probably because of a more controlled and protected environment provided by plants. Although less studied than bacteria, fungi are also crucial constituents of the microbiome whose diversity is also shaped by plants and HCs. Bell et al. (2014a) observed that fungal communities were even more sensitive than bacteria to HCs and that the introduction of willows (*Salix* spp.) promoted more diverse fungal communities, which diverged based on plant phylogeny.

Plant-microbiome interactions taking place in the endosphere and rhizosphere appear to enable much of the outcomes in polluted environments in terms of plant growth and HC degradation. Therefore, this book chapter examines the interactions between plants and their associated microbiomes in HC-contaminated soils. Recent studies are gathered together to shed light on how plants contribute to HC removal and by what means they are assisted by microorganisms and, the other way round, how microorganisms (i.e., bacteria and fungi) metabolize HCs and in what way they are supported by plants to do so.

10.2 The Active Role of Plants in the Metabolism of Hydrocarbons and How They Are Assisted by Microorganisms

10.2.1 Uptake and Degradation of Hydrocarbons by Plants

Although plants do not rely on an external supply of HCs for their metabolism, the uptake of these compounds can occur, taking place both in the phyllosphere and rhizosphere. HCs can volatilize (from the soil surface) to the leaf surface and be adsorbed and/or uptaken by plant leaves in the phyllosphere. Likewise, in the rhizosphere the process involves the desorption of HCs from soil followed by adsorption and/or uptake by plant roots from soil solution (Collins et al. 2006) (Fig. 10.1). Chemical properties of HCs definitely limit both plant foliar and root uptake. As the octanol-water partition coefficient (K_{ow}) and molecular weight of HCs increase, water solubility decreases hindering the transfer of HCs across biological membranes. In addition, soil properties (e.g., clay, soil organic matter content) govern the sorption of HCs to soil. Several studies report a successfully active role of plants to uptake, translocate, and/or degrade a wide variety of HCs. For instance, *Schefflera arboricola* and *Spathiphyllum wallisii*, two ornamental plants, were able to remove benzene from indoor air (Parseh et al. 2018). Similarly, *Scirpus grossus* showed the potential to withstand diesel in contaminated water with the ability to uptake and translocate the HC series C8–C32 (Al-Baldawi et al. 2015). Finally, ornamental *Tagetes patula* and *Mirabilis jalapa* demonstrated to have good

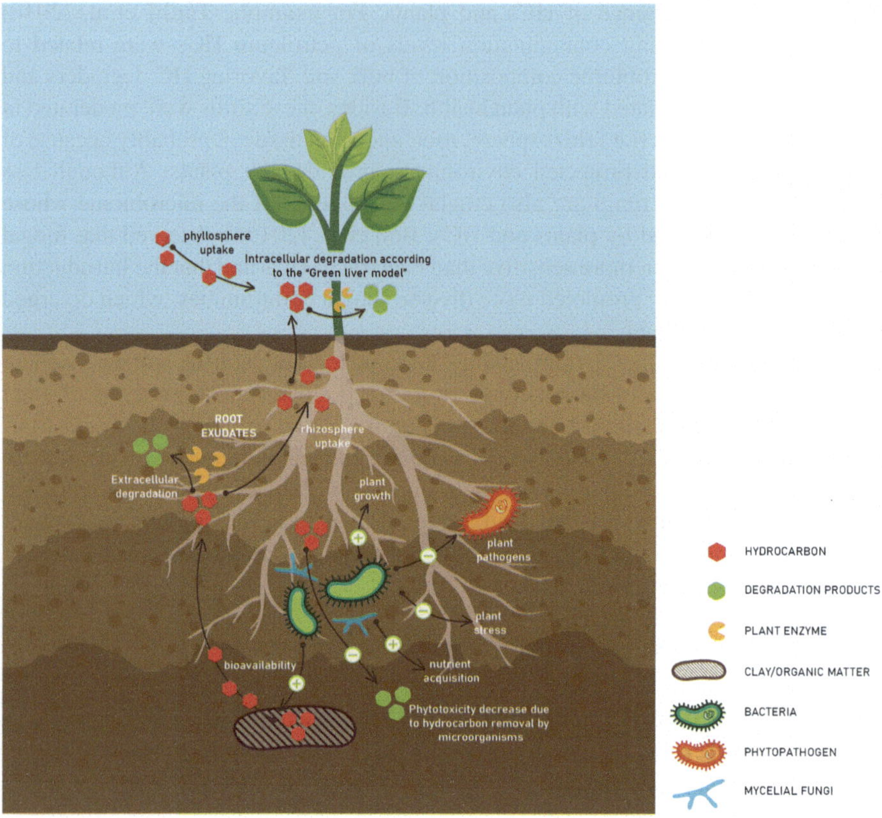

Fig. 10.1 Metabolism of hydrocarbons by plants and the role of microorganisms to assist them

ability to tolerate and accumulate benzo[a]pyrene from polluted soils (Sun and Zhou 2016), and PAHs were detected in alfalfa (*Medicago sativa* L.) tissues by fluorescence microscopy as well (Alves et al. 2017).

After being taken up by plants, HCs can be catabolized to non-toxic intermediates. The most accepted model that describes the metabolism of xenobiotics by plants considers plants as a 'green liver' due to the resemblance with the detoxification function of the mammalian liver. According to this model, the metabolism of xenobiotics by plants involves three steps: (1) chemical transformation through oxidation, reduction, or hydrolysis reactions, (2) conjugation to endogenous molecules (e.g., malonate, UDP-glucose, glutathione), and (3) internal compartmentalization and storage in the vacuole, incorporation into the cell wall, or excretion to the extracellular space (Sandermann 1992). For instance, *Clitoria ternatea* exhibited a high potential for airborne HC remediation. Ethylbenzene was not only taken up but also metabolized by the plant: 1-phenylethanol, acetophenone, and benzaldehyde were identified as metabolites from ethylbenzene degradation (Daudzai et al. 2018). Besides intracellular metabolism of HCs, plants may also have an active role in HC degradation via the root exudation of enzymes, which catalyze the oxidation of HCs and degrade them into intermediate products. In support of this, Barone et al. (2016) demonstrated that water-soluble protein extracts derived from maize (*Zea mays* L.) were able to degrade PAHs as a result of peroxidase, polyphenol oxidase, and catalase activities.

10.2.2 Phytotoxicity of Hydrocarbons

The balance between the uptake and the degradation of HCs by plants in the pathways explained above is the key of plant tolerance/sensitivity to HCs. The phytotoxicity of HCs can be manifested as a number of symptoms such as inhibition of germination, stunting of plant development, reduced plant growth, and tissue damage (Al-Baldawi et al. 2015; Chaîneau et al. 1997; Siddiqui et al. 2001). Mechanisms underlying HC phytotoxicity may be related both to direct effects on plant physiology (e.g., cell membrane disruption, damage of photosynthetic apparatus) or, indirectly, altering the physical and chemical properties of the soil where plants are growing. Moreover, the chemical structure of HCs, its concentration and bioavailability in soil, and the plant species are among the key factors that determine the severity of phytotoxicity (Efroymson et al. 2004). A typical example of phytotoxicity was surveyed by Somtrakoon and Chouychai (2013) who observed that germination of maize and rice (*Oryza sativa* L.) seeds was retarded by single or mixed PAHs. Similarly, the study carried out by Chaîneau et al. (1997) showed that growth of maize, wheat (*Triticum aestivum* L.), and bean (*Phaseolus vulgaris* L.) was reduced by the presence of fuel oil. Interestingly, growth inhibition increased with HC concentration but was not linearly proportional to the loading rates. Another remarkable example of phytotoxicity is described by Al-Baldawi et al. (2015) who observed that direct exposure to diesel-contaminated water caused severe damage to the root and stem structures, as demonstrated by SEM micrographs of *S. grossus* epidermis and cross-sections. Besides, it is important to highlight that biotransformation of HCs can lead to additional phytotoxic metabolites. This is exemplified in the recent work undertaken by Dubrovskaya et al. (2016) who found that some of the metabolites produced as a result of microbial degradation of phenanthrene (i.e., 9,10-phenanthrenequinone, 1-hydroxy-2-naphthoic and benzoic acids) are more toxic for plants than starting PAH molecules.

Because of the toxic effects of HCs on plants, performing a phytotoxicity assessment is an initial and essential step in phytoremediation trials. Phytotoxicity tests allow finding potential candidate species able to germinate and establish in HC-contaminated sites. In this context, several species of legumes, grasses, and crops have been tested for their ability to withstand the presence of HCs (Kirk et al. 2002; Muratova et al. 2008). The outcome of these screenings showed that the species that most frequently demonstrated a good performance to tolerate HCs, in terms of high germination and growth rates, were alfalfa and ryegrass (*Lolium perenne*). As a result, these species were used later on as candidates in phytoremediation trials with promising applications for the remediation of HC-contaminated soils (Bourceret et al. 2015).

10.2.3 The Role of Bacteria to Assist Plants in Hydrocarbon-Contaminated Soils

The term plant growth–promoting bacteria (PGPB) refers to a group of bacteria that are beneficial for plant development and can be found in close association with different plant tissues (e.g., roots, shoots, leaves, or even fruits and seeds). Therefore, the habitat of PGPB might be not only the rhizosphere but also internal tissues of plants colonized by endophytes with plant growth–promoting ability (Santoyo et al. 2016; Prasad et al. 2015). PGPB act both through (1) direct mechanisms: like the synthesis of phytohormones that enhance plant growth and the release of compounds that facilitate resource acquisition and (2) indirect mechanisms such as the competition with pathogens and the modulation of plant stress (Olanrewaju et al. 2017). A significant aspect of PGPB is that they act not only under normal conditions but also under environmental stress. PGPB can assist in overcoming the detrimental phytotoxic effects caused by organic pollutants promoting the establishment of plants in HC-contaminated soils by improving plant health and growth performance. The enhancement of a prolific root system by PGPB may benefit the uptake of water and nutrients, promote the rhizosphere effect, and increase the depth of the treatment zone, which often limits the success of phytoremediation. Moreover, bacteria may help plants to cope with pollutants regulating the stress induced by HCs (Fig. 10.1). This is exemplified in the work undertaken by Singha et al. (2018) who observed that rice stress response under the influence of pyrene was modulated by PGPB. Inoculation of rice with PGPB promoted not only the growth of rice (i.e., shoot and root length) but also improved rice antioxidant activity enhancing the levels of glutathione, glutathione-S-transferase, and superoxide dismutase activities. Likewise, PGPB have also demonstrated to influence phytodegradation of pollutants. For example, the inoculation of *C. ternatea* with plant growth–promoting endophytic bacteria *Bacillus cereus* modulated the expression of plant ethylbenzene degradation genes and increased ethylbenzene removal efficiency (Daudzai et al. 2018).

Besides the straight actions that PGPB have on plants, they can also have effects on soil pollutants. A distinct mechanism by which soil microorganisms may influence pollutant removal is the increase of HC bioavailability in the rhizosphere and plant uptake as a result. In support of this, Chen et al. (2017) observed that the inoculation of *Scirpus triqueter* with PGPB increased the amount of pyrene uptaken by the plant. Microorganisms may enhance desorption of HCs from soil by producing surface-active biomolecules termed biosurfactants. As a result of biosurfactant emulsifying action, HCs could be readily available not only for microorganisms but also for plants. In this sense, the degradation of HCs by PGPB may be a supplementary beneficial trait for plants. Bacteria being able to metabolize HCs (refer to Sect. 10.3.1) can reduce soil phytotoxicity via the effective removal of contaminants, which constitutes an additional gain for plants growing in polluted soils. In this sense, Baoune et al. (2018) isolated endophytic bacteria (*Streptomyces* genus) from roots of plants grown naturally in sandy contaminated soil that exhibited plant growth-promoting features and also could use petroleum as sole carbon and energy.

In view of the joint actions that PGPB may exhibit, Pacwa-Płociniczak et al. (2016) performed a broad screening to isolate bacteria strains that combine plant growth-promoting traits, HC-degrading ability, and biosurfactant/bioemulsifier production. Although from 42 HC-degrading isolates they could not identify a unique strain with a high performance for all the above-mentioned characteristics, they propose the application of a consortium composed of biosurfactant-producing strains together with plant growth-promoting strains as promising agents in microbe-assisted phytoremediation.

Extensive research has been conducted over the past years to develop bacteria-assisted phytoremediation as an efficient remedial strategy for petroleum HCs (Fatima et al. 2017). The key to enhance phytoremediation in this way is finding the suitable plant-bacteria partnerships, which can be accomplished through different approaches: native plant growth-promoting rhizobacteria (PGPR) (Gerhardt et al. 2015), colonizing endophytes (Syranidou et al. 2016), bioaugmentation with indigenous (Franchi et al. 2016) or allochthonous (Agnello et al. 2016) bacteria, biostimulation (Agarry et al. 2014), etc.

10.2.4 The Role of Fungi to Assist Plants in Hydrocarbon-Contaminated Soils

The interior of plants is an important habitat where colonizing fungi reside. Plants can live in symbiosis with non-pathogenic endophytic fungi like arbuscular mycorrhizae (AM) and ectomycorrhizae. To date, a number of plant-fungal interactions have been reported in contaminated soils, which may favor plants and, in turn, its phytoremediation potential. The main mechanisms by which fungal endophytes can assist plants in HC-contaminated soils are (1) improving plant growth, (2) modulating plant stress levels, (3) enhancing the adsorption and bioaccumulation of HCs by plants, and (4) reducing phytotoxicity by the removal of HCs provided that they possess the suitable degradation pathways (refer to Sect. 10.3.1) (Deng and Cao 2017; Rajtor and Piotrowska-Seget 2016). For example, it has been observed that AM inoculation alleviated diesel toxicity on *Melilotus albus*: plants had a better growth response and higher content of microelements than non-inoculated plants. Moreover, roots of inoculated plants had higher total antioxidant and nitrate reductase activities, indicating an improved physiological response (Hernández-Ortega et al. 2012). In what concerns the influence of fungi on HC bioavailability, it has been demonstrated that AM may facilitate the mobilization of HCs in soil enabling the adsorption and/or uptake by plants (Fig. 10.2). For instance, alkane bioaccumulation in roots of wheat was more important in AM-inoculated plants than in non-inoculated plants although this process accounted for only a small portion of the total HC removal, which was mainly due to biodegradation by bacteria and fungi (Lenoir et al. 2016). The authors hypothesized that the increased HC accumulation was related to a higher lipid content and volume of the root adsorption area in the

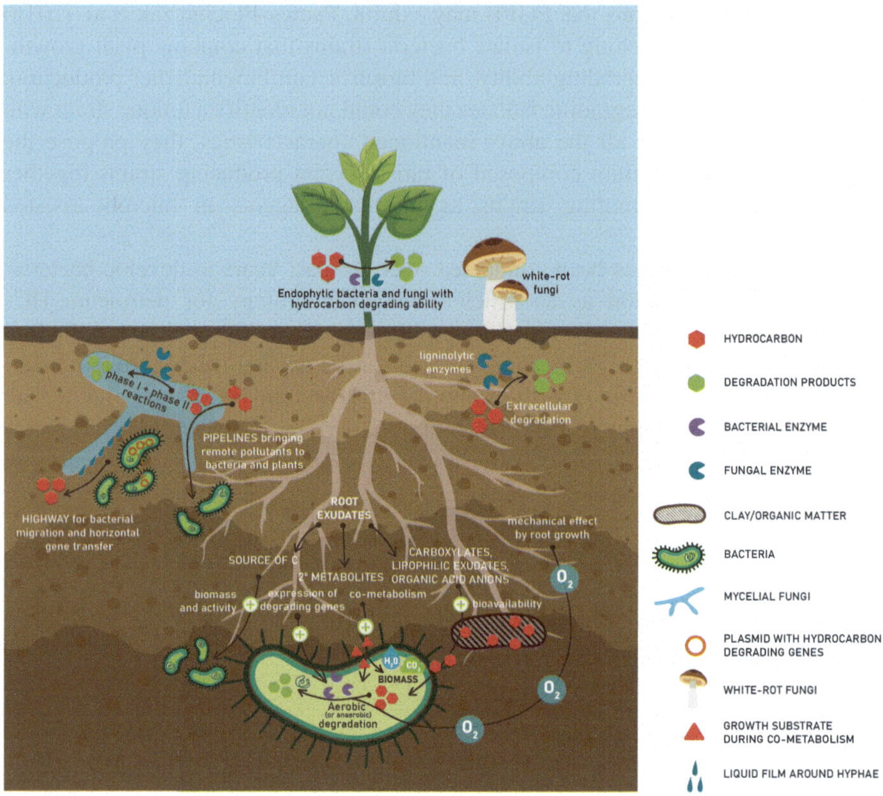

Fig. 10.2 Metabolism of hydrocarbons by microorganisms and the role of plants to assist them

presence of AM. In addition, the uptake of organic contaminants from soil by plants can be mediated by AM hyphae through a distinct mechanism. In a remarkable experiment using a compartmentalized cultivation system, Gao et al. (2010) observed that ryegrass (*Lolium multiflorum* Lam.) roots, which were grown in un-spiked clean soil, accumulated high concentrations of PAHs in the roots because abundant mycorrhizal hyphae extended from PAH-spiked soil, took PAHs and transported them to plants. Interestingly, AM acted as a pipeline dynamically linking soil pollutants, fungi, and plant roots. Furthermore, AM have been reported to synthesize compounds that alter HC bioavailability. This is supported by a recent study in which glomalin-related soil protein (GRSP), a *N*-linked glycoprotein produced by AM hyphae (Schindler et al. 2007), induced changes in roots that favored PAH adsorption and accumulation by ryegrass (Chen et al. 2018). Likewise, Gao et al. (2017) observed that inoculation with AM increased GRSP content and pyrene removal in soils planted with alfalfa.

The synergistic effect of plants and fungi has been used for the removal of organic contaminants in the context of microbe-assisted phytoremediation. For instance,

García-Sánchez et al. (2018) demonstrated that the combination of plants and white rot fungi (i.e., maize-*Crucibulum laeve* association) was more efficient than the individual use of plants or fungi for the treatment of aged PAH-polluted soils. Likewise, Asemoloye et al. (2017) reported that a synergistic approach that combined the joint action of guinea grass (*Megathyrsus maximus*) and rhizospheric fungi (i.e., *Aspergillus flavu*, *Aspergillus niger*, *Talaromyces purpurogenus*, and *Trichoderma harzianum*) isolated from a crude oil polluted site improved the soil nutrient content and sped up PAHs degradation rates. Finally, it is important to highlight that sole inoculation with AM may not be enough to achieve the presumed goals. Indeed, it will require the joint action of fungi and bacteria. For instance, Boldt-Burisch et al. (2018) demonstrated that mycorrhizal inoculation alone did not improve the growth of the legume *Lotus corniculatus* L. and the grass *Elymus trachycaulus* growing in oily substrates. By contrast, the inoculation with mycorrhizae plus bacteria led to a significantly positive response of both plant species.

10.3 The Active Role of Microorganisms in the Metabolism of Hydrocarbons and How They Are Assisted by Plants

10.3.1 Uptake and Degradation of Hydrocarbons by Bacteria

The metabolism of HCs by bacteria involves three fundamental steps: (1) access to the target HCs, (2) trans-membrane transport, and (3) enzymatic degradation.

Bacteria can gain access to the target HCs if they are dissolved in the aqueous phase or if small HC droplets are pseudo-solubilized (emulsified). Apart from water-soluble aromatics and short-chain HCs, most HCs are poorly soluble in water. As a result, the most common process to access them is the solubilization of little HC droplets. This can be achieved through the formation of micelles structured in the presence of biosurfactants. In contrast, large HC drops require the attachment of bacteria through direct surface contact, but this mechanism has been reported less frequently and the subsequent uptake mechanism remains poorly understood (Hua and Wang 2014). Indeed, *Sphingomonas paucimobilis* demonstrated to make phenanthrene bioavailable combining both mechanisms, i.e., the production of biosurfactants and the direct contact of cells adhering to phenanthrene crystals developing a biofilm over time (Coppotelli et al. 2010). Moreover, fungi have been suspected to facilitate bacterial access to hydrophobic substrates through direct bacterial-fungal interactions. Mycelial networks can act both as 'highways' that accelerate bacterial migration in the hydrophilic film around fungal hyphae as well as 'pipelines', which bring remote pollutants to bacteria by taking up and translocating them through their hyphae (Banitz et al. 2013; Harms and Wick 2006).

The transport of HCs across the membrane of bacteria can occur through passive diffusion and/or energy-dependent active transport, depending on HC type and concentration (Hua and Wang 2014). While most studies are devoted to low-molecular-weight HCs, the trans-membrane transport of high-molecular-weight HCs is rarely reported. Furthermore, it has been described from Gram-negative bacteria a system of outer membrane proteins with pore-like structure and a hydrophobic channel, which facilitate the passive diffusion of small HCs from the extracellular environment to the periplasm (Hearn et al. 2008). Once internalized, there are some reports that show the formation of lipid inclusion bodies inside the cell where HCs (e.g., octadecane) are deposited before being oxidized (Hua and Wang 2012).

The most documented pathways for intracellular HC degradation by bacteria occur under aerobic conditions, where molecular oxygen is critical to initiate the enzymatic attack (Fig. 10.2). The catabolism of aliphatic HCs involves a number of oxidation steps. The first key step consists in the hydroxylation of a terminal carbon, which is catalyzed by monooxigenases (e.g., alkane 1-monooxygenase, CYP153 alkane hydroxylase). Afterwards, the hydroxylated alkane is further oxidized to the corresponding aldehyde and carboxylic acid, which in turn enter in the β-oxidation route of fatty acids. The final product is acetyl-CoA, which is catabolized in the Krebs cycle, and fully oxidized to CO_2. The degradation of aromatic HCs requires, not only the initial hydroxylation of the aromatic ring, but also the opening of the hydroxylated aromatic ring (i.e., catechol or structurally related compounds) by aromatic-ring cleavage dioxygenases (e.g., intradiol or extradiol dioxygenases) following the ortho- or meta-cleavage of the ring. The resulting di- or trihydroxylated aromatic compounds can be introduced into the Krebs cycle and fully degraded to CO_2 (Das and Chandran 2011). Examples of aerobic bacteria such as *Pseudomonas*, *Alcaligenes*, *Sphingomonas*, *Rhodococcus*, and *Mycobacterium* have often been reported to degrade HCs. Indeed, biodegradation of HCs generally involves a number of different bacterial species within a consortium of microbes with broad enzymatic capacities rather than individual organisms (Santisi et al. 2015; Zafra et al. 2017). Recent studies demonstrate that functional bacterial communities co-acclimate to a changing environment of HC stress and are able to conduct biodegradation of HCs in a cooperative way creating interactive networks with each other (Wanapaisan et al. 2018; Wang et al. 2016).

Although the fastest and most complete degradation of HCs is performed in the presence of O_2, degradation of HCs is also possible under anaerobic conditions, but these pathways are less studied. Examples of such anaerobic reactions are the addition of toluene or *n*-alkanes to fumarate, the O_2-independent hydroxylation of ethylbenzene, and the reductive dearomatization of benzoyl-CoA (Rabus et al. 2016). In anaerobic and methanogenic environments, where HCs are biodegraded to methane, mutually beneficial interactions between syntrophic microorganisms play a key role. This implies that a cooperative action of mixed microbial populations is required for the ultimate removal of HCs (Gieg et al. 2014). Considering that O_2 is replenished in the rhizosphere by O_2 diffusion as a function of water/air-filled porosity, the anaerobic pathway is supposed to be less relevant than the aerobic route in the rhizosphere (Uteau et al. 2015).

Some key genes involved in HC degradation can be located on mobile elements. Moreover, the homology of DNA sequences and organization of degrading genes carried by conjugative plasmids may be indicators that horizontal gene transfers can occur between HC-degrading bacteria during microbial adaptation to xenobiotics (Abbasian et al. 2016). This is supported by in silico analysis, which demonstrated that HC-degrading genes *alkB* and *catA* can be subjected to horizontal transfer events among bacterial communities spreading the potential to degrade HCs (Rodrigues et al. 2018). In an interesting study, Taghavi et al. (2005) reported for the first time in planta horizontal gene transfer among plant-associated endophytic bacteria. They inoculated poplars with the endophyte *Burkholderia cepacia*, which contained a plasmid coding for toluene degradation. They observed that although the inoculated endophyte could not establish in the endophytic community, there was horizontal gene transfer of toluene degrading ability to different members of the endogenous endophytic community. Moreover, bacterial horizontal gene transfer can be facilitated by the network structures of mycelia. As described above, liquid films around hyphae constitute a continuous highway in which bacterial migration and contacts are favored (Berthold et al. 2016).

10.3.2 Degradation of Hydrocarbons by Fungi

Fungi are able to degrade HC both in an assimilative way to obtain energy and in a non-assimilative way through detoxification and co-metabolism pathways (Morelli et al. 2013). Fungi can be classified into ligninolytic and non-ligninolytic according to their ability to metabolize lignin in wood, and both types of fungi have a part in the degradation of PAHs (Aydin et al. 2017) (Fig. 10.2).

Major constituents of ligninolytic fungi are the white-rot fungi (WRF), i.e., wood-decaying basidiomycetes. Ligninolytic fungi are characterized for their ability to produce extracellular enzymes, which are responsible for the complete degradation of lignin. Ligninolytic enzymes encompass a vast array of enzymes such as peroxidases (e.g., lignin, manganese, and versatile peroxidases), laccases, and accessory enzymes (H_2O_2-generating enzymes and glioxal oxidase). A key characteristic of the complex ligninolytic enzymatic system is low substrate specificity. As a result, this feature can be exploited to extend the degrading ability of ligninolytic enzymes to break down other complex compounds (Kadri et al. 2017). This is exemplified in the work undertaken by Pozdnyakova et al. (2018) who studied the degradation of three-ringed PAHs by the white-rot fungus *Pleurotus ostreatus* and the litter-decomposing fungi *Agaricus bisporus*, demonstrating that the extracellular enzyme system of ligninolytic fungi plays a key role in the initial attack of PAH molecules yielding quinone metabolites. The degrading ability of ligninolytic fungi clearly represents a promising option for the bioremediation of HCs in soil. This is supported by a recent study performed by Košnář et al. (2018) that compared the removal of PAHs in soil after different bioremediation approaches in relation to extracellular enzyme activities. They observed that mycoremediation treatment with *P. ostreatus* outperformed natural attenuation and phytoremediation in terms of PAHs removal from soil.

The main weakness of WRF application for bioremediation is that they only grow under specific environmental conditions (e.g., on compact wood rich in lignocellulosic substrates, at acidic pH), which renders them inefficient to compete with non-ligninolytic fungi in soil and limits their contribution in the decomposition of HCs under natural conditions. In this sense, previous studies have demonstrated that is possible to isolate non-ligninolytic fungi (mainly belonging to the *Ascomycota* and *Zygomycota* phylum) from contaminated sites (Reyes-César et al. 2014). Likewise, highly diverse AM communities demonstrated to be able to colonize plants growing in weathered oil ponds indicating that AM are able to adapt to these harsh conditions (Garcés-Ruiz et al. 2017). This is particularly interesting because this kind of fungi exhibit tolerance to environmental pollutants as well as potential for their enzymatic transformation. The enzymatic transformation of HCs by non-ligninolytic fungi is typically slower than for ligninolytic fungi, and although is not fully understood, it is believed to use phase I (e.g., P450s and epoxide hydrolases) and phase II (e.g., glutathione S-transferases, NAD(P)H: quinine oxidoreductases and UDP-glucuronosyl transferases) intracellular enzymes (Marco-Urrea et al. 2015). These degradative pathways catalyze xenobiotic biotransformation and detoxification in most eukaryotes, thus being extensive to ligninolytic fungi too. Intracellular fungal degradation is exemplified in the study undertaken by Aranda et al. (2017) in which the Ascomycota fungi *Penicillium oxalicum* was found to exhibit a high and fast PAH degradation capability. *P. oxalicum* degradation of both anthracene and dibenzothiophene was mediated at the intracellular level by cytochrome P450 enzymes (CYPs). Indeed, the use of ^{13}C-anthracene enabled the identification of oxidized and hydroxylated derivatives, which are known as Phase I metabolites produced through CYP transformation. Additionally, this work highlighted that the presence of glucose was required to proceed with anthracene degradation, suggesting that fungi may not be able to use PAHs as a sole C and energy supply and may require additional C sources to co-metabolize xenobiotic compounds.

10.3.3 The Role of Plants in Assisting Hydrocarbon Uptake/Degradation by Microorganisms

Plants have a secondary supportive role in HC removal, both in the rhizosphere and endosphere, improving HC uptake/degradation by microorganisms through different mechanisms (Fig. 10.2).

The degradation of HCs by soil microorganisms is noteworthy under the root influence because of the *rhizosphere effect* to which the root surrounding microbiome is subject to. The rhizosphere effect is used to depict that, in comparison with bulk soil, the biomass and activity of microorganisms in the rhizosphere are enhanced (Warembourg 1997). This is the result of various processes driven by plants, but the most significant is probably root exudation (Rohrbacher and St-Arnaud 2016). Root exudates consist of a vast array of metabolites (e.g., organic, amino, and fatty acids, sugars, vitamins, nucleotides, flavonoids, phytohormones, etc.) released by plant roots to the surrounding media, many of which can be used as substrates for microbial metabolism leading to increased microbial biomass and/or activity. Moreover, the root architecture also shapes the rhizosphere microbiome by physical processes. The mechanical effect of growing roots comes with soil aeration, which influences the distribution of rhizosphere microorganisms, the concentration of O_2, and thus aerobic metabolism of HCs (van Dam and Bouwmeester 2016). Rhizodegradation takes advantage of the rhizosphere effect to achieve the removal of HCs in the rhizosophere by stimulating HC-degrader populations. This is evidenced in the study of Bourceret et al. (2018) that, using culture-independent methods, demonstrated that Gram-negative PAH-dioxygenase genes and transcripts were higher in the planted (alfalfa) than unplanted soil and were positively correlated to PAH degradation. Along with the enhancement of microbial biomass and/or activity of HC-degrading bacteria, root exudates can also promote HC biodegradation as a result of increasing HC bioavailability. Root-induced chemical changes in the rhizosphere by the release of organic anions (e.g., citrate, oxalate) contributes to HC desorption from the soil (e.g., organic matter and clay particles), improving the accessibility of degrading bacteria to HCs (Martin et al. 2014). This can be illustrated by a batch experiment showing that low-molecular-weight organic acids in aqueous solution could disrupt soil organic matter (SOM)-metal cation-mineral linkages in soils, resulting in the release of SOM from soil and simultaneous increase of dissolved organic carbon (DOC) in solution. The loss of SOM from soil and increase of DOC in solution were responsible for the enhanced PAH desorption from soil (Ling et al. 2015).

The so-called 'secondary compound hypothesis' states that plant secondary metabolites released into the rhizosphere can trigger the biodegradation of environmental pollutants. This may be explained by the fact that plant secondary metabolites can induce the expression of degradative genes or serve as primary substrates during the co-metabolism of HCs (Musilova et al. 2016). In support of the first mechanism, Yergeau et al. (2014) and Pagé et al. (2015) demonstrated through a metatranscriptomic approach an induced expression of several aliphatic- and aromatic-degrading genes in the rhizosphere of willows (*Salix purpurea*) growing in HC-contaminated soils as compared to bulk soils. The authors hypothesize that the secondary metabolite salicylate released by willows could mediate this process, as it has been reported to induce the transcription of PAH-degrading genes. Regarding the second mechanism, Rentz et al. (2005) corroborated that benzo[a]pyrene was removed from solution by *Sphingomonas yanoikuyae* while growing on root products as a primary carbon and energy source. This confirms the hypothesis that co-metabolism of xenobiotics, i.e., the transformation of a non-growth substrate (i.e., benzo[a]pyrene) in the obligate presence of a growth substrate (i.e., root extracts), can take place in the rhizosphere (Crowley et al. 2001; Dalton and Stirling 1982).

In addition to the processes taking place in the rhizosphere, it is important to take into consideration the endosphere as well. The endosphere is the interior of the plant that functions as a microbial habitat of endophytes where plants provide shelter and protection. Endophytic bacteria like those associated with *L. corniculatus* L. and *Oenothera biennis* L. collected in a long-term petroleum HC-polluted site have been shown to possess HC-degrading genes such as P450 gene, which encodes for cytochrome P450-type alkane hydroxylase (Pawlik et al. 2017). This confirms the importance of plants as the residence of endophytic bacteria with HC-degrading ability and thus with a high potential to improve phytoremediation of petroleum HCs. Moreover, plants can host fungi as well. In this sense, Fu et al. (2018) were able to demonstrate the biodegradation of phenanthrene by endophytic fungus *Phomopsis liquidambari* not only in vitro in liquid culture but also in vivo using rice seedlings.

10.4 The Construction of the Holobiont Concept Through Omics Approaches

The current trend toward developing reliable and predictable green remediation technologies puts the focus on the application of integrative omics tools to explore and harness the microbiome of polluted soils in order to fill in present knowledge gaps (Bell et al. 2014b; Quiza et al. 2015). In this sense, omics approaches are definitely enabling a deeper understanding of the complex relationships in the microbial network under the plant influence and subjected to HC stress (Table 10.1). For instance, metagenomics enables the prediction of the HC-degrading potential of rhizosphere microbial communities, metatranscriptomics can reveal the actual expression of HC-degrading genes under the plant influence, and metaproteomics/metabolomics makes possible the identification of the complete metabolic intermediates during HC degradation as well as the myriad of compounds released by root exudates. As a consequence, the use of such innovative technology platforms allows proposing new degradative pathways beyond the individual plant, bacterial, or fungal levels. Indeed, present research intends to understand not only how plants influence degrading microbial communities in the rhizosphere but also how they interplay as a metaorganism or holobiont (i.e., host and microbiome together) to degrade complex pollutants (El Amrani et al. 2015). Based on the latest evidence from the metagenomics level, Thijs et al. (2016) have recently proposed a competition-driven model to explain the establishment of a catabolic rhizosphere microbiome in contaminated soil. Furthermore, Yergeau et al. (2018) highlight the importance of considering plants and their associated microbiota as an 'interactome'. Performing simultaneous analysis of root and rhizosphere metatranscriptomes they found that plants and their associated microorganisms undergo a complete overhaul under HC stress modulating transcript abundances. Indeed, Gonzalez et al. (2018) conducted a complex metatranscriptomic study taking into consideration the entire microbiome and concluded that trees, fungi, and bacteria establish a tripartite mutualism in HC-polluted soils. They observed that while root and fungal expression patterns

Table 10.1 Summary of 'multi-omics'-based approaches to understand the role of the microbiome in hydrocarbon-contaminated (rhizosphere) soils

Type of omic-based approach	Technique	Brief description	Application	Example reference
Metagenomics	Amplicon sequencing	Next-generation sequencing of marker genes (typically the 16S rRNA gene and the ITS region), which are amplified from metagenomic DNA	Identify the composition and diversity of HC-contaminated soil's microbial community	Cecotti et al. (2018)
	Functional metagenomics	Large DNA fragments are inserted in vectors (e.g., cosmids, fosmids) and expressed in hosts (e.g., *Escherichia coli*) which are then screened for a particular function. Clones showing the desired activity are sequenced	Identify genes coding for specific functions (e.g., HC degrading genes such as alkB, nahAc, etc.) in the microbial community and link the function with the taxa responsible for it	Duarte et al. (2017)
	Shotgun metagenomics	Metagenomic DNA is sheared and sequenced. Both functions and taxonomy are derived from homology search in databases	Identify the microbial community composition, diversity, and functions in HC-contaminated soils (e.g., bioprospection of HC-degrading genes)	Vigneron et al. (2017)
	DNA or RNA SIP	A labeled substrate is used to feed a microbial community. Labeled (heavy) DNA or RNA is retrieved through centrifugation of nucleic acids in a density gradient. The labeled DNA or RNA is then sequenced directly or amplified to sequence marker genes	In combination with 16S rRNA gene amplicon sequencing it is possible to link taxonomic group identity to the metabolism of the substrate. This is particularly useful to study degradation of C-13 labeled HCs (e.g., C13-naphthalene)	Jiang et al. (2015)
Metatranscriptomics	Amplicon sequencing	Next-generation sequencing of marker genes (typically the 16S rRNA gene and the ITS region), which are amplified from cDNA retro-transcribed from total extracted RNA	Identify the composition and diversity of functionally active fraction of microflora in HC-contaminated soils	Bourceret et al. (2018)
	RNA-seq or whole-transcriptome shotgun sequencing	Sequencing of RNA: Not only mRNA transcripts but also miRNA, tRNA and rRNA	Microbial expression analysis of functionally active fraction of microflora in HC-contaminated soils. This represents accurately the activity and can provide a clue to the taxonomic groups that actively contribute to biodegradation and metabolism of HCs	Yergeau et al. (2018)

(continued)

Table 10.1 (continued)

Type of omic-based approach	Technique	Brief description	Application	Example reference
Metaproteomics	Electrophoresis and/or chromatography combined with chemical or metabolic labelling and MS	Gel-based (e.g., 2D-PAGE) or gel-free (e.g., nano-LC) approaches for protein separation coupled to protein identification by MS techniques (e.g., MALDI-TOF)	Identification of differential expression of proteins/enzymes involved in HC metabolism (alkB, nahAc, etc)	Festa et al. (2017)
Metabolomics	LC-MS GC-MS GC/LC-TOF-MS ^{1}H-NMR	Identification of different metabolites. Depending on their chemical properties (e.g., volatility, polarity, etc.) different types of metabolomics platforms are used	Study metabolic pathways of HC biodegradation: The presence of microbial metabolites can be considered as the 'ultimate proof' that a biochemical reaction has occurred. Analysis of metabolites in root exudates and changes in the root exudation patterns triggered by HCs in soils	Tian et al. (2018)

Abbreviations: *HC* hydrocarbons, *SIP* stable-isotope probing, *MS* mass spectrometry, *2D-PAGE* two-dimensional polyacrylamide gel electrophoresis, *LC* liquid chromatography, *GC* gas chromatography, *TOF* time-of-flight, ^{1}*H-NMR* nuclear magnetic resonance, *MALDI* matrix-assisted laser desorption/ionization nahAc gene coding for naphthalene 1,2-dioxygenase subunit alpha. alkB gene coding for alkane 1-monooxygenase enzyme responded to HC stress altering pathways associated to microbiome interactions, the apparatus necessary for the direct reduction of contamination stress came from bacteria. These results highlight how crucial it is to investigate the expression of the entire microbiome to have a full picture of the metaorganism responding to soil contamination.

10.5 Conclusion and Future Prospects

The evidence reviewed here supports the concept of a conjoint action of plants, bacteria, and fungi building an inseparable and highly dependent relationship. The particular approach of analyzing the contribution of plants to soil HC degradation with the assistance of soil microorganisms and, conversely, how the soil microorganisms could improve their degrading ability with the support of plants revealed a broad perspective of the multiple and diverse interactions that take place between plants and their associated microbiome. Moreover, it becomes manifest how the limitations of one actor could be overwhelmed by the abilities of the other. This leads to the accomplishment of a robust establishment in polluted soils and an effective HC degradation by the metaorganism or holobiont. In this sense, the current state of the art by means of a diffuse application of mixed omics approaches strengthens the idea that an integrated understanding of the relationships between plants, bacteria, and fungi determines the success of green remediation technologies. Furthermore, it is insinuated that green remediation is facing an auspicious transition moving from individual bio-, myco-, phyto-remediation toward the development of an integrative meta/holo-remediation notion.

Acknowledgments This review was supported by grant PICT-2016-0947. The authors would like to thank Dr. Sabrina Festa for proofreading the chapter. In addition, the authors are very grateful to industrial designer Carla Di Benedetto, who provided the expertise to create the illustrations presented in Figs. 10.1 and 10.2.

Glossary

Endophyte Microorganism residing within plant tissues (the endosphere). They may establish different types of interactions (e.g, mutualistic, pathogenic) with their plant hosts. For instance, they may improve the plant's ability to tolerate hydrocarbon stress

Endosphere Interior of the plant as a microbial habitat of endophytes. The term may refer to either the aerial (i.e., stems, leaves) and/or root components of a plant

Metaorganism/Holobiont Plant and its associated microorganisms

Microbiome Totality of microorganisms inhabiting a particular environment. For example, the rhizosphere microbiome refers to all microorganisms inhabiting the rhizosphere of a particular plant. The microbiome is a dynamic ecosystem driven by environmental changes (e.g., plant species, soil type, presence of pollutants, etc.)

Multi-omics Approach Combination of methods that use innovative technology platforms such as genomics, transcriptomics, proteomics, and metabolomics. Omics approaches typically generate large datasets to provide insight of genes, transcripts, proteins, and metabolites of a biological system. The prefix 'meta' is used when performed on all members of a mixed-species community as opposed to a single organism

Mycoremediation The use of fungi for soil remediation applications

Phyllosphere Surface area of the aerial portions of plants

Phytodegradation Breakdown of organic contaminants by plants through metabolic processes that occur within the plant; or through the effect of compounds, such as enzymes, produced by the plant

Rhizoremediation Degradation of pollutants by soil microorganisms, which are under the rhizosphere effect

Rhizosphere Zone in the soil under the direct influence of plant roots. This includes not only the surface of the roots (rhizoplane) but also any external region that is affected by root exudates

Rhizosphere Effect Phenomenon describing that in comparison with bulk soil, the biomass and activity of microorganisms in the rhizosphere are enhanced as a result of different mechanisms driven by plants (mainly root exudation)

Root Exudates Set of compounds (e.g., flavonoids, fatty acids, organic acids, aminoacids) produced by plants and secreted by roots into the soil or any other medium surrounding the roots. These molecules can be actively or passively released by plant roots. Root exudation patterns change under the influence of the plant (e.g., cultivar, plant species, developmental stage), environmental factors (e.g., soil type, pH, temperature, nutrient availability), and the presence of microorganisms.

References

Abatenh E, Gizaw B, Tsegaye Z, Wassie M (2017) The role of microorganisms in bioremediation: a review. Open J Environ Biol 2:38–46. https://doi.org/10.17352/ojeb.000007

Abbasian F, Lockington R, Megharaj M, Naidu R (2016) A review on the genetics of aliphatic and aromatic hydrocarbon degradation. Appl Biochem Biotechnol 178:224–250. https://doi.org/10.1007/s12010-015-1881-y

Abdel-Shafy HI, Mansour MSM (2016) A review on polycyclic aromatic hydrocarbons: source, environmental impact, effect on human health and remediation. Egypt J Pet 25:107–123. https://doi.org/10.1016/j.ejpe.2015.03.011

Agarry SE, Aremu MO, Aworanti OA (2014) Biostimulation and phytoremediation treatment strategies of gasoline-nickel co-contaminated soil. Soil Sediment Contam 23:227–244. https://doi.org/10.1080/15320383.2014.812612

Agnello AC, Bagard M, van Hullebusch ED, Esposito G, Huguenot D (2016) Comparative bioremediation of heavy metals and petroleum hydrocarbons co-contaminated soil by natural attenuation, phytoremediation, bioaugmentation and bioaugmentation-assisted phytoremediation. Sci Total Environ 563–564:693–703. https://doi.org/10.1016/j.scitotenv.2015.10.061

Ahmed F, Fakhruddin ANM (2018) A review on environmental contamination of petroleum hydrocarbons and its biodegradation. Int J Environ Sci Nat Resour 11:1–7. https://doi.org/10.19080/IJESNR.2018.11.555811

Al-Baldawi IA, Abdullah SRS, Anuar N, Suja F, Mushrifah I (2015) Phytodegradation of total petroleum hydrocarbon (TPH) in diesel-contaminated water using *Scirpus grossus*. Ecol Eng 74:463–473. https://doi.org/10.1016/j.ecoleng.2014.11.007

Alves WS, Manoel EA, Santos NS, Nunes RO, Domiciano GC, Soares MR (2017) Detection of polycyclic aromatic hydrocarbons (PAHs) in *Medicago sativa* L. by fluorescence microscopy. Micron 95:23–30. https://doi.org/10.1016/j.micron.2017.01.004

Aranda E, Godoy P, Reina R, Badia-Fabregat M, Rosell M, Marco-Urrea E, García-Romera I (2017) Isolation of *Ascomycota* fungi with capability to transform PAHs: insights into the biodegradation mechanisms of *Penicillium oxalicum*. Int Biodeterior Biodegrad 122:141–150. https://doi.org/10.1016/j.ibiod.2017.05.015

Asemoloye MD, Ahmad R, Jonathan SG (2017) Synergistic action of rhizospheric fungi with *Megathyrsus maximus* root speeds up hydrocarbon degradation kinetics in oil polluted soil. Chemosphere 187:1–10. https://doi.org/10.1016/j.chemosphere.2017.07.158

Aske N, Kallevik H, Sjöblom J (2001) Determination of saturate, aromatic, resin, and asphaltenic (SARA) components in crude oils by means of infrared and near-infrared spectroscopy. Energy Fuels 15:1304–1312. https://doi.org/10.1021/ef010088h

Aydin S, Karaçay HA, Shahi A, Gökçe S, Ince B, Ince O (2017) Aerobic and anaerobic fungal metabolism and Omics insights for increasing polycyclic aromatic hydrocarbons biodegradation. Fungal Biol Rev 31:61–72. https://doi.org/10.1016/j.fbr.2016.12.001

Banitz T, Johst K, Wick LY, Schamfuß S, Harms H, Frank K (2013) Highways versus pipelines: contributions of two fungal transport mechanisms to efficient bioremediation. Environ Microbiol Rep 5:211–218. https://doi.org/10.1111/1758-2229.12002

Baoune H, Ould El Hadj-Khelil A, Pucci G, Sineli P, Loucif L, Polti MA (2018) Petroleum degradation by endophytic *Streptomyces* spp. isolated from plants grown in contaminated soil of southern Algeria. Ecotoxicol Environ Saf 147:602–609. https://doi.org/10.1016/j.ecoenv.2017.09.013

Barone R, de Biasi MG, Piccialli V, de Napoli L, Oliviero G, Borbone N, Piccialli G (2016) Degradation of some representative polycyclic aromatic hydrocarbons by the water-soluble protein extracts from *Zea mays* L. cv PR32-B10. Chemosphere 160:258–265. https://doi.org/10.1016/j.chemosphere.2016.06.069

Bell TH, El-Din Hassan S, Lauron-Moreau A, Al-Otaibi F, Hijri M, Yergeau E, St-Arnaud M (2014a) Linkage between bacterial and fungal rhizosphere communities in hydrocarbon-contaminated soils is related to plant phylogeny. ISME J 8:331–343. https://doi.org/10.1038/ismej.2013.149

Bell TH, Joly S, Pitre FE, Yergeau E (2014b) Increasing phytoremediation efficiency and reliability using novel omics approaches. Trends Biotechnol 32:271–280. https://doi.org/10.1016/j.tibtech.2014.02.008

Berthold T, Centler F, Hübschmann T, Remer R, Thullner M, Harms H, Wick LY (2016) Mycelia as a focal point for horizontal gene transfer among soil bacteria. Sci Rep 6:1–8. https://doi.org/10.1038/srep36390

Boldt-Burisch K, Naeth MA, Schneider U, Schneider B, Hüttl RF (2018) Plant growth and arbuscular mycorrhizae development in oil sands processing by-products. Sci Total Environ 621:30–39. https://doi.org/10.1016/j.scitotenv.2017.11.188

Bourceret A, Leyval C, de Fouquet C, Cébron A (2015) Mapping the centimeter-scale spatial variability of PAHs and microbial populations in the Rhizosphere of two plants. PLoS One 10: e0142851. https://doi.org/10.1371/journal.pone.0142851
Bourceret A, Leyval C, Faure P, Lorgeoux C, Cébron A (2018) High PAH degradation and activity of degrading bacteria during alfalfa growth where a contrasted active community developed in comparison to unplanted soil. Environ Sci Pollut Res 25(29):29556–29571. https://doi.org/10.1007/s11356-018-2744-1
Cecotti M, Coppotelli BM, Mora VC, Viera M, Morelli IS (2018) Efficiency of surfactant-enhanced bioremediation of aged polycyclic aromatic hydrocarbon-contaminated soil: link with bioavailability and the dynamics of the bacterial community. Sci Total Environ 634:224–234. https://doi.org/10.1016/j.scitotenv.2018.03.303
Chaîneau CH, Morel JL, Oudot J (1997) Phytotoxicity and plant uptake of fuel oil hydrocarbons. J Environ Qual 26:1478–1483. https://doi.org/10.2134/jeq1997.00472425002600060005x
Chen X, Liu X, Zhang X, Cao L, Hu X (2017) Phytoremediation effect of *Scirpus triqueter* inoculated plant-growth-promoting bacteria (PGPB) on different fractions of pyrene and Ni in co-contaminated soils. J Hazard Mater 325:319–326. https://doi.org/10.1016/j.jhazmat.2016.12.009
Chen S, Wang J, Waigi MG, Gao Y (2018) Glomalin-related soil protein influences the accumulation of polycyclic aromatic hydrocarbons by plant roots. Sci Total Environ 644:465–473. https://doi.org/10.1016/j.scitotenv.2018.06.370
Collins CD, Martin I, Fryer M (2006) Evaluation of models for predicting plant uptake of chemicals from soil. Science Report – SC050021/SR 1, Environment Agency
Coppotelli BM, Ibarrolaza A, Dias RL, Del Panno MT, Berthe-Corti L, Morelli IS (2010) Study of the degradation activity and the strategies to promote the bioavailability of phenanthrene by *Sphingomonas paucimobilis* strain 20006FA. Microb Ecol 59:266–276. https://doi.org/10.1007/s00248-009-9563-3
Correa-García S, Pande P, Séguin A, St-Arnaud M, Yergeau E (2018) Rhizoremediation of petroleum hydrocarbons: a model system for plant microbiome manipulation. Microb Biotechnol 11:819–832. https://doi.org/10.1111/1751-7915.13303
Crowley DE, Luepromechai E, Singer AC (2001) Metabolism of Xenobiotics in the Rhizosphere. In: Hall JC, Hoagland RE, Zablotowicz RM (eds) Pesticide biotransformation in plants and microorganisms: similarities and divergences, ACS symposium series, vol 777. American Chemical Society, Washington, DC, pp 333–352. https://doi.org/10.1021/bk-2001-0777.ch018
Dalton H, Stirling DI (1982) Co-metabolism. Philos Trans R Soc Lond Ser B Biol Sci 297:481–496. https://doi.org/10.1098/rstb.1982.0056
Das N, Chandran P (2011) Microbial degradation of petroleum hydrocarbon contaminants : an overview. Biotechnol Res Int 2011:1–13. https://doi.org/10.4061/2011/941810
Daudzai Z, Thiravetyan P, Treesubsuntorn C (2018) Inoculated *Clitoria ternatea* with *Bacillus cereus* ERBP for enhancing gaseous ethylbenzene phytoremediation: plant metabolites and expression of ethylbenzene degradation genes. Ecotoxicol Environ Saf 164:50–60. https://doi.org/10.1016/j.ecoenv.2018.07.121
Deng Z, Cao L (2017) Fungal endophytes and their interactions with plants in phytoremediation: a review. Chemosphere 168:1100–1106. https://doi.org/10.1016/j.chemosphere.2016.10.097
Duarte M, Nielsen A, Camarinha-Silva A, Vilchez-Vargas R, Bruls T, Wos-Oxley ML, Jauregui R, Pieper DH (2017) Functional soil metagenomics: elucidation of polycyclic aromatic hydrocarbon degradation potential following 12 years of in situ bioremediation. Environ Microbiol 19:2992–3011. https://doi.org/10.1111/1462-2920.13756
Dubrovskaya EV, Pozdnyakova NN, Muratova AY, Turkovskaya OV (2016) Changes in phytotoxicity of polycyclic aromatic hydrocarbons in the course of microbial degradation. Russ J Plant Physiol 63:172–179. https://doi.org/10.1134/S1021443716010052
Efroymson RA, Sample BE, Peterson MJ (2004) Ecotoxicity test data for total petroleum hydrocarbons in soil: plants and soil-dwelling invertebrates. Hum Ecol Risk Assess 10:207–231. https://doi.org/10.1080/10807030490438175

El Amrani A, Dumas AS, Wick LY, Yergeau E, Berthomé R (2015) 'Omics' insights into PAH degradation toward improved green remediation biotechnologies. Environ Sci Technol 49:11281–11291. https://doi.org/10.1021/acs.est.5b01740

Fatima K, Imran A, Naveed M, Afzal M (2017) Plant-bacteria synergism: an innovative approach for the remediation of crude oil-contaminated soils. Soil Environ 36:93–113. https://doi.org/10.25252/SE/17/51346

Festa S, Coppotelli BM, Madueño L, Loviso CL, Macchi M, Neme Tauil RM, Valacco MP, Morelli IS (2017) Assigning ecological roles to the populations belonging to a phenanthrene-degrading bacterial consortium using omic approaches. PLoS One 12:1–21. https://doi.org/10.1371/journal.pone.0184505

Franchi E, Agazzi G, Rolli E, Borin S, Marasco R, Chiaberge S, Conte A, Filtri P, Pedron F, Rosellini I, Barbafieri M, Petruzzelli G (2016) Exploiting hydrocarbon-degrading indigenous Bacteria for bioremediation and phytoremediation of a multicontaminated soil. Chem Eng Technol 39:1676–1684. https://doi.org/10.1002/ceat.201500573

Fu W, Xu M, Sun K, Hu L, Cao W, Dai C, Jia Y (2018) Biodegradation of phenanthrene by endophytic fungus *Phomopsis liquidambari* in vitro and in vivo. Chemosphere 203:160–169. https://doi.org/10.1016/j.chemosphere.2018.03.164

Gao Y, Cheng Z, Ling W, Huang J (2010) Arbuscular mycorrhizal fungal hyphae contribute to the uptake of polycyclic aromatic hydrocarbons by plant roots. Bioresour Technol 101:6895–6901. https://doi.org/10.1016/j.biortech.2010.03.122

Gao Y, Zong J, Que H, Zhou Z, Xiao M, Chen S (2017) Inoculation with arbuscular mycorrhizal fungi increases glomalin-related soil protein content and PAH removal in soils planted with *Medicago sativa* L. Soil Biol Biochem 115:148–151. https://doi.org/10.1016/j.soilbio.2017.08.023

Garcés-Ruiz M, Senés-Guerrero C, Declerck S, Cranenbrouck S (2017) Arbuscular mycorrhizal fungal community composition in *Carludovica palmata*, *Costus scaber* and *Euterpe precatoria* from weathered oil ponds in the Ecuadorian Amazon. Front Microbiol 8:1–13. https://doi.org/10.3389/fmicb.2017.02134

García-Sánchez M, Košnář Z, Mercl F, Aranda E, Tlustoš P (2018) A comparative study to evaluate natural attenuation, mycoaugmentation, phytoremediation, and microbial-assisted phytoremediation strategies for the bioremediation of an aged PAH-polluted soil. Ecotoxicol Environ Saf 147:165–174. https://doi.org/10.1016/j.ecoenv.2017.08.012

Gerhardt KE, Gerwing PD, Huang XD, Greenberg BM (2015) Microbe-assisted phytoremediation of petroleum impacted soil: a scientifically proven green technology. In: Fingas M (ed) Handbook of oil spill science technology. Wiley, Hoboken, pp 407–427. https://doi.org/10.1002/9781118989982.ch16

Gieg LM, Fowler SJ, Berdugo-Clavijo C (2014) Syntrophic biodegradation of hydrocarbon contaminants. Curr Opin Biotechnol 27:21–29. https://doi.org/10.1016/j.copbio.2013.09.002

Gonzalez E, Pitre FE, Pagé AP, Marleau J, Nissim WG, St-Arnaud M, Labrecque M, Joly S, Yergeau E, Brereton NJB (2018) Trees, fungi and bacteria: tripartite metatranscriptomics of a root microbiome responding to soil contamination. Microbiome 6:53. https://doi.org/10.1186/s40168-018-0432-5

Harms H, Wick LY (2006) Dispersing pollutant-degrading bacteria in contaminated soil without touching it. Eng Life Sci 6:252–260. https://doi.org/10.1002/elsc.200620122

Hearn EM, Patel DR, van den Berg B (2008) Outer-membrane transport of aromatic hydrocarbons as a first step in biodegradation. Proc Natl Acad Sci 105:8601–8606. https://doi.org/10.1073/pnas.0801264105

Hernández-Ortega HA, Alarcón A, Ferrera-Cerrato R, Zavaleta-Mancera HA, López-Delgado HA, Mendoza-López MR (2012) Arbuscular mycorrhizal fungi on growth, nutrient status, and total antioxidant activity of *Melilotus albus* during phytoremediation of a diesel-contaminated substrate. J Environ Manag 95:S319–S324. https://doi.org/10.1016/j.jenvman.2011.02.015

Hua F, Wang H (2012) Uptake modes of octadecane by *Pseudomonas* sp. DG17 and synthesis of biosurfactant. J Appl Microbiol 112:25–37. https://doi.org/10.1111/j.1365-2672.2011.05178.x

Hua F, Wang HQ (2014) Uptake and trans-membrane transport of petroleum hydrocarbons by microorganisms. Biotechnol Biotechnol Equip 28:165–175. https://doi.org/10.1080/13102818.2014.906136

Jiang L, Song M, Luo C, Zhang D, Zhang G (2015) Novel phenanthrene-degrading bacteria identified by DNA-stable isotope probing. PLoS One 10:1–14. https://doi.org/10.1371/journal.pone.0130846

Kadri T, Rouissi T, Kaur Brar S, Cledon M, Sarma S, Verma M (2017) Biodegradation of polycyclic aromatic hydrocarbons (PAHs) by fungal enzymes: a review. J Environ Sci (China) 51:52–74. https://doi.org/10.1016/j.jes.2016.08.023

Kirk JL, Klirnomos JN, Lee H, Trevors JT (2002) Phytotoxicity assay to assess plant species for phytoremediation of petroleum-contaminated soil. Biorem J 6:57–63. https://doi.org/10.1080/10889860290777477

Košnář Z, Částková T, Wiesnerová L, Praus L, Jablonský I, Koudela M, Tlustoš P (2018) Comparing the removal of polycyclic aromatic hydrocarbons in soil after different bioremediation approaches in relation to the extracellular enzyme activities. J Environ Sci 76:249. https://doi.org/10.1016/j.jes.2018.05.007

Krahforst K, Healey TE (2017) Unraveling the complexities of upland spilled fuels: selected case studies. In: Stout SA, Wang Z (eds) Oil spill environmental forensics case studies. Elsevier, Saint Louis, pp 201–238. https://doi.org/10.1016/B978-0-12-804434-6.00010-0

Lenoir I, Anissa L-HS, Laruelle F, Dalpé Y, Fontaine J (2016) Arbuscular mycorrhizal wheat inoculation promotes alkane and polycyclic aromatic hydrocarbon biodegradation: microcosm experiment on aged-contaminated soil. Environ Pollut 213:549–560. https://doi.org/10.1016/j.envpol.2016.02.056

Ling W, Sun R, Gao X, Xu R, Li H (2015) Low-molecular-weight organic acids enhance desorption of polycyclic aromatic hydrocarbons from soil. Eur J Soil Sci 66:339–347. https://doi.org/10.1111/ejss.12227

Marco-Urrea E, García-Romera I, Aranda E (2015) Potential of non-ligninolytic fungi in bioremediation of chlorinated and polycyclic aromatic hydrocarbons. New Biotechnol 32:620–628. https://doi.org/10.1016/j.nbt.2015.01.005

Martin BC, George SJ, Price CA, Ryan MH, Tibbett M (2014) The role of root exuded low molecular weight organic anions in facilitating petroleum hydrocarbon degradation: current knowledge and future directions. Sci Total Environ 472:642–653. https://doi.org/10.1016/j.scitotenv.2013.11.050

Morelli IS, Saparrat MCN, Del Panno MT, Coppotelli BM, Arrambari A (2013) Bioremediation of PAH-contaminated soil by fungi. In: Goltapeh EM, Danesh YR, Varma A (eds) Fungi as bioremediators. Springer, Heidelberg, pp 159–179. https://doi.org/10.1007/978-3-642-33811-3

Mullins OC (2008) Review of the molecular structure and aggregation of asphaltenes and petroleomics. SPE J 13:48–57. https://doi.org/10.2118/95801-PA

Muratova AY, Dmitrieva T, Panchenko LV, Turkovskaya OV (2008) Phytoremediation of oil-sludge – contaminated soil. Int J Phytoremediation 10:486–502. https://doi.org/10.1080/15226510802114920

Musilova L, Ridl J, Polivkova M, Macek T, Uhlik O (2016) Effects of secondary plant metabolites on microbial populations: changes in community structure and metabolic activity in contaminated environments. Int J Mol Sci 17:1–31. https://doi.org/10.3390/ijms17081205

Newman LA, Reynolds CM (2004) Phytodegradation of organic compounds. Curr Opin Biotechnol 15:225–230. https://doi.org/10.1016/j.copbio.2004.04.006

Olanrewaju OS, Glick BR, Babalola OO (2017) Mechanisms of action of plant growth promoting bacteria. World J Microbiol Biotechnol 33:1–16. https://doi.org/10.1007/s11274-017-2364-9

Pacwa-Płociniczak M, Płociniczak T, Iwan J, Zarska M, Chorazewski M, Dzida M, Piotrowska-Seget Z (2016) Isolation of hydrocarbon-degrading and biosurfactant-producing bacteria and assessment their plant growth-promoting traits. J Environ Manag 168:175–184. https://doi.org/10.1016/j.jenvman.2015.11.058

Pagé AP, Yergeau É, Greer CW (2015) *Salix purpurea* stimulates the expression of specific bacterial xenobiotic degradation genes in a soil contaminated with hydrocarbons. PLoS One 10:1–16. https://doi.org/10.1371/journal.pone.0132062

Parseh I, Teiri H, Hajizadeh Y, Ebrahimpour K (2018) Phytoremediation of benzene vapors from indoor air by *Schefflera arboricola* and *Spathiphyllum wallisii* plants. Atmos Pollut Res 9:1083. https://doi.org/10.1016/j.apr.2018.04.005

Pawlik M, Cania B, Thijs S, Vangronsveld J, Piotrowska-Seget Z (2017) Hydrocarbon degradation potential and plant growth-promoting activity of culturable endophytic bacteria of *Lotus corniculatus* and *Oenothera biennis* from a long-term polluted site. Environ Sci Pollut Res 24:19640–19652. https://doi.org/10.1007/s11356-017-9496-1

Pozdnyakova N, Dubrovskaya E, Chernyshova M, Makarov O, Golubev S, Balandina S, Turkovskaya O (2018) The degradation of three-ringed polycyclic aromatic hydrocarbons by wood-inhabiting fungus *Pleurotus ostreatus* and soil-inhabiting fungus *Agaricus bisporus*. Fungal Biol 122:363–372. https://doi.org/10.1016/j.funbio.2018.02.007

Prasad R (2017) Mycoremediation and environmental sustainability, vol 1. Springer, Cham. ISBN 978-3-319-68957-9. https://link.springer.com/book/10.1007/978-3-319-68957-9

Prasad R (2018) Mycoremediation and environmental sustainability, vol 2. Springer, Cham. ISBN 978-3-319-77386-5. https://www.springer.com/us/book/9783319773858

Prasad R, Kumar M, Varma A (2015) Role of PGPR in soil fertility and plant health. In: Egamberdieva D, Shrivastava S, Varma A (eds) Plant Growth-Promoting Rhizobacteria (PGPR) and medicinal plants. Springer, Cham, pp 247–260

Quiza L, St-Arnaud M, Yergeau E (2015) Harnessing phytomicrobiome signaling for rhizosphere microbiome engineering. Front Plant Sci 6:1–11. https://doi.org/10.3389/fpls.2015.00507

Rabus R, Boll M, Heider J, Meckenstock RU, Buckel W, Einsle O, Ermler U, Golding BT, Gunsalus RP, Kroneck PMH, Krüger M, Lueders T, Martins BM, Musat F, Richnow HH, Schink B, Seifert J, Szaleniec M, Treude T, Ullmann GM, Vogt C, Von Bergen M, Wilkes H (2016) Anaerobic microbial degradation of hydrocarbons: from enzymatic reactions to the environment. J Mol Microbiol Biotechnol 26:5–28. https://doi.org/10.1159/000443997

Rajtor M, Piotrowska-Seget Z (2016) Prospects for arbuscular mycorrhizal fungi (AMF) to assist in phytoremediation of soil hydrocarbon contaminants. Chemosphere 162:105–116. https://doi.org/10.1016/j.chemosphere.2016.07.071

Rentz JA, Alvarez PJJ, Schnoor JL (2005) Benzo[a]pyrene co-metabolism in the presence of plant root extracts and exudates: implications for phytoremediation. Environ Pollut 136:477–484. https://doi.org/10.1016/j.envpol.2004.12.034

Reyes-César A, Absalón ÁE, Fernández FJ, González JM, Cortés-Espinosa DV (2014) Biodegradation of a mixture of PAHs by non-ligninolytic fungal strains isolated from crude oil-contaminated soil. World J Microbiol Biotechnol 30:999–1009. https://doi.org/10.1007/s11274-013-1518-7

Rodrigues EM, Freitas F d S, Siqueira T d P (2018) Detection of horizontal transfer of housekeeping and hydrocarbons catabolism genes in bacterial genus with potential to application in bioremediation process. OALib 05:1–10. https://doi.org/10.4236/oalib.1104454

Rohrbacher F, St-Arnaud M (2016) Root exudation: the ecological driver of hydrocarbon Rhizoremediation. Agronomy 6:1–27. https://doi.org/10.3390/agronomy6010019

Sandermann H (1992) Plant metabolism of xenobiotics. Trends Biochem Sci 17:82–84. https://doi.org/10.1016/0968-0004(92)90507-6

Santisi S, Cappello S, Catalfamo M, Mancini G, Hassanshahian M, Genovese L, Giuliano L, Yakimov MM (2015) Biodegradation of crude oil by individual bacterial strains and a mixed bacterial consortium. Brazilian. J Microbiol 46:377–387. https://doi.org/10.1002/clen.200700042

Santoyo G, Moreno-Hagelsieb G, del Carmen Orozco-Mosqueda M, Glick BR (2016) Plant growth-promoting bacterial endophytes. Microbiol Res 183:92–99. https://doi.org/10.1016/j.micres.2015.11.008

Schindler FV, Mercer EJ, Rice JA (2007) Chemical characteristics of glomalin-related soil protein (GRSP) extracted from soils of varying organic matter content. Soil Biol Biochem 39:320–329. https://doi.org/10.1016/j.soilbio.2006.08.017

Siddiqui S, Adams WA, Schollion J (2001) The phytotoxicity and degradation of diesel hydrocarbons in soil. J Plant Nutr Soil Sci 164:631–635. https://doi.org/10.1002/1522-2624(200112)164:6<631::AID-JPLN631>3.0.CO;2-E

Singh M, Pant G, Hossain K, Bhatia AK (2017) Green remediation. Tool for safe and sustainable environment: a review. Appl Water Sci 7:2629–2635. https://doi.org/10.1007/s13201-016-0461-9

Singha LP, Sinha N, Pandey P (2018) Rhizoremediation prospects of polyaromatic hydrocarbon degrading rhizobacteria, that facilitate glutathione and glutathione-S-transferase mediated stress response, and enhance growth of rice plants in pyrene contaminated soil. Ecotoxicol Environ Saf 164:579–588. https://doi.org/10.1016/j.ecoenv.2018.08.069

Somtrakoon K, Chouychai W (2013) Phytotoxicity of single and combined polycyclic aromatic hydrocarbons toward economic crops. Russ J Plant Physiol 60:139–148. https://doi.org/10.1134/S1021443712060155

Sun Y, Zhou Q (2016) Uptake and translocation of benzo[a]pyrene (B[a]P) in two ornamental plants and dissipation in soil. Ecotoxicol Environ Saf 124:74–81. https://doi.org/10.1016/j.ecoenv.2015.09.037

Syranidou E, Christofilopoulos S, Gkavrou G, Thijs S, Weyens N, Vangronsveld J, Kalogerakis N (2016) Exploitation of endophytic bacteria to enhance the phytoremediation potential of the wetland helophyte *Juncus acutus*. Front Microbiol 7:1–15. https://doi.org/10.3389/fmicb.2016.01016

Taghavi S, Barac T, Greenberg B, Vangronsveld J, Van Der Lelie D, Borremans B (2005) Horizontal gene transfer to endogenous endophytic bacteria from poplar improves phytoremediation of toluene. Appl Environ Microbiol 71:8500–8505. https://doi.org/10.1128/AEM.71.12.8500

Tardif S, Yergeau E, Tremblay J, Legendre P, Whyte LG, Greer CW (2016) The willow microbiome is influenced by soil petroleum-hydrocarbon concentration with plant compartment-specific effects. Front Microbiol 7:1–14. https://doi.org/10.3389/fmicb.2016.01363

Thijs S, Sillen W, Rineau F, Weyens N, Vangronsveld J (2016) Towards an enhanced understanding of plant-microbiome interactions to improve phytoremediation: engineering the metaorganism. Front Microbiol 7:1–15. https://doi.org/10.3389/fmicb.2016.00341

Tian Z, Vila J, Yu M, Bodnar W, Aitken MD (2018) Tracing the biotransformation of polycyclic aromatic hydrocarbons in contaminated soil using stable isotope-assisted metabolomics. Environ Sci Technol Lett 5:103–109. https://doi.org/10.1021/acs.estlett.7b00554

Tormoehlen LM, Tekulve KJ, Nañagas KA (2014) Hydrocarbon toxicity: a review. Clin Toxicol 52:479–489. https://doi.org/10.3109/15563650.2014.923904

USEPA (2008) Green remediation: incorporating sustainable environmental practices into remediation of contaminated sites. USEPA, Washington, DC

Uteau D, Hafner S, Pagenkemper SK, Peth S, Wiesenberg GLB, Kuzyakov Y, Horn R (2015) Oxygen and redox potential gradients in the rhizosphere of alfalfa grown on a loamy soil. J Plant Nutr Soil Sci 178:278–287. https://doi.org/10.1002/jpln.201300624

van Dam NM, Bouwmeester HJ (2016) Metabolomics in the rhizosphere: tapping into belowground chemical communication. Trends Plant Sci 21:256–265. https://doi.org/10.1016/j.tplants.2016.01.008

Varjani SJ (2017) Microbial degradation of petroleum hydrocarbons. Bioresour Technol 223:277–286. https://doi.org/10.1016/j.biortech.2016.10.037

Vigneron A, Alsop EB, Cruaud P, Philibert G, King B, Baksmaty L, Lavallée D, Lomans BP, Kyrpides NC, Head IM, Tsesmetzis N (2017) Comparative metagenomics of hydrocarbon and methane seeps of the Gulf of Mexico. Sci Rep 7:1–12. https://doi.org/10.1038/s41598-017-16375-5

Wanapaisan P, Laothamteep N, Vejarano F, Chakraborty J, Shintani M, Muangchinda C, Morita T, Suzuki-Minakuchi C, Inoue K, Nojiri H, Pinyakong O (2018) Synergistic degradation of pyrene

by five culturable bacteria in a mangrove sediment-derived bacterial consortium. J Hazard Mater 342:561–570. https://doi.org/10.1016/j.jhazmat.2017.08.062
Wang H, Wang B, Dong W, Hu X (2016) Co-acclimation of bacterial communities under stresses of hydrocarbons with different structures. Sci Rep 6:34588. https://doi.org/10.1038/srep34588
Warembourg FR (1997) The 'rhizosphere effect': a plant strategy for plants to exploit and colonize nutrient-limited habitats. Bocconea 7:187–194
Yergeau E, Sanschagrin S, Maynard C, St-Arnaud M, Greer CW (2014) Microbial expression profiles in the rhizosphere of willows depend on soil contamination. ISME J 8:344–358. https://doi.org/10.1038/ismej.2013.163
Yergeau E, Tremblay J, Joly S, Labrecque M, Maynard C, Pitre FE, St-Arnaud M, Greer CW (2018) Soil contamination alters the willow root and rhizosphere metatranscriptome and the root-rhizosphere interactome. ISME J 12:869–884. https://doi.org/10.1038/s41396-017-0018-4
Zafra G, Absalón ÁE, Anducho-Reyes MÁ, Fernandez FJ, Cortés-Espinosa DV (2017) Construction of PAH-degrading mixed microbial consortia by induced selection in soil. Chemosphere 172:120–126. https://doi.org/10.1016/j.chemosphere.2016.12.038

Chapter 11
Rhizoremediation: A Unique Plant Microbiome Association of Biodegradation

Arvind Kumar, Sruchi Devi, Himanshu Agrawal, Simranjeet Singh, and Joginder Singh

Abstract Microbes can be incorporated into a particular area where it can interact with selected pollutants and convert them into nontoxic or relatively less toxic compounds by excellent colonization in plant roots, thereby helping in the degradation of pollutants. This process is called 'rhizoremediation' which serves to emphasize the role of rhizosphere having competent microbes. These days majority of the process involving the degradation of environmental pollutants occurs through rhizospheric microbes. Root exudates can be taken as the best food source that is available in the soil for these microbes. The plants' uptake of heavy metals from soils in high concentrations negatively influences the interaction of microbes with plants, its growth and consequently the crop's production and yield. Heavy metals behave as genotoxic substances, and they disintegrate different cell organelles, rupture the cell membranes and disturb the physiological process like carbohydrate metabolism, protein synthesis and respiration photosynthesis. Some of the species of the pseudomonad family are root colonizer, having high efficiency for remediating pollutants via phytoremediators. In the previous two decades, various research articles on rhizodegradation of different toxicants utilizing diverse plants or potentially microbial inoculants have been documented.

11.1 Introduction

By the advancement in the industrial sector and globalization, there has been a lot of increase in pollutants from the past few decades, and they have impacted the environment massively (Singh et al. 2017a, b). To maintain the equilibrium of environment, it is necessary to utilize the sources, contaminants and their pathways in big cities. Due to a massive increase in the population and anthropogenic cultures, cities are indulging themselves as the producer of contaminants (Kaur et al. 2018; Kumar et al. 2017). The contaminants showing bad effects on the environment are

A. Kumar (✉) · S. Devi · H. Agrawal
Department of Biochemistry, Lovely Professional University, Phagwara, Punjab, India

S. Singh · J. Singh
Department of Biotechnology, Lovely Professional University, Phagwara, Punjab, India

A. Varma et al. (eds.), *Plant Microbe Symbiosis*,
https://doi.org/10.1007/978-3-030-36248-5_11

polycyclic aromatic hydrocarbon (PAH), petroleum hydrocarbons (PHCs), heavy metals, pesticides and salts (Kumar et al. 2016; Singh et al. 2016; Mishra et al. 2016; CCME 2001). The polycyclic aromatic hydrocarbon family or complex is said to contain more than 100 organic compounds (Kumar et al. 2014a, 2015a, b). These hydrocarbons are generated in the environment by inadequate combustion of forest fires, heating in homes, traffic pollution and release of organic waste (Johnsen et al. 2005). These compounds are categorized as hydrophobic in nature (Cerniglia 1993). The increased fused benzene ring brings an increase in hydrophobicity and solubility and decrease in the volatility of PAHs (Wilson and Jones 1993). Inadequate combustion of organic matter causes PAH increase (Guerin and Jones 1988). PAHs are produced naturally during burning of bushes and vegetation of forest and in the thermal geologic production (Blumer 1976). It has been found that some of the PAHs and their alkyl homologous are derived from biogenic precursors (Wakeham et al. 1980). The outstanding sources instigating PAH production are petroleum product spillage, anthropogenic sources, fuel combustion, waste incineration, pyrolytic processes and domestic heaters (Kumar et al. 2013, 2014b; Freeman and Cattell 1990). Phenanthrene (1.4 mg/kg), dibenzoanthracene (0.8 mg/kg), pyrene (4.0 mg/kg) and chrysene (0.9 mg/kg) were found at 19–35 cm in depth (Oviasogie et al. 2006). PAH concentration shows variation in accordance with the level of industrialization, the proximity of production source contamination and PAH release in the environment. Kanaly and Harayama (2000) documented the PAH content in soil and sediment which ranges from 1 μg/Kg. A number of environment samples give rise to PAHs, for example, foodstuff (Lijinsky 1991), air (Freeman and Cattell 1990), sediments (Shiaris and Jambard 1986), soil (Jones et al. 1989), water (Cerniglia and Heitkamp 1989), oils and tars (Nishioka et al. 1986). The PAHs contamination in industrial spillage and leakage of fuel/oil storage product undertreated water product disposals arises from a very common medium (Wilson and Jones 1993).

Creosote and anthracene oil account for their 85% utilization as pesticides in treating wood for PAHs (Walter et al. 1991). PAH transportation is via air through volatilization (Ramadah et al. 1982). Similarly, the earth acts as the focal point for the variety of PAH. The major discharge occurs in the winter season by burning. PAHs undergoes contamination by fume condensation, engine vehicles, including flash outflow and diesel autos, trucks and transport, tire particles and greasing up oils and oil expanding barometrical PAHs (Juhasz and Naidu 2000). There are prospect chances where the lighter PAHs can be separated and then excessive burning of petroleum derivatives like coal, diesel and oil are substantial products of lighter PAHs since fuel is still considered the only significant PAH (Guoa et al. 2003). The PAHs nature holds physical and concoction attributes of PAH. PAHs are exploited by photooxidation and compound oxidation (Shiaris and Jambard 1986). Therefore, the organic changes lead to PAH misfortune (Mueller et al. 1990). The naphthalene, phenanthrene, anthracene and fluorene are the rings that have broadly contemplated in microbial digestion of PAHs. In 2003, Comprehensive Environment Response, Compensation and Liability Act ranked lead, cadmium and mercury on the basis of their hazardous nature in the second, seventh and third position. The soil is exceptionally filled with substantial metals (Peters 1999).

Such method is followed to bring advancement in the remediation process of natural soil complementing the landfill transport, unearthing, adjustment and cremation and takes charge of particle trade or coagulation filtration that is costly and disturbs the life of locals (Haritash and Kaushik 2009). All the bioremediation innovations are subjected to the utilization of plants and eco-friendly microbes in the termination of harmful contaminants and converting them into less poisonous and less hazardous substances infertile natural soil (Shukla et al. 2010). With the assistance of in situ methods, the dirt and groundwater is dealt with set up without removal, while it is exhumed before treatment with ex-situ applications (McGuinness and Dowling 2009). In biosorption, the microbes are capable of heavy metal binding affinity for phytochelatins (PCs) and metallothioneins (MTs), and both animals and plants utilize the immobilization as major mechanisms for inhibition of the concentrations of internal reactive metal species (Francova et al. 2003).

In phytoremediation, the plants are used to degrade the toxic contaminants by phytoextraction, phytotransformation, phytovolatilization, rhizofiltration and phytostabilization (Prasad 2011). In phytoextraction, uptake and concentration of pollutants into harvestable biomass are sequestrated or incinerated. In phytotransformation, the enzymatic modification is inactivated which degrades (phytodegradation) or immobilizes (phytostabilization) the pollutants. In phytovolatilization, the soil pollutants are removed and are released through leaves via a process called evapotranspiration, and in rhizofiltration, the mass of roots removes pollutants and filters the water. Bioremediation through plants is the most successful method as reported; for rhizoremediation, the use of plant-associated bacteria provides a good potential (Mejare and Bulow 2001). The association between plants and microbes plays a significant role in the remediation of organic pollutant or contaminants which have been proved by studies through rhizosphere levels in the environment (Gerhardt et al. 2009; Ho et al. 2007), the phyllosphere and the internal area of plants (Kidd et al. 2008; Sandhu et al. 2007). Rhizoremediation is the most potent approach for PAH remediation in soil (Mohan et al. 2006). Soil microflora has a significant role in rhizoremediation of xenobiotics (Barac et al. 2009). In soil the interaction between microbes act as degrader, plants and PAHs is controlled or regulated during the process of rhizosphere (Ma et al. 2011). PAH rhizoremediation depended on coordination between specific plants and microbial communities that are present near the root area (Barac et al. 2009). Rhizosphere mechanism facilitates the degradation of PAHs. Plants exude the harmful organic pollutant through their roots and increase the activity of microorganism hydrocarbon degraders in the surrounding area of roots (de Carcer et al. 2007). The abilities of microorganisms like bacteria for biodegradation, expression and their maintenance in the rhizosphere are extremely important for the effective removal of contaminants (Phillips et al. 2012). Thus, bioremediation has a great contribution to rhizoremediation and phytoremediation. Significantly to the outcome of hazardous desecrate and can be used for the removal of contaminants from the environment (Mohan et al. 2006; Ma et al. 2011). Rhizoremediation of PAHs is the most potent approach for PAH remediation in soil proposed by researchers (Ma et al. 2009). Rhizospheric bacteria are designated as plant-associated bacteria having a fair contribution in biodegradation of toxic organic compounds from contaminated soil and could also have the potential to improve phytoremediation (David and Sharon 2009).

The discharge of root exudates a variety of organic acids, organic compounds and amino acids etc., responsible for the creation of rhizophores, phytohormone, HCN and enzyme phosphatases combination of different activities of roots of the plant and microbial communities of rhizosphere by plant growth promote rhizobacteria (PGPR) (Prasad et al. 2015). These associations are also effective for the process of ecorestoration on polluted areas or sites (Lee et al. 2012). *B. aryabhattai* strains may be utilized as an environment-friendly means of revegetating barren lands (Lee et al. 2012). The rhizobacteria have many valuable effects on the growth of plants and on the bacteria which promote plant growth such as rhizobacteria, utilized for several degradations, even though their mechanism to promote the plant growth has not been completely established (Babalola 2010). Straight phytohormonal accomplishments like availability boost-up of nutrients of plant and the improvement of beneficial plant microorganisms are important mechanisms (Dodd et al. 2010). When an appropriate rhizospheric isolated strain interacts which a particular plant, it develops bioremediation process along with the native population (Bisht et al. 2010). In addition, bacteria degrading the pollutants use the rising root system through root colonization, and hence the bacteria are now spread in the soil (Harms and Wick 2006). Plant roots have performed some positive roles, together with the ability to synthesize, roots may control the community of soil microbes in their instant locality, and as a result exude a variety of nutrients, interact with herbivores and help to support beneficial symbiosis, modify the chemical and physical properties of the soil and act as a growth inhibitor for challenging plant species (Walker et al. 2003). Plant growth may be facilitated by PGP bacteria either directly or indirectly (Glick 1995). The rhizoremediation is a natural process, but it can be customized by the deliberate development of the well-equipped rhizospheric microorganisms, while it can adapt by using an appropriate association of plant and microbe. Degradation of pollutant may be performed by integration of plant and PGPR.

Similarly, a naphthalene-degrading microbe interacts to grass species which confines the seeds of different grasses from the toxic activity of naphthalene and breaks the newly growing roots (Kuiper et al. 2004). According to previous studies, it has been reported that the symbiotic association of plant and microbes helps to degrade the hazardous and xenobiotic compounds like PCBs, PAHs and TCE (Kamaludeen and Ramasamy 2008). Degrading bacteria injected mechanically to contaminated sites to show their action.

11.2 Bioremediation Approaches

The poisonous quality, mutagenicity and cancer-causing nature properties are a matter of concern for the polycyclic sweet-smelling hydrocarbons (PAHs) (Guijian et al. 2008). Consequently, there is an investigation for the microorganisms as a method for bioremediation shown in polluted situations Prasad and Aranda (2018). PAHs and other natural contaminants are related to both abiotic and biotic procedures in the earth; these procedures are microbial change, photooxidation, volatilization, bioaccumulation and substance oxidation. As indicated by an investigation, the microbial movement has been accounted for as the most compelling and noteworthy in the evacuation of PAHs (Cerniglia 1993; Cerniglia and Heitkamp 1989). A few microorganisms may likewise debase PAHs in the dirt (Abd et al. 2009). The term bioremediation alludes to the debasement of ecological poisons with the assistance of living beings (Barea et al. 2005). In bioremediation, to quicken the synergist capacities through large common lessons like biostimulation or bioaugmentation and expansion of regular or designed microorganisms. As indicated by a report of Environmental Protection Agency in the United States (USEPA 1999), characteristic weakening procedures through ruinous procedures, for example, biodegradation and concoction changes may decrease contaminant mass, through straightforward weakening or scattering diminish contaminant focuses. The contaminants are biodegraded by micro-organisms or a few microscopic organisms that are bound to soil particles in nature.

11.3 Bioaugmentation Approaches

In Bioaugmentation, the microorganisms are brought into tainted condition, which has particular catabolic capacities. These microorganisms supplement the indigenous populace and accelerate the debasement of toxins (Kuiper et al. 2004; Louisa 2010). A few investigations have exhibited that bioaugmentation which did not improve biodegradation related to normal lessening (Cosgrove et al. 2010). Bioaugmentation is demonstrated as a fruitful method for remediation of PAHs in silt with poor or lacking inborn corruption potential (Juhasz and Naidu 2000). Survival and action of the life forms are the fundamental issues in applying bioaugmentation in the earth (Louisa 2010). An assortment of variables counting pH and redox, the proximity of harmful pollutant or contaminants, focus and bioavailability of contaminants or nonattendance of key co-substrates are restrained Bioaugmentation (Kuiper et al. 2004). In any case, the choice of a suitable bacterial strain acts as a crucial part for the achievement of bioaugmentation process. Among choice of the strain for enlargement purpose, the type of microbial groups that near to specifically source natural surroundings ought to be viewed as most promising micro-organisms (Tyagi et al. 2011). If there should be an occurrence of remediation of man-made contaminants, Bioaugmentation techniques effectively pertinent when particular microorganisms with the suitable catabolic pathways may not be available in the tainted condition (Louisa 2010).

11.4 Phytoremediation Approaches

In a research centre trial, a few agents have made similar examinations among measurement and the reactions of soil microbial groups among the phytoremediation of PAHs (Walter et al. 1991). The improvement of transgenic poplars (*Populus* sp.) by communicating a cytochrome P450 has been found by a few analysts and a group of catalysts ordinarily associated with the digestion of harmful mixes. A few labourers recommended that transgenic plants may have the capacity to add to the more extensive and more secure use of phytoremediation (Frerot et al. 2006). The designed plants indicated upgraded execution about the digestion of trichloroethylene and the expulsion of a scope of other poisonous unstable natural contaminations, including carbon tetrachloride, vinyl chloride, benzene and chloroform. Beforehand, various natural contaminations, for example, TCE (trichloroethylene), herbicides, for example, atrazine, explosives, PHC, mono fragrant hydrocarbons (BTEX) and PAHs, methyl tertiary butyl ether (MTBE) and polychlorinated biphenyls (PCBs) have been effectively phytoremediated (Pilon 2005). In vitro state and a large number of the mechanisms investigated in phytoremediation field trials examine the impacts of plants on expulsion of contaminants from spiked soil and soil unearthed from debased locales (Wakeham et al. 1980), and the greater part of these examinations gave important bits of knowledge into the particular systems of phytoremediation of natural contaminants (Ho et al. 2007). Price of the phytoremediation is lesser than conventional procedures both in situ and ex situ; plants can be effortlessly checked, and the plausibility of recuperation and re-utilization of important items are fundamental focal points of phytoremediation. Utilization of normally happening living beings and conservation in condition and minimal effort of phytoremediation are likewise two primary significances (Shukla et al. 2010).

11.5 Rhizoremediation Approaches

Analysts depicted a critical strategy for the separation of microorganisms (Kuiper et al. 2001). Microorganisms can join to chose poison with great root colonization and help in the debasement of pollutant. They have named this procedure ‘rhizoremediation’ rather than phytoremediation to accentuate the parts of the root exudates and the capable organism found in the rhizosphere. The larger part of natural toxins and corruption process happen through rhizospheric organisms. Root exudates are the best sustenance source accessible in the soil for these organisms (Bais et al. 2006). The plants’ uptake of overwhelming metals in high fixation from soils hurtfully impacts the advantageous interaction, development and thus the yield creation (Wani et al. 2007). These overwhelming metals are used as genotoxic substance and deteriorating cell organelles, to crack the cell films (Sharma and Talukdar 1987) and bother the physiological procedure, similar to the mechanism of photosynthesis or by inhibiting the respiration rate, protein blend and digestion of sugar. *Pseudomonas putida* acts as a root colonizer, and it is active for rhizoremediation of poisons and the natural control for bugs (Lazaro et al. 2000). In earlier two decades, a substantial amount of distributions on rhizodegradation of different natural toxicants utilizing distinctive plants as well as microbial inoculants have been distributed (Mohan et al. 2006; Wenzel 2009).

As indicated by a theory, when an appropriate microbial or rhizosphere strain is joined together with a reasonable plant, at that point these adjusted microscopic organisms with the ordinary indigenous populace can make a bond on the root together and help to upgrade the bioremediation procedure. Pioneer work about the debasement of mixes in the rhizosphere was chiefly engaged for herbicides and pesticides (Eerd et al. 2003). The areas or field are debased with contamination in soils that have experienced delayed times of maturing, for the most part, giving off an impression of being substantially less receptive to rhizodegradation than natural soil (Phillips et al. 2006; Olson et al. 2007; Dams et al. 2007). In 2009, Wenzel has inferred that short bioavailability is a primary cause for the disappointment of rhizodegradation in field-tainted and matured spiked soils (Wenzel 2009). There is some critical utilization of crisp and quickly matured, spiked soil material for the materialness of rhizodegradation and in addition for the assessment of information acquired on it. Vaccination of a few strains that go about as degrader can likewise improve rhizodegradation. Prior the researchers have detailed that the microbial medicines had all the earmarks of being effective in in vitro explorings (Child et al. 2007); however this analysis is fizzled when connected to long-haul defiled soil (Van Dillewijn et al. 2007). This again demonstrates the significance of the bioavailability and lab tests. A normally happened rhizoremediation process could rely upon the expanded root arrangement of the plant species where countless harbour, the foundation of essential and auxiliary digestion, survival and natural connections with different living beings (Kuiper et al. 2004). Some dubious consequences of conventional immunization baffling the examinations (Dzantor 2007), refined methodologies are the principal prerequisite for the upgrade of rhizodegradation in the earth. Presentation of healthful inclination in the mix with immunized strains helps to improve the corruption capacities of microorganisms of strains.

Specialists have recognized phenylpropanoids, a root exudate natural exacerbates that made a dietary predisposition and upgraded PCB debasement (Narasimhan et al. 2003). Plant roots working in the dirt with consolidating added substances (supplements) enhance air circulation in soil (Kuiper et al. 2004). An assortment of photosynthetic natural aggravates is discharged from plants which help in the debasement of poisons (Pilon 2005). The root exudates comprise of an assortment of some essential constituent like water dissolvable, insoluble, and unstable mixes including alcohols, sugars, amino acids, proteins, nucleotides, flavonones, phenolic mixes, natural acids and certain catalysts (Anderson et al. 1993). Ordinarily, in a harmonious relationship, there is an equivalent development or advancement of plant and soil microorganisms in the rhizosphere; in this connection plants give off important supplements and space for the organisms to develop, while consequently, the microorganisms give a solid domain to the soil where plant roots can develop legitimately.

In particular, in the rhizosphere, the transportation of oxygen and water happened through the plant. Furthermore, sugars, alcohols and starch are the principle phytochemicals of plants radiate and are essential wellsprings of starch (sustenance) for the particular soil microorganisms that giving a solid soil condition (Shukla et al. 2011). Yet, on the other hand, these phytochemicals stifle the development of different plants that are developing in a similar soil and which might be an allelopathic operator. Consequently, plants are shielded from soil pathogens, poisons and other unsafe chemicals by trading these phytochemicals; these hurtful conditions are normally present or developing in the dirt condition (Prasad 2011). In correlation of a vegetated soil, the microbial population can be a few requests of greatness higher than an unvegetated soil. At times called rhizodegradation or phytostimulation, rhizosphere biodegradation of plant helped bioremediation. Rhizodegradation improves the breakdown of contaminants by expanding the bioactivity and animate the microbial populaces in plant rhizosphere condition.

11.6 Factor Influencing PAH Degradation

Soil sort, surface, molecule size, supplements and natural issue content are the fundamental variables which can constrain the bioavailability of contaminations and impact the rate of rhizoremediation of PAHs in soil (Wenzel 2009). The procedure of corruption is impacted by few conditions, for example, low atomic weight PAHs, emanation or statement PAHs that has occurred late, direct estimation of pH in soil and the nearness of proper PAHs debasing microbes and plants with huge root surface zone that help to encourage deterioration (USEPA 2008). In the interactions among roots and microorganisms, root organisms are considered essential for PAH phytoremediation (Rugh et al. 2005). Common constriction or corruption happened in three chains PAHs in the period in vegetated settings is in 4 months (Parrish et al. 2006). The corrupted items are less dangerous for other soil living beings and furthermore, fill in as a vitality source. A few research show that PAHs comprise fewer benzene rings that can be effectively processed by microorganisms in the soil. In 2005, Johnson proposed that microbial debasement of PAHs and extra hydrophobic substrates is constrained by the sums broke up in water stage, with crystalline, sorbed and non-watery stage fluid disintegrated PAHs being inaccessible to PAH corrupting creatures (Johnsen et al. 2005). The bioavailability of contamination is a primary issue for soil bioremediation. The idea of the vast majority of the natural toxins is diverse like some natural contamination break down inadequately in water since they are profoundly hydrophobic mixtures and numerous natural poisons shape buildings with soil molecule, this absence of bioavailability frequently diminishes the evacuation efficiencies (Prasad 2011).

11.7 Bioavailability Approaches

Bioavailability is an energetic procedure, dictated by the time of substrate mass exchange to microbes with respect to their natural catabolism (Prasad 2011). Bioavailability alludes as parts of chemicals that are able to change or uptake by living beings from the encompassing bio-improved condition where life form interceded biochemical changes happen (Semple et al. 2003; Harmsen et al. 2005). The bioavailability of the particular contamination, organic action, microbial alterations of roots to build their dissolvability, physical and concoction properties of the toxin, types of soil, natural conditions and compound speciation in the rhizosphere increment the appropriateness of rhizoremediation process (Pilon 2005; Wenzel 2009). Biosurfactants increment the bioavailability of hydrocarbons bringing about improved development and debasement of contaminants by hydrocarbon-corrupting microorganisms introduced in contaminated soil (Płociniczak et al. 2011).

The vapour weight and Henry's consistent are two critical toxin properties controlling the destiny of poison in the earth (Wenzel 2009). The instability of toxin is shown by vapour weight when there is a lack of water in the soil, yet the Henry's consistent gives a superior estimation for unpredictability of poison in wet and overwhelmed soil. Some very unstable mixes, for example, chloroethene, remain in the soil for a brief period; these unpredictable mixes are not an essential focus for rhizodegradation. Dissolvability of poison is additionally altered by nature of soil particles. Natural problems like the quality of substance, dirt substance and synthesis of minerals, redox potential and pH are known as essential regulators of natural poison dissolvability, with the hydrophobic, nonpolar natural issue being of specific significance for restricting natural toxins (Reid et al. 2000). Natural contaminants become bound to soil particles and build the contact time. The rendering contaminations are less bioavailable in condition (Wenzel 2009). The authoritative of the natural toxin to soil particles for a long stretch is exceptionally hurtful and "maturing" the procedure recognized sorption against minerals and natural mixes in the soil, and ensuring interparticle dispersion of minerals and nano microspores (Semple et al. 2003; Reid et al. 2000).

In nature, the microorganisms have a retention ability, and the bioavailability of poisons in soil relies upon the solvency of toxin, as well as the dissemination of contamination and mass transportation towards the destinations where degrader populaces are copious (Semple et al. 2003). Bioavailability is a standout among the most constraining components in bioremediation of relentless natural contaminations in soil (Mohan et al. 2006; Reid et al. 2000). In bioreactor frameworks, this issue is regularly tended to by unsettling and blending and expansion of surfactants (Kamaludeen and Ramasamy 2008). In the later past, a few microorganisms have been accounted for to be chemotactic towards various natural contaminations, for instance, toluene going about as chemoattractant to *Pseudomonas putida* (Paul et al. 2006). Chemotactic microscopic organisms may be more able for bioremediation than their non-chemotactic partners (Paul et al. 2006).

11.8 Biodegradation of PAH

The most generally revealed bacterial species incorporate *Acinetobacter calcoaceticus*, *Pseudomonas vesicularis*, *Pseudomonas fluorescens*, *Pseudomonas cepacia*, *Corynebacterium renale*, *Alcaligens denitrificans*, *Rhodococcus* sp., *Mycobacterium* sp., *Moraxella* sp., *Pseudomonas putida*, *Beijerinckia sp.*, *Bacillus cereus*, *Micrococcus* sp., *Pseudomonas paucimobilis* and *Sphingomonas* sp. (Cao et al. 2009). Numerous bacterial, contagious and algal strains have been built up to corrupt a broad assortment of PAHs (Jain et al. 2005). *Pseudomonas putida* upgrades the metabolic building and hereditary control applications for the articulation of qualities encoding in a few degradative catalysts (Jain et al. 2005). Hence, a *P. putida* strain builds the productivity of corruption of naphthalene and salicylate (Jain et al. 2005). Likewise, another examination showed *Pseudomonas* and *Burkholderia* build the productivity of the naphthalene debasement process performed by various microbial strains in demonstrated soil frameworks (Filonov et al. 2006). At the point when microscopic organisms are developed on an option carbon source, can corrupt BaP in fluid culture tests (Ye et al. 1996). Secluded spore-forming PAHs corrupt microbes and revealed the result of the hereditary investigations of their debasement pathways which may prompt the revelation of novel qualities (Sorkhoh et al. 2011). The rhizospheres demonstrated a huge impact on the debasement of natural poison (Chaudhry et al. 2005). Recently, four microorganisms are purportedly disengaged from *P. deltoides*, a non-sullied rhizosphere which could corrupt 80–90% of anthracene and naphthalene within 1 week (Bisht et al. 2010). *Mycobacterium vanbaalenii* is ready to debase an incredible assortment of low and high subatomic weight PAHs in the soil. The flexibility of this species makes it plausible inoculants in the remediation of PAH's defiled destinations (Kim et al. 2005). *Calotropis* sp. can take substantial metals into its tissues because of their capacities to assimilate and endure overwhelming metals. It is prevailing and a basic forsaken plant that develops generally in warm and urbanizing locales (Al-Yemni et al. 2011). The leaf biomass of *Calotropis procera* can be utilized as great bio-sorbent for the evacuation of Cr (III) from watery arrangements and as an optional technique for their expulsion from mechanical profluent (Overah 2011).

11.9 The Rate of PAH Biodegradation

As the level and rate of PCB debasement diminish, the degree of chlorination will increase. Among the 2-month dynamic treatment stage 18% 5-Cl-PCBs, 24% 4-Cl-PCBs, 28% 3-Cl-PCBs and 62% 2-Cl-PCBs biodegradation will be more effective (Liu et al. 2007). Among uninvolved stage, the reversibly sorbed PCBs will be bio balanced outside and inside in 15 years, separately. The biodegradation rate of PAHs is very unusual and relies upon the physical and chemical parameters of the area and also the number and sorts of microorganism displayed and PAH structure. As the rate and level of PAH debasement diminish, the quantity of benzene ring increments. In strong and dregs, PAHs sorb to natural issue and the rate of their sorption unequivocally controls the rate of which microorganism can corrupt the contamination. Momentum looks into on PAH concentrates on procedures that improve the ability and along these lines, the debasement rate of PAHs at dirtied site upgraded. For land treatment, a successive dynamic distant biotreatment approach is a viable plan for corruption of both PAHs and PCBs. The quantitative model, together with a research centre and field testing, can be a valuable apparatus for the arrangement, plan and operation of comparable land treatment frameworks.

11.10 Microbial Enzymes Involved in PAH Degradation Process

Microbes start PAH debasement during the activity of intracellular oxygenase, dioxygenases, phosphatases, dehydrogenase, dehalogenases, nitroreductases, nitrilases and lignolytic compounds (Johnsen et al. 2005) (Table 11.1). Multicomponent compounds dioxygenases are those which comprise of a ferredoxin and a reductase and a terminal dioxygenase present in electron transport chain (Chadhain et al. 2006). The naphthalene dioxygenase is best examined through the PAH dioxygenase from *Pseudomonas putida* encoded by the NAH plasmid pDTG1 (Dennis and Zylstra 2004).

11.11 Improvement in Rhizoremediation

An optional approach to enhance the rhizoremediation is the determination of microscopic organisms in the rhizosphere of plants which can create biosurfactants (Płociniczak et al. 2011). Rhizoremediation can enhance a few viewpoints like bioavailability of contaminant atoms and articulation and upkeep of hereditarily designed plant microbial frameworks and exudates of root for the feasibility of procedure. Recognized microscopic organisms are developing in PAHs' despoiled zone that produces biosurfactants which encourage solubilization of PAHs and subsequently biodegradation with microorganisms (Kuiper et al. 2004).

A variety of biodegradative microbes display helpful chemotaxis towards contaminations in this property (Bisht et al. 2010). Consequently, biosurfactant and chemotaxis played out a consolidated activity to bacterial multiplication, and organisms are spread in contaminated soils, with a specific end goal to clean nature (Gerhardt et al. 2009).

As microbial corruption of contaminants in the rhizosphere provides a beneficial conclusion for the plant, the convergence of poison is diminished in the territory of roots so the plant can develop superior to those in sullied territories (Natsch et al.

Table 11.1 Significant enzymes related to bioremediation

Enzyme	Target pollutants
Aromatic dehalogenase	Chlorinated aromatics (DDT, PCBs, etc.)
Carboxyl esterases	Xenobiotics
Cellulases	Complex cellulosic materials
Cytochrome P450	Xenobiotics (PCBs)
Dehalogenase	Chlorinated solvents and ethylene
Dioxygenases	Aromatic compounds
Glutathione	Xenobiotics
Haloalkane dehalogenases	Halogenated aliphatic compounds
Horseradish peroxidase	Chlorophenol, phenol
Hydroxylases	Hydrocarbon (aromatic and aliphatic)
Peroxygenases	Xenobiotics
Peroxidases	Xenobiotics; phenols; PAHs
Laccase	Oxidative step in the degradation of explosives
Lipases	Polyaromatic hydrocarbon; triglycerol
N-Glycosyltransferases	Xenobiotics
Nitrilase	Herbicides
Nitroreductase	Explosives (RDX and TNT)
N-Malonyltransferases	Xenobiotics
Monooxygenase	Heterocyclic hydrocarbons
O-Demethylase	Alachlor, metolachlor
O-Glucosyltransferases	Xenobiotics
O-Malonyltransferases	Xenobiotics
Oxidoreductase	Xenobiotics (phenols and aniline)
Oxygenase	Chlorinated biphenyls, aliphatic olefins
Phosphatase	Organophosphates
Phosphotriesterases	Organophosphates
Phytase	Organophosphates
Peroxidase	Lignin, phenolic compounds

1996). This relationship of plant and organisms is mutually benefiting both. It has been suggested that plants of exacting genotypes are chosen to exhibit their underlying foundations. A trial exhibited for alkane monooxygenase feature is very common in endophytic and rhizosphere microbial community than soil despoiled with hydrocarbons (Siciliano et al. 2003). Nonetheless, the outcomes observed in the investigation of the predominance of the xylene monooxygenase or naphthalene dioxygenase qualities were the polar opposite; the nearness of these catalysts was higher in mass soil microbial groups than close to plant. This recommends impacting the rhizosphere by plants which rely upon the contaminant existing in the soil. Among some examination, it has been reasoned that this impact relies upon the type of plant. It has driven the theory of viability of rhizoremediation procedures which are acknowledged with the purpose of some best plant bacterium mix. HPLC investigation verified that the rhizospheric *Pseudomonas* sp. of *Calotropis* plant is a decent degrader for anthracene (63.53%) and naphthalene (78.44%). Rhizosphere of *Calotropis* sp. is a source of *Pseudomonas* sp. that has strong PGP traits, PAH debasement and natural exercises to control phytopathogenic growths. Many examinations are used to affirm their adequacy in field conditions (Shukla et al. 2012).

Researchers built up chemotaxis of PAHs, debasing rhizosphere microscopic organisms to naphthalene, phenanthrene and root exudates (Shukla et al. 2011). Anthracene and pyrene are microscopic organism-repellent. The cooperation among microbes and roots may enhance bioavailability and increment PAH corruption in the rhizosphere. The phenanthrene-corrupting movement of *P. putida* by root concentrates and exudates prescribed that presentation of the catalyst may not happen by rhizodegradation of PAHs (Rentz et al. 2004). The expansion of hereditarily built plant-microorganism enhance the rhizoremediation procedure e.g. the quality cloning of plants containing bacterial quality for the debasement of natural toxins, root-colonizing microscopic organisms (e.g. *P. fluorescens*) communicating compounds for degradation e.g. orthomonooxygenase act on toluene debasement (Francova et al. 2003). The investigations infer that the rhizospheric microorganisms related with the particular plant are yet not settled in the accessible report. Despite the fact that a lot of tests have been performed on bioremediation, numerous researchers are concentrating on plant growth promoting (PGP) movement of the rhizosphere of various plants. But it is quite evident from the literature that there is no information accessible about rhizosphere group of a particular plant. Its subatomic portrayal and use in economical horticulture, biofertilization and ecorestoration are accounted for. The rebuilding of debased locales through rhizoremediation can be effectively utilized by selecting a particular kind of plant cultivar for particular rhizobacteria or by vaccination of proficient strain from rhizobacteria on plant seeds (Kamaludeen and Ramasamy 2008).

11.12 Conclusion

Heavy metals are released into the environment in various ways directly or indirectly. It is released as a by-product in wastewater or is directly introduced into the environment. These heavy metals are highly persistent in nature and have a negative impact on microflora, soil fertility and human health. The natural flow of contaminants is altered by the current scenario of industrial activity as some of the novel metals are introduced into the environment. The discharge rate of these effluents in soil and water has been boosted up by the increase in the industrialization and urbanization including many other activities like mining, farming, military activities, waste practices, etc. Heavy metals are added permanently to the soil as they are not subjected to microbial attack or degradation; therefore they are considered the most conventional pollutants in the environment. Rhizoremediation is a plant-based technique in which the pollution caused by metals is reduced by stabilizing them in the plant's rhizosphere by the process of sorption and binding (sequestration) where the availability of metals to the livestock, human and wildlife is lowered by immobilizing them in plant roots. The main aim of this technique is to stabilize the metals rather than to remove them from site, unlike other phytoremediation techniques so that the risk to human health and environment is reduced. Plants that help in phytoremediation acquire many features such as easy and quick to grow, easy to establish and care for, form thick canopies along with dense root system and magnanimous to the high concentration of metal and the site conditions. In future studies, this technique is considered more advantageous as compared to other techniques as it is less extortionate, less environmentally indistinct and easily implemented.

References

Abd EHE, Hafez EE, Hussain AA, Ali AG, Hanafy AA (2009) Isolation and identification of three rings polyaromatic hydrocarbons (anthracene and phenanthrene) degrading bacteria. Am Eurasian J Agric Environ Sci 5:31–38

Al-Yemni MN, Sher H, Sheikh MA, Eid EM (2011) Bioaccumulation of nutrient and heavy metals by *Calotropis procera* and *Citrullus colocynthis* and their potential use as contamination indicators. Sci Res Essays 6:966–976

Anderson TA, Guthrie EA, Walton BT (1993) Bioremediation in the rhizosphere, plant roots and associated microbes clean contaminated soil. Environ Sci Technol 27:2630–2636

Babalola OO (2010) Beneficial bacteria of agricultural importance. Biotechnol Lett 32:1559–1570

Bais HT, Perry LG, Simon G, Vivanco JM (2006) The role of root exudates in rhizosphere interactions with plants and other organisms. Plant Biol 57:233–266

Barac T, Weyens N, Oeyen L, Taghavi S, van der Lelie D, Dubin D, Spliet M, Vangronsveld J (2009) Field note: hydraulic containment of a BTEX plum using poplar trees. Int J Phytoremediation 11:416–424

Barea JM, Pozo MJ, Azcon R, Azcon AC (2005) Microbial cooperation in the rhizosphere. J Exp Bot 56:1761–1778

Bisht S, Pandey P, Sood A, Sharma S, Bisht NS (2010) Biodegradation of naphthalene and anthracene by chemotactically active rhizobacteria of *Populus deltoides*. Braz J Microbiol 41:922–930

Blumer M (1976) Polycyclic aromatic compounds in nature. Sci Am 3(1976):35–45

Canadian Council of Ministers of the Environment (CCME) (2001) Canada wide standards for petroleum hydrocarbons (PHC) in soil. Canadian Council of Ministers of the Environment, Winnipeg

Cao B, Nagarajan K, Kai CL (2009) Biodegradation of aromatic compounds: current status and opportunities for biomolecular approaches. Appl Microbiol Biotechnol 85:207–228

Cerniglia CE (1993) Biodegradation of polycyclic aromatic hydrocarbons. Curr Opin Biotechnol 4:331–338

Cerniglia CE, Heitkamp MA (1989) Microbial degradation of polycyclic aromatic hydrocarbons in the aquatic environment. In: Varanasi U (ed) Metabolism of polycyclic aromatic hydrocarbons in the aquatic environment. CRC Press, Boca Raton, pp 42–64

Chadhain SMN, Norman RS, Pesce KV, Kukor JJ, Zylstra GJ (2006) Microbial (phytoremediation) of soils. Plant Soil 321:385–408

Chaudhry Q, Blom-Zandstra M, Gupta S, Joner EJ (2005) Utilising the synergy between plants and rhizosphere microorganisms to enhance breakdown of organic pollutants in the environment. Environ Sci Pollut Res 12:34–48

Child R, Miller CD, Liang Y, Sims RC, Anderson AJ (2007) Pyrene mineralization by Mycobacterium sp. strain in a barley rhizosphere. J Environ Qual 36:1260–1265

Cosgrove L, McGeechan PL, Handley PS, Robson GD (2010) Effect of biostimulation and bioaugmentation on degradation of polyurethane buried in soil. Appl Environ Microbiol 76:810–819

Dams RI, Paton GI, Killham K (2007) Rhizoremediation of pentachlorophenol by Sphingobium chlorophenolicum. Chemosphere 68:864–870

David ND, Sharon LD (2009) Improving phytoremediation through biotechnology. Curr Opin Biotechnol 20:1–3

de Carcer DA, Martin M, Mackova M, Macek T, Karlson U, Rivilla R (2007) The introduction of genetically modified microorganisms designed for rhizoremediation induces changes on native bacteria in the rhizosphere but not in the surrounding soil. ISME J 1:215–223

Dennis JJ, Zylstra GJ (2004) Complete sequence and genetic organization of pDTG1, the 83 kilobase naphthalene degradation plasmid from *Pseudomonas putida* strain. J Mol Biol 341:753–768

Dodd I, Zinovkina N, Safronova V, Belimov A (2010) Rhizobacterial mediation of plant hormone status. Ann Appl Biol 157:361–379

Dzantor EK (2007) Phytoremediation: the state of rhizosphere "engineering" for accelerated rhizodegradation of xenobiotic contaminants. J Chem Technol Biotechnol 82:228–232

Eerd LLV, Hoagland RE, Zablotowicz RM, Hall JC (2003) Pesticide metabolism in plants and microorganisms. Weed Sci 51:472–495

Filonov AE, Puntus IF, Karpov AV, Kosheleva IA, Akhmetov LI, Yonge DR, Petersen JN, Boronin AM (2006) Assessment of naphthalene biodegradation efficiency of *Pseudomonas* and *Burkholderia* strains tested in soil model systems. J Chem Technol Biotechnol 18:216–224

Francova K, Sura M, Macek T, Szekeres M, Bancos S, Demnerova K (2003) Preparation of plants containing bacterial enzyme for the degradation of polychlorinated biphenyls. Fresenius Environ Bull 12:309–313

Freeman DJ, Cattell FCR (1990) Wood burning as a source of atmospheric polycyclic aromatic hydrocarbons. Environ Sci Technol 24:1581–1585

Frerot H, Lefebvre C, Gruber W, Collin C, Dos SA, Escarre J (2006) Specific interactions between local metallicolous plants improve the phytostabilization of mine soils. Plant Soil 282:53–65

Gerhardt KE, Huang XD, Glick BR, Greenberg BM (2009) Phytoremediation and rhizoremediation of organic soil contaminants: potential and challenges. Plant Sci 176:20–30

Glick BR (1995) The enhancement of plant growth by free living bacteria. Can J Microbiol 41:109–114

Guerin WF, Jones GE (1988) Mineralisation of phenanthrene by a *Mycobacterium sp.* Appl Environ Microbiol 54:937–944

Guijian L, Zhiyuan N, Daniel VN, Jian X, Liugen Z (2008) Polycyclic aromatic hydrocarbons (PAHs) from coal combustion: Emissions, analysis, and toxicology. Rev Environ Contam Toxicol 192:1–28

Guoa H, Leea SC, Hoa KF, Wangb XM, Zou SC (2003) Particle associated polycyclic aromatic hydrocarbons in urban air of Hong Kong. Atmos Environ 37:5307–5317

Haritash AK, Kaushik CP (2009) Biodegradation aspects of polycyclic aromatic hydrocarbons (PAHs): a review. J Hazard Mater 169:1–15

Harms H, Wick LY (2006) Dispersing pollutant degrading bacteria in contaminated soil without touching it. Eng Life Sci 6:252–260

Harmsen J, Rulkens W, Eijsackers H (2005) Bioavailability: concept for understanding or tool for predicting. Land Contam Reclamat 13:161–171

Ho CH, Applegate B, Banks MK (2007) Impact of microbial plant interactions on the transformation of polycyclic aromatic hydrocarbons in rhizosphere of *Festuca arundinacea*. Int J Phytoremediation 9:107–114

Jain RK, Kapur M, Labana S, Lal B, Sarma PM, Bhattacharya D, Thakur IS (2005) Microbial diversity: application of microorganisms for the biodegradation of xenobiotics. Curr Sci 89:101–112

Johnsen AR, Wickb LY, Harmsb H (2005) Principles of microbial PAH degradation in soil. Environ Pollut 133:71–84

Jones KC, Stratford JA, Waterhouse KS, Furlong ET, Giger W, Hites RA, Schaffner C, Johbston AE (1989) Increases in the polynuclear aromatic hydrocarbon content of an agricultural soil over the last century. Environ Sci Technol 23:95–101

Juhasz AL, Naidu R (2000) Bioremediation of high molecular weight polycyclic aromatic hydrocarbons: a review of the microbial degradation of benzo[a]pyrene. Int Biodeterior Biodegrad 45:57–88

Kamaludeen PB, Ramasamy K (2008) Rhizoremediation of metals: harnessing microbial communities. Indian J Microbiol 48:80–88
Kanaly RA, Harayama S (2000) Biodegradation of high molecular weight polycyclic aromatic hydrocarbons by bacteria. J Bacteriol 182:2059–2067
Kaur P, Singh S, Kumar V, Singh N, Singh J (2018) Effect of rhizobacteria on arsenic uptake by macrophyte Eichhornia *crassipes* (Mart.) Solms. Int J Phytoremediation 20(2):114–120
Kidd P, Prieto FA, Monterroso C, Acea M (2008) Rhizosphere microbial community and hexachlorocyclohexane degradative potential in contrasting plant species. Plant Soil 302:233–247
Kim YH, Freeman JP, Moody JD, Engesser KH, Cerniglia CE (2005) Effects of pH on the degradation of phenanthrene and pyrene by *Mycobacterium vanbaalenii*. Appl Microbiol Biotechnol 67:275–285
Kuiper I, Bloemberg GV, Lugtenberg BJJ (2001) Selection of a plant bacterium pair as a novel tool for rhizostimulation of polycyclic aromatic hydrocarbon degrading bacteria. Mol Plant-Microbe Interact 14:1197–1205
Kuiper I, Lagendijk EL, Bloemberg GV, Lugtenberg BJ (2004) Rhizoremediation: a beneficial plant microbe interaction. Mol Plant-Microbe Interact 17:6–15
Kumar V, Upadhyay N, Singh S, Singh J, Kaur P (2013) Thin-layer chromatography: comparative estimation of soil's atrazine. Curr World Environ 8(3):469–472
Kumar V, Upadhyay N, Kumar V, Kaur S, Singh J, Singh S, Datta S (2014a) Environmental exposure and health risks of the insecticide monocrotophos—a review. J Biodivers Environ Sci 5:111–120
Kumar V, Singh S, Manhas A, Singh J, Singla S, Kaur P (2014b) Bioremediation of petroleum hydrocarbon by using Pseudomonas species isolated from petroleum contaminated soil. Orient J Chem 30(4):1771–1776
Kumar V, Singh S, Kashyap N, Singla S, Bhadrecha P, Kaur P (2015a) Bioremediation of heavy metals by employing resistant microbial isolates from agricultural soil irrigated with industrial waste water. Orient J Chem 31(1):357–361
Kumar V, Singh S, Singh J, Upadhyay N (2015b) Potential of plant growth promoting traits by bacteria isolated from heavy metal contaminated soils. Bull Environ Contam Toxicol 94:807–815
Kumar V, Kaur S, Singh S, Upadhyay N (2016) Unexpected formation of N′-phenyl-thiophosphorohydrazidic acid O, S-dimethyl ester from acephate: chemical, biotechnical and computational study. 3 Biotech 6(1):1
Kumar V, Singh S, Singh R, Upadhyay N, Singh J (2017) Design, synthesis, and characterization of 2, 2-bis (2, 4-dinitrophenyl)-2-(phosphonatomethylamino) acetate as a herbicidal and biological active agent. J Chem Biol 10(4):179–190
Lazaro M, Cayo R, Estrella D, Ronchel MC, Garcia JM, Lene W, Ramos JL (2000) Survival of *Pseudomonas putida* KT2440 in soil and in the rhizosphere of plants under greenhouse and environmental conditions. Soil Biol Biochem 32:315–321
Lee S, Ka JO, Gyu SH (2012) Growth promotion of *Xanthium italicum* by application of rhizobacterial isolates of *Bacillus aryabhattai* in microcosm soil. J Microbiol 50:45–49
Lijinsky W (1991) The formation and occurrence of polynuclear aromatic hydrocarbons associated with food. Mutat Res 259:251–262
Liu L, Tindall JA, Friedel MJ (2007) Biodegradation of PAHs and PCBs in soils and Sludges. Water Air Soil Pollut 181(1–4):281–296
Louisa WP (2010) Review: in situ and bioremediation of organic pollutants in aquatic sediments. J Hazard Mater 177:81–89
Ma B, He Y, Chen HH, Xu JM (2009) Dissipation of polycyclic aromatic hydrocarbons (PAHs) in the rhizosphere: synthesis through metal analysis. Environ Pollut 159:855–861
Ma Y, Prasad MNV, Rajkumar M, Freitas H (2011) Plant growth promoting rhizobacteria and endophytes accelerate phytoremediation of metalliferous soils. Biotechnol Adv 29(2):248–258
McGuinness M, Dowling D (2009) Plant associated bacterial degradation of toxic organic compounds in soil. Int J Environ Res Public Health 6:2226–2247

Mejare M, Bulow L (2001) Metal-binding proteins and peptides in bioremediation and phytoremediation of heavy metals. Trends Biotechnol 19:67–73
Mishra V, Gupta A, Kaur P, Singh S, Singh N, Gehlot P, Singh J (2016) Synergistic effects of Arbuscular mycorrhizal fungi and plant growth promoting rhizobacteria in bioremediation of iron contaminated soils. Int J Phytoremediation 18(7):697–703
Mohan SV, Kisa T, Ohkuma T, Kanaly RA, Shimizu Y (2006) Bioremediation technologies for treatment of PAH contaminated soil and strategies to enhance process efficiency. Rev Environ Sci Biotechnol 5:347–374
Mueller JG, Chapman PJ, Pritchard PH (1990) Action of a fluoranthene utilizing bacterial community on polycyclic aromatic hydrocarbon components of creosote. Appl Environ Microbiol 55:3085–3090
Narasimhan K, Basheer C, Bajic VB, Swarup S (2003) Enhancement of plant microbe interactions using a rhizosphere metabolomics driven approach and its application in the removal of polychlorinated biphenyls. Plant Physiol 132:146–153
Natsch AC, Keel J, Troxler M, Zala N, Albertini V, Defago G (1996) Importance of preferential flow and soil management in vertical transport of a biocontrol strain of *Pseudomonas fluorescens* in structured field soil. Appl Environ Microbiol 62:33–40
Nishioka M, Chang HC, Lee V (1986) Structural characteristics of polycyclic aromatic hydrocarbon isomers in coal tars and combustion products. Environ Sci Technol 20:1023–1027
Olson PE, Castro A, Joern M, DuTeau NM, Pilon SEAH, Reardon KF (2007) Comparison of plant families in a greenhouse phytoremediation study on an aged polycyclic aromatic hydrocarbon contaminated soil. J Environ Qual 36:1461–1469
Overah LC (2011) Biosorption of Cr (III) from aqueous solution by the leaf biomass of *Calotropis procera*-'bombom'. J Appl Sci Environ Manag 15(1):87–95
Oviasogie PO, Ukpebor EE, Omoti U (2006) Distribution of polycyclic aromatic hydrocarbons in rural agricultural wetland soils of the Niger Delta region. Afr J Biotechnol 5:1415–1421
Parrish Z, White J, Isleyen M, Gent M, Iannucci BW, Eitzer B, Kelsey J, Mattina M (2006) Accumulation of weathered polycyclic aromatic hydrocarbons (PAHs) by plant and earthworm species. Chemosphere 64:609–618
Paul D, Singh R, Jain RK (2006) Chemotaxis of *Ralstonia sp.* towards p-nitrophenol in soil. Environ Microbiol 8:1797–1804
Peters RW (1999) Chelant extraction of heavy metals from contaminated soils. J Hazard Mater 66:151–210
Phillips LA, Greer CW, Germida JJ (2006) Culture based and culture independent assessment of the impact of mixed and single plant treatments on rhizosphere microbial communities in hydrocarbon contaminated flare pit soil. Soil Biol Biochem 38:2823–2833
Phillips LA, Greer CW, Richard EF, James JG (2012) Plant root exudates impact the hydrocarbon degradation potential of a weathered hydrocarbon contaminated soil. Appl Soil Ecol 52:56–64
Pilon SEAH (2005) Phytoremediation. Annu Rev Plant Biol 56:15–39
Płociniczak MP, Płaza GA, Seget ZP, Cameotra SS (2011) Environmental applications of biosurfactants: recent advances. Int J Mol Sci 12:633–654
Prasad MNV (2011) A state of the art report on bioremediation, its applications to contaminated sites in India. Ministry of Environment and Forests, Government of India, 2011
Prasad R, Aranda E (2018) Approaches in bioremediation: the new era of environmental microbiology and nanobiotechnology. Springer, Cham. ISBN 978-3-030-02369-0. https://www.springer.com/gp/book/9783030023683
Prasad R, Kumar M, Varma A (2015) Role of PGPR in soil fertility and plant health. In: Egamberdieva D, Shrivastava S, Varma A (eds) Plant Growth-Promoting Rhizobacteria (PGPR) and medicinal plants. Springer, Cham, pp 247–260
Ramadah T, Alfheim I, Rustad S, Olsen T (1982) Chemical and biological characterization of emissions from small residential stoves burning wood and charcoal. Chemosphere 11:601–611
Reid BJ, Jones KC, Semple KT (2000) Bioavailability of persistent pollutants in soils and sediments a perspective on mechanisms, consequences and assessment. Environ Pollut 108:103–112

Rentz JA, Alvarez PJJ, Schnoor JL (2004) Repression of *Pseudomonas putida* phenanthrene degrading activity by plant root extracts and exudates. Environ Microbiol 6:574–583
Rugh C, Susilawati E, Kravchenko A, Thomas J (2005) Biodegrader metabolic expansion during polyaromatic hydrocarbons rhizoremediation. Z Naturforsch 60:331–339
Sandhu A, Halverson LJ, Beattie GA (2007) Bacterial degradation of airborne phenol in the phyllosphere. Environ Microbiol 9:383–392
Semple KT, Morriss AWJ, Paton GI (2003) Bioavailability of hydrophobic contaminants in soils: fundamental concepts and techniques for analysis. Eur J Soil Sci 54:809–818
Sharma A, Talukdar G (1987) Effects of metals on chromosomes of higher organisms. Environ Mutagen 9:191–226
Shiaris MP, Jambard SD (1986) Polycyclic aromatic hydrocarbons in surficial sediments of Boston Harbour, MA, USA. Mar Pollut Bull 17:469–472
Shukla KP, Singh NK, Sharma S (2010) Bioremediation: developments, current practices and perspectives. Genet Eng Biotechnol J 3:1–20
Shukla KP, Sharma S, Singh NK, Singh V, Tiwari K, Singh S (2011) Nature and role of root exudates: efficacy in bioremediation. Afr J Biotechnol 10:9717–9724
Shukla KP, Sharma S, Singh NK, Singh V (2012) Deciphering rhizosphere soil system for strains having plant growth promoting and bioremediation traits. Agric Res 1(3):251–257
Siciliano SD, Germida JJ, Banks K, Greer CW (2003) Changes in microbial community composition and function during a polyaromatic hydrocarbon phytoremediation field trial. Appl Environ Microbiol 69:483–489
Singh S, Singh N, Kumar V, Datta S, Wani AB, Singh D, Singh J (2016) Toxicity, monitoring and biodegradation of the fungicide carbendazim. Environ Chem Lett 14:317–329
Singh S, Kumar V, Upadhyay N, Singh J, Singla S, Datta S (2017a) Efficient biodegradation of acephate by *Pseudomonas pseudoalcaligenes* PS-5 in the presence and absence of heavy metal ions [Cu(II) and Fe(III)], and humic acid. 3 Biotech 7(4):262
Singh S, Kumar V, Chauhan A, Datta S, Wani AB, Singh N, Singh J (2017b) Toxicity, degradation and analysis of the herbicide atrazine. Environ Chem Lett 16:1–27
Sorkhoh NA, Al-Mailem DM, Ali N, Al-Awadhi H, Salamah S, Eliyas M, Radwan SS (2011) Bioremediation of volatile oil hydrocarbons by epiphytic bacteria associated with American grass (*Cynodon sp.*) and broad bean (*Vicia faba*) leaves. Int Biodeterior Biodegrad 65:797–802
Tyagi M, da Fonseca MR, de Carvalho CR (2011) Bioaugmentation and biostimulation strategies to improve the effectiveness of bioremediation processes. Biodegradation 22:231–241
U.S. Environmental Protection Agency (1999) Use of monitored natural attenuation at superfund
US Environmental Protection Agency (EPA) (2008) Integrated risk information. http://www.epa.gov/iris
Van Dillewijn P, Caballero A, Paz JA, Gonzales MM, Oliva JM, Ramos JL (2007) Bioremediation of 2,4,6-trinitrotoluene under field conditions. Environ Sci Technol 41:1378–1383
Wakeham SG, Schaffner C, Giger W (1980) Polycyclic aromatic hydrocarbons in recent lake sediments II compound having anthropogenetic origins. Geochim Cosmochim Acta 44:403–413
Walker TS, Bais HP, Grotewold E, Vivanco JM (2003) Root exudation and rhizosphere biology. Plant Physiol 132:44–51
Walter U, Beyer M, Klein J, Rehm HJ (1991) Degradation of pyrene by *Rhodococcus sp.* Appl Microbiol Biotechnol 34:671–676
Wani PA, Khan MS, Zaidi A (2007) Cadmium, chromium and copper in green gram plants. Agron Sustain Dev 27:145–153
Wenzel WW (2009) Rhizosphere processes and management in plant assisted bioremediation. Plant Soil 321(1–2):385–408
Wilson SC, Jones KC (1993) Bioremediation of soils contaminated with polynuclear aromatic hydrocarbons (PAHs): a review. Environ Pollut 88:229–249
Ye D, Siddiqi MA, Maccubbin AE, Kumar S, Sikka HC (1996) Degradation of polynuclear aromatic hydrocarbons by *Sphingomonas paucimobilis*. Environ Sci Technol 30:136–142

Chapter 12
Pesticide Tolerant Rhizobacteria: Paradigm of Disease Management and Plant Growth Promotion

Tina Roy, Nirmalendu Das, and Sukanta Majumdar

Abstract Plant growth-promoting rhizobacteria (PGPR) are soil bacteria, colonizing rhizospheric region of plants, which have the ability to enhance plant health and promote plant growth by increasing seed emergence, plant weight, and yields to a wide variety of crops either through direct action or via biological control of plant diseases. PGPR improve plant growth by either fixing atmospheric nitrogen; solubilizing insoluble phosphates and iron and producing plant growth regulators (PGRs) like auxins, gibberellins, cytokinins, etc.; or suppression of deleterious root-colonizing microorganisms including plant pathogens through antibiosis, i.e., production of fungitoxic compounds, competition with pathogenic microorganisms for nutrients by producing siderophores, or niche exclusion. Indiscriminate use of different chemicals in the form of fertilizers and pesticides targeting to increase the agricultural produce for ever-increasing population outburst has led to the contamination of the groundwater, soil, and sediments. Accretion of diversified range of chemicals in significant quantities has a direct impact not only on the living beings but also on the environment. The ecological balance of the soil microorganisms has been distorted which show the negative impact on their rhizospheric competence. When exogenous PGPR are applied in this pesticide-infested soil, they not only hardly show their plant growth-promoting or disease-suppressing activities but also might not survive at all. Isolation of native PGPR from the pesticide-challenged rhizospheric soils mostly shows pesticide-tolerant/degrading properties. These PGPR might show the rhizospheric competence in similar pesticide-infested soil. These strains easily acclimatize in the pesticide-contaminated microenvironment in soil and show their plant growth-promoting and pathogen-suppressive activities.

T. Roy
Microbiology and Microbial Biotechnology Laboratory, Department of Botany, University of Gour Banga, Malda, West Bengal, India

Post Graduate Department of Botany, Barasat Government College, Barasat, West Bengal, India

N. Das
Post Graduate Department of Botany, Barasat Government College, Barasat, West Bengal, India

S. Majumdar (✉)
Microbiology and Microbial Biotechnology Laboratory, Department of Botany, University of Gour Banga, Malda, West Bengal, India

A. Varma et al. (eds.), *Plant Microbe Symbiosis*,
https://doi.org/10.1007/978-3-030-36248-5_12

12.1 Introduction

Plant growth-promoting rhizobacteria (PGPR) are the soil bacteria dwelling around the root surface and are directly or indirectly involved in promoting plant growth and development via manufacturing and secreting various regulatory substances in the vicinity of rhizosphere (Ahemad and Kibret 2014). The PGPR may facilitate the plant growth by either directly supporting acquirement of resources (phosphorus, nitrogen, and essential minerals) or increasing plant hormone levels or indirectly by decreasing the deleterious effects of various pathogens on growth and development of plant in the forms of biocontrol agents (BCA) (Hariprasad et al. 2013; Bhatt and Vyas 2014; Garcia et al. 2015). They can improve plant's tolerance to stresses, such as drought (Ngumbi and Kloepper 2016), salinity, metal toxicity, and pesticide load as well as play a role in bioremediation of different xenobiotics present in contaminated soils (Jiang et al. 2008a).

The term (PGPR) was first introduced by Joseph W. Kloepper during the late 1970s (Kloepper and Schroth 1978; Suslow et al. 1979). The soil surrounding the roots called rhizosphere, is rich in various plant exudates that provide nutrients to soil bacteria so that they are attracted toward that zone (Dobbelaere et al. 2003; Gray and Smith 2005; Prasad et al. 2015). The nutrient-rich zone of root surroundings may attract different soil bacteria irrespective of their beneficial or harmful nature to the plants. The microorganisms compete for the resources, and finally, those who are able to colonize may show the positive, negative, or neutral effect on plant growth and development depending upon soil quality and other factors (Singh and Varaprasad 2008; Shrivastava et al. 2014). Mainly leguminous plants are symbiotically associated with the endophytes and form bead- or knot-like formation in the root surface called nodules (Sturz et al. 1997). Nodule formers are mainly integrated into the plant tissues (Sturz et al. 2000). Non-nodule formers or free living bacteria compete with other soil inhabitants and colonize the rhizosphere depending on plant species and various root secretions (Kang et al. 2010a, b).

The unwarranted use of pesticides for better production of agricultural produce to meet up the global demand has led to the accretion of their gigantic residual amounts in the environment. It causes disturbances in the microbial community particularly in the soil in addition to other environmental hazards. Some microorganisms build up resistance after a long-term exposure to agrochemicals by using them as the source of nutrient and energy and can successfully be used for bioremediation of pesticide-contaminated soils (Khan et al. 2009). These microbes may also have other plant growth-promoting features in addition to pesticide degradation and can be used to enhance the remediation process (Shahgoli and Ahangar 2014). The use of pesticide-tolerant plant growth-promoting microbial agents established better in the pesticide-contaminated soil when exogenously applied, together with their bioremediation capabilities.

This chapter accentuates the latest paradigms of applicability of pesticide-tolerant/degrading rhizobacteria in different agroecosystems particularly in pesticide-stress conditions to minimize the global addiction on hazardous synthetic pesticides to stabilize the agroecosystems for sustainable agriculture.

12.2 Mechanism of Plant Growth Promotion

12.2.1 Direct Mechanism of Plant Growth Promotion

12.2.1.1 N_2 Fixation

Nitrogen is a very much essential element for living organisms; it is required in the synthesis of amino acids, DNA, RNA, etc. The atmospheric nitrogen is 78% of total air which is huge, but plants are unable to get access to it. They only get nitrogen in the form of ammonia or nitrate, which is not abundant, so external application of urea is a very common practice, and non-judicial uses of urea made the soil more acidic. The soil bacteria may be free living like *Azotobacter* or symbiotic with some plants like *Rhizobium* that can fix atmospheric nitrogen by nitrogenase enzyme which converts N_2 to NH_3. Symbiotic nitrogen fixers or biological nitrogen fixers (BNF) transport nitrogen in the form of amides and ureides. Symbiotic nitrogen fixers, viz., *Rhizobium leguminosarum bv. trifolii*, fix atmospheric nitrogen (N_2) into ammonia (NH_3) and export them to the host plants (Hoffman et al. 2014; Ohyama 2010).

Owens (1973) first time reported that certain strains of nitrogen-fixing nodulating bacterium *Rhizobium japonicum* were able to produce a toxic metabolite in the culture filtrate which was phytotoxic in nature. He referred the properties of the toxin to "rhizobiotoxine." Tu (1978, 1979) in consecutive publications demonstrated successful parasitism of *Rhizobium japonicum* on *Fusarium* and *Phytophthora*. Dobereiner and Day (1976) in Brazil rediscovered that *Azospirillum* is capable of enhancing non-legume plant growth. The best known among the nonsymbiotic PGPB are bacteria of the genus *Azospirillum*. Other than *Azospirillum*, several non-rhizobial isolates, viz., *Acetobacter*, *Achromobacter*, *Alcaligenes*, *Arthrobacter*, *Azotobacter*, *Azomonas*, *Bacillus*, *Beijerinckia*, *Clostridium*, *Corynebacterium*, *Derxia*, *Enterobacter*, *Herbaspirillum*, *Klebsiella*, *Pseudomonas*, *Rhodospirillum*, *Rhodopseudomonas*, and *Xanthobacter*, have been reported to fix atmospheric nitrogen non-symbiotically.

12.2.1.2 Plant Hormones

Indole acetic acid (IAA), one of the most active auxins, is a product of L-tryptophan metabolism produced by most of the PGPR (Lynch 1985). It stimulates cell elongation by increasing osmotic contents and permeability of water and by decreasing cell wall pressure. IAA also delayed leaf abscission and induced fruiting (Zhao et al. 2010). The production of longer root and increased number of root hair by IAA help the plant to uptake more nutrients from the soil, and thus it helps in growth. There are two pathways of IAA synthesis. The first one is tryptophan-dependent, in which L-tryptophan converts to indole-3-acetamide by the enzyme tryptophan 2-monooxygenase in the first step, and the second step is conversion of indole-3-

acetamide (IAM) to IAA by IAM hydrolase (Mano and Nemoto 2012; Zakharova et al. 1999). The second pathway is tryptophan-independent pathway and is still undefined (Zhang et al. 2008).

Gibberellin (GA) is another plant hormone well reported in PGPR. It is important for seed germination, stem elongation, leaf enlargement, trichome development, pollen maturation, and flowering also (Achard et al. 2009). Some physicochemical methods demonstrate the presence of different GAs like GA_1, GA_4, GA_9, and GA_{20} in gnotobiotic culture of *Rhizobium meliloti* (Atzorn et al. 1988). Later different GAs are established in different microorganisms like *Azospirillum* sp., *Bradyrhizobium japonicum, Bacillus* sp., *Herbaspirillum seropedicae*, etc. (Bottini et al. 1989; Piccoli et al. 1996, 1997; Gutiérrez-Mañero et al. 2001) which helps in the improvement of plant growth.

Cytokinin and other plant hormones also played important role in plant growth and development. Ryu et al. (2003) reported that some PGPR strains release a blend of volatile organic compounds (VOCs) that promote growth in *Arabidopsis* seedlings and induce resistance against *Erwinia caratovora* subsp. *carotovora*. Using transgenic and mutant lines of *Arabidopsis*, they provided evidence that the signal pathway activated by volatiles from one PGPR strain is dependent on cytokinin activation for growth promotion and dependent on an ethylene-signaling pathway for induced pathogen resistance.

12.2.1.3 Phosphate Solubilization

Phosphate (P) plays an important role in most of the major metabolic processes in plants like photosynthesis, energy transfer, signal transduction, macromolecular biosynthesis and respiration, (Khan et al. 2010) and also nitrogen fixation in legumes (Saber et al. 2005). Though phosphate is available in soils in both inorganic and organic forms, there are some limitations for plants to avail it for their growth. Inorganic phosphates are mostly insoluble mineral complexes; some of them are deposited after frequent application of chemical fertilizers. These insoluble, precipitated forms of phosphates cannot be absorbed by plants (Rengel and Marschner 2005). Only 0.1% of phosphate in the soil is being uptaken by plants in soluble form (Zhou et al. 1992) because it fixes in soil in insoluble form by the process P-fixation. Repeated use of phosphate fertilizers and phosphate-containing pesticides increases insoluble phosphate in soil by the abovementioned process (Goldstein et al. 1993; Khan et al. 2009).

Phosphatase (EC 3.1.3.1)-producing soil microbes including bacteria, fungi, actinomycetes, and even algae are capable to degrade insoluble phosphate into soluble form. *Pseudomonas, Bacillus, Rhodococcus, Arthrobacter, Serratia, Chryseobacterium, Gordonia, Phyllobacterium, Delftia* sp. all are phosphate solubilizers (Wani et al. 2005; Chen et al. 2006), *Azotobacter* (Kumar et al. 2001), *Xanthomonas* (De Freitas et al. 1997), *Enterobacter, Pantoea*, and *Klebsiella* (Chung et al. 2005). A halophilic P-solubilizer *Kushneria sinocarni* isolated from the sediment of Daqiao saltern of the eastern coast of China showed potentiality as salt-tolerant PGPR in salt-stressed soils (Zhu et al. 2011). Ghosh et al. (2016) reported the role of phosphate-solubilizing *Burkholderia* spp. for successful colonization and growth promotion of *Lycopodium cernuum* L. in lateritic belt of Birbhum district of West Bengal, India.

12.2.1.4 ACC Deaminase (EC 4.1.99.4)

ACC deaminase is an enzyme that cleaves the 1-aminocyclopropane-1-carboxylate (ACC) into ammonia and α-ketobutyrate (Honma and Shimomura 1978). ACC is the precursor of plant hormone ethylene, responsible for leaf abscission, fruit ripening, flowering and flower wilting, *Rhizobia* nodule formation, seed germination, root elongation, and branching, and it is a stress-related hormone (Abeles et al. 1992; Goodlass and Smith 1979; Glick et al. 2007). High level of ACC produces much more ethylene and leads the plant toward growth inhibition or death. ACC deaminase-producing microorganisms decrease plant ethylene level by decreasing ACC level in plants (Glick et al. 1998, 2007). Some ACC deaminase-producing microorganisms are *Pseudomonas putida*, *Burkholderia cepacia*, *Citrobacter freundii*, *Serratia marcescens*, *Achromobacter* sp., *Rhizobium* sp., *Bacillus anthracis*, etc. (Jacobson et al. 1994; Maxton et al. 2017; Glick et al. 1995; Belimov et al. 2001, 2005; Ma et al. 2003a, b; Read et al. 2002)

12.2.2 Indirect Mechanism of Plant Growth Promotion

12.2.2.1 Induced Systemic Resistance (ISR)

Rhizobacteria can induce systemic resistance (ISR) in plants that is analogous to pathogen-induced systemic acquired resistance (SAR). SAR develops when plants successfully trigger their defense mechanism following an earlier localized exposure to a pathogen. SAR induces a hypersensitive reaction through which it becomes limited in a local necrotic lesion of brown, desiccated tissue (Van Loon et al. 1998). Both SAR and ISR are effective against different types of pathogens. However, ISR can be differentiated from SAR in that the rhizobacteria do not cause any visible symptoms on the host plant (Van Loon et al. 1998). Bacterial determinants of ISR include lipopolysaccharides, siderophores, and salicylic acid (SA). Whereas some of the rhizobacteria induce resistance through the SA-dependent SAR pathway, others prefer jasmonic acid (JA) and ethylene (ET) perception by the plant for ISR to develop (Fig. 12.1). ISR offers a natural mechanism for biological control of plant disease. It has been speculated that the successful colonization of plant roots by rhizobacteria, in turn, aggravates a signal, which spreads systemically within the plant and increases the synthesis of defense enzymes and proteins and thus protects the host from subsequent infection. ISR thus extended the protective action of PGPR

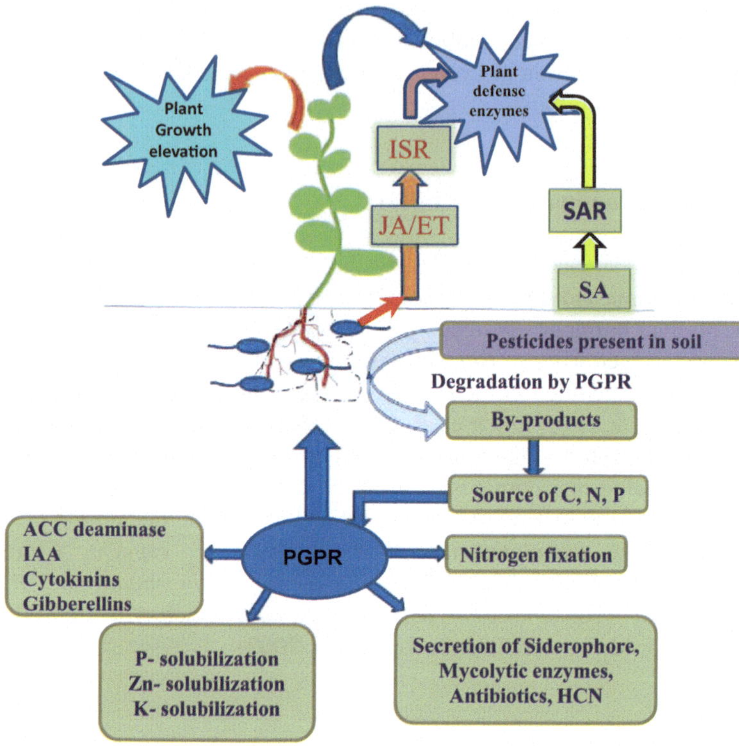

Fig. 12.1 Schematic representation of mechanism of plant growth promotion and plant defence by pesticide-degrading/tolerant bacteria

from their antagonistic activity against soil-borne pathogens in the rhizosphere to a defense-stimulating effect above the surface of the ground tissues against foliar pathogens (Van Loon and Bakker 2006).

Induction of resistance in plants following the application of PGPR can be correlated with the accumulation of defense-related enzymes like phenylalanine ammonia lyase, peroxidase, and polyphenol oxidase.

Phenylalanine ammonia lyase (PAL) [EC 4.3.1.3] converts L-phenylalanine into trans-cinnamic acid which is the precursor of flavonoid pigments, lignin, and phytoalexins by phenylpropanoid pathway (Massala et al. 1980). Increase in PAL activity subsequently increases the phenolic contents leading to disease resistance (Klessig and Malamy 1994).

Peroxidase (PO) [EC 1.11.1.7] is one of the key enzymes involved in phenylpropanoid pathway, and it is associated with disease resistance in plants (Hammerschmidt et al. 1982). PO is a component of an early response during infection and plays a major role in the biosynthesis of lignin, which limits the extent of pathogen spread (Bruce and West 1989). PGPR like *Bacillus megaterium*, *B. pumilus*, *Ochrobactrum anthropi*, and *Serratia marcescens* were successfully utilized to overcome several root diseases of tea (Chakraborty et al. 2006).

Polyphenol oxidase (PPO) [EC 1.10.3.2] is a member of oxidoreductase group of enzymes which catalyzes the oxidation of monophenolic and orthophenolic compounds using molecular oxygen. Higher PPO activity found in BCA-pretreated plants challenged inoculated with pathogens (Anand et al. 2007; Selvaraj and Ambalavanan 2013). PPO genes are upregulated during any kind of wound, infections, etc., and some specific PPO gene family will be expressed during infection (Constabel and Ryan 1998; Richter et al. 2012). It has been also reported that overexpression of PPO decreases disease susceptibility in tomato and potato plants (Li and Steffens 2002).

Elevation of all the above three enzymes in tomato plants treated with *P. fluorescens* upon challenged inoculation with *Alternaria solani* and *Septoria lycopersici* was reported by Anand et al. (2007). *B. amyloliquefaciens*, *B. subtilis*, *B. pasteurii*, *B. cereus*, *B. pumilus*, *B. mycoides*, and *B. sphaericus*, significantly reduced the severity of several diseases in tomato, bell pepper, muskmelon, watermelon, sugar beet, tobacco, *Arabidopsis* sp., and cucumber (Choudhary and Johri 2009).

12.2.2.2 Antibiosis

Plant growth promoters help in plant growth indirectly by controlling diseases of the host plant by virtue of their antagonistic effect against pathogens. They secrete some metabolites that are constitutive or may be produced by the induction of other competitors for resources¸ such as antibiotics, cell wall-degrading enzymes, siderophores, and HCN (Olanrewaju et al. 2017).

Iron is an essential element for living organisms irrespective of plant or animals. It may be a part of enzyme like oxidoreductase, also required for electron transporters like ferredoxins, and can transport oxygen like hemoglobin. But maximum irons in soil are available in the form of ferric oxide or hydroxides which are stable compounds. When any microorganism needs iron, then its physical signals induce production of an iron carrier molecule called siderophore, which has high affinity for irons and chelated iron for them. Some specific signaling molecule and receptors are involved in these mechanisms (Wandersman et al. 2004; Andrews et al. 2003). Microbial siderophores are typically classified into three types like catecholates, hydroxamates, and α-carboxylates depending on the chemical nature and their iron binding sites (Winkelmann 2002).

Some rhizobacteria are capable of producing HCN (hydrogen cyanide, also known as cyanide) (Rezzonico et al. 2007). It is a volatile, secondary metabolite that suppresses the development of microorganisms and that also affects negatively the growth and development of plants (Siddiqui et al. 2006). Cyanide is toxic to plants capable of disrupting enzyme activity involved in major metabolic processes; its role as a biocontrol substance is overwhelming (Devi et al. 2007; Voisard et al.

1989). Hydrogen cyanide (HCN) among cyanogenic compounds effectively blocks the cytochrome oxidase pathway and is highly toxic to all aerobic microorganisms at very low concentrations. However, the microbes, which produce the compound, mainly pseudomonads, are reported to be resistant (Bashan and de-Bashan 2005). HCN is formed from glycine through the action of HCN synthetase enzyme, which is a membrane-bound flavoenzyme that oxidizes glycine, producing HCN and CO_2. Production of HCN by rhizobacteria has been reported to inhibit the growth of *M. phaseolina* (Reetha et al. 2014). *P. fluorescence* can successfully eradicate some soil pathogen by producing HCN (Voisard et al. 1989).

Antibiotics like 2,4-diacetylphloroglucinol, phenazine-1-carboxylic acid, phenazine-1-carboxamide, pyoluteorin, pyrrolnitrin, butyrolactones, kanosamine, zwittermicin A, aerugine, rhamnolipids, cepaciamide A, pseudomonic acid, azomycin, antitumor antibiotics FR901463, cepafungins, and antiviral antibiotic karalicin are produced by some plant growth-promoting rhizobacteria. All of these antibiotics have antiviral, antimicrobial, insecticidal, antihelminthic, phytotoxic, antioxidant, and cytotoxic properties (Dilantha Fernando et al. 2005). All have different modes of actions. Some attack the cellular membranes; others have inhibitory effects on the ribosome or other cellular activities (Reid et al. 2002). Different members of *Pseudomonas*, a very well-known PGPR, produce antibiotics like pyrrolnitrin used as biopesticide named fludioxonil derived from pyrrolnitrin (Sturz 2006). Some other antibiotics characterized are agrocin 84 from *Agrobacterium* sp.; herbicolin A from *Erwinia* sp.; iturin A, surfactin, and zwittermicin A from *Bacillus* sp., and xanthobacin from *Stenotrophomonas* sp., which also have biocontrol abilities (Gardi and Jeffery 2009).

Many soil inhabitants are capable of producing different hydrolytic enzymes like chitinase, glucanase, protease, cellulase (endoglucanases, exo-cellobiohydrolase, exoglucanases, and β-glucosidases) which inhibit the fungal pathogens by damaging the integrity of fungal cell wall (Majumdar and Chakraborty 2017).

12.3 Pesticide Degradation/Toleration by Soil Bacteria

In the last 30 years, a number of pesticide-degrading/tolerating microbial strains have been reported. Most of the studies were done on the pesticide degradations in surface soils by different bacterial strains only. The pesticide biodegradation in the rhizosphere of plants and in the soil of subsurface layers has not been studied extensively (Linn et al. 1993). The microbial activity, in the plant root surroundings, is more, and the population is also greater as the region is rich in nutrients. Possibly that is the reason why p-nitrophenol, degraded product of parathion, is mineralized faster in the rhizosphere of rice than in unplanted soil under non-flooded and flooded conditions (Reddy and Sethunathan 1994). The degradation rate of carbofuran in cornfield is studied by Parkin and Shelton (1992). The degradation rates are much higher in the planted furrow than between the rows of corn which indicates that differences in degradation rates may be due to the augmented availability of carbon in the plant rhizosphere and thus enhances microbial activity under nutrient-rich conditions. Synthetic pesticides are mainly used against pests and pathogens during conventional agricultural practices and crop protection. Different groups of pesticides are commercially available and give quick result, and residue of these remains year after year. Bioremediation of contaminated soils with PGPR is now an emerging technique and getting the attention of the scientific community (Huang et al. 2004; Jiang et al. 2008a, b).

Pseudomonas, *Azospirillum*, *Agrobacterium*, *Bacillus*, *Enterobacter*, and *Flavobacterium* are some of the genera of known PGPR strains and also able to degrade organic and inorganic pesticide contaminants in soils as well (Zhuang et al. 2007). The inoculum density of bacteria is an important factor for biodegradation of applied pesticides (Karpouzas et al. 2005; Ramadan et al. 1990). Apart from that, many biotic and abiotic factors have shown their effect on the degradation process. Properties of soil like organic matter content, pH, temperature, soil texture, and pesticide concentration are also important for the inoculant (Liu et al. 1990; Singh et al. 2006). Lakshmi et al. (2009) during their experiment showed that the populations of *B. cereus*, *Klebsiella* sp., *Serratia marcescens*, and *P. aeruginosa* increased in soil samples after 30 days of incubation period with chlorpyrifos (50 mg kg^{-1}). Genetic studies of pesticide-degrading bacteria showed involvement of some specific enzymes in the degradation process. Involvement of plasmid in pesticide degradation has also been documented (Sayler et al. 1990).

Bacterial transformation of organochlorine like DDT under anaerobic conditions by the process of dechlorination has been reported (McRae 1989). *Alcaligenes eutrophus* A5 has been reported to grow on 4-chlorobiphenyl and used it as the sole carbon source. It can tolerate and degrade up to 1 ppm of both the ortho and para isomers of DDT when incubated with sufficient inoculum (Nadeau et al. 1994). The mechanism for attack on DDT by this bacterium appears to be analogous to the 4-chlorobiphenyl degradation pathway and probably results from the actions of the enzymes specific for 4-chlorobiphenyl degradation. A dioxygenase enzyme converts DDT to a dihydroxy derivative that undergoes meta-cleavage and ultimately produces 4-chlorobenzoic acid. This is the first report of a bacterium metabolizing DDT aerobically, though the environmental conditions are unclear (Aislabie and Lloyd-Jones 1995). Few bacterial strains are able to metabolize atrazine, an herbicide of s-triazine family, in batch culture. *Nocardia* and *Pseudomonas* species were isolated from atrazine-contaminated soil that can utilize one or more of the side chains of atrazine aerobically as the sole source of carbon. The metabolites deethylatrazine and deisopropylatrazine were shown to accumulate in soil (Cook 1987). According to Mandelbaum et al. (1995), a *Pseudomonas* species is able to mineralize atrazine from soil and used it as the sole source of nitrogen with sodium citrate as the carbon source. *Rhodococcus* spp. also can degrade some herbicides including atrazine (Behki et al. 1993; Behki and Khan 1994). Another strain of *Rhodococcus* TE1 has been found to degrade other triazine herbicides like simazine, propazine, and cyanazine. Plasmid genes were involved in the ability of *Rhodococcus* TE1 to dealkylate atrazine (Behki et al. 1993). Two bacterial isolates from agricultural field, *Stenotrophomonas* and *Arthrobacter*, have shown triazine-degrading capacity. Genes like *atzA* and *atzD* are involved in upper and lower catabolic pathway (Garcia et al. 2009). Major enzymes for hydrolysis, dealkylation, deamination, and ring cleavage are involved in atrazine degradation, and the ultimate products are cyanuric acid, ammonia, and carbon dioxide (Govantes et al. 2009; Rehan et al. 2011). The genera *Achromobacter*, *Pseudomonas*, and *Flavobacterium* are able to degrade carbofuran, the carbamate pesticide, isolated from carbofuran-treated soils. Some isolates can utilize methylamine, which is a by-product after hydrolysis of the N-methylcarbamate ester linkage of carbofuran by carbofuran hydrolase. Carbofuran phenol was also seen to accumulate as a by-product, and some isolates can mineralize carbofuran but were unable to use carbofuran phenol. So the researchers suggested an alternative pathway for degradation of carbofuran phenol (Karns et al. 1986; Chaudhry and Ali 1988). A plasmid-encoded carbofuran hydrolase gene *mcd* was first isolated by Tomasek and Karns (1989) from *Achromobacter* WM111. Some other reports of carbaryl hydrolase gene *cehA* and *cahA* were isolated from *Rhizobium* sp. (Hashimoto 2002). *cehA gene* was also found in *Pseudomonas* sp. which can degrade oxamyl (Rousidou et al. 2016). Molecular characterization of a *Rhodococcus* sp. strain NI86/21 shows a cluster of gene that codes for an aldehyde dehydrogenase and cytochrome P450 specific for thiocarbamate degradation (Nagy et al. 1995).

A *Pseudomonas* species isolated from rhizosphere of brinjal was able to degrade carbosulfan. Parathion hydrolase (phosphotriesterase) cleaves the phosphodiester linkage of organophosphate pesticide parathion to form DETP (O, O-diethylthiophosphoric acid) and p-nitrophenol (Brown 1980). The gene-encoding phosphotriesterase, designated as *opd* (organophosphate degradation), was found in strains like *Pseudomonas diminuta* and *Flavobacterium* sp. from different geographic regions (Chaudhry et al. 1988; Serdar and Gibson 1985). The breakdown product is carbon, phosphorus, and nitrogen. Six bacterial isolates from agricultural soil like *Stenotrophomonas maltophilia*, *Proteus vulgaris*, *Vibrio metschnikovii*, *Serratia ficaria*, *Serratia* spp., and *Yersinia enterocolitica* can degrade tetrachlorvinphos, as confirmed by GC-MS, when applied in consortium (Ortiz-Hernández and Sánchez-Salinas 2010). *Alcaligenes*, *Arthrobacter*, *Flavobacterium*, and *Pseudomonas* have the capacity to convert 2,4-dichlorophenoxyacetic acid, a phenoxyacetate group of pesticide, to 2,4-dichlorophenol and then hydroxylated to 3,5-dichlorocatechol which is subsequently metabolized by a modified ortho-cleavage pathway to chloromaleylacetic acid (Evans et al. 1971).

12.4 Plant Growth-Promoting Activities of Pesticide-Degrading/Tolerating Strains

A number of pesticide-tolerant/degrading PGPR strains were isolated from agricultural field which showed biocontrol activity against certain plant pathogenic fungi (Table 12.1). The fipronil- and pyriproxyfen-tolerant *Rhizobium* sp. strain MRL3 is able to exhibit all PGP traits in the absence as well as in presence of the insecticides.

Table 12.1 Different pesticide-tolerant/degrading PGPR strains with their plant growth-promoting activities

PGPR strains	Name of the pesticide	Host plant	PGP activity	References
Mesorhizobium (MRC4)	Fipronil and pyriproxyfen	Chickpea	IAA, siderophore, EPS, HCN, ammonia, catalase	Ahemad and Khan (2010a, b)
Mesorhizobium, *Rhizobium*, *Bradyrhizobium*, *Pseudomonas*, *Bacillus*, *Enterobacter*, and *Klebsiella*	Quizalofop-p-ethyl, clodinafop, metribuzin, glyphosate, fipronil, pyriproxyfen, imidacloprid, thiamethoxam, tebuconazole, hexaconazole, metalaxyl, and ketazin	Chickpea, pea, green gram, lentil, and mustard	IAA, siderophore, EPS, HCN, ammonia production; phosphate solubilization	Ahemad and Khan (2011b)
Bacillus subtilis, *Bacillus amyloliquefaciens*, *Bacillus pumilus*	Acibenzolar-s-methyl, metribuzin, napropamide, propamocarb hydrochloride, and thiamethoxam	Rice	NS	
Azotobacter vinelandii, *Azotobacter salinestris*, *Azotobacter* sp., *Azotobacter nigricans* subsp. *nigricans*, and *Azotobacter tropicalis*	Pendimethalin, glyphosate, chlorpyrifos, and phorate (5%)	Rice	IAA, GA production; phosphate solubilization	Chennappa et al. (2014)
Burkholderia sp.	Phorate, mancozeb, chlorpyrifos, and endosulfan	Tomato	IAA, phosphate solubilization	Tripti et al. (2015)
Pseudomonas spp., *Bacillus* spp.	Methomyl, imidacloprid, and carbendazim	Cowpea	IAA, siderophore, HCN, chitinase production; biocontrol of *Macrophomina* sp.	Bandopadhyay et al. (2018)
Bacillus cereus and *Bacillus safensis*	Methomyl, imidacloprid, and carbendazim	Lentil, cowpea	IAA, siderophore, ammonia, chitinase, phosphate solubilization; biocontrol of *Alternaria* sp.	Roy et al. (2018)

NS not studied

Both insecticides at recommended dose reduced plant dry weight, symbiotic properties, nutrient uptake, and yield of the lentil. The application of *Rhizobium* sp. strain MRL3 significantly reduces the harmful effect of pesticides on the lentil plants (Ahemad and Khan 2011a). Application of *Mesorhizobium* isolate MRC4 in fipronil- and pyriproxyfen-treated soil also increased symbiotic properties (nodulation and leghemoglobin content), root N, shoot N, root P, shoot P, total yield, and seed protein content of chickpea compared to the un-inoculated control (Ahemad and Khan 2010a, b). Out of the 14 *Azotobacter* strains isolated from different paddy cultivating soils, Chennappa et al. (2014) identified the presence of five *Azotobacter* species, viz., *A. vinelandii*, *A. salinestris*, *Azotobacter* sp., *A. nigricans* subsp. *nigricans*, and *A. tropicalis*. Thirteen strains out of 14 were able to grow on media supplemented with different pesticides such as pendimethalin, glyphosate (herbicides), and chlorpyrifos and phorate (insecticides) commonly used for the paddy. Five *Azotobacte*r strains showed their ability to grow at more than 5% pesticide-supplemented media, without affecting their growth rate and metabolic activities including IAA production. *Burkholderia* sp. strain L2 showed resistance against four commercially used pesticides and also showed PGP activities. *Burkholderia* sp. strain L2 isolated from rhizosphere of *L. esculentum* showed P solubilization and IAA production efficiency even in the presence of higher concentration pesticide (Tripti et al. 2015). Out of the 20 isolates isolated from rhizosphere of okra plants, 2 pesticide-tolerant strains were able to survive at 500 mg L^{-1} of bifenthrin. The isolates showed efficient Zn solubilization, catalase activity, and root colonization abilities (Najam-ul-Sehar et al. 2015). Romeh and Hendaw (2014) studied bioremediation of certain organophosphorus pesticides by two biofertilizers, *Paenibacillus (Bacillus) polymyxa* and *Azospirillum lipoferum*.

Two methomyl-degrading *Bacillus* strains, *B. cereus* and *B. safensis*, were isolated from pesticide-infested soil which showed growth-promoting activities on lentil and suppressing the leaf spot and blight pathogen *Alternaria* sp. (Roy and Das 2017; Roy et al. 2018). They grown in methomyl-supplemented minimal salt media, using it as sole carbon source as well as also used carbendazim and imidacloprid. Both these isolates displayed different plant growth promoting features in presence of all the three pesticides. They produced improved amount of chitinase and more or less similar amount of phosphate solubilization in presence of all the three pesticides (Roy et al. 2018). Bandopadhyay et al. (2018) showed five pesticide-tolerant strains that produced different fungitoxic compounds to protect *Vigna unguiculata* against some seed- and soilborne diseases caused by *Macrophomina phaseolina* and promote the growth of the plant in vitro.

12.5 Conclusion

To fight against the hazardous effect of synthetic pesticides, biological approach is by far more important in the niche of biodiversity; henceforth PGPR are inevitable. Though global consumption of biopesticides is increasing day by day, the use of synthetic pesticides is still in its climax. Use of PGPR is increasing day by day as they may be exploited as both biofertilizer and biopesticides as well. However, as our agricultural fields are already saturated with chemical pesticides, exogenous PGPR might not survive or barely perform their plant growth-promoting (PGP) and biocontrol activities in this soil. Pesticide-tolerant/degrading PGPR might show advantages in this transitional stage and might act as paradigm of disease management and plant growth promotion in the present circumstances. The use of up-to-date approaches and techniques like nanoencapsulation in combination with other multidisciplinary approaches like biotechnology, nanotechnology, material science, chemical engineering, and genetic engineering as well as different ecological and functional biological approaches might provide new formulations and openings which have the potentiality to diminish the loopholes of PGPR research.

References

Abeles FB, Morgan PW, Saltveit ME Jr (1992) Ethylene in plant biology, 2nd edn. Academic, New York

Achard P, Gusti A, Cheminant S, Alioua M, Dhondt S, Coppens F, Beemster GT, Genschik P (2009) Gibberellin signaling controls cell proliferation rate in *Arabidopsis*. Curr Biol 19 (14):1188–1193

Ahemad M, Khan MS (2010a) Growth promotion and protection of lentil (*Lens esculenta*) against herbicide stress by *Rhizobium* species. Ann Microbiol 60(4):735–745

Ahemad M, Khan MS (2010b) Ameliorative effects of *Mesorhizobium* sp. MRC4 on chickpea yield and yield components under different doses of herbicide stress. Pestic Biochem Physiol 98:183–119

Ahemad M, Khan MS (2011a) Insecticide-tolerant and plant-growth-promoting *Rhizobium* improves the growth of lentil (*Lens esculentus*) in insecticide-stressed soils. Pest Manag Sci 67(4):423–429

Ahemad M, Khan MS (2011b) Assessment of pesticide tolerant and functional diversity of bacterial strains isolated from rhizosphere of different crops. Insight Microbiol 1(1):8–19

Ahemad AB, Kibret M (2014) Mechanisms and applications of plant growth promoting rhizobacteria: current perspective. J King Saud Univ Sci 26:1–20

Aislabie J, Lloyd-Jones G (1995) A review of bacterial-degradation of pesticides. Soil Res 33:925–942

Anand T, Chandrasekaran A, Kuttalam S, Raguchander T, Prakasam V, Samiyappan R (2007) Association of some plant defense enzyme activities with systemic resistance to early leaf blight and leaf spot induced in tomato plants by azoxystrobin and *Pseudomonas fluorescens*. J Plant Interact 2(4):233–244

Andrews SC, Robinson AK, Rodríguez-Quiñones F (2003) Bacterial iron homeostasis. FEMS Microbiol Rev 27(2–3):215–237

Atzorn R, Crozier A, Wheeler CT, Sandberg G (1988) Production of gibberellins and indole-3-acetic acid by *Rhizobium phaseoli* in relation to nodulation of *Phaseolus vulgaris* roots. Planta 175:532–538

Bandopadhyay A, Roy T, Das N (2018) Isolation of some soil bacteria showing potentiality for disease control, growth enhancement and pesticide degradation in *Vigna unguiculata* L. Plant Archives 18:79–88

Bashan Y, de-Bashan LE (2005) Bacteria. In: Hillel D, Hillel D (eds) Encyclopaedia of soils in the environment, vol 1. Elsevier, Oxford, pp 103–115

Behki RM, Khan SU (1994) Degradation of atrazine, propazine, and simazine by *Rhodococcus* strain B-30. J Agril Food Chem 42:1237–1241

Behki RM, Topp E, Dick W, Germon P (1993) Metabolism of the herbicide atrazine by *Rhodococcus* strains. Appl Environ Microbiol 59:1955–1959

Belimov AA, Safronova VI, Sergeyeva TA, Egorova TN, Matveyeva VA, Tsyganov VE, Borisov AY, Tikhonovich IA, Kluge C, Preisfeld A, Dietz KJ, Stepanok VV (2001) Characterisation of plant growth-promoting rhizobacteria isolated from polluted soils and containing 1-aminocyclopropane-1-carboxylate deaminase. Canadian J Microbio 47:642–652

Belimov AA, Hontzeas N, Safronova VI, Demchinskaya SV, Piluzza G, Bullitta S, Glick BR (2005) Cadmium-tolerant plant growth-promoting bacteria associated with the roots of Indian mustard (*Brassica juncea* L. Czern.). Soil Biol Biochem 37:241–250

Bhatt PV, Vyas BRM (2014) Screening and characterization of plant growth and health promoting rhizobacteria. Int J Curr Microbiol App Sci 3:139–155

Bottini R, Fulchieri M, Pearce D, Pharis RP (1989) Identification of gibberellins A1, A3, and iso-A3 in cultures of *Azospirillum lipoferum*. Plant Physiol 90:45–47

Brown KA (1980) Phosphotriesterases of *Flavobacterium* sp. Soil Biol Biochem 12:105–112

Bruce RJ, West CA (1989) Elicitation of lignin biosynthesis and isoperoxidase activity by pectic fragments in suspension cultures of castor bean. Plant Physiol 91(3):889–897

Chakraborty U, Chakraborty BN, Basnet M (2006) Plant growth promotion and induction of resistance in *Camellia sinensis* by *Bacillus megaterium*. J Basic Microbiol 46:186–195

Chaudhry GR, Ali AN (1988) Bacterial metabolism of carbofuran. Appl Environ Microbiol 54:1414–1419

Chaudhry GR, Ali AN, Wheeler WB (1988) Isolation of a methyl parathion degrading *Pseudomonas* sp. that possesses DNA homologous to the *opd* gene from a *Flavobacterium* sp. Appl Environ Microbiol 54:288–293

Chen YP, Rekha PD, Arun AB, Shen FT, Lai WA, Young CC (2006) Phosphate solubilizing bacteria from subtropical soil and their tricalcium phosphate solubilizing abilities. Appl Soil Ecol 34:33–41

Chennappa G, Adkar-Purushothama CR, Naik MK, Suraj U, Sreenivasa MY (2014) Impact of pesticides on PGPR activity of *Azotobacter* sp. isolated from pesticide flooded paddy soils. Greener J Agril Sci 4(4):117–129

Choudhary DK, Johri BN (2009) Interactions of *Bacillus* spp. and plants—with special reference to induced systemic resistance (ISR). Microbiol Res 164(5):493–513

Chung H, Park M, Madhaiyan M, Seshadri S, Song J, Cho H, Sa T (2005) Isolation and characterization of phosphate solubilizing bacteria from the rhizosphere of crop plants of Korea. Soil Biol Biochem 37:1970–1974

Constabel CP, Ryan CA (1998) A survey of wound- and methyl jasmonate-induced leaf polyphenol oxidase in crop plants. Phytochemistry 47:507–511

Cook AM (1987) Biodegradation of s-triazine xenobiotics. FEMS Microbiology Rev 46:93–116

de Freitas JR, Banergee MR, Jermida JJ (1997) Phosphate solubilizing bacteria enhance the growth and yield but not phosphorus uptake of Canola (*Brassica napus* L.). Biol Fertil Soils 24:358–364

Devi KK, Seth N, Kothamasi S, Kothamasi D (2007) Hydrogen cyanide-producing rhizobacteria kill subterranean termite *Odontotermes obesus* (rambur) by cyanide poisoning under in vitro conditions. Curr Microbiol 54:74–78

Dilantha Fernando WG, Nakkeeran S, Zhang Y (2005) Biosynthesis of antibiotics by PGPR and its relation in biocontrol of plant diseases. In: Siddiqui ZA (ed) PGPR: biocontrol and biofertilization. Springer, Dordrecht, pp 67–109

Dobbelaere S, Vanderleyden J, Okon Y (2003) Plant growth promoting effects of diazotrophs in the rhizosphere. CRC Crit Rev Plant Sci 22:107–149

Dobereiner J, Day JM (1976) Associative symbiosis and free-living systems. In: Newton WE, Nyman CJ (eds) Proceedings of the 1st international symposium on nitrogen fixation. Washington State University Press, Pullman, pp 518–538

Evans WC, Smith BSW, Fernley HN, Davies JI (1971) Bacterial metabolism of 2,4-dichlorophenoxyacetic acid. Biochem J 122:543–551

Garcia PV, Pereira N, Oliveira LM (2009) Side-effects of organic and synthetic pesticides on cold-stored diapausing prepupae of *Trichogramma cordubensis*. BioControl 54:451–458

Garcia FP, Menendez E, Rivas R (2015) Role of bacterial bio fertilizers in agriculture and forestry. AIMS Bioeng 2:183–205

Gardi C, Jeffery S (2009) Soil biodiversity. Joint Research Center, European Commission, Luxembourg, p 27

Ghosh R, Barman S, Mukherjee S, Mandal NC (2016) Role of phosphate solubilizing *Burkholderia* spp. for successful colonization and growth promotion of *Lycopodium cernuum* L. (Lycopodiaceae) in lateritic belt of Birbhum district of West Bengal, India. Microbiol Res 183:80–91

Glick BR, Karaturovíc D, Newell P (1995) A novel procedure for rapid isolation of plant growth-promoting rhizobacteria. Can J Microbiol 41:533–536

Glick BR, Penrose DM, Li J (1998) A model for the lowering of plant ethylene concentrations by plant growth promoting bacteria. J Theor Biol 190:63–68

Glick BR, Cheng Z, Czarny J, Duan J (2007) Promotion of plant growth by ACC deaminase containing soil bacteria. Eur J Plant Pathol 119:329–339

Goldstein H, Rasbash J, Yang M, Woodhouse G, Nuttall D, Thomas S (1993) A multilevel analysis of school examination results. Oxford Rev of Edu 19:425–433

Goodlass G, Smith KA (1979) Effects of ethylene on root extension of pea (*Pisum sativum* L.) and white clover (*Trifolium repens* L.). Plant Soil 51:387–395

Govantes F, Porrúa O, Garcia V, Santero E (2009) Atrazine biodegradation in the lab and in the field: enzymatic activities and gene regulation. Microbial Biotech 2(2):178–185

Gray EJ, Smith DL (2005) Intracellular and extracellular PGPR: commonalities and distinctions in the plant– bacterium signaling processes. Soil Biol Biochem 37:395–412

Gutiérrez-Mañero F, Ramos-Solano B, Probanza A, Mehouachi J, Tadeo FR, Talon M (2001) The plant-growth-promoting rhizobacteria *Bacillus pumilus* and *Bacillus licheniformis* produce high amounts of physiologically active gibberellins. Physiol Plant 111:206–211

Hammerschmidt R, Nuckles EM, Kuc J (1982) Association of enhanced peroxidase activity with induced systemic resistance of cucumber of *Colletotrichum lagenarium*. Physiol Plant Pathol 20:73–82

Hariprasad P, Chandrashekar S, Singh SB, Niranjana SR (2013) Mechanisms of plant growth promotion and disease suppression by *Pseudomonas aeruginosa* strain 2apa. J Basic Microbiol 54(8):792–801

Hashimoto Y (2002) Study of the bacteria pathogenic for aphids, isolation of bacteria and identification of insecticidal compound. Rep Hokkaido Prefectural Agric Exp Station 102:1–48

Hoffman BM, Lukoyanov D, Yang ZY, Dean DR, Seefeldt LC (2014) Mechanism of nitrogen fixation by nitrogenase: the next stage. Chem Rev 114(8):4041–4062

Honma M, Shimomura T (1978) Metabolism of 1-Aminocyclopropane-1-carboxylic. Acid Agrl Bio Chem 42(10):1825–1831

Huang XD, El-Alawi Y, Penrose DM, Glick BR, Greenberg BM (2004) A multi-process phytoremediation system for removal of polycyclic aromatic hydrocarbons from contaminated soils. Environ Pollut 130:465–476

Jacobson CB, Pasternak JJ, Glick BR (1994) Partial purification and characterization of ACC deaminase from the plant growth-promoting rhizobacterium *Pseudomonas putida* GR12-2. Can J Microbiol 40:1019–1025

Jiang X, Li D, Xu X, Ying Y, Li Y, Ye Z, Wang J (2008a) Immunosensors for detection of pesticide residues. Biosens Bioelectron 23(11):1577–1587

Jiang CY, Sheng XF, Qian M, Wang QY (2008b) Isolation and characterization of a heavy metal-resistant *Burkholderia* sp. from heavy metal-contaminated paddy field soil and its potential in promoting plant growth and heavy metal accumulation in metal-polluted soil. Chemosphere 72:157–164

Kang HM, Sul JH, Service SK, Zaitlen NA, Kong SY, Freimer NB, Sabatti C, Eskin E (2010a) Variance component model to account for sample structure in genome-wide association studies. Nat Genet 42:348–354

Kang BG, Kim WT, Yun HS, Chang SC (2010b) Use of plant growth-promoting rhizobacteria to control stress responses of plant roots. Plant Biotechnol Rep 4:179–183

Karns JS, Mulbry W, Nelson JO, Kearney PC (1986) Metabolism of carbofuran by pure bacterial culture. Pesticide Biochem Physio 25(2):211–217

Karpouzas DG, Fotopoulou A, Menkissoglu-Spiroudi U, Singh BK (2005) Non-specific biodegradation of the organophosphorus pesticides, cadusafos and ethoprophos, by two bacterial isolates. FEMS Microbiol Ecol 53:369–378

Khan MS, Zaidi A, Wani PA, Oves M (2009) Role of plant growth promoting rhizobacteria in the remediation of metal contaminated soils. Environ Chem Lett 7:1–19

Khan MS, Zaidi A, Ahemad M, Oves M, Wani PA (2010) Plant growth promotion by phosphate solubilizing fungi – current perspective. Arch Agron Soil Sci 56(1):73–98

Klessig DF, Malamy J (1994) The salicylic acid signals in plants. Plant Mol Biol 26:1439–1458

Kloepper JW, Schroth MN (1978) Plant growth-promoting rhizobacteria on radishes. In: Proceedings of the 4th international conference on plant pathogenic bacteria. Gilbert-Clarey, Tours, pp 879–882

Kumar A, Roach C, Hirsh IS, Turley S, de Walque S, Michels PA, Hol WG (2001) An unexpected extended conformation for the third TPR motif of the peroxin PEX5 from *Trypanosoma brucei*. J Mol Biol 307(1):271–282

Lakshmi CV, Kumar M, Khanna S (2009) Biodegradation of chlorpyrifos in soil by enriched cultures. Curr Microbiol 58:35–38

Li L, Steffens JC (2002) Over expression of polyphenol oxidase in transgenic tomato plants results in enhanced bacterial disease resistance. Planta 215(2):239–247

Linn DM, Carski TH, Brusseau ML, Chang FH (1993) Sorption and degradation of pesticides and organic chemicals in soils, SSSA Special Publication No. 32. Soil Science Society of America

Liu SY, Lu MH, Bollag JM (1990) Transformation of metolachlor in soil inoculated with a *Streptomyces* sp. Biodegradation 1:9–17

Lynch JM (1985) Origin, nature and biological activity of aliphatic substances and growth hormones found in soil. In: Vaughan D, Malcom RE (eds) Soil organic matter and biological activity. Dr. W. Junk Publishers, Dordrecht, pp 151–174

Ma W, Guinel FC, Glick BR (2003a) The *Rhizobium leguminosarum* bv *viciae* ACC deaminase protein promotes the nodulation of pea plants. Appl Environ Microbiol 69:4396–4402

Ma W, Sebestianova S, Sebestian J, Burd GI, Guinel F, Glick BR (2003b) Prevalence of 1-aminocyclopropaqne-1-carboxylate in deaminase in *Rhizobia* spp. Anton Leeuw 83:285–291

Majumdar S, Chakraborty U (2017) Optimization of protease production from plant growth promoting *Bacillus amyloliquefaciens* showing antagonistic activity against phytopathogens. Int J Pharm Bio Sci 8(2):635–642

Mandelbaum RT, Allan DL, Wackett LP (1995) Isolation and characterization of a *Pseudomonas* sp. that mineralizes the s-triazine herbicide atrazine. Appl Environ Microbiol 61:1451–1457

Mano NM, Nemoto RA (2012) The pathway of Auxin biosynthesis in plants. J Exp Bot 63:2853–2872

Massala R, Legrand M, Fritig B (1980) Effect of amino oxyacetate, a competitive inhibitor of phenylalanine ammonia lyase, on the hypersensitive resistance of tobacco to tobacco mosaic virus. Physiol Plant Pathol 16:213–226

Maxton SP, Prasad SM, Andy A, Masih SA (2017) Characterization of ACC deaminase producing *B. cepacia*, *C. freundii* and *S. marcescens* for plant growth promoting activity. Int J Curr Microbiol App Sci 6(8):883–897

McRae IC (1989) Microbial metabolism of pesticides and structurally related compounds. Rev Environ Contam Toxicol 109:1–87

Nadeau LJ, Menn FM, Breen A, Sayler GS (1994) Aerobic degradation of l,l,l-trichloro-2,2-bis (4-chloropheny1) ethane (DDT) by *Alcaligenes eutrophus* A5. Appl Environ Microbiol 60:51–55

Nagy I, Schoofs G, Compernolle F, Proost P, Vanderleyden J, de Mot R (1995) Degradation of the thiocarbamate herbicide EPTC (8-ethyl dipropylcarbamothioate) and biosafening by *Rhodococcus* sp. strain NI86/21 involve an inducible cytochrome P-450 system and aldehyde dehydrogenase. J Bacteriol 177:676–687

Najam-ul-Sehar, Ahmad M, Akhtar MF, Jamil M, Latif M, Ahmad I (2015) Pesticide tolerant plant growth promoting rhizobacteria isolated from rhizosphere of okra. Soil Environ 34(2):111–118

Ngumbi E, Kloepper J (2016) Bacterial-mediated drought tolerance: current and future prospects. Appl Soil Ecol 105:109–125

Ohyama T (2010) Nitrogen as a major essential element of plants. In: Ohyama T, Sueyoshi K (eds) Nitrogen assimilation in plants, 1st edn. Research Signpost, Trivandrum, pp 1–17

Olanrewaju OS, Glick BR, Babalola OO (2017) Mechanisms of action of plant growth promoting bacteria. World J Microbiol Biotechnol 33(11):197

Ortiz-Hernández ML, Sánchez-Salinas E (2010) Biodegradation of the organophosphate pesticide tetrachlorvinphos by bacteria isolated from agricultural soils in México. Rev Int Contam Ambie 26:27–38

Owens LD (1973) Herbicidal potential of rhizobiotoxine. Weed Sci 21:63–66

Parkin TB, Shelton DR (1992) Spatial and temporal variability of carbofuran degradation in soil. J Environ Qual 21:672–678

Piccoli P, Masciarelli O, Bottini R (1996) Metabolism of 17, 17 [2 H2]-Gibberellins A4, A9, and A20 by *Azospirillum lipoferum* in chemically-defined culture medium. Symbiosis 21:167–178

Piccoli P, Lucangeli D, Schneider G, Bottini R (1997) Hydrolysis of [17,17-2 H2]Gibberellin A20-glucoside and [17,17-2 H2]Gibberellin A20-glucosyl ester by *Azospirillum lipoferum* cultured in a nitrogen-free biotin-based chemically-defined medium. Plant Growth Regul 23:179–182

Prasad R, Kumar M, Varma A (2015) Role of PGPR in soil fertility and plant health. In: Egamberdieva D, Shrivastava S, Varma A (eds) Plant Growth-Promoting Rhizobacteria (PGPR) and medicinal plants. Springer, Cham, pp 247–260

Ramadan MA, EL-Tayeb OM, Alexander M (1990) Inoculum size as a factor limiting success of inoculation for bioremediation. Appl Environ Microbiol 56:1392–1396

Read TD, Salzberg SL, Pop M, Shumway M, Umayam L, Jiang L, Holtzapple E, Busch JD, Smith KL, Schupp JM, Solomon D, Keim P, Fraser CM (2002) Comparative genome sequencing for discovery of novel polymorphisms in *Bacillus anthracis*. Science 296:2028–2033

Reddy R, Sethunathan N (1994) Mineralization of p-nitrophenol in the rhizosphere of rice. Agric Ecosyst Environ 47:313–317

Reetha S, Bhuvaneswari G, Thamizhiniyan P, Mycin TR (2014) Isolation of indole acetic acid (IAA) producing rhizobacteria of *Pseudomonas fluorescens* and *Bacillus subtilis* and enhance growth of onion (*Allium cepa* L). Int J Curr Microbiol App Sci 3(2):568–574

Rehan A, Saleem MA, Freed S (2011) Baseline susceptibility and stability of insecticide resistance of *Spodoptera litura* (F.) (Lepidoptera: Noctuidae) in the absence of selection pressure. Pak J Zool 43(5):973–978

Reid TC, Hausbeck MK, Kizilkaya K (2002) Use of fungicides and biological controls in the suppression of *Fusarium* crown and root rot of asparagus under greenhouse and growth chamber conditions. Plant Dis 86:493–498

Rengel Z, Marschner P (2005) Nutrient availability and management in the rhizosphere: exploiting genotypic differences. New Phytol 168(2):305–312

Rezzonico F, Zala M, Keel C, Duffy B, Moënne-Loccoz Y, Défago G (2007) Is the ability of biocontrol fluorescent pseudomonads to produce the antifungal metabolite 2,4- diacetylphloroglucinol really synonymous with higher plant protection? New Phytol 173:861–872

Richter C, Dirks ME, Gronover CS, Pruefer D, Moerschbacher BM (2012) Silencing and heterologous expression of ppo-2 indicate a specific function of a single polyphenol oxidase isoform in resistance of dandelion (*Taraxacum officinale*) against *Pseudomonas syringae* pv. *tomato*. Mo Plant Microbe Interact 25:200–210

Romeh AA, Hendaw MY (2014) Bioremediation of certain organophosphorus pesticides by two biofertilizers, *Paenibacillus* (*Bacillus*) polymyxa (Prazmowski) and *Azospirillum lipoferum* (Beijerinck). J Agric Sci Technol 16:265–276

Rousidou K, Chanika E, Georgiadou D, Soueref E, Katsarou D, Kolovos P, Ntougias S, Tourna M, Tzortzakakis EA, Karpouzas DG (2016) Isolation of Oxamyl-degrading bacteria and identification of cehA as a novel Oxamyl hydrolase gene. Front Microbiol 7:616

Roy T, Das N (2017) Isolation, characterization and identification of two methomyl degrading Bacteria from a pesticide treated crop field in West Bengal, India. Microbiology (Moscow) 86:753–764

Roy T, Bandopadhyay A, Sonawane JP, Majumdar S, Mahapatra RN, Alam S, Das N (2018) Bio-effective disease control and plant growth promotion in lentil by two pesticide degrading strains of *Bacillus* sp. Biol Control 127:55–63

Ryu CM, Farag MA, Hu CH, Reddy MS, Wei HX, Pare PW, Kloepper JW (2003) Bacterial volatiles promote growth in *Arabidopsis*. Proc Natl Acad Sci 100:4927–4932

Saber K, Nahla L, Ahmed D, Chedly A (2005) Effect of P on nodule formation and N fixation in bean. Agron Sustain Dev 25(3):389–393

Sayler GS, Hooper SW, Layton AC, King JMH (1990) Catabolic plasmids of environmental and ecological significance. Microb Ecol 19:1–20

Selvaraj T, Ambalavanan S (2013) Induction of defense-related enzymes in anthurium by application of fungal and bacterial biocontrol agents against *Colletotrichum gloeosporioides*. J Curr Microbiol App Sci 2(12):661–670

Serdar CM, Gibson DT (1985) Enzymic hydrolysis of organophosphates: cloning and expression of a parathion hydrolase gene from *Pseudomonas diminuta*. Bio/Technology 3:567–571

Shahgoli H, Ahangar AG (2014) Factors controlling degradation of pesticides in the soil environment: a review. Agril Sci Dev 3:273–278

Shrivastava S, Prasad R, Varma A (2014) Anatomy of root from eyes of a microbiologist. In: Morte A, Varma A (eds) Root engineering, vol 40. Springer-Verlag, Berlin, pp 3–22

Siddiqui IA, Shaukat SS, Hussain Sheikh I, Khan A (2006) Role of cyanide production by *Pseudomonas fluorescens* CHA0 in the suppression of root-knot nematode, *Meloidogyne javanica* in tomato. World J Microbiol Biotechnol 22:641–650

Singh AK, Varaprasad KS (2008) Criteria for identification and assessment of agro-biodiversity heritage sites: evolving sustainable agriculture. Curr Sci 94(9):1131–1138

Singh BK, Walker A, Denis J, Wright DJ (2006) Bioremedial potential of fenamiphos and chlorpyrifos degrading isolates: influence of different environmental conditions. Soil Biol Biochem 38:2682–2693

Sturz AV (2006) Bacterial root zone communities, beneficial allelopathies and plant disease control. In: Mukerji KG (ed) Allelochemical biological control of plant pathogens and diseases. Springer, Dordrecht, pp 123–142

Sturz AV, Christie BR, Matheson BG, Nowak J (1997) Biodiversity of endophytic bacteria which colonize red clover nodules, roots, stems and foliage and their influence on host growth. Biol Fertil Soils 25:13–19

Sturz AV, Christie BR, Nowak J (2000) Bacterial endophytes: potential role in developing sustainable systems of crop production. Crit Rev Plant Sci 19:1–30

Suslow TV, Kloepper JW, Schroth MN, Burr TJ (1979) Beneficial bacteria enhance plant growth rhizobacteria. Calif Agric Exp Stn 33:15–17

Tomasek PH, Karns JS (1989) Cloning of a carbofuran hydrolase gene from *Achromobacter* sp. WM111 and its expression in gram-negative bacteria. J Bacteriol 171:4038–4044

Tripti KA, Kumar V, Anshumali (2015) Effect of commercial pesticides on plant growth-promoting activities of *Burkholderia* sp. strain L2 isolated from rhizosphere of *Lycopersicon esculentum* cultivated in agricultural soil. Toxicol Environ Chem 97(9):1180–1189

Tu JC (1978) Protection of soybean from severe *Phytophthora* root rot by *Rhizobium*. Physiol Plant Pathol 12:233–240

Tu JC (1979) Evidence of differential tolerance among some root rot fungi to rhizobial parasitism in vitro. Physiol Plant Pathol 14:171–177

Van Loon LC, Bakker PAHM (2006) Induced systemic resistance as a mechanism of disease suppression by rhizobacteria. In: Siddiqui ZA (ed) PGPR: biocontrol and biofertilization. Springer, Dordrecht, pp 39–66

Van Loon LC, Bakker PAHM, Pieterse CMJ (1998) Systemic resistance induced by rhizosphere bacteria. Annu Rev Phytopathol 36:453–483

Voisard C, Keel C, Haas D, Defago G (1989) Cyanide production by *Pseudomonas fluorescens* helps suppress black root of tobacco under gnotobiotic conditions. EMBO J 8:351–358

Wandersman A, Keener DC, Snells-John J, Miller R, Flaspohler P, Dye M (2004) Empowerment evaluation: principles and action. In: Jason LA, Keys CB, Suarez-Balcazar Y, Taylor RR, Davis M, Durlak J et al (eds) Participatory community research: theories and methods in action. American Psychological Association, Washington, DC, pp 139–156

Wani SP, Singh P, Dwivedi RS, Navalgund RR, Ramakrishna A (2005) Biophysical indicators of agro-ecosystem services and methods for monitoring the impacts of NRM technologies at different scale. In: Shiferaw B, Freemen HA, Swinton SM (eds) Natural resource Management in Agriculture: methods for assessing economic and environmental impacts. CAB International, Wallingford, pp 97–123

Winkelmann G (2002) Microbial siderophore-mediated transport. Biochem Soc Trans 30:691–696

Zakharova E, Shcherbakov A, Brudnik V, Skripko N, Bulkhin N, Ignatov V (1999) Biosynthesis of indole-3-acetic acid in *Azospirillum brasilense*. Insights from quantum chemistry. Eur J Biochem 259:572–576

Zhang R, Wang B, Ouyang J, Li J, Wang Y (2008) Arabidopsis indole synthase, a homolog of tryptophan synthase alpha, is an enzyme involved in the Trp-independent indole-containing metabolite biosynthesis. J Integr Plant Biol 50:1070–1077

Zhao J, Zhou L, Wub J (2010) Promotion of *Salvia miltiorrhiza* hairy root growth and tanshinone production by polysaccharide–protein fractions of plant growth-promoting rhizobacterium *Bacillus cereus*. Process Biochem 45:1517–1522

Zhou K, Binkley D, Doxtader KG (1992) A new method estimating gross phosphorus mineralization and immobilisation rates in soils. Plant Soil 147:243–250

Zhu L, Huang X, Shi H, Cai X, Song Y (2011) Transport pathways and potential sources of PM10 in Beijing. Atmos Environ 45:594–604

Zhuang X, Chen J, Shim H, Bai Z (2007) New advances in plant growth-promoting rhizobacteria for bioremediation. Environ Int 33:406–413

Chapter 13
Structure and Function of Rhizobiome

Raja V. N. R. Vukanti

Abstract Plant roots can select for certain microbial species from soil microbiome and interact with them. As a consequence, the structure (or composition) of root-associated microbiome (here after referred to as rhizobiome) is significantly different from that of soil microbiome. Although, it is widely accepted that rhizobiome positively influences plant growth and health, relatively less is known about its complete structure and function. High-resolution and large-scale studies are essential to unravel the structure and function of rhizobiome. Moreover, identification of "core rhizobiome" or "heritable rhizobiome" of different crop plants is a top priority for accelerating translational research toward improving crop productivity in an environmentally sustainable manner. Here, I summarize information about the structure and function of various rhizobiomes that is recently made available using culture-independent technologies. I also review the factors that regulate composition of rhizobiome. Specifically, I discuss the role of root exudates and plant immune system in shaping rhizobiome.

13.1 Introduction

With the greatest diversity of microbes on Earth estimated existence of a trillion (10^{12}) species (Locey and Lennon 2016), only a small number of microbial species that exist in any environment have been described (Gilbert et al. 2014). Likewise, less is known about microbes that associate with plant roots or "rhizobiome" (Bisseling et al. 2009; O'Brien et al. 2018; Walters et al. 2018).

Rhizobiome is found in three microhabitats of roots (or rhizocompartments), viz., endorhizosphere (area within root cells and between root cells), rhizoplane (surface of root), and rhizosphere (area of soil that is being influenced by plant root secretions) (Reinhold-Hurek et al. 2015). Rhizobiome is actually a part of "plant microbiome," which includes microbes that inhabit the three root microhabitats,

R. V. N. R. Vukanti (✉)
Department of Microbiology, Bhavan's Vivekananda College, Sainikpuri, Secunderabad, Telangana, India
e-mail: rvukanti@kent.edu

A. Varma et al. (eds.), *Plant Microbe Symbiosis*,
https://doi.org/10.1007/978-3-030-36248-5_13

leaves, shoots, flowers, and seeds of plants (Turner et al. 2013a). However, among the different plant organs, the microhabitats of root are the prime sites for plant-microbe and microbe-microbe interaction; this is mainly because of extensive root exudation (Bais et al. 2006; Jones et al. 2009; Singh et al. 2019). Root exudates cause rhizosphere to become a nutrient-rich environment when compared to surrounding bulk or root-free soil (Demoling et al. 2007). Accordingly, the microbes are more abundant and more active in, on, and around the roots, when compared to those in bulk soil (Heijnen et al. 1995; Semenov et al. 1999; De Angelis et al. 2009).

Although diverse groups of organisms such as bacteria, fungi, protozoa, and algae are found to be associated with plant roots, bacteria are the most abundant organisms. One gram of plant root contains up to 1 billion bacterial cells (Reinhold-Hurek et al. 2015) and more than 30,000 bacterial species (Mendes et al. 2011). Rhizosphere microbiome is considered as the second genome of a plant (Berendsen et al. 2012). Rhizobiome is an important component of global biogeochemical cycling (Philippot et al. 2009). Together, because of the enormous energy flux within the rhizosphere and volume of rhizosphere soil, along with myriad types of organisms and the galaxy of interactions, it is considered as the largest ecosystem on Earth (Barriuso and Solano 2008). Therefore, it is of fundamental importance that the structure and function of rhizobiome are comprehensively understood.

Rhizobiome can positively influence plant growth and health (Podile and Kishore 2007; Berendsen et al. 2012; Turner et al. 2013a; Kwak et al. 2018). An understanding of the overall structure and function of rhizobiome can be translated into a solution for global food production problem. It is estimated that human population of the world may reach 10 billion by 2050 (www.unfpa.org). For improving food production, if more agricultural land is created, it will destroy the biodiversity. Similarly, if more fertilizers and pesticides are utilized, they will pollute the ecosystems and atmosphere. Excessive application of phosphorous and nitrogen fertilizers forces these chemicals to leach from the field and appear in nearby groundwater, rivers, and streams. These chemicals in natural waters cause algal blooms; nitrates can be further converted into greenhouse gas—nitrous oxide (Nosengo 2003; Reay 2004). Therefore, production of sufficient food in an environmentally sustainable manner is a major challenge in the twenty-first century. While, plant biotechnologists focus on creating plant varieties that have enhanced resistance to diseases and pests and greater tolerance to abiotic stresses such as drought and salinity, microbiologists should explore the role of bacteria that live in, on, and around the plant roots to enhance nutrient supply for plants and also to protect plants from pathogens.

Overall, because of the emerging functional importance for the rhizobiome, high-resolution and large-scale studies for characterization of structure and function of rhizobiome are warranted. These studies unravel the composition, lifestyles, and ecological roles of rhizobiome. In addition, this information is useful for favoring translational research that is aimed to increase crop productivity while promoting sustainable agriculture. Here, we summarize the information about the structure and function of various rhizobiomes that is recently made available using culture-independent technologies. In addition, we review the factors, including the role of root exudates and plant immune system, which regulate composition of rhizobiome.

13.2 Structure of Rhizobiome

Before the advent of genomic technologies, majority of researchers have screened rhizobiomes using culture-dependent methods for bacteria (and fungi) that have potential for commercial exploitation as bio-fertilizers and biocontrol agents. Because, only a minority (less than 1%) of environmental bacteria can be cultured in a laboratory (Rappé and Giovannoni 2003), culture-dependent studies are insufficient for complete characterization of rhizobiomes. In recent years, using culture-independent methods such as 16S rRNA gene sequencing and 16S rRNA microarray (PhyloChip), a few researchers have characterized the structure of rhizobiomes of different plants including various food crops (see Table 13.1, for the references).

Rhizobiome is less diverse and different when compared to the bulk soil microbiome (Lundberg et al. 2012), and it is assembled from the bulk soil microbiome (Edwards et al. 2015; Fig. 13.1). The main phyla detected in soil microbiome are *Proteobacteria*, *Acidobacteria*, *Bacteroidetes*, *Chloroflexi*, *Actinobacteria*, and *Gemmatimonadetes* (Kwak et al. 2018). The rhizobiome of various plants is

Table 13.1 Dominant bacterial phyla found in rhizobiome of various plants including the important food crops (see text for more details)

Plant	Dominant phyla	References
Barley	*Actinobacteria, Bacteroidetes, Proteobacteria*	Bulgarelli et al. (2015)
Maize[a]	*Actinobacteria, Bacteroidetes, Proteobacteria*	Niu et al. (2017)
Maize[a]	*Proteobacteria*	Walters et al. (2018)
Mustard	*Actinobacteria, Bacteroidetes, Proteobacteria*	Wagner et al. (2016)
Potato[b]	*Actinobacteria, Bacteroidetes, Firmicutes, Proteobacteria*	Weinert et al. (2011)
Rice	*Bacteroidetes, Firmicutes, Proteobacteria, Chloroflexi, Fibrobacteres, and Spirochaetes*	Edwards et al. (2015)
Soybean	*Firmicutes, Proteobacteria*	Sugiyama et al. (2017)
Sugarcane	*Bacteroidetes, Proteobacteria*	de Souza et al. (2016)
Thale cress	*Actinobacteria, Bacteroidetes, Proteobacteria*	Bulgarelli et al. (2012)
Thale cress	*Actinobacteria, Proteobacteria, Cyanobacteria*	Lundberg et al. (2012)
Tomato	*Bacteroidetes, Firmicutes, Proteobacteria, Verrucomicrobia, Ignavibacteriae*	Kwak et al. (2018)
Wheat	*Actinobacteria, Firmicutes, Proteobacteria*	Velazquez-Sepulveda et al. (2012)
Wild oat[b]	*Actinobacteria, Bacteroidetes, Firmicutes, Proteobacteria, Verrucomicrobia, and Nitrospira*	De Angelis et al. (2009)

[a]Phyla belonging to "core rhizobiome" or "heritable rhizobiome" of maize are represented
[b]Except these two studies, all of the above studies characterized rhizobiomes using high-throughput 16S rRNA gene amplicon sequencing. The studies on rhizobiomes of wild oat, sugar beet, and potato were carried out using 16S rRNA microarray (PhyloChip)

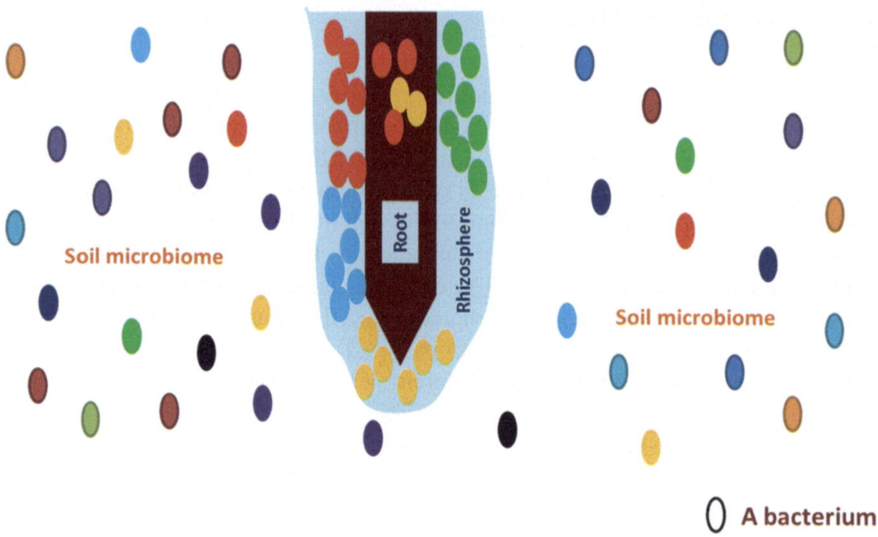

Fig. 13.1 Cartoon of rhizobiome (microbes present in the three microhabitats of root such as endorhizosphere, rhizoplane, and rhizosphere) emphasizing that only a subset of soil microbiome is able to dominate rhizobiome (picture is not drawn to scale)

dominated by members of four bacterial phyla: *Actinobacteria*, *Bacteroidetes*, *Firmicutes*, and *Proteobacteria* (Table 13.1 and Niu et al. 2017). In other words, the relative abundance of members of the above phyla increases in the rhizobiomes, when compared to that of bulk soil, although the structure of rhizobiome varies with plant type at lower taxonomic levels (e.g., family and genus). Among the four most abundant phyla in rhizobiomes, members of *Proteobacteria* were always found. *Bacteroidetes* can get involved in denitrification (Van Spanning et al. 2005). Whereas *Proteobacteria*, *Bacteroidetes*, and *Firmicutes* represent copiotrophs (r-strategists), *Actinobacteria* represent oligotrophs (k-strategists) and are prolific producers of diverse antimicrobial compounds (De Angelis et al. 2009; Chaparro et al. 2014). *Actinobacteria* are found mostly in soils, rhizosphere, and endorhizosphere (Turner et al. 2013a). The abundant representation of both copiotrophs and oligotrophs in the rhizobiome suggests that roots selectively recruit and associate with microbes using the mechanisms which deviate from the statement that "hungry soil microbes migrate toward root because of its exudation."

In barley rhizobiome, when compared to the bulk soil, enrichment of actinobacterial family, *Microbacteriaceae*; proteobacterial families, *Comamonadaceae*, *Oxalobacteraceae*, *Xanthomonadaceae*, *Rhizobiaceae*, and *Myxococcaceae*; and *Bacteroidetes* family, *Flavobacteriaceae*, was observed (Bulgarelli et al. 2015).

The simplified core microbiome associated with maize roots was identified (Niu et al. 2017), and it is represented by seven bacterial strains: *Enterobacter cloacae*, *Stenotrophomonas maltophilia*, *Ochrobactrum pituitosum*, *Herbaspirillum frisingense*, *Pseudomonas putida*, *Chryseobacterium indologenes*, and *Curtobacterium pusillum*.

All of these bacteria except *Chryseobacterium indologenes* and *Curtobacterium pusillum* belong to *Proteobacteria. Chryseobacterium indologenes* belongs to *Bacteroidetes. Curtobacterium pusillum* belongs to *Actinobacteria. E. cloacae* interacted positively with *S. maltophilia*, *O. pituitosum*, and *C. indologenes* and negatively with *C. pusillum.* Similarly, maize core microbiome of rhizosphere (plus, rhizoplane) was identified (Walters et al. 2018) from a very large field study ($n = 4911$) that included 27 maize inbred lines in five fields across three states of the United States and with partial replication of the experiment after 5 years. The maize core rhizobiome of rhizosphere consisted of seven operational taxonomic units (OTUs) all belonging to phylum *Proteobacteria*, including three α-proteobacteria (*Agrobacterium*, *Bradyrhizobiaceae*, *Devosia*), two β-proteobacteria (*Comamonadaceae*), and two γ-proteobacteria (*Pseudomonas* and *Sinobacteraceae*). Of these bacteria, *Pseudomonas* was abundantly associated with maize roots (Niu et al. 2017; Walters et al. 2018).

In rice rhizobiome, when compared to bulk soil, OTUs belonging to phyla *Bacteroidetes*, *Firmicutes*, *Proteobacteria*, *Chloroflexi*, *Fibrobacteres*, and *Spirochaetes* were differentially enriched (Edwards et al. 2015). Specifically, OTUs belonging to β-proteobacterial families, *Rhodocyclaceae* and *Comamonadaceae*, and α-proteobacterial genus *Pleomorphomonas* were enriched in all three microhabitats (endorhizosphere, rhizoplane, and rhizosphere) of rice roots. OTUs belonging to *Bacteroidetes*, *Proteobacteria*, *Chloroflexi*, *Fibrobacteres*, and *Spirochaetes* were enriched in rhizoplane and endorhizosphere of rice. Only 17 OTUs mainly belonging to *Proteobacteria* and *Acidobacteria* were significantly depleted in rhizosphere of rice. 730 OTUs mainly belonging to *Acidobacteria* and *Planctomycetes* were depleted in rhizoplane. 1961 OTUs mainly belonging to *Acidobacteria*, *Planctomycetes*, *Chloroflexi*, and *Verrucomicrobia* were reduced in endorhizosphere. Since rhizoplane shares 713 of the 1961 OTUs that were depleted in endorhizosphere, it was suggested that rhizoplane acts as a gate as it controls the entry of microbes into the endorhizosphere. Accordingly, the diversity of rice microbiome decreased in endorhizosphere when compared to the rhizosphere.

In soybean rhizosphere, when compared to the bulk soil, relative abundance of *Proteobacteria* and *Firmicutes* was higher and *Acidobacteria* was lower (Sugiyama et al. 2017). Significantly higher abundance of bacteria belonging to taxonomic families such as the *Bradyrhizobiaceae* and *Bacillaceae* and lower abundance of families such as *Gemmatimonadaceae* and *Chitinophagaceae* were observed in the soybean rhizosphere, when compared to the bulk soil.

In the sugarcane rhizobiome, when compared to the bulk soil, enrichment of *Proteobacteria* families, *Hyphomicrobiaceae*, *Rhodospirillaceae*, and *Sinobacteraceae*; *Bacteroidetes* family, *Cytophagaceae*; and *Verrucomicrobia* family, *Chthoniobacteraceae*, was observed (de Souza et al. 2016). Specifically, *Proteobacteria* genera, *Azospirillum*, *Beijerinckia*, *Bradyrhizobium*, *Burkholderia*, *Herbaspirillum*, and *Gluconacetobacter*, were found in the sugarcane rhizobiome.

In thale cress (*Arabidopsis thaliana*) endorhizosphere microbiome, when compared to the bulk soil, significant enrichment of an actinobacterial family, *Streptomycetaceae*; a *Bacteroidetes* family, *Flavobacteriaceae*; and proteobacterial families, *Rhizobiaceae*, *Comamonadaceae*, and *Oxalobacteraceae*, was observed (Bulgarelli et al. 2012). In another study of thale cress rhizobiome, when compared to the bulk soil, significant enrichment of an actinobacterial family, *Streptomycetaceae*, and proteobacterial families, *Rhizobiaceae*, *Methylobacteriaceae*, *Pseudomonadaceae*, and *Moraxellaceae*, and depletion of *Acidobacteria*, *Verrucomicrobia*, *Gemmatimonadetes*, and *Cyanobacteria* and various proteobacterial families (*Sphingomonadaceae*, *Phyllobacteriaceae*, *Xanthomonadaceae*) were observed (Lundberg et al. 2012). A subset of soil bacteria is recruited into the endorhizosphere. When compared to soil communities, the endorhizosphere communities were less diverse.

In tomato rhizosphere, when compared to the bulk soil, significant enrichment of *Bacteroidetes*, *Firmicutes*, *Proteobacteria*, *Verrucomicrobia*, *Ignavibacteriae*, and BRC1 and lower abundance of *Acidobacteria* and *Gemmatimonadetes* were observed (Kwak et al. 2018). The tomato rhizosphere had lesser number of OTUs compared to bulk soil and indicates that the species richness in the rhizosphere is lesser than that of bulk soil. In wheat rhizosphere, 30 OTUs belonging to *Proteobacteria* (α-proteobacteria, β-proteobacteria, γ-proteobacteria, and δ-proteobacteria), *Firmicutes* (*Bacillus* spp. and *Clostridium* spp.), and *Actinobacteria* and uncultivable bacteria were observed (Velazquez-Sepulveda et al. 2012). Wild oat rhizosphere, when compared to bulk soil, had higher abundance of *Proteobacteria*, *Firmicutes*, *Actinobacteria*, *Bacteroidetes*, *Verrucomicrobia*, and *Nitrospira* (De Angelis et al. 2009). Relative abundance for 7% of oat rhizosphere microbial community varied when compared to the bulk soil.

13.3 Functions of Rhizobiome

Plants depend on rhizobiome for a variety of biochemical functions that improve plant growth and health (Fig. 13.2; Podile and Kishore 2007; Berendsen et al. 2012; Turner et al. 2013a; Kwak et al. 2018; Singh et al. 2019). Rhizobiome contributes to (1) plant growth, by increasing availability of limiting nutrients to plant and by producing plant growth hormones, ACC deaminase, and/or volatile organic compounds, and (2) plant health, by controlling pathogens using antimicrobials and other mechanisms and by stimulating plant immunity.

13.3.1 Rhizobiome Contributes to Plant Nutrition by Increasing Availability of the Limiting Nutrients

Certain members of rhizobiome, often referred to as plant growth-promoting rhizobacteria (PGPR), have the capacity to fix the atmospheric N_2, thereby improving the availability of nitrogen to the plant (James 2000; Prasad et al. 2015). Both symbiotic N_2-fixing rhizobia (located in endorhizosphere) such as *Rhizobium* sp., *Sinorhizobium* sp., *Mesorhizobium* sp., *Bradyrhizobium* sp., *Azorhizobium* sp., and *Allorhizobium* sp. and free-living N_2-fixing microbes (located in rhizosphere) such

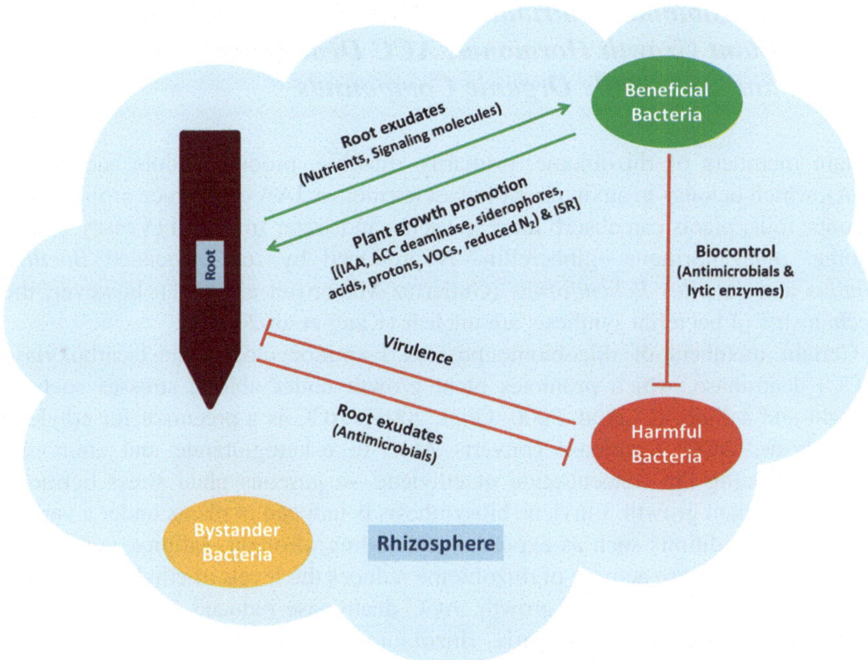

Fig. 13.2 Scheme of plant-microbe interactions that occur within rhizosphere (picture not drawn to scale). Root exudates can both recruit and defend certain members of soil microbiome. Beneficial rhizobiome promotes plant growth by a variety of direct and indirect mechanisms. *IAA* indole acetic acid, *ACC* deaminase, 1-aminocyclopropane-1-carboxylase deaminase, *VOCs* volatile organic compounds, *ISR* induced systemic resistance

as *Azospirillum* sp., *Herbaspirillum* sp., *Acetobacter* sp., *Azotobacter* sp., *Azoarcus* sp., *Bacillus polymyxa*, *Burkholderia* sp., and *Gluconacetobacter diazotrophicus*, significantly improve nitrogen nutrition in plants (Vessey 2003).

Less than 5% of phosphorus in soils is directly available to plants, and rhizobiome has the capacity to make it bioavailable to plants (Rodriguez et al. 2006). Microbes solubilize phosphorus from organic and inorganic sources by secreting phosphatases or organic acids such as acetate, succinate, citrate, and gluconate. Although organic acids in plant root exudates can also contribute to phosphate solubilization, phosphate-solubilizing microbes have a greater role in this regard.

Rhizobiome can also improve availability of iron to plants. Like phosphorus, iron is less soluble in soil and it is the third limiting plant nutrient. Availability of iron is increased by microbes and plant roots through the release of organic acids or protons and a variety of siderophores (Ahmed and Holmstrom 2014). While the organic acids decrease soil pH and increase the solubility of iron, siderophores chelate the iron; the iron-siderophore complex is then taken up by root cells. *Bacillus subtilis* GB03 promoted acquisition of iron by thale cress when it is grown in iron-limited soil (Zhang et al. 2009). Further, bacterial siderophores efficiently scavenge iron and make it unavailable for fungal plant pathogens in rhizosphere (Duijff et al. 1999).

13.3.2 Rhizobiome Contributes to Plant Growth by Producing Plant Growth Hormones, ACC Deaminase, and/or Volatile Organic Compounds

Certain members of rhizobiome, primarily rhizobia, produce indole acetic acid (IAA), which belongs to auxin class of plant hormones. IAA can induce proliferation of roots; thus, plants can absorb more nutrients and water from soil (Vessey 2003). Another plant hormone—gibberellin—is produced by root-associated *Bacillus pumilus* and *Bacillus licheniformis* (Gutierrez-Manero et al. 2001); however, the mechanisms of bacterial synthesis are unclear (Kang et al. 2009).

Certain members of rhizobiome possess 1-aminocyclopropane-1-carboxylase (ACC) deaminase, which promotes plant growth under abiotic stresses such as drought and salinity (Li et al. 2000; Glick 2005). ACC is a precursor for ethylene biosynthesis. ACC deaminase converts ACC to α-ketoglutarate and ammonia, thereby reducing the concentration of ethylene—a gaseous plant stress hormone that inhibits plant growth. Ethylene biosynthesis is induced in plants under a variety of stressful conditions such as exposure to flooding, drought, salinity, and pathogens. ACC deaminase activity of rhizobiome reduces the levels of ethylene in plants and, therefore, promotes plant growth. ACC deaminase reduced salt stress in pea plants (Wang et al. 2016). Similarly, rhizobiome assembled under drought stress improved plant resistance to drought stress (Rolli et al. 2015).

Certain microbes produce volatile organic compounds (VOC) such as 1-butanone (acetoin) and 2,3-butanediol, which can promote plant growth (Ryu et al. 2003).

13.3.3 Rhizobiome Directly Contributes to Plant Health by Controlling the Pathogens

Plants grown in disease-suppressive soils are protected from certain pathogens such as *Streptomyces* scab disease of potato, *Fusarium* wilt of melons, *Thielaviopsis* black rot of tobacco, *Rhizoctonia* damping-off of sugar beet, and take-all disease of wheat (Weller et al. 2002). Pasteurization of disease-suppressive soil eliminates its ability to suppress the diseases (Mendes et al. 2011; Yin et al. 2013); this suggests that native soil microbiome—especially rhizobiome—is responsible for disease suppression. Repeated cultivation of a single crop plant causes the soil to become disease-suppressive by allowing the enrichment of antagonistic microbiome (Raaijmakers and Weller 1998). In sugar beet, the relative abundance of certain rhizospheric bacteria belonging to *Proteobacteria* (*Pseudomonadaceae*, *Burkholderiaceae*, *Xanthomonadales*), *Firmicutes* (*Lactobacillaceae*), and *Actinobacteria* was associated with suppression of root disease caused by *Rhizoctonia solani* (Mendes et al. 2011). Similarly, *Rhizoctonia* patch disease was suppressed by collective action of three bacterial species: *Pantoea agglomerans*, *Exiguobacterium acetylicum*, and *Microbacteria* sp. (Barnett et al. 2006). The core microbiome of maize roots that comprised of seven bacterial strains protected the plant from colonization by *Fusarium verticillioides* (formerly *Fusarium moniliforme*), the causative agent of seedling blight disease (Niu et al. 2017). Further, when native rhizobiome of a tomato cultivar that was resistant to soilborne pathogen *Ralstonia solanacearum* was transplanted to a susceptible tomato cultivar, it suppressed the disease in the susceptible cultivar (Kwak et al. 2018). In this tomato study, a *Flavobacterium* sp. was identified to offer protection from *R. solanacearum.* These results suggest that disease suppression in certain soils is due to either diverse microbial communities or a single microbial species. Certain members of rhizobiome produce a variety of antimicrobial compounds such as hydrogen cyanide, phenazines, 2,4-diacetylphloroglucinol (DAPG), pyoluteorin, pyrrolnitrin, cyclic lipopeptide surfactants, zwittermicin A, and bacteriocins as well as lytic enzymes (Lugtenberg and Kamilova 2009). Using these antimicrobials and lytic enzymes, rhizobiome can directly protect plants from pathogens. In addition, rhizobiome can indirectly affect soilborne pathogens by competing for micronutrients.

13.3.4 Rhizobiome Indirectly Contributes to Plant Health by Stimulating the Plant Resistance

Beneficial members of rhizobiome provide indirect plant protection by inducing systemic resistance. Induced systemic resistance (ISR) primes plants to have higher and faster defensive capacity (Zamioudis and Pieterse 2012). ISR depends on jasmonate and ethylene signaling, and it is observed when the plants were exposed to beneficial bacteria or their conserved cellular constituents such as flagella, cell wall components, O-antigen of lipopolysaccharide, and siderophores, which are often referred to as microbe-associated molecular patterns (MAMPs). A rhizobacterium *Pseudomonas putida* KT2440 protected thale cress from *Pseudomonas syringae* pv. *tomato* DC3000—a phytopathogen; the mechanism involved was ISR (Matilla et al. 2010). A rhizobacterium *Pseudomonas aureofaciens* 63-28 induced the defense system in soybean seedlings that led to improved resistance to *Rhizoctonia solani* AG-4 (Jung et al. 2011). Similarly, some isolates of *Mitsuaria* and *Burkholderia* inhibited fungal and oomycetal diseases in tomato and soybean (Benítez and Gardener 2009). Together, rhizobiome protects plants from pathogens and improve their fitness (Haney et al. 2015).

13.3.5 Functional Genomics of Barley Rhizobiome

When compared to the information about the structure of rhizobiomes, not much is known about global function of rhizobiome. Bulgarelli et al. (2015), using metatranscriptomics, identified the 12 biological functional categories that were significantly enriched in barley rhizobiome. The 12 categories of biological function, listed in ascending order of p-value, were type III protein secretion system, adhesion, regulation of virulence, siderophores, secretion, transposable elements, periplasmic stress, sugar phosphotransferase systems, bacteriophage integration-excision-lysogeny, invasion and intracellular resistance, type VI protein secretion system, and detoxification. Majority of these functions such as adhesion, detoxification, stress responses, sugar transport, and secretion are required for survival of rhizobiome. Siderophores are useful for iron mobilization. The rest of the functions were responsible for microbe-microbe interactions (type V1 secretion system), host-pathogen interactions (type III secretion system, regulation of virulence, invasion, and intracellular resistance), and microbe-phage interactions (transposable elements and bacteriophage integration).

13.4 Factors That Affect Structure of Rhizobiome

Structure (or composition) of rhizobiome is governed by various abiotic and biotic factors. For example, soil type (Bulgarelli et al. 2012; Lundberg et al. 2012; Edwards et al. 2015), plant species (Berg and Smalla 2009; Turner et al. 2013b), plant genotype (Edwards et al. 2015; Wagner et al. 2016), plant age (Kwak et al. 2018; O'Brien et al. 2018; Walters et al. 2018), plant developmental stage (Berg et al. 2014; Chaparro et al. 2014), cultivar type (Kwak et al. 2018), maternal effect (Hardoim et al. 2012), and fertilizer amendment (O'Brien et al. 2018) affect the structure of the rhizobiome. Soil type is identified as the main driver for altering the composition of thale cress core rhizobiome (Bulgarelli et al. 2012; Lundberg et al. 2012). Interestingly, the plant cell wall (lignocellulose) serves as a cue for colonization by certain root-associated microbiota, especially some *Proteobacteria* (Bulgarelli et al. 2012). Further, amoeba—a protozoan predator of bacteria—altered the composition of rhizobiome of thale cress (Rosenberg et al. 2009). It was suggested that the bacterial groups that have avoided predation by amoeba have enriched in the rhizosphere.

How plants assemble certain microbial species and communities is an active area in plant microbiology. Mechanistic explanation for the several biotic factors (such as plant species, plant genotype, plant age, plant developmental stage, and cultivar type) that determine the structure of rhizobiomes can be likely that these factors differentially alter the composition of root exudates (Haichar et al. 2008; Walters et al. 2018). The two-component system involved in chemotaxis is significantly expressed during later stages of plant development, and its expression is correlated with root exudates—glycine and xylitol (Chaparro et al. 2014). Glycine is a chemoattractant for several PGPR and endophytic bacteria (de Weert et al. 2002), and xylitol is a chemoattractant for nonsymbiotic N_2-fixing *Azotobacter vinelandii*. Soils exposed to specific fractions of root exudates significantly altered the composition of their microbiome (Badri et al. 2013). Thale cress mutant that secretes more phenolic compounds (and less sugars), when compared to the wild type, was found to enrich beneficial microbes such as PGPR and those involved in N_2 fixation in its rhizobiome (Badri et al. 2009). Glucosinolate, a secondary metabolite produced by brassicaceous plants, altered the composition of thale cress rhizobiome (Bressan et al. 2009). Rice root exudates, in the absence of plant, have prepared *Azoarcus* sp. BH72 to get into roots by upregulating its genes responsible for rhizosphere competence and endophytic colonization (Shidore et al. 2012). A leguminous plant, pea, drastically affected its rhizobiome, when compared to that of cereal plants such as wheat and oat (Turner et al. 2013b). The rate of root exudation varies with plant species, the age of plant, and environmental conditions (Nguyen 2003) including biotic stress (Matilla et al. 2010). These observations underscore that root exudates alter the structure of rhizobiome. Therefore, the role of root exudates in shaping the rhizobiome is discussed.

13.4.1 Root Exudates Determine the Structure of Rhizobiome

13.4.1.1 Root Exudates

Roots of land plants, apart from performing general functions such as acquisition of water and minerals as well as anchorage to the aboveground plant parts, have a special function—synthesis and secretion of root exudates (Flores et al. 1999). Depending on their age, plant roots secrete up to 10–40% of carbon and 10–16% of nitrogen in the form of root exudates (Bais et al. 2006; Jones et al. 2009). Root exudates are chemicals secreted into the nearby soil by roots. Root exudates contain a variety of simple chemical (or low-molecular-weight) compounds such as simple carbohydrates, amino acids, organic acids, vitamins, phenolics (phenylpropanoids and flavonoids), and plant hormones and, sometimes, complex (or high-molecular-weight) substances such as phytosiderophores, polysaccharides, and proteins (de Weert et al. 2002; Badri et al. 2013). About 100 different chemicals were found in the root exudate of a simple plant: thale cress—*Arabidopsis thaliana* (Strehmel et al. 2014). Both active and passive mechanisms of transport are utilized for the secretion of chemicals by root cells. Passive processes such as vesicular transport (Battey and Blackbourn 1993) and diffusion (Sanders and Bethke 2000) participate in the exudation of a few low-molecular-weight (small polar and uncharged molecules) and high-molecular-weight chemicals, respectively. Similarly, facilitated diffusion is responsible for secretion of other low-molecular-weight compounds such as sugars, amino acids, and carboxylic acids; for this, specific transporter proteins are involved in secretion of specific exudates (Svennerstam et al. 2007). Further, it is suggested that secretion of certain secondary metabolites by plant roots is an active process. The involvement of ABC transporter is shown for the secretion of genistein from soybean plant roots; genistein is a chemical signal for establishing symbiosis by *Rhizobium* sp. (Peters et al. 1986; Redmond et al. 1986; Sugiyama et al. 2007). Although it is not clear which of the root cells secrete root exudates, some evidence suggests that root cap and root hair cells are involved in secretion of compounds (Czarnota et al. 2003). Root hairs comprise about 77% of the total root surface area and are important for anchorage and uptake of water and minerals and root exudation.

Root exudation is a costly process for plants (Badri and Vivanco 2009) and it adds organic carbon into the soil; soil respiration returns the carbon to the atmosphere. Previously, it was thought that the release of organic carbon from roots had no purpose toward development of the plant; however, it is now well established that the root exudates support biological activity in the rhizosphere, and the activity is mostly beneficial to the plant. In addition, root exudates mediate acquisition of nutrients such as iron, manganese, copper, zinc, and phosphorus by plants growing in the environments that have these nutrients in less available forms. Some root exudates act as chelators of iron (phytosiderophores) or phosphorus (organic acids such as citrate, malate, and oxalate) that increase nutrient availability in soils with high pH (Dakora and Phillips 2002). On the other hand, roots of some plants also release root cap "border" cells into rhizosphere (Iijima et al. 2000); these cells are often referred to as rhizodeposits. In thale cress, the cells shed by roots have been shown to attach rhizobia to root cells (Vicré et al. 2005). Further, rhizodeposits including root exudates serve as nutrients for rhizosphere bacteria (Heijnen et al. 1995; Knee et al. 2001). Root exudates along with "border cells" not only influence biology in the soil or rhizosphere but also physicochemical properties of soil. Roots, along with their exudates, influence biological activity in the soil, especially in the rhizosphere.

13.4.1.2 Root Exudates Recruit Certain Microbial Species into Rhizocompartments

The root exudates mediate interactions between root and rhizobiome (Fig. 13.2). Legume plant roots secrete a flavonoid—genistein—to attract and to associate with N_2-fixing rhizobia that are present in the soil; genistein upregulates *Rhizobium meliloti* genes responsible for root nodulation (Peters et al. 1986; Redmond et al. 1986). Strigolactones present in root exudates of Fabaceae plants attract arbuscular mycorrhizae and induce its hyphal branching, which is essential for colonization of root (Akiyama et al. 2005). Benzoxazinoids in root exudates of maize attract *Pseudomonas putida* into the rhizosphere from the bulk soil (Neal et al. 2012). Citric acid and fumaric acid released from tomato plant roots attract plant growth-promoting *Pseudomonas fluorescens* into the rhizosphere (de Weert et al. 2002). Root-secreted malic acid recruits plant growth-promoting *Bacillus subtilis* into rhizosphere upon infection with foliar pathogen (Rudrappa et al. 2008). Further, malic acid (an exudate of tomato plant) was shown to induce biofilm formation by *Bacillus subtilis* on tomato root surface (Chen et al. 2012). Thale cress root-secreted polysaccharides such as arabinogalactan, pectin, and xylan have induced biofilm formation in plant growth-promoting *B. subtilis*, and these polysaccharides were incorporated into the bacterial biofilm matrix (Beauregard et al. 2013). Root colonization by bacteria is an important trait for plant growth promotion; therefore, plants, via their root exudates, can trigger biofilm formation by a PGPR (Chen et al. 2012; Beauregard et al. 2013). Further, root exudates collected from rice have been shown to alter the expression of genes responsible for endophytic colonization in *Azoarcus* sp. strain BH72 (Shidore et al. 2012). Certain chemicals present in root exudates can alter community composition of soil microbiome, in the absence of plant (Badri et al. 2013). Along these lines, root exudates were implicated in the development of plant type and developmental stage-specific microbiomes (Berg et al. 2014; Chaparro et al. 2014).

13.4.1.3 Root Exudates Defend Plant Roots from Natural Enemies, Thereby Indirectly Altering Rhizobiome

Because of higher nutrient availability in the rhizosphere, a variety of detrimental organisms such as pathogenic fungi, oomycetes, bacteria, viruses, nematodes, and root-feeding arthropods invade rhizosphere. Antimicrobials in the root exudates defend plant roots from these natural enemies (Baetz and Martinoia 2014). While some of these antimicrobials are constitutively produced by the plant root (Vaughan et al. 2013), some are inducible under specific conditions such as under pathogen attack (Brigham et al. 1999; Bais et al. 2002) and at certain stage of development (Park et al. 2004; Chaparro et al. 2014). Constitutively made antimicrobials and inducible antimicrobials in plant roots are referred to as phytoanticipins and phytoalexins, respectively (Van Etten et al. 1994). The semivolatile diterpene rhizathalene was continuously secreted by thale cress roots (Vaughan et al. 2013). Amino acid canavanine present in root exudates of certain legumes acts as an antimicrobial agent against a broad range of microbes without affecting rhizobia (Cai et al. 2009). Rosmarinic acid, an antimicrobial, is secreted by sweet basil roots when induced by the cell wall extracts of a *Phytophthora* sp. or a *Pythium* sp. (Bais et al. 2002). Rosmarinic acid is also produced by the roots of *Coleus blumei* when induced by beta-cryptogein, which is an oomycetal elicitor that mimics pathogen attack (Vukovic' et al. 2013). Antimicrobial naphthoquinones were secreted by roots of *Lithospermum erythrorhizon* when induced by fungal elicitors (Brigham et al. 1999). Similarly, roots of barley exuded five antifungal phenylpropanoids (phenolics) when these were attacked by *Fusarium graminearum* (Lanoue et al. 2010). Further, the composition of root exudate in terms of antimicrobials was shown to be altered by exogenous addition of signaling molecules such as salicylic acid, methyl jasmonate, and nitric oxide suggesting that plants do alter their exudation profile in response to biotic stress (Badri et al. 2008). Root exudates contain allelochemicals—the chemicals that repel the roots of other plant species (Callaway and Aschehoug 2000). Root exudates mediate communication between not only root and microbes but also between roots of two different plants. Antimicrobials in the root exudates not only defend plants against their natural enemies but also indirectly promote abundance of beneficial rhizobiome through their antagonistic effects on pathogenic microbes. Rice and bean plant roots secrete compounds that mimic acylated homoserine lactones, which interfered with biofilm formation by root-associated bacteria (Pérez-Montaño et al. 2013).

13.4.2 Plant Immune System Also Shapes Rhizobiome

Mutants of thale cress that are deficient in systemic acquired resistance (SAR) had an altered rhizobiome (Hein et al. 2008). Salicylic acid (SA) is a plant defense hormone that mediates SAR. SA, generally present in leaves, under the condition of either modified production or exogenous application, modulated the composition of rhizobiome; SA was also shown to be a carbon substrate or signal for growth of root-associated microbiota (Lebeis et al. 2015). These results point toward the role of plant immunity in modulating the composition of rhizobiome.

Together, root secretions (Chaparro et al. 2012, 2014; Walters et al. 2018) and plant immunity (Hein et al. 2008; Lebeis et al. 2015; Poole et al. 2018) are the main mechanisms for regulating the composition of rhizobiome. The detailed mechanisms for acquisition of microbiome at community level by plants are yet to be elucidated. Possibly, each microhabitat of the root has a role in the selection of its microbiome (Niu et al. 2017).

13.5 Methods for Studying Rhizobiome Structure and Function

The first step toward the characterization of rhizobiome is to isolate microbes belonging to different microhabitats of root such as endorhizosphere, rhizoplane, and rhizosphere. Soil-attached roots (rhizosphere) can be obtained by washing the roots and then collected by centrifugation; microbes or their DNA can be extracted from the soil (Schlaeppi et al. 2014). Microbes from rhizoplane can be collected by mechanical removal methods such as rigorous shaking of roots with glass beads or by ultrasonication; microbial cells are collected from the supernatant (Reinhold-Hurek et al. 2015). However, these methods for removing microbes may not be completely successful. Microbes from endorhizosphere can be obtained by surface sterilization of roots with ethanol or sodium hypochlorite and then maceration of the roots (Gyaneshwar et al. 2001).

Previously, culture-based methods were utilized for the determination of structure of rhizobiome; however, only a small percentage of environmental bacteria can be grown in pure culture largely owing to the inability to recreate their original environment in the laboratory (Rappé and Giovannoni 2003). Subsequently, the dynamics of rhizobiomes was studied using fingerprinting methods such as denaturing gradient gel electrophoresis (DGGE) and terminal restriction fragment length polymorphism (T-RFLP) (Muyzer et al. 1993; De Angelis et al. 2009). These fingerprinting methods do not depend on culturing microbes but are low-resolution techniques for the characterization of highly diverse microbes of soil, which has at least 10^6 different genomes per gram of soil (Torsvik et al. 2002). DGGE and T-RFLP are capable of resolving 10^2 operational taxonomic units (OTUs); OTUs are assumed to be distinct taxa or phylotypes (Osborn et al. 2000). Clone library of

16S rRNA gene offers higher resolution than DGGE and T-RFLP but is limited to describing the abundant taxa in the sample. Further, these three methods introduce polymerase chain reaction (PCR) bias, which is the preferential amplification of DNA from the most abundant bacteria; the PCR bias is also due to primer design because primers are designed based on available sequence data only (Hawkes et al. 2007). High-density 16S rRNA PhyloChip can resolve up to 10^4 OTUs or taxa (Brodie et al. 2006). Similarly, pyrosequencing of 16S rRNA gene amplicons was utilized as a high-resolution method (Bulgarelli et al. 2012; Lebeis et al. 2012; Lundberg et al. 2012); however, this method is also criticized for its inability to provide biologically meaningful resolution and PCR bias (Pinto and Raskin 2012). 16S rRNA gene sequences are inadequate for identification of bacteria at the subspecies or strain level. Metagenomic sequencing of microbiomes has the capacity to provide strain-level resolution and information (Tett et al. 2012). Stable isotope probing when combined with DNA sequencing can facilitate understanding interactions between plants and rhizobiome (Bressan et al. 2009). Metatranscriptomics can be used to simultaneously characterize structures and functions of active rhizobiomes (Turner et al. 2013b). Further, metatranscriptomics overcomes the limitation of PCR bias and characterizes organisms belonging to all domains of life. The latter is important, because rhizosphere is a complex environment that is also inhabited by eukaryotes such as fungi, protozoa, oomycetes, and nematodes. Although it is accepted that cultivation-independent methods provide more insights into microbiomes associated with plants, large-scale cultivation method was employed to understand microbiome associated with thale cress (Bai et al. 2015).

13.6 Summary

Because of the functional importance of rhizobiomes, it is fundamental to understand their structure and function. In general, relative abundance of certain taxa belonging to four major phyla, viz., *Actinobacteria*, *Bacteroidetes*, *Firmicutes*, and *Proteobacteria*, is higher in rhizobiomes, when compared to that of bulk soil. Plants recruit certain microbial species and communities. Root exudation and plant immunity are the two main mechanisms that regulate composition of rhizobiome. Multiple mechanisms are employed by the members of rhizobiome toward improving plant growth and health. Both culture-dependent and culture-independent methods are required to understand the structure and function of rhizobiome. Detailed investigations of rhizobiome and its interaction with plant contribute to enhanced food production and reduced negative effects on environment.

References

Ahmed E, Holmstrom SJ (2014) Siderophores in environmental research: roles and applications. Microb Biotechnol 7:196–208

Akiyama K, Matsuzaki K, Hayashi H (2005) Plant sesquiterpenes induce hyphal branching in arbuscular mycorrhizal fungi. Nature 435:824–827

Badri DV, Vivanco JM (2009) Regulation and function of root exudates. Plant Cell Environ 32:666–681

Badri DV, Loyola-Vargas VM, Du J, Stermitz FR, Broeckling CD, Iglesias-Andreu L, Vivanco JM (2008) Transcriptome analysis of *Arabidopsis* roots treated with signaling compounds: a focus on signal transduction, metabolic regulation and secretion. New Phytol 179:209–223

Badri DV, Weir TL, van der Lelie D, Vivanco JM (2009) Rhizosphere chemical dialogues: plant-microbe interactions. Curr Opin Biotechnol 20:642–650

Badri DV, Chaparro JM, Zhang R, Shen Q, Vivanco JM (2013) Application of natural blends of phytochemicals derived from the root exudates of *Arabidopsis* to the soil reveal that phenolic-related compounds predominantly modulate the soil microbiome. J Biol Chem 288:4502–4512

Baetz U, Martinoia E (2014) Root exudates: the hidden part of plant defense. Trends Plant Sci 19:90–98

Bai Y, Müller DB, Srinivas G, Garrido-Oter R, Potthoff E, Rott M, Dombrowski N, Münch PC, Spaepen S, Remus-Emsermann M, Hüttel B, McHardy AC, Vorholt JA, Schulze-Lefert P (2015) Functional overlap of the *Arabidopsis* leaf and root microbiota. Nature 528:364–369

Bais HP, Walker TS, Schweizer HP, Vivanco JM (2002) Root specific elicitation and antimicrobial activity of rosmarinic acid in hairy root cultures of sweet basil (*Ocimum basilicum* L.). Plant Physiol Biochem 40:983–995

Bais HP, Weir TL, Perry LG, Gilroy S, Vivanco JM (2006) The role of root exudates in rhizosphere interactions with plants and other organisms. Annu Rev Plant Biol 57:233–266

Barnett SJ, Roget DK, Ryder MH (2006) Suppression of *Rhizoctonia solani* AG-8 induced disease on wheat by the interaction between *Pantoea*, *Exiguobacterium*, and *Microbacteria*. Aust J Soil Res 44:331–342

Barriuso J, Solano BR (2008) Ecology, genetic diversity and screening strategies of plant growth promoting rhizobacteria (PGPR). J. Plant Nutrition:1–17. https://doi.org/10.1002/9783527621989

Battey NH, Blackbourn HD (1993) The control of exocitosis in plant cells. New Phytol 125:307–308

Beauregard PB, Chai YR, Vlamakis H, Losick R, Kolter R (2013) *Bacillus subtilis* biofilm induction by plant polysaccharides. Proc Natl Acad Sci U S A 110:1621–1630

Benítez MS, Gardener BB (2009) Linking sequence to function in soil bacteria: sequence-directed isolation of novel bacteria contributing to soilborne plant disease suppression. Appl Environ Microbiol 75:915–924

Berendsen R, Pieterse C, Bakker P (2012) The rhizosphere microbiome and plant health. Trends Plant Sci 17:478–486

Berg G, Smalla K (2009) Plant species and soil type cooperatively shape the structure and function of microbial communities in the rhizosphere. FEMS Microbiol Ecol 68:1–13

Berg G, Grube M, Schloter M, Smalla K (2014) Unraveling the plant microbiome: looking back and future perspectives. Front Microbiol 5:148

Bisseling T, Dangl JL, Schulze-Lefert P (2009) Next-generation communication. Science 324:691

Bressan M, Roncato MA, Bellvert F, Comte G, Haichar FZ, Achouak W, Berge O (2009) Exogenous glucosinolate produced by *Arabidopsis thaliana* has an impact on microbes in the rhizosphere and plant roots. ISME J 3:1243–1257

Brigham LA, Michaels PJ, Flores HE (1999) Cell-specific production and antimicrobial activity of naphthoquinones in roots of *Lithospermum erythrorhizon*. Plant Physiol 119:417–428

Brodie EL, Desantis TZ, Joyner DC, Baek SM, Larsen JT, Andersen GL, Hazen TC, Richardson PM, Herman DJ, Tokunaga TK, Wan JM, Firestone MK (2006) Application of a high density oligonucleotide microarray approach to study bacterial population dynamics during uranium reduction and reoxidation. Appl Environ Microbiol 72:6288–6298

Bulgarelli D, Rott M, Schlaeppi K, Ver Loren van Themaat E, Ahmadinejad N, Assenza F, Rauf P, Huettel B, Reinhardt R, Schmelzer E, Peplies J, Gloeckner FO, Amann R, Eickhorst T, Schulze-Lefert P (2012) Revealing structure and assembly cues for *Arabidopsis* root-inhabiting bacterial microbiota. Nature 488:91–95

Bulgarelli D, Garrido-Oter R, Münch PC, Weiman A, Dröge J, Pan Y, McHardy AC, Schulze-Lefert P (2015) Structure and function of the bacterial root microbiota in wild and domesticated barley. Cell Host Microbe 17(3):392–403

Cai T, Cai W, Zhang J, Zheng H, Tsou AM, Xiao L, Zhong Z, Zhu J (2009) Host legume-exuded antimetabolites optimize the symbiotic rhizosphere. Mol Microbiol 73:507–517

Callaway RM, Aschehoug ET (2000) Invasive plants versus their new and old neighbors: a mechanism for exotic invasion. Science 90:521–523

Chaparro J, Sheflin A, Manter D, Vivanco J (2012) Manipulating the soil microbiome to increase soil health and plant fertility. Biol Fertil Soils 48:489–499

Chaparro JM, Badri DV, Vivanco JM (2014) Rhizosphere microbiome assemblage is affected by plant development. ISME J 8:790–803

Chen Y, Cao S, Chai Y, Clardy J, Kolter R, Guo JH, Losick R (2012) A *Bacillus subtilis* sensor kinase involved in triggering biofilm formation on the roots of tomato plants. Mol Microbiol 85 (3):418–430

Czarnota MA, Paul RN, Weston LA, Duke SO (2003) Anatomy of sorgoleone-secreting root hairs of *Sorghum* species. Int J Plant Sci 164:861–866

Dakora FD, Phillips DA (2002) Root exudates as mediators of mineral acquisition in low-nutrient environments. Plant Soil 245:35–47

De Angelis KM, Brodie EL, De Santis TZ, Andersen GL, Lindow SE, Firestone MK (2009) Selective progressive response of soil microbial community to wild oat roots. ISME J 3:168–178

de Souza RS, Okura VK, Armanhi JS, Jorrín B, Lozano N, da Silva MJ, González-Guerrero M, de Araújo LM, Verza NC, Bagheri HC, Imperial J, Arruda P (2016) Unlocking the bacterial and fungal communities assemblages of sugarcane microbiome. Sci Rep 6:28774

de Weert S, Vermeiren H, Mulders IHM, Kuiper I, Hendrickx N, Bloemberg GV, Vanderleyden J, De Mot R, Lugtenberg BJJ (2002) Flagella-driven chemotaxis towards exudate components is an important trait for tomato root colonization by *Pseudomonas fluorescens*. Mol Plant-Microbe Interact 15:1173–1180

Demoling F, Figueroa D, Baath E (2007) Comparison of factors limiting bacterial growth in different soils. Soil Biol Biochem 39:2485–2495

Duijff BJ, Recorbet G, Bakker PAHM, Loper JE, Lemanceau P (1999) Microbial antagonism at the root level is involved in the suppression of Fusarium wilt by the combination of nonpathogenic *Fusarium oxysporum* Fo47 and *Pseudomonas putida* WCS358. Phytopathology 89:1073–1079

Edwards J, Johnson C, Santos-Medellín C, Lurie E, Podishetty NK, Bhatnagar S, Eisen JA, Sundaresan V (2015) Structure, variation, and assembly of the root-associated microbiomes of rice. Proc Natl Acad Sci U S A 112:911–920

Flores HE, Vivanco JM, Loyola-Vargas VM (1999) "Radicle" biochemistry: the biology of root-specific metabolism. Trends Plant Sci 4:220–226

Gilbert JA, Jannson JK, Knight R (2014) The Earth microbiome project: successes and aspirations. BMC Biol 12:69

Glick BR (2005) Modulation of plant ethylene levels by the bacterial enzyme ACC deaminase. FEMS Microbiol Lett 251:1–7

Gutierrez-Manero FJ, Ramos-Solano B, Probanza A, Mehouachi J, Tadeo FR, Talon M (2001) The plant-growth-promoting rhizobacteria *Bacillus pumilus* and *Bacillus licheniformis* produce high amounts of physiologically active gibberellins. Physiol Plant 111:206–211

Gyaneshwar P, James EK, Mathan N, Reddy PM, Reinhold-Hurek B, Ladha JK (2001) Endophytic colonization of rice by a diazotrophic strain of *Serratia marcescens*. J Bacteriol 183:2634–2645

Haichar FZ, Marol C, Berge O, Rangel-Castro JI, Prosser JI, Balesdent J, Heulin T, Achouak W (2008) Plant host habitat and root exudates shape soil bacterial community structure. ISME J 2:1221–1230

Haney CH, Samuel BS, Bush J, Ausubel FM (2015) Associations with rhizosphere bacteria can confer an adaptive advantage to plants. Nat Plants 1(6):15051

Hardoim PR, Hardoim CCP, van Overbeek LS, van Elsas JD (2012) Dynamics of seed-borne rice endophytes on early plant growth stages. PLoS One 7(2):e30438

Hawkes CV, DeAngelis KM, Firestone MK (2007) Root interactions with soil microbial communities and processes. In: Cardon Z, Whitbeck J (eds) The rhizosphere. Elsevier, New York

Heijnen CE, Page S, Vanelsas JD (1995) Metabolic activity of *Flavobacterium* strain P25 during starvation and after introduction into bulk soil and the rhizosphere of wheat. FEMS Microbiol Ecol 18:129–138

Hein JW, Wolfe GV, Blee KA (2008) Comparison of rhizosphere bacterial communities in *Arabidopsis thaliana* mutants for systemic acquired resistance. Microb Ecol 55:333–343

Iijima M, Griffiths B, Bengouch AG (2000) Sloughing of cap cells and carbon exudation from maize seedling roots in compacted sand. New Phytol 145:477–482

James EK (2000) Nitrogen fixation in endophytic and associative symbiosis. Field Crop Res 65:197–209

Jones DL, Nguyen C, Finlay RD (2009) Carbon flow in the rhizosphere: carbon trading at the soil-root interface. Plant Soil 321:5–33

Jung WJ, Park RD, Mabood F, Souleimanov A, Smith D (2011) Effects of *Pseudomonas aureofaciens* 63-28 on defense responses in soybean plants infected by *Rhizoctonia solani*. J Microbiol Biotechnol 21:379–386

Kang SM, Joo GJ, Hamayun M, Na CI, Shin DH, Kim HY, Hong JK, Lee IJ (2009) Gibberellin production and phosphate solubilization by newly isolated strain of *Acinetobacter calcoaceticus* and its effect on plant growth. Biotechnol Lett 31:277–281

Knee EM, Gong FC, Gao M, Teplitski M, Jones AR, Foxworthy A, Mort AJ, Bauer WD (2001) Root mucilage from pea and its utilization by rhizosphere bacteria as a sole carbon source. Mol Plant-Microbe Interact 14:775–784

Kwak MJ, Kong HG, Choi K, Kwon SK, Song JY, Lee J, Lee PA, Choi SY, Seo M, Lee HJ, Jung EJ, Park H, Roy N, Kim H, Lee MM, Rubin EM, Lee SW, Kim JF (2018) Rhizosphere microbiome structure alters to enable wilt resistance in tomato. Nat Biotechnol 36:1100–1109

Lanoue A, Burlat V, Henkes GJ, Koch I, Schurr U, Röse US (2010) De novo biosynthesis of defense root exudates in response to *Fusarium* attack in barley. New Phytol 185:577–588

Lebeis SL, Rott M, Dangl JL, Schulze-Lefert P (2012) Culturing a plant microbiome community at the cross-rhodes. New Phytol 196:341–344

Lebeis SL, Herrera Paredes S, Lundberg DS, Breakfield N, Gehring J, McDonald M, Malfatti S, Glavina del Rio T, Jones CD, Tringe SG, Dangl JL (2015) Plant microbiome: salicylic acid modulates colonization of the root microbiome by specific bacterial taxa. Science 349:860–864

Li J, Ovakim DH, Charles TC, Glick BR (2000) An ACC deaminase minus mutant of *Enterobacter cloacae* UW4 no longer promotes root elongation. Curr Microbiol 41:101–105

Locey KJ, Lennon JT (2016) Scaling laws predict global microbial diversity. Proc Natl Acad Sci U S A 113:5970–5975

Lugtenberg B, Kamilova F (2009) Plant-growth-promoting Rhizobacteria. Annu Rev Microbiol 63:541–556

Lundberg DS, Lebeis SL, Paredes SH, Yourstone S, Gehring J, Malfatti S, Tremblay J, Engelbrektson A, Kunin V, del Rio TG, Edgar RC, Eickhorst T, Ley RE, Hugenholtz P, Tringe SG, Dangl JL (2012) Defining the core *Arabidopsis thaliana* root microbiome. Nature 488:86–90

Matilla MA, Ramos JL, Bakker PAHM, Doornbos R, Badri DV, Vivanco JM, Ramos-González MI (2010) *Pseudomonas putida* KT2440 causes induced systemic resistance and changes in *Arabidopsis* root exudation. Environ Microbiol Rep 2:381–388

Mendes R, Kruijt M, de Bruijn I, Dekkers E, van der Voort M, Schneider JH, Piceno YM, DeSantis TZ, Andersen GL, Bakker PA, Raaijmakers JM (2011) Deciphering the rhizosphere microbiome for disease-suppressive bacteria. Science 332:1097–1100

Muyzer G, de Waal EC, Uitterlinden AG (1993) Profiling of complex microbial populations by denaturing gradient gel electrophoresis analysis of polymerase chain reaction-amplified genes coding for 16S rRNA. Appl Environ Microbiol 59:695–700

Neal AL, Ahmad S, Gordon-Weeks R, Ton J (2012) Benzoxazinoids in root exudates of maize attract *Pseudomonas putida* to the Rhizosphere. PLoS One 7(4):e35498

Nguyen C (2003) Rhizodeposition of organic C by plants: mechanisms and controls. Agronomie 23:375–396

Niu B, Paulson JN, Zheng X, Kolter R (2017) Simplified and representative bacterial community of maize roots. Proc Natl Acad Sci U S A 114:2450–2459

Nosengo N (2003) Fertilized to death. Nature 425:894–895

O'Brien FJM, Dumont MG, Webb JS, Poppy GM (2018) Rhizosphere bacterial communities differ according to fertilizer regimes and cabbage (*Brassica oleracea* var. *capitata* L.) harvest time, but not aphid herbivory. Front Microbiol 9:1620

Osborn AM, Moore ERB, Timmis KN (2000) An evaluation of terminal-restriction fragment length polymorphism (T-RFLP) analysis for the study of microbial community structure and dynamics. Environ Microbiol 2:39–50

Park WJ, Hochholdinger F, Gierl A (2004) Release of the benzoxazinoids defense molecules during lateral- and crown root emergence in *Zea mays*. J Plant Physiol 161:981–985

Pérez-Montaño F, Jiménez-Guerrero I, Contreras Sánchez-Matamoros R, López-Baena FJ, Ollero FJ, Rodríguez-Carvajal MA, Bellogín RA, Espuny MR (2013) Rice and bean AHL-mimic quorum-sensing signals specifically interfere with the capacity to form biofilms by plant-associated bacteria. Res Microbiol 164:749–760

Peters NK, Frost JW, Long SR (1986) A plant flavone, luteolin, induces expression of *Rhizobium meliloti* nodulation genes. Science 233:977–980

Philippot L, Hallin S, Borjesson G, Baggs EM (2009) Biochemical cycling in the rhizosphere having an impact on global change. Plant Soil 321:61–81

Pinto AJ, Raskin L (2012) PCR biases distort bacterial and archaeal community structure in pyrosequencing datasets. PLoS One 7(8):e43093

Podile AR, Kishore K (2007) Plant growth-promoting rhizobacteria. In: Plant associated bacteria. Springer, Dordrecht, pp 195–230

Poole P, Ramachandran V, Terpolilli J (2018) Rhizobia: from saprophytes to endosymbionts. Nat Rev Microbiol 16:291–303

Prasad R, Kumar M, Varma A (2015) Role of PGPR in soil fertility and plant health. In: Egamberdieva D, Shrivastava S, Varma A (eds) Plant Growth-Promoting Rhizobacteria (PGPR) and medicinal plants. Springer, Cham, pp 247–260

Raaijmakers JM, Weller DM (1998) Natural plant protection by 2,4-diacetylphloroglucinol-producing *Pseudomonas* spp. in take-all decline soils. Mol Plant-Microbe Interact 11:144–152

Rappé MS, Giovannoni SJ (2003) The uncultured microbial majority. Annu Rev Microbiol 57:369–394

Reay DS (2004) Fertiliser 'solution' could turn local problem global. Nature 427:485

Redmond JW, Batley M, Djordjevic MA, Innes RW, Kuempel PL, Rolfe BG (1986) Flavones induce expression of nodulation genes in *Rhizobium*. Nature 323:632–635

Reinhold-Hurek B, Bünger W, Burbano CS, Sabale M, Hurek T (2015) Roots shaping their microbiome: global hotspots for microbial activity. Annu Rev Phytopathol 53:403–424

Rodriguez H, Fraga R, Gonzalez T, Bashan Y (2006) Genetics of phosphate solubilization and its potential applications for improving plant growth-promoting bacteria. Plant Soil 287:15–21

Rolli E, Marasco R, Vigani G, Ettoumi B, Mapelli F, Deangelis ML, Gandolfi C, Casati E, Previtali F, Gerbino R, Pierotti Cei F, Borin S, Sorlini C, Zocchi G, Daffonchio D (2015) Improved plant resistance to drought is promoted by the root-associated microbiome as a water stress-dependent trait. Environ Microbiol 17:316–331

Rosenberg K, Bertaux J, Krome K, Hartmann A, Scheu S, Bonkowski M (2009) Soil amoebae rapidly change bacterial community composition in the rhizosphere of *Arabidopsis thaliana*. ISME J 3:675–684

Rudrappa T, Czymmek KJ, Pare PW, Bais HP (2008) Root-secreted malic acid recruits beneficial soil bacteria. Plant Physiol 148:1547–1556

Ryu CM, Farag MA, Hu CH, Reddy MS, Wei HX, Paré PW, Kloepper JW (2003) Bacterial volatiles promote growth in *Arabidopsis*. Proc Natl Acad Sci U S A 100:4927–4932

Sanders D, Bethke P (2000) Membrane transport. In: Buchanan BB, Gruisham W, Jones RL (eds) Biochemistry and molecular biology of plants. ASPP, Rockville, MD, pp 110–158

Schlaeppi K, Dombrowski N, Oter RG, Ver Loren van Themaat E, Schulze-Lefert P (2014) Quantitative divergence of the bacterial root microbiota in *Arabidopsis thaliana* relatives. Proc Natl Acad Sci U S A 111(2):585–592

Semenov AM, van Bruggen AHC, Zelenev VV (1999) Moving waves of bacterial populations and total organic carbon along roots of wheat. Microb Ecol 37:116–128

Shidore T, Dinse T, Öhrlein J, Becker A, Reinhold-Hurek B (2012) Transcriptomic analysis of responses to exudates reveal genes required for rhizosphere competence of the endophyte *Azoarcus* sp. strain BH72. Environ Microbiol 14:2775–2787

Singh D, Raina TK, Kumar A, Singh J, Prasad R (2019) Plant microbiome: a reservoir of novel genes and metabolites. Plant Gene 18:100177. https://doi.org/10.1016/j.plgene.2019.100177

Strehmel N, Böttcher C, Schmidt S, Scheel D (2014) Profiling of secondary metabolites in root exudates of *Arabidopsis thaliana*. Phytochemistry 108:35–46

Sugiyama A, Shitan N, Yazaki K (2007) Involvement of a soybean ATP-binding cassette-type transporter in the secretion of genistein, a signal flavanoid in legume–*Rhizobium* symbiosis. Plant Physiol 144:2000–2008

Sugiyama A, Unno Y, Ui O, Yoshikawa E, Suzuki H, Minamisawa K, Yazaki K (2017) Assessment of bacterial communities of black soybean grown in fields. Commun Integr Biol 10(5–6): e1378290

Svennerstam H, Ganeteg U, Bellini C, Nasholm T (2007) Comprehensive screening of *Arabidopsis* mutants suggests the lysine histidine transporter 1 to be involved in plant uptake of amino acids. Plant Physiol 143:1853–1860

Tett AJ, Turner TR, Poole PS (2012) Genomics and the rhizosphere. Wiley, New York

Torsvik V, Ovreas L, Thingstad TF (2002) Prokaryotic diversity–magnitude, dynamics, and controlling factors. Science 296:1064–1066

Turner TR, James EK, Poole PS (2013a) The plant microbiome. Genome Biol 14:209

Turner TR, Ramakrishnan K, Walshaw J, Heavens D, Alston M, Swarbreck D, Anne Osbourn A, Alastair Grant A, Philip S, Poole PS (2013b) Comparative metatranscriptomics reveals kingdom level changes in the rhizosphere microbiome of plants. ISME J 7:2248–2258

Van Etten HD, Mansfield JW, Bailey JA, Farmer EE (1994) Two classes of plant antibiotics: phytoalexins versus "phytoanticipins". Plant Cell 6:1191–1192

Van Spanning RJM, Delgado MJ, Richardson DJ (2005) The nitrogen cycle: denitrification and its relationship to N_2 fixation. In: Werner D, Newton WE (eds) Nitrogen fixation in agriculture, forestry, ecology, and the environment. Springer, Dordrecht, pp 277–342

Vaughan MM, Wang Q, Webster FX, Kiemle D, Hong YJ, Tantillo DJ, Coates RM, Wray AT, Askew W, O'Donnell C, Tokuhisa JG, Tholl D (2013) Formation of the unusual semivolatile diterpene rhizathalene by the *Arabidopsis* class I Terpene synthase TPS08 in the root stele is involved in defense against belowground herbivory. Plant Cell 25:1108–1125

Velazquez-Sepulveda I, Orozco-Mosqueda MC, Prieto-Barajas CM, Santoyo G (2012) Bacterial diversity associated with the rhizosphere of wheat plants (*Triticum aestivum*): toward a metagenomic analysis. Phyton Int J Exp Bot 81:81–87

Vessey JK (2003) Plant growth promoting rhizobacteria as biofertilizers. Plant Soil 255:571–586

Vicré M, Santaella C, Blanchet S, Gateau A, Driouich A (2005) Root border-like cells of *Arabidopsis*. Microscopical characterization and role in the interaction with rhizobacteria. Plant Physiol 138:998–1008

Vukovic' R, Bauer N, Curković-Perica M (2013) Genetic elicitation by inducible expression of b-cryptogein stimulates secretion of phenolics from *Coleus blumei* hairy roots. Plant Sci 199–200:18–28

Wagner MR, Lundberg DS, del Rio TG, Tringe SG, Dangl JL, Mitchell-Olds T (2016) Host genotype and age shape the leaf and root microbiomes of a wild perennial plant. Nat Commun 7:12151

Walters WA, Jin Z, Youngblut N, Wallace JG, Sutter J, Zhang W, González-Peña A, Peiffer J, Koren O, Shi Q, Knight R, del Rio TG, Tringe SG, Buckler ES, Dangl JL, Ley RE (2018) Large-scale replicated field study of maize rhizosphere identifies heritable microbes. Proc Natl Acad Sci U S A 115:7368–7373

Wang QY, Dodd IC, Belimov AA, Jiang F (2016) Rhizosphere bacteria containing 1-aminocyclopropane-1-carboxylate deaminase increase growth and photosynthesis of pea plants under salt stress by limiting Na^+ accumulation. Funct Plant Biol 43:161–172

Weinert N, Piceno Y, Ding GC, Meincke R, Heuer H, Berg G, Schloter M, Andersen G, Smalla K (2011) PhyloChip hybridization uncovered an enormous bacterial diversity in the rhizosphere of different potato cultivars: many common and few cultivar-dependent taxa. FEMS Microbiol Ecol 75:497–506

Weller DM, Raaijmakers JM, McSpadden Gardener BB, Thomashow LS (2002) Microbial populations responsible for specific soil suppressiveness to plant pathogens. Annu Rev Phytopathol 40:309–348

Yin C, Hulbert SH, Schroeder KL, Mavrodi O, Mavrodi D, Dhingra A, Schillinger WF, Paulitz TC (2013) Role of bacterial communities in the natural suppression of *Rhizoctonia solani* bare patch disease of wheat (*Triticum aestivum* L.). Appl Environ Microbiol 79:7428–7438

Zamioudis C, Pieterse CMJ (2012) Modulation of host immunity by beneficial microbes. Mol Plant-Microbe Interact 25:139–150

Zhang H, Sun Y, Xie X, Kim MS, Dowd SE, Paré PW (2009) A soil bacterium regulates plant acquisition of iron via deficiency-inducible mechanisms. Plant J 58:568–577

Chapter 14
Soil Microbes-Medicinal Plants Interactions: Ecological Diversity and Future Prospect

Ramesh Kumar Kushwaha, Vereena Rodrigues, Vinay Kumar, Himani Patel, Meenakshi Raina, and Deepak Kumar

Abstract Plants live in association with microbes in both above- and belowground part, as some are beneficial and some are harmful to the plant. Microbes which are found within the plant tissue, namely, endophytes, can have beneficial, neutral, or detrimental effects on plant health and development. Several works have been done on plant-microbe interactions and microbial diversity of rhizospheric region of medicinal plants. Therefore, plant secondary metabolite and root exudates which include various sugars and organic acids influence biogeochemical reactions and thus plant metabolism. Signaling molecules like strigolactones induce the colonization of the mycorrhiza fungi with plant root and stimulate the germination of the parasitic plant such as *Striga*. Similarly, the flavonoids secreted by leguminous roots increase the growth of symbiotic and nonsymbiotic nitrogen-fixing bacteria and also attract pathogenic oomycetes as well. Root-associated microflora and endophytes (fungi or bacteria) help plant growth by secreting the plant hormone (auxin/cytokinin) and nutrients like phosphorus, nitrogen, and iron. Microbial association with

---

R. K. Kushwaha
Microbial Technology Laboratory, CSIR – Central Institute of Medicinal and Aromatic Plants, Research Centre, Bangalore, Karnataka, India

Present address: School of Biochemistry, REVA University, Rukmini Knowledge Park, Kattigenahalli, Yalahanka, Bangalore, Karnataka, India

V. Rodrigues
Plant Biology and Systematics, CSIR – Central Institute of Medicinal and Aromatics Plants, Research Center, Bangalore, Karnataka, India

V. Kumar
Genetics and Plant Molecular Biology, National Botanical Research Institute, Lucknow, Uttar Pradesh, India

H. Patel
Ancient DNA Lab, Birbal Sahni Institute of Plaeosciences, Lucknow, Uttar Pradesh, India

M. Raina · D. Kumar (✉)
Department of Botany, Central University of Jammu, Samba, Jammu and Kashmir, India
e-mail: deepakkumar@cujammu.ac.in

A. Varma et al. (eds.), *Plant Microbe Symbiosis*,
https://doi.org/10.1007/978-3-030-36248-5_14

root may induce plant resistance against the several biotic and abiotic stresses, such as toxic metals, pathogens, drought, high temperature, saline soils, adverse soil pH, and transplant shock. Study the plant-microbe interaction in the era of next-generation sequencing opens a new way to understand their association as well as help in improvement of sustainable agriculture. Finding answers of these questions "Who is there?" and "What are they doing?" extended by "How do they live under given conditions?", "How do they respond to environmental changes and perturbations?", "How do they interact with each other?", and "How do they affect plant growth and development?" may be used in the future to support plant growth and improve crop yield. Exploration of endophytic or rhizospheric microbes in the future for enhancement of secondary metabolites in medicinal plants might be a new vista opened for the sustainable agriculture practices. In this chapter, we will focus our attention to the role of medicinal plant-microbe interaction to root and shoot in positive and negative aspect.

14.1 Introduction

Rhizospheric soil contains numerous bacteria, fungi, protozoa, and nematodes, the diversity and functions of which are influenced by the root exudates, respiration, and biogeochemical reactions (Narula et al. 2009; Shrivastava et al. 2014). Some of the nematodes consume bacteria and fungi as food for their survival and growth. The exudates secreted by roots contain numerous compounds that facilitate the multiple functions such as disease suppression and help in nutrient cycling. Moreover, the genetics of the plant species also influence on the diversity of rhizospheric microorganisms (Peiffer et al. 2013; Chaparro et al. 2014). For example, in root exudate, flavonoids secreted by the leguminous plants increase the growth of symbiotic and nonsymbiotic nitrogen-fixing bacteria, root nodules formation, and nitrogen intake by plants. Strigolactones are a class of phytohormone secreted by the plants root to stimulate the growth of symbiotic arbuscular mycorrhizal (AM) fungi and help in germination of the parasitic plant *Striga*. However, allelochemicals can inhibit the growth and proliferation of other microorganisms in the rhizosphere or negatively influence on same plants (Zhang et al. 2010).

Plant roots secrete a wide range of organic compounds in the rhizospheric region which act as food source for microbial communities to increase microorganism density and other activities in the rhizospheric region compared to bulk soil (the soil away from the rhizosphere is known as bulk soil). The important activities of the microorganisms in the rhizosphere are (1) solubilization of inorganic P and organic P sources (unavailable to plants) into the soluble P form (available to plants); (2) nitrogen fixation, that is, conversion of free nitrogen into available form of nitrogen compound; (3) phytohormone production; (4) ACC deaminase activity; (5) siderophore production; (6) antipathogenic activity; etc. (Ahmed et al. 2014; Abeer et al. 2016; Arora et al. 2001). The rhizobacteria having the above plant growth-promoting activity are called plant growth-promoting bacteria (PGPB) such as *Azotobacter chroococcum*, *Bacillus pumilus*, *Bacillus licheniformis*, *Bacillus megaterium*, *Bacillus mucilaginous*, *Bacillus*

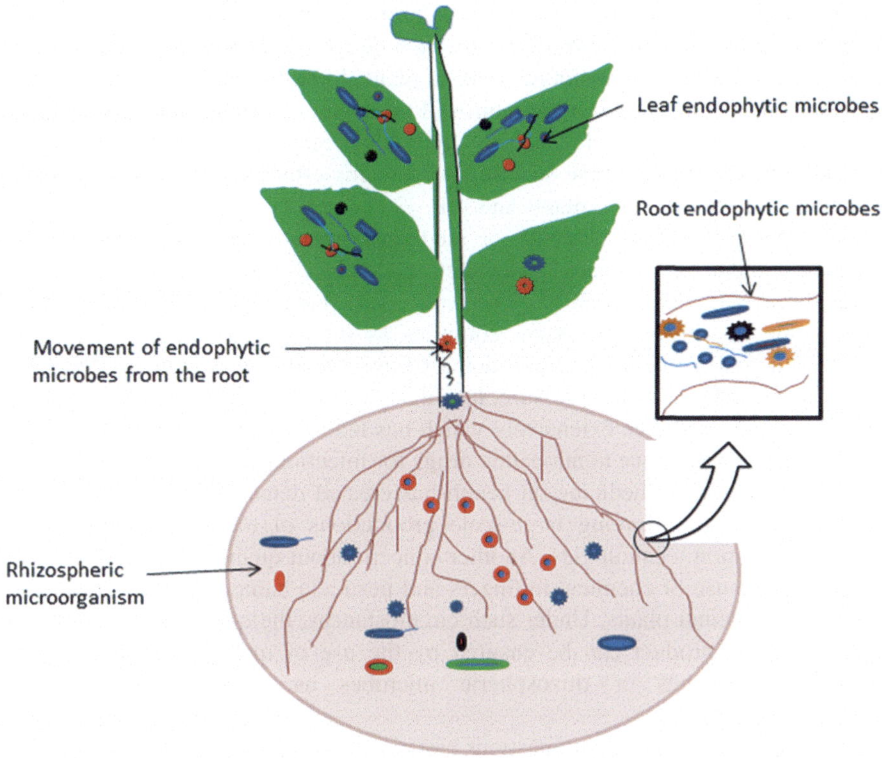

Fig. 14.1 Diagrammatic representation of rhizospheric microorganism, microbes inside the root, movement of endophytic microbes, and colonization within leaves

firmus, *Burkholderia cepacia*, *Pseudomonas aureofaciens*, *Pseudomonas chlororaphis*, *Pseudomonas fluorescens*, *Pseudomonas solanacearum*, *Pseudomonas syringae*, *Serratia entomophila*, *Streptomyces griseoviridis*, *Streptomyces lydicus*, and various *Rhizobia* spp. (Souza et al. 2015). Other than the rhizobacteria, mycorrhiza also helps in plant growth through the same mechanisms; some of those fungi are *Penicillium pinophilum*, *Penicillium oxalicum*, *Aspergillus niger*, *A. fumigates*, and *Trichoderma* spp. (Nadeem et al. 2014).

Apart from root exudates, types and properties of soil also influence the diversity and the composition of both free-living rhizospheric microorganisms and endophytic microorganisms (Fig. 14.1) (Lakshmanan et al. 2014; Bulgarelli et al. 2012). The endophytic bacteria and fungi enter through the root terminal and are transported to different tissues which are also influenced by the geographical soil type. The microbial communities analyzed through the culture-independent technique from different geographical rhizospheric soils of cactus differed according to the soil type (Andrew et al. 2012). In another study, the actinobacterial population was different around the rhizosphere of strawberry plants grown in different soils and also differed much from the bulk soil. This result proves the role of plants in modulating the microbial richness around the rhizosphere (Costa et al. 2006). Varieties of fungal and bacterial species are reported from the different geographical regions and have been isolated from the root and shoot of soybean plant (Impullitti and Malvick 2013; de Almeida Lopes et al. 2016). Apart from soil type, other extrinsic abiotic and biotic factors also modulate the rhizospheric microbial communities of the host plant (Vacheron et al. 2015). These soil microorganisms establish the relationship and communication between the plants and soil.

The earth's enormous diversity of medicinal plants is a rich source of safe bioactive compounds compared to synthetic medicine for the treatment of various human diseases (Nema et al. 2013). The use of medicinal plants has been adopted for a long time in Europe and Asian countries as tradition like Indian Ayurvedic medicine, herbal medicine, and traditional Chinese medicine (Joy et al. 1998). Of course, increased population pressure, fast lifestyle, and cost-effectiveness tuned the use of synthetic medicine extensively which has led to various side effects and the development of resistance to allopathic drugs for infectious diseases. The detrimental effect of synthetic medicine on health has created demand for medicinal plant products which is enforcing large-scale productions of medicinal plants using modern cultivation technologies. Another concern about quality of medicinal plant products is the use of chemical fertilizers and pesticide hampering the growth and quality of medicinal plants. Under such circumstances, the quality and quantity of medicinal plant product can be ensured by the use of medicinal plants-specific endophytic microbes or rhizospheric microbes as plant growth-regulating biofertilizers.

Therefore, the fundamental questions associated with plants and their associated microorganisms are "Who is there?", "How do they respond with different environmental conditions?", and "Do they impact plant health and growth?". Genomic studies of the plant-associated microorganisms are capable of revealing answers to the issue. Metagenomic analysis of the entire rhizospheric microbiome or DNA-based analyses of individual microorganism species reveal the composition of rhizospheric microbiome and functional potential of individual microbes (Knief 2014). Currently several molecular techniques are available, while next-generation sequencing (NGS) techniques have the greatest impact on DNA- and RNA-based analysis techniques; both rhizospheric and endophytic bacterial diversities in coastal halophyte *Messerschmidia sibirica* were analyzed through the illumina-based technique (Tian and Zhang 2017).

In this chapter, we collected the information about the medicinal plants and their rhizospheric microorganisms (bacteria and fungi) in context of understanding the impact on plant growth, secondary metabolite accumulation, to help in abiotic and biotic stress. This information will provide the further aspect of rhizospheric microorganism concern with medicinal plant growth and their product in context of organic agriculture cultivation.

14.2 Diversity of Rhizospheric Bacteria Associated with Medicinal Plant

Rhizospheric bacteria associated with medicinal plants are of immense significance and therefore widely studied through the culture-dependent and culture-independent metagenomics sequence (Table 14.1). They are well-known to induce plant growth and improve secondary metabolite accumulation as well and also recognized as a source of numerous bioactive compounds (Bafana and Lohiya 2013). The diversity of rhizosphere microbial communities is always plant-specific in terms of both structural and functional diversities. Because of plant specificity, the rhizosphere bacterial diversity of *Derris elliptica*, *Pueraria mirifica*, and *Indigofera tinctoria* are significantly different from one another (Nimnoi et al. 2011). Similarly, the diazotrophic communities of medicinal plant *Matricaria chamomilla* L., *Calendula officinalis* L., and *Solanum distichum* are different and were dominated by Gram-positive *Bacillus* spp. which have prime importance to suppress the pathogen (Koeberl et al. 2013). The rhizosphere of *Ocimum sanctum* aromatic plants retains the maximum population of *Klebsiella pneumoniae*, *Staphylococcus pasteuri*, *Bacillus cereus*, *Brevibacillus agri*, *Bacillus subtilis*, *Pseudomonas putida*, *Bacillus megaterium*, *Enterobacter* sp., *Cronobacter sakazakii*, *Pantoea agglomerans*, *Alcaligenes* sp., *Micrococcus* sp., *Bacillus thuringiensis*, *Bacillus firmus*, *Pseudomonas rhizosphaerae*, and *Flexibacter* sp. (Singh et al. 2015). Zhao and his group isolated 50 bacterial strains from rhizosphere of some medicinal plant, belonging to the genus *Cystobacter*, *Archangium*, *Corallococcus*, *Myxococcus*, *Pyxidicoccus*, *Stigmatella*, and *Chondromyces* in which *Myxococcus* was dominant over to others (Zhao et al. 2013). Two rod-shaped exopolysaccharides producing rhizospheric bacterial strain DRP 35 and DR-9 were isolated from rhizosphere of *Angelica sinensis* belonging to genera *Terriglobus* and *Mucilaginibacter*, respectively (Whang et al. 2014; Lee et al. 2013).

The continuous cultivation of any medicinal plant and non-medicinal plant decrease the land productivity which negatively affect the plant growth and quality as well. For example, the continuous cultivation of medicinal plant *Rehmannia glutinosa* influenced on the microbial community and rhizospheric activities which leads to decreased productivity (Qi et al. 2009). Further study of Qi et al. (2012) proved the relative proportions of the bacterial communities around the wild-type rhizosphere of *Rumex patientia* are different in *Bacteroidetes*, *Firmicutes*, *Gemmatimonadetes*, and *Verrucomicrobia* bacterial group than in non-rhizosphere soils. However, the intercropping of peanut with traditional Chinese medicinal plant (*Atractylodes lancea* and *Euphorbia pekinensis*) increased the productivity by increasing the Gram-negative bacterial population and reducing the phenolic allelochemicals (Dai et al. 2009, 2013).

Very similar phenotypic and chemotaxonomic properties to genus *Pontibacter*, a novel Gram-negative, pink-pigmented bacterium is isolated from rhizosphere of medicinal plant *Nerium indicum* (*Chuvanna arali*) (Raichand et al. 2011). The different proportion of rhizospheric bacteria such as *Actinobacteria* (12%),

Table 14.1 Medicinal plants and their rhizospheric bacteria

Plant species	Microorganisms	References
Ferula species	Bacterial species	Wang et al. (2018)
Glycyrrhiza inflata	*Actinomycete* strain BMP B8152	Cao et al. (2018)
Limonium sinense	*Glutamicibacter halophytocola* strain KLBMP 5180	Qin et al. (2018)
Radix pseudostellariae	*Pseudomonas* spp. and *Fusarium* spp.	Chen et al. (2017a, b)
Anoectochilus roxburghii	*Bacillus* sp. FJAT-14262	Chen et al. (2016)
Echinacea purpurea	*Rheinheimera* sp. EpRS3	Presta et al. (2016)
Sapindus saponaria	Actinobacteria, Acidobacteria, and Proteobacteria	Garcia et al. (2016)
Rhododendron arboreum	Actinobacteria, Acidobacteria, Proteobacteria, and Chloroflexi	Debnath et al. (2016)
Ocimum sanctum	*Klebsiella pneumoniae*, *Staphylococcus pasteuri*, *Bacillus cereus*, *Brevibacillus agri*, *Bacillus subtilis*, *Pseudomonas putida*, *Bacillus megaterium*, *Enterobacter* sp.,*Cronobacter sakazakii*, *Pantoea agglomerans*, *Alcaligenes* sp., *Micrococcus* sp., *Bacillus thuringiensis*, *Bacillus firmus*, *Pseudomonas rhizosphaerae*, *Flexibacter* sp.	Singh et al. (2015)
Fraxinus chinensis	*Pseudoxanthomonas humi*	Akter et al. (2015)
Echinacea purpurea and *Echinacea angustifolia*	*Pseudomonas*, *Actinobacteria*, and *Bacillus* sp.	Chiellini et al. (2014)
Angelica sinensis	*Mucilaginibacter polysacchareus*, *M. myungsuensis*, *M. ximonensis*, *Terriglobus saanensis*	Whang et al. (2014) Lee et al. (2013)
Calendula officinalis, *Matricaria chamomilla*, *Solanum distichum*	*Bacillus* sp.	Koeberl et al. (2013)
Rumex patientia	*Actinobacteria*, *Firmicutes*, *Chloroflexi*, *Bacteroidetes*, *Acidobacteria*, *Proteobacterium*, *Gemmatimonadetes*, *Verrucomicrobia*, *Planctomycetes*	Qi et al. (2012)
Atractylodes lancea	Gram-negative bacteria	Dai et al. (2013)
Plectranthus tenuiflorus	*Micrococcus luteus*, *Paenibacillus* sp., *Pseudomonas* sp., *Acinetobacter calcoaceticus*, *Bacillus* sp., *B. megaterium*, *B. pumilus*, *B. licheniformis*	El-Deeb et al. (2013)
Origanum vulgare	*Pseudomonas*, *Stenotrophomonas*	Bafana and Lohiya (2013)
Typhonium giganteum	*Kribbella flavida*, *K. karoonensis*, *K. alba*	Xu et al. (2012)
Ginseng plants	Actinomycetes	Zhang et al. (2013)
Hypericum silenoides	*Sphingobium*, *Stenotrophomonas*, *Agrobacterium*, *Pantoea*, *Acinetobacter*, *Enterobacter*, *Pseudomonas*, *Serrati*	Lopez-Fuentes et al. (2012)

(continued)

Table 14.1 (continued)

Plant species	Microorganisms	References
Ajuga bracteosa	*Pseudomonas*	Kumar et al. (2012)
Nerium indicum	*Pontibacter*	Raichand et al. (2011)
Fritillaria thunbergii	*Actinobacteria, Bacteroidetes, Proteobacteria, Acidobacteria*	Shi et al. (2011)
Astragalus membranaceus	*Geodermatophilus obscurus*	Zhang et al. (2011a)
Phytolacca acinosa	*Aspergillus fumigatus*	Guo et al. (2010)
Agathosma betulina	*Cryptococcus laurentii*	Cloete et al. (2009)
Catharanthus roseus, Aloe vera, Ocimum sanctum, Coleus forskohlii	*Pseudomonas, Azospirillum, Azotobacter*	
Cassia auriculata, Annona squamosa L., *Eclipta alba*	*Corynebacterium, Micrococcus, Serratia, Bacillus, Pseudomonas, Enterobacter*	Tamilarasi et al. (2008)

Bacteroidetes (18%), *Proteobacteria* (55%), and *Acidobacteria* (12%) are reported from the medicinal plant *Fritillaria thunbergii* (Shi et al. 2011). Both non-rhizospheric and rhizospheric bacterial communities were studied in medicinal plant *Ocimum sanctum*, *Coleus forskohlii*, *Catharanthus roseus*, and *Aloe vera*, showed the maximum population in rhizospheric soil and was mostly dominated by *Azospirillum*, *Azotobacter*, and *Pseudomonas* spp.

Rhizospheric and root endophytic bacteria were isolated from medicinal plant such as *Origanum vulgare*, *Hypericum silenoides*, and *Ajuga bracteosa*. *Origanum vulgare* was surrounded with a total of 120 morphologically different kind of bacteria, which was mostly belongs to *Gammaproteobacteria* and *Firmicutes*; however, dominating species were belonging to *Pseudomonas* and *Stenotrophomonas* (Bafana and Lohiya 2013). In case of *Ajuga bracteosa*, a total of 123 morphologically different bacteria were isolated from roots and rhizospheric region and were mostly belonging to alpha- and *Gammaproteobacteria* (Kumar et al. 2012). A total of 103 bacterial isolates from *Hypericum silenoides* root and rhizosphere were isolated which belong to genera *Pseudomonas*, *Sphingobium*, *Stenotrophomonas*, *Pantoea*, *Serratia*, *Acinetobacter*, and *Agrobacterium* (Lopez-Fuentes et al. 2012). Both endophytic and rhizospheric bacterial communities were analyzed from the same species of medicinal plant *Echinacea purpurea* and *Echinacea angustifolia*, observing the maximum dominance of *Pseudomonas*, *Actinobacteria*, and *Bacillus* spp. (Chiellini et al. 2014). Biocontrol is a part of PGP activity, so the presence of *Actinobacteria* may play as a biocontrol because it is a potential source of antibiotics which are reported from the several medicinal plants. *Gemmatimonadetes*, *Acidobacteria*, *Actinobacteria*, *Proteobacteria*, *Planctomycetes*, and *Bacteroidetes* bacterial phyla are abundant in the rhizosphere of *Boswellia sacra* (Khan et al. 2017b). The dominance of three bacterial phyla *Actinobacteria*, *Acidobacteria*, and *Proteobacteria* were observed in *Sapindus saponaria* and *Rhododendron arboreum* medicinal plants (Garcia et al. 2016;

Debnath et al. 2016). Two bacterial species *Bacillus* sp. FJAT-14262 and *Rheinheimera* sp. EpRS3 are reported from the medicinal plants *Anoectochilus roxburghii* and *Echinacea purpurea*, respectively, characterized as antimicrobial producer (Chen et al. 2016; Presta et al. 2016).

14.3 Diversity of Rhizosphere Fungi Associated with Medicinal Plant

A wide variety of fungi species are associated with different medicinal plants that depend upon the plant host, edaphic factor, climate, season, and abiotic and biotic factor. Table 14.2 presented the various rhizospheric-associated fungi concern with different medicinal plants. The variety of secondary metabolites or root volatiles produced by medicinal plants carefully carve a niche of species-specific rhizospheric microbiome. The metagenomic analysis of rhizosphere reveals the dominant species of *Phoma*, *Volutella*, *Pachycudonia*, *Heterodermia*, *Gibberella*, *Cladosporium*, *Trichocladium*, and *Sporothrix* fungal community around *Coptis chinensis* root. The populations of AM fungi largely differ despite having similar growth conditions in medicinal plants *Trachyspermum copticum*, *Smilax* spp, *Euphoria longan*, *Rauvolfia serpentina*, *Rauvolfia tetraphylla*, *Centella asiatica*, *Emblica officinalis*, *Aloe barbadensis*, and *Sapindus trifoliatus* (Hussain and Srinivas 2013). Rhizospheric fungi from different species of the *Cassia* plants such as *C. alata*, *C. occidentalis*, and *C. sophera* have been reported. Moreover, most of the isolated fungi were belonging to *Globus* genus that was dominant in *C. alata* followed by *C. occidentalis* and *C. sophera* (Chatterjee et al. 2010).

The associations of rhizospheric fungi to the plants are important regarding to plant growth as well as considerably influence on the secondary metabolite accumulation within host plant (Shaikh and Mokat 2018). Total 11 fungal species were isolated from *Santalum album* rhizospheric soil, in which mostly were belonging to *Hyphomycetes* such as *Aspergillus restrictus*, *A. fumigatus*, *A. terricola*, *A. niger*, *A. funiculosus*, *A. flavus*, *A. terreus*, *A. flavipes*, *Fusarium oxysporum*, and *Penicillium* spp. and one only belonged to *Basidiomycetes* (*Mycelia sterilia*) (Thombre et al. 2016). From 14 cultivar of medicinal plant *Paeonia suffruticosa*, total 31 AM fungi were isolated in which mostly was belonged to *Glomus* genus, followed by *Acaulospora* and *Scutellospora* genera (Shi et al. 2013). However, from rhizosphere of Paris-type *Magnolia cylindrical* medicinal plant, 17 species of AM were isolated that belonged to genera *Glomus* (8 species), *Acaulospora* (6 species), *Scutellospora* (2 species), and *Gigaspora* (1 species) (Yang et al. 2011). Two AM fungi *Glomus mosseae* and *Glomus intraradices* are isolated from rhizosphere of *Bacopa monnieri* that increased the plant growth and induced the salinity tolerance by various mechanisms (Khaliel et al. 2011).

The colonization and diversity pattern of AM fungi is dependant on edaphic factors and type of vegetation. AM fungi such as *Acaulospora delicata*, *Glomus*

Table 14.2 Medicinal plants and their rhizospheric fungi

Plant species	Rhizospheric fungi	References
Coptis chinensis	*Phoma, Volutella, Pachycudonia, Heterodermia, Gibberella, Cladosporium, Trichocladium*, and *Sporothrix*	Song et al. (2018)
Boswellia sacra	Ascomycota and Basidiomycota	Khan et al. (2017b)
Taxus	301 species of fungi	Hao et al. (2018)
Lilium davidii var.unicolor	*Fusarium, Rhizoctonia, Verticillium, Penicillium, and Ilyonectria*	Shang et al. (2016)
Sanctum album	*Aspergillus restrictus, A. fumigatus, A. terricola, A. niger, A. funiculosus, A. flavus, A. terreus, A. flavipes, Fusarium oxysporum, Penicillium* spp., *and Mycelia sterilia*	Thombre et al. (2016)
Pinellia ternata, Atractylodes lancea, Ophiopogon platyphyllum, Dioscorea zingiberensis, Euphorbia pekinensis	*Verticillium* sp.; *Fusarium* sp.	Dai et al. (2009)
Andrographis paniculata	*Glomus aggregatum, Acaulospora scrobiculata*	Radhika and Rodrigues (2010)
Hemidesmus indicus	*Glomus multicaule, G. maculosum, G. geosporum, G. fasciculatum, Ambispora leptoticha*	Radhika and Rodrigues (2010)
Aloe vera	*Glomus multicaule, G. maculosum, G. geosporum*	Radhika and Rodrigues (2010)
Azadirachta indica	*A. Scrobiculata, S. calospora, G. fasciculatum, G. albida*	Radhika and Rodrigues (2010)
Naregamia alata	*G. rubiforme, G. maculosum, G. fasciculatum, A. scrobiculata, A. leptoticha, A. nicolsonii, S. verrucosa*	Radhika and Rodrigues (2010)
Physalis minima	*G. maculosum, G. geosporum, G. rubiforme, G. fasciculatum, G. multicaule, A. rehmi*	Radhika and Rodrigues (2010)
Panax ginseng	*G. fasciculatum, G. geosporum, G. macrocarpum, G. microaggregatum, G. Mosseae, A. cavernata, A. Spinosa, Sordariomycetes, Alatospora, Eurotiomycetes, Leotiomycetes, Saccharomycetes, Mucorales*, and *Pezizomycetes*	Cho et al. (2009); Dong et al. (2017)
Panax notoginseng	*G. mosseae, G. versiforme, G. claroideum, G. monosporum, G. constrictum, Absidia panacisoli*	Zhang et al. (2011b); Zhang et al. (2018)
Arnica montana	*G. versiforme, G. macrocarpum, G. fasciculatum, G. geosporum, G. constrictum, G. intraradices, G. mosseae*	Jurkiewicz et al. (2010)

(continued)

Table 14.2 (continued)

Plant species	Rhizospheric fungi	References
Echinacea purpurea	*G. intraradices*	Araim et al. (2009)
Cercidiphyllum japonicum	*S. aurigloba*, *Archaeospora leptoticha*, *G. aggregatum*, *G. constrictum*, *G. dimorphicum*, *G. fasciculatum*, *G. flavisporum*, *G. intraradices*, *G. mosseae*	Wang et al. (2008)
Hippophae rhamnoides	*G. albidum*, *G. claroideum*, *G. constrictum*, *G. coronatum*, *G. intraradices*	Tang et al. (2004)
Ziziphus jujuba Mill. var. inermi	*G. monosporum*, *G. reticulatum*, *G. coronatum*, *G. intraradices*	Tang et al. (2004)
Lycium barbarum	*G. margarita*, *G. albidum*	Tang et al. (2004)
Taxus chinensis	*A. denticulate*, *G. reticulatum*, *G. verruculosum*, *G. viscosum*, *G. fasciculatum*, *G. aggregatum*, *G. ambisporum*, *G. clarum*, *G. constrictum*, *G. geosporum*, *G. magnicaule*	Wang et al. (2008)
Euptelea pleiosperma	*S. verrucosa*, *G. ambisporum*, *G. hyderabadensis*, *G. constrictum*, *G. geosporum*, *G. fasciculatum*, *G. intraradices*	Wang et al. (2008)
Cassia alata, *C. occidentalis*, *C. sophera*	*Glomus* spp.	Chatterjee et al. (2010)
Curcuma mangga	*Penicillium digitatum*, *Fusarium oxysporum*, *Sclerotium rolfsii*, *Alternaria brassicicola*, *Colletotrichum gloeosporioides*	Khamna et al. (2009)
Ocimum sanctum, *Centella asiatica*	AM and endophytic fungi	
Paeonia suffruticosa	*Acaulospora*, *Glomusm*, *Scutellospora*, *Curvularia*, *Guehomyces*, *Exophiala*, and *Fusarium*	Shi et al. (2013), Zhang et al. (2018)
Artemisia annua	*Glomusm osseae*, *Glomus intraradices*, *G. aggregatum*, *G. fasciculatum*	Awasthi et al. (2011)
Magnolia cylindrica	*Scutellospora*, *Glomus*, *Acaulospora*, *Gigaspora*	Yang et al. (2011)
Bacopa monnieri	*G. constrictum*, *G. fasciculatum*, *G. geosporum*, *G. intraradices*, *G. mosseae*, *G. rubiforme*	Panwar and Tarafdar (2006)
Sorghum bicolor	*G. mosseae*, *G. intraradices*	Sun and Tang (2013)
Curculigo orchioides	*G. microcarpum*, *G. geosporum*	Sharma et al. (2008)
Ginseng plants	Soil fungi	Zhang et al. (2013)

aggregatum, *G. fasciculatum*, *G. geosporum*, *G. intraradices*, and *G. mosseae* are observed in the roots of *Indigofera tinctoria*, *I. aspalathoides*, and *Eclipta prostrata* (Sundar et al. 2011). Because of the significant role in the abiotic stress-relieving potential, five genera of AM fungi were isolated from Indian Thar Desert habituated medicinal plant *Leptadenia reticulata*, *Mitragyna parvifolia*, and *Withania coagulans* (Panwar and Tarafdar 2006). Awasthi and his group showed the compatibility and synergy between AM fungus *Glomus mosseae* and rhizobacterium *Bacillus subtilis* and suggested the use of consortia for *Artemisia annua* cultivation to achieve maximum herbage and artemisinin content in leaf. Zubek and Blaszkowski (2009) and Zubek et al. (2011) identified 33 genera of fungi including AM fungi and dark septate endophyte (DSE) associations in 36 medicinal plant species. Thirty-four out of 36 medicinal plants are dominated by the colonization of *Convallaria majalis* (77.9%) followed by *Helianthus tuberosus* (2.5%); however, mycelium of DSE was only observed in 13 plant species, i.e., percentage of root colonization by DSE was less than AM. Gorsi (2002) reported several AM fungi around 76 medicinal plants collected from the Azad Jammu and Kashmir. From Western Ghats of India, 36 medicinal plants were taken under consideration for the AM fungi isolation by the Radhika and Rodrigues (2010); most of medicinal plants are associated with AM fungi. Rhizospheric community of *Dichanthelium lanuginosum* medicinal plant is *Acaulospora*, *Archaeospora*, *Glomus*, *Paraglomus*, and *Scutellospora*; however, the rhizosphere of *Phellodendron amurense* is dominated by *Glomus*, *Scutellospora*, and *Hyponectria* (Appoloni et al. 2008; Cai et al. 2009). Rhizospheric community of medicinal plant *Boswellia sacra* is dominated by *Aspergillus*, *Coprinopsis*, *Chaetomium*, *Exophiala*, *Glomus*, *Haematonectria*, *Rhizophagus*, *Spizellomyces*, *Veronaea*, etc. (Khan et al. 2017b). Johnson and Stephan (2016) studied the rhizospheric community of the following medicinal plants *Achyranthes aspera* L., *Aristolochia indica* L., *Cleome viscosa* L., *Catharanthus roseus* L., *Gynandropsis pentaphylla* L., and *Gymnema sylvestre* which are dominated by genera *Aspergillus*, *Alternaria*, *Trichoderma*, *Penicillium*, *Cladosporium*, and *Fusarium*. From *Panax notoginseng*, six rhizospheric fungi are reported such as *G. mosseae*, *G. versiforme*, *G. claroideum*, *G. monosporum*, *G. constrictum*, and *Absidia panacisoli* (Zhang et al. 2011b, 2018).

From *Panax ginseng* the following fungi *G. fasciculatum*, *G. geosporum*, *G. macrocarpum*, *G. microaggregatum*, *G. Mosseae*, *A. cavernata*, and *A. Spinosa* are reported. Because of consistent cultivation of *P. ginseng* obstacle for the change in diversity and composition of fungal communities in rhizosphere, the continuous cultivation leads to increase the diversity of *Sordariomycetes*, *Alatospora*, *Eurotiomycetes*, *Leotiomycetes*, *Saccharomycetes*, *Mucorales*, and *Pezizomycetes* species (Cho et al. 2009; Dong et al. 2017). Metagenomics sequencing analysis of *Taxus* rhizosphere revealed the presence of 301 species of nine fungi phyla (Hao et al. 2018).

14.4 Effect of Rhizospheric Microbiome on Medicinal Plant Growth and Nutrient Uptake

The microbial association in the rhizosphere of plant is positively influences on the improvement of microbial diversity, influencing production of growth-promoting auxins and cytokinins, aiding nutrient availability and uptake, activating host defense mechanisms, providing tolerance to stress, and controlling pathogens through antagonism. These beneficial plant growth-promoting (PGP) microbes can be used in the formulation of biofertilizers and biocontrol agents.

Rhizospheric bacteria and fungi produce siderophores which are compounds that chelate iron (Fe) with high affinity and aid transport across membranes (Das et al. 2007). Iron is mostly present in the form of insoluble complexes in the soil which makes it less available to the plants. Iron is an important component in the synthesis of chlorophyll, maintaining the structure and function of chloroplast, a prosthetic group for the functioning of several enzymes including cytochromes in electron transport. Iron deficiency in plants leads to low photosynthetic efficiency, chlorosis, etc. Sufficient availability of iron is known to promote growth and development in plants. *Azotobacter vinelandii*, *Bacillus subtilis*, *Pseudomonas fluorescens*, *Magnetospirillum magneticum*, *Staphylococcus hyicus*, etc. produce siderophores. Siderophores from *Rhizobium meliloti* and *B. thuringiensis* SMA5 promote plant growth in *Mucuna pruriens* and *Aloe vera*, respectively (Arora et al. 2001; Meena et al. 2018). Colonization by *Glomus fasciculatum*, *G. versiforme*, *G. clarum*, *G. mosseae*, and *G. etunicatum* in *Piper longum* increases shoot length, nutrient content, biomass, chlorophyll content, etc. (Gogoi and Singh 2011). *Panax ginseng* seedlings treated with *Pseudomonas simiae* N3, *Burkholderia ginsengiterrae* N11-2, and *Chryseobacterium polytrichastri* N10 show higher chlorophyll and higher biomass than untreated seedlings (Farh et al. 2017). *Piriformospora indica* shows growth-promoting effect in a number of plants, viz., *Artemisia annua*, *Azadirachta indica*, *Abrus precatorius*, *Bacopa monnieri*, *Withania somnifera*, *Curcuma longa*, *Trigonella fornum-graecum*, *Stevia rebaudiana*, etc. (Kumar et al. 2017; Bagde et al. 2010; Prasad et al. 2008; Das et al. 2013). Medicinal plants (*Withania somnifera*, *Psorelea corylifolia*, *Clitoria ternatea*, *Plumbago zeylanica*, *Abelmoschus moschatus*) grown in various soil types accumulate increased dry matter on mycorrhizal inoculation (Chandra et al. 2010). *Azotobacter chrooccocum* promotes plant growth in *Calendula officinalis* (Hosseinzadah et al. 2011), *Adhatoda vasica* (Naik 2006), and *Ocimum basilicum* (Ordookhan et al. 2011). These PGP microbes bring about plant growth through various mechanisms such as production of metabolites like hydrogen cyanide (HCN) and 2,4-diacetylphloroglucinol (DAPG); production of bioactive volatiles like acetoin and 2,3-butanediol; production of plant hormones like auxins, cytokinins, gibberellins, and abscisic acid; synthesis of 1-aminocyclopropane-1-carboxylate (ACC) deaminase that metabolizes ethylene precursor ACC; etc. It has been observed that seed germination and seedling growth improvement are the influence of gibberellins synthesized by the PGP microbes. *Poncirus trifoliata* L. Raf seedlings inoculated with *Rhizophagus irregularis* BGC JX04B enhance lateral root formation (Chen et al. 2017a, b). The AM fungus was found to influence the upregulation of genes involved in auxin signaling among the various differentially expressed genes (DEGs). Indoleacetic acid (IAA) produced by PGP microbes result in root proliferation, development of root hair, metabolism, etc. Bacteria like *Azospirillum* sp., *Bacillus megaterium*, *Bradyrhizobium* sp., *Rhizobium* sp., *Klebsiella* sp., *Jeotgalicoccus huakuii*, *Agrobacterium* sp., *Pseudomonas* sp., *Erwinia herbicola*, etc. are capable of producing IAA (Misra et al. 2017; Katiyar et al. 2016). Rhizospheric microbes also increase the nutrient absorption such as Ca, Cu, Zn, Fe, Mn, Mg, Na, etc. AM fungal hyphae can aid in uptake and assimilation of NH_4^+ and NO_3^- into amino acids. They increase the root surface area and produce nitrogenases and phosphatases, other beneficial metabolites, and enzymes. Bacteria such as *Acetobacter*, *Agrobacterium radiobacter*, *Azospirillum lipoferum*, *Arthrobacter mysorens*, *Azoarcus*, *Burkholderia*, *Bacillus polymyxa*, *Diazotrophicus*, *Paenibacillus polymyxa*, *Gluconacetobacter*, *Pseudomonas putida*, *Herbaspirillum*, *Rhodobacter capsulatus*, etc. increase plant growth through nitrogen fixation (Adesemoye and Egamberdieva 2013; Backer et al. 2018). Unlike the symbiotic *Rhizobium* species, these free-living nitrogen-fixing microbes are beneficial to a wide range of plants. Further few microbes that are incapable of fixing atmospheric nitrogen aid in the efficient uptake of nitrogen, likely by the proliferation of roots allowing access to large area of soil (Beattie 2015). In *Chlorophytum borivilianum*, *Glomus fasciculatum* increases P and K uptake, and *Azotobacter chroococcum* increases nitrogen uptake (Solanki et al. 2011). *P. extremorientalis* strain TSAU20 and *Mesorhizobium* sp. strain NWXJ31 enhanced N, P, K, and Mg uptake as well as produced higher biomass in *Glycyrrhiza uralensis* Fisch. (Egamberdieva et al. 2017).

Phosphate (P) deficiency in plants leads to stunting, poor root development, flowering inhibition, etc. Low amount of P is available to plants as they take up P in the ionic form ($H_2PO_4^-$ and $H_2PO_4^{2-}$) unlike the insoluble complexes present in soil. A number of rhizospheric bacteria and fungi are capable of solubilizing phosphates by producing organic acids which decrease the pH or chelate the mineral ions, releasing P into solution; by the secretion of phosphatases, phytases; thereby increase P uptake by plants and promote plant growth. To name a few, bacteria like *Bacillus stratosphericus*, *B. marisflavi*, *Burkholderia gladioli*, *Serratia marcescens*, *Enterobacter erogenes*, etc. and fungi *Penicillium pinophilum*, *Aspergillus fumigatus*, *Aspergillus niger*, etc. are P-solubilizers (Misra et al. 2017; Gupta et al. 2011; Wahid and Mehana 2000). *B. gladiola* and *E. aerogenes* aid growth of *Stevia rebaudiana* by increasing P availability (Mamta et al. 2010). A consortium of phosphate-solubilizing bacteria (PSB) enhances plant growth and accumulation of biomass and stevioside as well as rebaudioside-A in *Stevia rebaudiana* (Gupta et al. 2011). Phosphate-solubilizing microbes *Pseudomonas synxantha*, *Serratia marcescens*, *Burkholderia gladioli*, and *Enterobacter hormaechei* increase leaf length, root length, number of leaves, and total gel volume in *Aloe vera* (Gupta et al. 2012).

The microbial diversity associated with medicinal plants influences and enhances the accumulation of pharmaceutically important secondary metabolites in plants. Plant growth and yield of antimalarial compound artemisinin in *Artemisia annua* L. is enhanced by a consortium of *Glomus mosseae* and *Bacillus subtilis* (Awasthi et al. 2011). A comparative study of the mycorrhizal communities of *Cassia alata*, *C. occidentalis*, and *C. sophera* revealed highest root colonization by *Glomus* species in *C. alata*, and interestingly the species is considered most potent for antimicrobial activity among the three (Chatterjee et al. 2010). *Glomus mosseae* or *G. intraradices* colonized *Sorghum* plants that contain higher alcohols, ethers, acids, and alkenes as compared to the non-colonized plants (Sun and Tang 2013). *Glomus intraradices* inoculation in *Salvia officinalis* enhances essential oil yield and quality (Geneva et al. 2010). *Plectranthus amboinicus* seedlings when primed with indigenous fungi, viz., *Acaulospora scrobiculata*, *A. bireticulata*, *Glomus mosseae*, *G. aggregatum*, *G. geosporum*, *Gigaspora margarita*, and *Scutellospora heterogama*, improved plant growth and phytochemical constituents like alkaloids, tannins, flavonoids, and saponins (Rajeshkumar et al. 2008). *Glomus* species enhances production of anticancer alkaloid vinblastine in *Catharanthus roseus* (Rosa-Mera et al. 2011); *Acaulospora mellea* and *Glomus intraradices* enhance camptothecin in *Camptotheca acuminata* (Yang et al. 2012); forskolin in *Coleus forskohlii* by *Glomus bagyarajii* and *Scutellospora calospora* (Sailo and Bagyaraj 2005); antioxidant compounds rosmarinic acid and caffeic acid in *Ocimum basilicum* by *Glomus caledonium* (Toussaint et al. 2007); andrographolide in *Andrographis paniculata* by *Gigaspora albida* (Radhika and Rodrigues 2011); essential oil content in *O. basilicum* by *Glomus mosseae* BEG 12, *Gigaspora margarita* BEG 34, and *Gigaspora rosea* BEG 9 (Copetta et al. 2006); and *Glomus fasciculatum* in *Mentha arvensis* (Gupta et al. 2002).

14.5 Effect of Rhizospheric Microbiome on Biotic and Abiotic Stress Tolerance

The sessile nature of plants exposes them to a broad range of environmental stresses (abiotic) as well as stresses induced by living entities (biotic) that they cannot escape. Environmental stresses such as drought, low/high temperature, salt stress, acidic conditions, heavy metal stress, nutrient stress, etc. are considered. While biotic stress induced by living organisms like bacteria, viruses, fungi, parasites, beneficial and harmful insects that may damage to the plant. AM fungi are known to enhance tolerance to several biotic and abiotic stresses. Salt stress inhibits plant growth by reducing uptake of water and essential nutrients and excessive salt uptake. Salinity also leads to reduced photosynthetic ability. The resulting decreased NADP yield in turn causes increased transfer of excess photoexcitation energy to oxygen producing reactive oxygen species (ROS). ROS cause a number of metabolic disorders due to its oxidizing nature. Plant yield is drastically reduced under salinity stress in *Satureja hortensis*, *Citronella*, *Hyoscyamus niger*, *Matricaria chamomilla*, etc. (Teixeira da Silva and Egamberdieva 2013). PGPRs like *Rhizobium*, *Azospirillum*, *Arthrobacter*, *Flavobacterium*, *Bacillus*, *Pseudomonas*, etc. improve plant metabolic processes, osmoregulation, and resistance to starvation to impart tolerance to drought and salinity (Egamberdieva and Islam 2008). *Pseudomonas putida* imparts drought tolerance by accumulation of osmolytes like proline and glycine, changing membrane integrity and ROS scavenging. Tolerance to stress is also provided by the modulation of genes associated with salicylic acid (PR1), jasmonate transcription activation (MYC2), dehydration responsive element binding (DREB1A), dehydrins (DHN), late embryogenesis abundant proteins (LEA), and antioxidant enzymes (Tiwari et al. 2016). Microbial ACC deaminase aids plant growth in adverse conditions like poor soil aeration, waterlogging, excessive toxins, drought, and salinity by counteracting the inhibitory effects of ethylene. On the other hand, ABA from microbial colonization is involved in ABA-dependent stomatal movements, ABA-mediated stress response signaling aiding in salt and drought stress tolerance (Sah et al. 2016). Under nutrient-poor soils, *Cryptococcus laurentii* is capable of aiding growth of *Agathosma betulina* (Berg) Pillans (Cloete et al. 2009). *Streptomyces pactum* (Act12) increases resistance, yield, and quality in ginseng (Zhang et al. 2013). *P. extremorientalis* improves plant health under salt stress in *Silybum marianum* (Egamberdieva et al. 2013a). *Diversispora versiformis* increases root length and total dry weight under salinity stress in *Chrysanthemum morifolium* (Wang et al. 2018). *Glomus mosseae*, *G. intraradices*, and *G. etunicatum* are effective in mitigating adverse effects of salinity by reducing oxidative damage in *O. basilicum* (Abeer et al. 2016). *Glomus mosseae* and *G. intraradices* AM fungi improve the growth and salinity tolerance in *Bacopa monnieri* (Khaliel et al. 2011). AMF colonization improves growth and adaption to drought stress in *Rosa damascena* Mill. (Abdel-Salam et al. 2017). Under drought conditions *Pseudomonas* inoculation increases biomass and growth of *Catharanthus roseus* (Jaleel et al. 2007). *Rhizobium galegae* and *Pseudomonas extremorientalis* alleviate salt stress in *Galega officinalis* (Egamberdieva et al. 2013b). *Glomus intraradices* increases yield of methyl chavicol, methyl euginol, etc., in *Ocimum basilicum* under metal contamination, viz., Cd, Pb, and Ni (Prasad et al. 2011). *Bacillus megaterium* alleviates nickel (Ni) stress by enhanced antioxidant enzyme activity of ascorbate peroxidase (APX), catalase (CAT), peroxidase (POD), and superoxide dismutase (SOD); increased production of flavonoids, phenols, and proline; and reduction in malondialdehyde (MDA) and hydrogen peroxide (H_2O_2) in *Vinca rosea* (Khan et al. 2017a, b). *Pseudomonas simiae* N3, *Burkholderia ginsengiterrae* N11-2, and *Chryseobacterium polytrichastri* N10 from the rhizosphere confer aluminum stress resistance in *Panax ginseng* (Farh et al. 2017). Plants in association with these rhizobacteria showed higher Al stress linked gene expression as compared to the non-bacterized plants.

A number of microbes are found to be potential in both plant growth improvement and disease resistance. PGPRs show antagonism to pathogenic microbes through antibiotic and hydrogen cyanide production; secretion of chitinases, proteases, beta-1, and 3-glucanase; parasitism, competition for nutrition or colonization, etc., thereby conferring disease resistance in plants (Ahmed et al. 2014). In addition, PGP microbes establish induced systemic resistance (ISR) in plants; induce production of pathogenesis-related proteins (PR proteins); and increase activity of phenylalanine ammonia-lyase (PAL) and polyphenol oxidase (PPO) which are enzymes associated with lignin and phytoalexin formation. *Bacillus subtilis* Jdm2 from the rhizosphere of *Trichosanthes kirilowii* is antagonistic to nematode and enhances plant growth (Wei et al. 2014). *Fusarium oxysporum*, *F. solani*, *Macrophomina phaseolina*, and *Rhizoctonia solani*-mediated diseases in mung bean can effectively be controlled by the application of *Pseudomonas aeruginosa* and *Paecilomyces lilacinus* along with medicinal plant *Launaea nudicaulis* (Mansoor et al. 2007). Root-rot diseases caused by *Fusarium chlamydosporum* and *Ralstonia solanacearum* in *C. forskohlii* can be controlled by AM colonization of *Pseudomonas monteilii* and *Glomus fasciculatum* (Singh et al. 2013). *Bacillus* and *Pseudomonas* isolates from *Thymus vulgaris*, *Majorana hortensis*, *Matricaria chamomilla*, *Cymbopogon citratus*, and *Melissa officinalis* show antagonism against *F. oxysporum* and *R. solani* (Ahmed et al. 2014). Rhizobacteria *Staphylococcus saprophyticus*, *Serratia marcescens*, *Brevibacillus agri*, *Pseudomonas hunanensis*, etc. show antagonism against fungal pathogens *Nigrospora sphaerica*, *Curvularia eragrostidis*, *R. solani*, *F. oxysporum*, and *Pestalotiopsis theae* (Dutta and Thakur 2017). Rhizospheric bacteria associated with arid soil-grown *Calendula officinalis* L., *Matricaria chamomilla* L., and *Solanum distichum* are of value in suppression of pathogens (Koeberl et al. 2013). Among the various isolates, *Bacillus subtilis* subsp. *subtilis* Co1-6, *Paenibacillus polymyxa* Mc5Re-14, and *Streptomyces subrutilus* Wb2n-11 are found promising for their antagonistic activity against plant pathogenic nematode *Meloidogyne incognita* and bacterium *Ralstonia solanacearum.* These isolates also improve stress tolerance and plant growth. Further, *Bacillus subtilis* subsp. *subtilis* Co1-6 and *Paenibacillus polymyxa* Mc5Re-14 increase the production of major flavonoids like apigenin-7-*O*-glucoside and apigenin in *M. chamomilla. Streptomyces* isolates from rhizosphere soils of medicinal plants (*Curcuma mangga* Val. and Zijp., *Ocimum sanctum* L., *Cymbopogon citratus* Stapf., *Achyranthes aspera* L., *Zingiber officinale* Rose., etc.) show inhibitory activity against phytopathogenic fungi *Fusarium oxysporum*, *Colletotrichum gloeosporioides*, *Alternaria brassicicola*, *Penicillium digitatum*, and *Sclerotium rolfsii* (Khamna et al. 2009). *Delftia tsuruhatensis*, a rhizospheric bacterium from *Rauwolfia serpentina*, suppresses pathogenic fungi like *Fusarium oxysporum*, *Phomopsis vexans*, *Alternaria solani*, *Rhizoctonia solani*, *Pythium aphanidermatum*, *Colletotrichum capsici*, etc. (Prasannakumar et al. 2015). Some rhizospheric bacteria are capable of controlling pathogenesis in plants by utilizing the pathogenic toxins as an energy source. For example, *Burkholderia ambifaria*, *B. cepacia*, *Klebsiella oxytoca*, etc. used fusaric acid produced by the wilt-causing pathogens *Fusarium oxysporum*, *F. solani*, and *F. verticillioides* as carbon source, thus reducing its pathogenic effects (Simonetti et al. 2018).

14.6 Conclusions

The medicinal plant products and yield are influenced by the presence of microorganism around the root. The rhizospheric microbes help in plant growth via solubilization of mineral phosphate, nitrogen fixation, and growth hormone secretion and provide immunity against the abiotic and biotic stress. So, the application of rhizospheric bacteria or fungi to medicinal plant improves the quality and yield of medicinal product. In the current scenario of increasing interest toward organic cultivation, there is an immediate need for crop-specific potential rhizospheric microbes for the improvement in cultivation of medicinal plants. A wide variety of bacterial and fungal communities are recognized in the rhizosphere which has high significance in plant nutrient acquisition and secondary metabolite alteration. The application of AM fungi or plant growth-promoting rhizobacteria is a sustainable and environment-friendly technique to enhance the quantity and quality of the medicinal plant compounds. Apart from the medicinal plants, numerous rhizospheric bacteria or fungi are also recognized as an alternative source of bioactive compound. Research focusing on understanding the diversity and function of rhizospheric microbes with medicinal plants will be prerequisite in the future to better understand the secondary metabolite accumulation within plants.

References

Abdel-Salam E, Alatar A, El-Sheikh MA (2017) Inoculation with arbuscular mycorrhizal fungi alleviates harmful effects of drought stress on damask rose. Saudi J Biol Sci 25:1772–1780. https://doi.org/10.1016/j.sjbs.2017.10.015

Abeer H, Salwa AA, Alqarawi AA, Allah EE, Egamberdieva D (2016) Arbuscular mycorrhizal fungi enhance basil tolerance to salt stress through improved physiological and nutritional status. Pak J Bot 48:37–45

Adesemoye AO, Egamberdieva D (2013) Beneficial effects of plant growth-promoting rhizobacteria on improved crop production: prospects for developing economies. In: Maheshwari DK, Saraf M, Aeron A (eds) Bacteria in agrobiology: crop productivity. Springer, Berlin, pp 45–63

Ahmed EA, Hassan EA, El Tobgy KM, Ramadan EM (2014) Evaluation of rhizobacteria of some medicinal plants for plant growth promotion and biological control. Ann Agric Sci 59:273–280

Akter S, Jo H, Du J, Won K, Yin CS, Kook M, Yu H, Choi HS, Kim MK, Yi TH (2015) *Pseudoxanthomonashumi* sp. nov., a bacterium isolated from rhizospheric soil of *Fraxinus chinensis* in Gyeonggi Province, South Korea. Arch Microbiol 197:1165–1172

Andrew DR, Fitak RR, Munguia-Vega A, Racolta A, Martinson VG, Dontsova K (2012) Abiotic factors shape microbial diversity in Sonoran Desert soils. Appl Environ Microbiol 78:7527–7537

Appoloni S, Lekberg Y, Tercek MT, Zabinski CA, Redecker D (2008) Molecular community analysis of arbuscular mycorrhizal fungi in roots of geothermal soils in Yellowstone National Park (USA). Microb Ecol 56:649–659

Araim G, Saleem A, Arnason JT, Charest AC (2009) Root colonization by an arbuscular mycorrhizal (AM) fungus increases growth and secondary metabolism of purple coneflower. *Echinacea purpurea* L. Moench. J Agric Food Chem 57:2255–2258

Arora NK, Kang SC, Maheshwari DK (2001) Isolation of siderophore-producing strains of *Rhizobium meliloti* and their biocontrol potential against *Macrophomina phaseolina* that causes charcoal rot of groundnut. Curr Sci 81:673–677

Awasthi A, Bharti N, Nair P, Singh R, Shukla AK, Gupta MM, Darokar MP, Kalra A (2011) Synergistic effect of *Glomus mosseae* and nitrogen fixing *Bacillus subtilis* strain Daz26 on artemisinin content in *Artemisia annua* L. Appl Soil Ecol 49:125–130

Backer R, Rokem JS, Ilangumaran G, Lamont J, Praslickova D, Ricci E, Subramanian S, Smith DL (2018) Plant growth-promoting rhizobacteria: context, mechanisms of action, and roadmap to commercialization of biostimulants for sustainable agriculture. Front Plant Sci 9:1473

Bafana A, Lohiya R (2013) Diversity and metabolic potential of culturable root-associated bacteria from *Origanum vulgare* in sub-Himalayan region. World J Microbiol Biotechnol 29:63–74

Bagde US, Prasad R, Varma A (2010) Interaction of *Piriformospora indica* with medicinal plants and of economic importance. Afr J Biotechnol 9:9214–9226

Beattie GA (2015) Microbiomes: curating communities from plants. Nature 528:340–341

Bulgarelli D, Rott M, Schlaeppi K, Ver Loren van Themaat E, Ahmadinejad N, Assenza F, Rauf P, Huettel B, Reinhardt R, Schmelzer E, Peplies J, Gloeckner FO, Amann R, Eickhorst T, Schulze-Lefert P (2012) Revealing structure and assembly cues for *Arabidopsis* root-inhabiting bacterial microbiota. Nature 488:91–95

Cai BY, Ge QP, Jie WG, Yan XF (2009) The community composition of the arbuscular mycorrhizal fungi in the rhizosphere of *Phellodendron amurense*. Mycosystema 28:512–520

Cao C, Sun Y, Wu B, Zhao S, Yuan B, Qin S, Jiang J, Huang, Y (2018) Actinophytocola glycyrrhizae sp. nov. isolated from the rhizosphere of *Glycyrrhiza inflata*. Int J Syst Evol Microbiol 68:2504–2508

Chandra KK, Kumar N, Chand G (2010) Studies on mycorrhizal inoculation on dry matter yield and root colonization of some medicinal plants grown in stress and forest soils. J Environ Biol 31:975–979

Chaparro JM, Badri DV, Vivanco JM (2014) Rhizosphere microbiome assemblage is affected by plant development. ISME J 8:790–803

Chatterjee S, Chatterjee S, Dutta S (2010) A survey on VAM association in three different species of Cassia and determination of antimicrobial property of these phytoextracts. J Med Plant Res 4:286–292

Chen Q, Liu B, Wang J, Che J, Liu G, Guan X (2016) Antifungal lipopeptides produced by *Bacillus* sp. FJAT-14262 isolated from rhizosphere soil of the medicinal plant *Anoectochilus roxburghii*. Appl Biochem Biotechnol 182:155–167

Chen W, Li J, Zhu H, Xu P, Chen J, Yao Q (2017a) Arbuscular mycorrhizal fungus enhances lateral root formation in *Poncirus trifoliata* (L.) as revealed by RNA-Seq analysis. Front Plant Sci 8:2039. https://doi.org/10.3389/fpls.2017.02039

Chen J, Wu L, Xiao Z, Wu Y, Wu H, Qin X, Wang J, Wei X, Khan MU, Lin S, Lin W (2017b) Assessment of the diversity of *Pseudomonas* spp. and *Fusarium* spp. in *Radix pseudostellariae* rhizosphere under monoculture by combining DGGE and quantitative PCR. Front Microbiol 8:1748

Chiellini C, Maida I, Emiliani G, Mengoni A, Mocali S, Fabiani A, Biffi S, Maggini V, Gori L, Vannacci A, Gallo E, Firenzuoli F, Fani R (2014) Endophytic and rhizospheric bacterial communities isolated from the medicinal plants *Echinacea purpurea* and *Echinacea angustifolia*. Int Microbiol 17:165–174

Cho EJ, Lee DJ, Wee CD, Kim HL, Cheong YH, Cho JS, Sohn BK (2009) Effects of AM fungi inoculation on growth of *Panax ginseng* C.A. Meyer seedlings and on soil structures in mycorrhizosphere. Sci Hortic 122:633–637

Cloete KJ, Valentine AJ, Stander MA, Blomerus LM, Botha A (2009) Evidence of symbiosis between the soil yeast *Cryptococcus laurentii* and a *Sclerophyllous* medicinal shrub, *Agathosma betulina* (Berg.) Pillans. Microb Ecol 57:624–632

Copetta A, Lingua G, Berta G (2006) Effects of three AM fungi on growth, distribution of glandular hairs, and essential oil production in *Ocimum basilicum* L. var. Genovese. Mycorrhiza 16:485–494

Costa R, Gotz M, Mrotzek N, Lottmann J, Berg G, Smalla K (2006) Effects of site and plant species on rhizosphere community structure as revealed by molecular analysis of microbial guilds. FEMS Microbiol Ecol 56:236–249

Dai CC, Xie H, Wang XX, Li PD, Zhang TL, Li YL, Tan X (2009) Intercropping peanut with traditional Chinese medicinal plants improves soil microcosm environment and peanut production in subtropical China. Afr J Biotechnol 8:3739–3746

Dai CC, Chen Y, Wang XX, Li PD (2013) Effects of intercropping of peanut with the medicinal plant *Atractylodes lancea* on soil microecology and peanut yield in subtropical China. Agrofor Syst 87:417–426

Das A, Prasad R, Srivastava A, Giang PH, Bhatnagar K, Varma A (2007) Fungal siderophores: structure, functions and regulations. In: Varma A, Chincholkar SB (eds) Microbial siderophores, vol 12. Springer-Verlag, Berlin, pp 1–42

Das A, Prasad R, Srivastava RB, Deshmukh S, Rai MK, Varma A (2013) Co-cultivation of *Piriformospora indica* with medicinal plants: case studies. In: Varma A, Kost G, Oelmuller R (eds) *Piriformospora indica*: Sebacinales and their biotechnological applications. Springer-Verlag, Berlin, pp 149–171

de Almeida Lopes KB, Carpentieri-Pipolo V, Oro TH, Stefani Pagliosa E, Degrassi G (2016) Culturable endophytic bacterial communities associated with field-grown soybean. J Appl Microbiol 120:740–755

Debnath R, Yadav A, Gupta VK, Singh BP, Handique PJ, Saikia R (2016) Rhizospheric bacterial community of endemic *Rhododendron arboreum* Sm. Ssp. delavayi along Eastern Himalayan Slope in Tawang. Front Plant Sci 7:1345

Dong LL, Niu WH, Wang R, Xu J, Zhang LJ, Zhang J, Chen SL (2017) Changes of diversity and composition of fungal communities in rhizosphere of *Panax ginseng*. Zhongguo Zhong Yao Za Zhi 42:443–449

Dutta J, Thakur D (2017) Evaluation of multifarious plant growth promoting traits, antagonistic potential and phylogenetic affiliation of rhizobacteria associated with commercial tea plants grown in Darjeeling, India. PLoS One 12(8):e0182302

Egamberdieva D, Islam KR (2008) Salt tolerant rhizobacteria: plant growth promoting traits and physiological characterization within ecologically stressed environment. Wiley, Weinheim, pp 257–281

Egamberdieva D, Jabborova D, Mamadalieva N (2013a) Salt tolerant *Pseudomonas extremorientalis* able to stimulate growth of *Silybum marianum* under salt stress. Med Aromat Plant Sci Biotechnol 7:7–10

Egamberdieva D, Berg G, Lindström K, Räsänen LA (2013b) Alleviation of salt stress of symbiotic *Galega officinalis* L. (goat's rue) by co-inoculation of rhizobium with root-colonizing *Pseudomonas*. Plant Soil 369:453–465

Egamberdieva D, Wirth S, Li L, Abd-Allah EF, Lindström K (2017) Microbial cooperation in the rhizosphere improves liquorice growth under salt stress. Bioengineered 8:433–438

El-Deeb B, Fayez K, Gherbawy Y (2013) Isolation and characterization of endophytic bacteria from *Plectranthus tenuiflorus* medicinal plant in Saudi Arabia desert and their antimicrobial activities. J Plant Interact 8:56–64

Farh ME, Kim YJ, Sukweenadhi J, Singh P, Yang DC (2017) Aluminium resistant, plant growth promoting bacteria induce overexpression of aluminium stress related genes in *Arabidopsis thaliana* and increase the ginseng tolerance against aluminium stress. Microbiol Res 200:45–52

Garcia A, Polonio J, Polli A, Santos C, Rhoden S, Quecine M, Azevedo JL, Pamphile JA (2016) Rhizosphere bacteriome of the medicinal plant *Sapindus saponaria* L. revealed by pyrosequencing. Genet Mol Res 15:1–9

Geneva MP, Stancheva IV, Boychinova MB, Mincheva NH, Yonova PA (2010) Effects of foliar fertilization and arbuscular mycorrhizal colonization on *Salvia officinalis* L. growth, antioxidant capacity, and essential oil composition. J Sci Food Agric 90:696–702

Gogoi P, Singh RK (2011) Differential effect of some arbuscular mycorrhizal fungi on growth of *Piper longum* L. (Piperaceae). Ind J Sci Technol 4:119–125

Gorsi MS (2002) Studies on mycorrhizal association in some medicinal plants of Azad Jammu and Kashmir. Asian J Plant Sci 1:383–387

Guo DZ, Chen J, Du XP, Han BX (2010) Screening of molluscicidal strain against *Oncomelania hupensis* from the rhizosphere of medicinal plant *Phytolacca acinosa* Roxb. Pharmacogn Mag 6:159–165

Gupta ML, Prasad A, Ram M, Kumar S (2002) Effect of the vesicular-arbuscular mycorrhizal (VAM) fungus *Glomus fasciculatum* on the essential oil yield related characters and nutrient acquisition in the crops of different cultivars of menthol mint (*Mentha arvensis*) under field conditions. Bioresour Technol 81:77–79

Gupta M, Bisht S, Singh B, Gulati A, Tewari R (2011) Enhanced biomass and steviol glycosides in *Stevia rebaudiana* treated with phosphate-solubilizing bacteria and rock phosphate. Plant Growth Regul 65:449–457

Gupta M, Kiran S, Gulati A, Singh B, Tewari R (2012) Isolation and identification of phosphate solubilizing bacteria able to enhance the growth and aloin-A biosynthesis of *Aloe barbadensis* Miller. Microbiol Res 167:358–363

Hao DC, Zhang CR, Xiao PG (2018) The first Taxus rhizosphere microbiome revealed by shotgun metagenomic sequencing. J Basic Microbiol 58:501–512

Hosseinzadah F, Satei A, Ramezanpour (2011) Effects of mycorrhiza and plant growth promoting rhizobacteria on growth, nutrient uptake and physiological characteristics in *Calendula officinalis* L. Middle East J Sci Res 8(5):947–953

Hussain SA, Srinivas P (2013) Association of arbuscular mycorrhizal fungi and other rhizosphere microbes with different medicinal plants. Res J Biotechnol 8:24–28

Impullitti AE, Malvick DK (2013) Fungal endophyte diversity in soybean. J Appl Microbiol 114:1500–1506

Jaleel CA, Manivavannan P, Sankar P, Krishnakumar B, Gopi AR, Somasundaram R, Pannerselvam (2007) *Pseudomonas fluorescens* enhances biomass yield and Ajmalicine production in *Catharanthus roseus* under water deficit stress. Colloid Surf B Biointerfaces 60:7–11

Johnson MP, Stephan R (2016) Association of arbuscular mycorrhizal fungi and other rhizosphere microbes with different medicinal plants in the calcareous soil of Ariyalur District, India. Int J Curr Mirobiol App Sci 5(9):659–666

Joy P, Thomos J, Mathew S, Skaria BP (1998) Medicinal plants. Kerala Agricultural University Press, Kerala

Jurkiewicz A, Ryszka P, Anielska T, Waligorski P, Białon'ska D, Goralska K, Michael MT, Turnau K (2010) Optimization of culture conditions of *Arnica montana* L: effects of mycorrhizal fungi and competing plants. Mycorrhiza 20:293–306

Katiyar D, Hemantaranjan A, Singh B (2016) Plant growth promoting rhizobacteria-an efficient tool for agriculture promotion. Adv Plants Agric Res 4:426–434

Khaliel AS, Shine K, Vijayakumar K (2011) Salt tolerance and mycorrhization of *Bacopa monneiri* grown under sodium chloride saline conditions. Afr J Microbiol Res 5:2034–2040

Khamna S, Yokota A, Lumyong S (2009) *Actinomycetes* isolated from medicinal plant rhizosphere soils: diversity and screening of antifungal compounds, indole-3-acetic acid and siderophore production. World J Microbiol Biotechnol 25:649–655

Khan WU, Ahmad SR, Yasin NA, Ali A, Ahmad A, Akram W (2017a) Application of *Bacillus megaterium* MCR-8 improved phytoextraction and stress alleviation of nickel in *Vinca rosea*. Int J Phytoremediation 19:813–824

Khan AL, Asaf S, Al-Rawahi A, Lee I-J, Al-Harrasi A (2017b) Rhizospheric microbial communities associated with wild and cultivated frankincense producing *Boswellia sacra* tree. PLoS One 12(10):e0186939

Knief C (2014) Analysis of plant microbe interactions in the era of next generation sequencing technologies. Front Plant Sci 5:216

Koeberl M, Schmidt R, Ramadan EM, Bauer R, Berg G (2013) The microbiome of medicinal plants: diversity and importance for plant growth, quality, and health. Front Microbiol 4:400

Kumar G, Kanaujia N, Bafana A (2012) Functional and phylogenetic diversity of root-associated bacteria of *Ajuga bracteosa* in Kangra valley. Microbiol Res 167:220–225

Kumar P, Pagano M, O'donovan A (2017) Mycosphere essay 18: biotechnological advances of beneficial fungi for plants. Mycosphere 8:445–455

Lakshmanan V, Selvaraj G, Bais HP (2014) Functional soil microbiome: belowground solutions to an aboveground problem. Plant Physiol 166:689–700

Lee HR, Han SI, Rhee KH, Whang KS (2013) *Mucilaginibacter herbaticus* sp. nov., isolated from the rhizosphere of the medicinal plant *Angelica sinensis*. Int J Syst Evol Microbiol 63:2787–2793

Lopez-Fuentes E, Ruiz-Valdiviezo VM, Martinez-Romero E, Gutierrez-Miceli FA, Dendooven L, Rincon-Rosales R (2012) Bacterial community in the roots and rhizosphere of *Hypericum silenoides* Juss. 1804. Afr J Microbiol Res 6:2704–2711

Mamta RP, Pathania V, Gulati A, Singh B, Bhanwra RK, Tewari R (2010) Stimulatory effect of phosphate-solubilizing bacteria on plant growth, stevioside and rebaudioside-A contents of *Stevia rebaudiana* Bertoni. Appl Soil Ecol 46:222–229

Mansoor F, Sultana V, Ehteshamul-Haque S (2007) Enhancement of biocontrol potential of *Pseudomonas aeruginosa* and *Paecilomyces lilacinus* against root rot of mungbean by a medicinal plant *Launaea nudicaulis* L. Pak J Bot 39:2113–2119

Meena NK, Tara N, Saharan BS (2018) Review on PGPR: an alternative for chemical fertilizers to promote growth in *Aloe vera* plants. Int J Curr Microbiol App Sci 7:3546–3551

Misra S, Dixit VK, Khan MH, Kumar Mishra S, Dviwedi G, Yadav S, Singh Chauhan P (2017) Exploitation of agro-climatic environment for selection of 1-aminocyclopropane-1-carboxylic acid (ACC) deaminase producing salt tolerant indigenous plant growth promoting rhizobacteria. Microbiol Res 205:25–34

Nadeem SM, Ahmad M, Zahir ZA, Javaid A, Ashraf M (2014) The role of mycorrhizae and plant growth promoting rhizobacteria (PGPR) in improving crop productivity under stressful environments. Biotechnol Adv 32:429–448

Naik AT (2006) Biological and Molecular characterization of *Azotobacter chroococcum* isolated from different agroclimatic zones of Karnataka and their influences on growth and biomass of *Adhatoda vasica* Nees. MSc, (Agri.) thesis, University of Agricultural Sciences, Bangalore, India

Narula N, Kothe E, Behl RK (2009) Role of root exudates in plant-microbe interactions. J Appl Bot Food Qual 82:122–130

Nema R, Khare S, Jain P, Pradhan A, Gupta A, Singh D (2013) Natural products potential and scope for modern cancer research. Am J Plant Sci 4:1270–1277

Nimnoi P, Lumyong S, Pongsilp N (2011) Impact of rhizobial inoculants on rhizosphere bacterial communities of three medicinal legumes assessed by denaturing gradient gel electrophoresis (DGGE). Ann Microbiol 61:237–245

Ordookhan K, Sharafzadeh S, Zare M (2011) Influence of PGPR on growth, essential oil and nutrients uptake of sweet basil. Adv Environ Biol 5:672–677

Panwar J, Tarafdar JC (2006) Distribution of three endangered medicinal plant species and their colonization with arbuscular mycorrhizal fungi. J Arid Environ 65:337–350

Peiffer JA, Spor A, Koren O, Jin Z, Tringe SG, Dangl JL, Buckler ES, Ley RE (2013) Diversity and heritability of the maize rhizosphere microbiome under field conditions. Proc Natl Acad Sci USA 110:6548–6553

Prasad R, Sharma M, Kamal S, Rai MK, Rawat AKS, Pushpangdan P, Varma A (2008) Interaction of *Piriformospora indica* with medicinal plants. In: Varma A (ed) Mycorrhiza. Springer-Verlag, Berlin, pp 655–678

Prasad A, Kumar S, Khaliq A, Pandey A (2011) Heavy metals and arbuscular mycorrhizal (AM) fungi can alter the yield and chemical composition of volatile oil of sweet basil (*Ocimum basilicum* L.). Biol Fertil Soils 47:853–861

Prasannakumar SP, Gowtham HG, Hariprasad P, Shivaprasad K, Niranjana SR (2015) *Delftia tsuruhatensis* WGR–UOM–BT1, a novel rhizobacterium with PGPR properties from *Rauwolfia serpentina* (L.) Benth. ex Kurz also suppresses fungal phytopathogens by producing a new antibiotic—AMTM. Lett Appl Microbiol 61:460–468

Presta L, Bosi E, Fondi M, Maida I, Perrin E, Miceli E, Maggini V, Bogani P, Firenzuoli F, Di Pilato V, Rossolini GM, Mengoni A, Fani R (2016) Phenotypic and genomic characterization of the antimicrobial producer *Rheinheimera* sp. EpRS3 isolated from the medicinal plant *Echinacea purpurea*: insights into its biotechnological relevance. Res Microbiol 168:293–305

Qi JJ, Yao HY, Ma XJ, Zhou LL, Li XN (2009) Soil microbial community composition and diversity in the rhizosphere of a Chinese medicinal plant. Commun Soil Sci Plant Anal 40:1462–1482

Qi X, Wang E, Xing M, Zhao W, Chen X (2012) Rhizosphere and non-rhizosphere bacterial community composition of the wild medicinal plant *Rumex patientia*. World J Microbiol Biotechnol 28:2257–2265

Qi X, Wang E, Chen X (2012) Molecular characterization of bacterial population in the *Rumex patientia* rhizosphere soil of Jilin, China. Res J Biotechnol 8:64–71

Qin S, Feng WW, Zhang YJ, Wang TT, Xiong YW, Xing K (2018) Diversity of bacterial microbiota of coastal halophyte *Limonium sinense* and amelioration of salinity stress damage by symbiotic plant growth-promoting *Actinobacterium Glutamicibacter halophytocola* KLBMP 5180. Appl Environ Microbiol 84:e01533–18

Radhika KP, Rodrigues BF (2010) Arbuscular mycorrhizal fungal diversity in some commonly occurring medicinal plants of Western Ghats, Goa region. J For Res 21:45–52

Radhika KP, Rodrigues BF (2011) Influence of arbuscular mycorrhizal fungi on andrographolide concentration in *Andrographis paniculata*. Aust J Med Herbal 23:34–36

Raichand R, Kaur I, Singh NK, Mayilraj S (2011) *Pontibacterrhizosphera* sp. nov., isolated from rhizosphere soil of an Indian medicinal plant *Nerium indicum*. Antonie Van Leeuwenhoek 100:129–135

Rajeshkumar S, Nisha MC, Selvaraj T (2008) Variability in growth, nutrition and phytochemical constituents of *Plectranthus amboinicus* (Lour) Spreng. as influenced by indigenous arbuscular mycorrhizal fungi. Mj Int J Sci Tech 2:431–439

Rosa-Mera CJDA, Ferrera-Cerrato R, Alarc_on A, Sánchez-Colín MDJ, Muñoz-Muñiz OD (2011) Arbuscular mycorrhizal fungi and potassium bicarbonate enhance the foliar content of the vinblastine alkaloid in *Catharanthus roseus*. Plant Soil 349:367–376

Sah SK, Reddy KR, Li J (2016) Abscisic acid and abiotic stress tolerance in crop plants. Front Plant Sci 7:571

Sailo G, Bagyaraj DJ (2005) Influence of different AM-fungi on the growth, nutrition and forskolin content of *Coleus forskohlii*. Mycol Res 109:795–798

Shaikh NM, Mokat ND (2018) Role of rhizosphere fungi associated with commercially explored medicinal and aromatic plants: a review. Curr Agric Res J 6(1):72–77

Shang Q, Yang G, Wang Y, Wu X, Zhao X, Hao H, Li Y, Xie Z, Zhang Y, Wang R (2016) Illumina-based analysis of the rhizosphere microbial communities associated with healthy and wilted Lanzhou lily (*Lilium davidii* var. unicolor) plants grown in the field. World J Microbiol Biotechnol 32:95

Sharma D, Kapoor R, Bhatnagar AK (2008) Arbuscular mycorrhizal (AM) technology for the conservation of *Curculigo orchioides* Gaertn.: an endangered medicinal herb. World J Microbiol Biotechnol 24:395–400

Shi JY, Yuan XF, Lin HR, Yang YQ, Li ZY (2011) Differences in soil properties and bacterial communities between the rhizosphere and bulk soil and among different production areas of the medicinal plant *Fritillaria thunbergii*. Int J Mol Sci 12:3770–3785

Shi ZY, Chen YL, Hou XG, Gao SC, Wang F (2013) Arbuscular mycorrhizal fungi associated with tree peony in 3 geographic locations in China. Turk J Agric For 37:726–733

Shrivastava S, Prasad R, Varma A (2014) Anatomy of root from eyes of a microbiologist. In: Morte A, Varma A (eds) Root engineering, vol 40. Springer-Verlag, Berlin, pp 3–22

Simonetti E, Roberts IN, Montecchia MS, Gutierrez-Boem FH, Gomez FM, Ruiz JA (2018) A novel *Burkholderia ambifaria* strain able to degrade the mycotoxin fusaric acid and to inhibit *Fusarium* spp. growth. Microbiol Res 206:50–59

Singh R, Soni SK, Kalra A (2013) Synergy between *Glomus fasciculatum* and a beneficial *Pseudomonas* in reducing root diseases and improving yield and forskolin content in *Coleus forskohlii* Briq. under organic field conditions. Mycorrhiza 23:35–44
Singh M, Awasthi A, Soni SK, Singh R, Verma RK, Kalra A (2015) Complementarity among plant growth promoting traits in rhizospheric bacterial communities promotes plant growth. Sci Rep 5:15500
Solanki AS, Kumar V, Sharma S (2011) AM fungi and *Azotobacter chroococcum* affecting yield, nutrient uptake and cost efficacy of *Chlorophytum borivilianum* in Indian Arid Region. J Agric Technol 7:983–991
Song X, Pan Y, Li L, Wu X, Wang Y (2018) Composition and diversity of rhizosphere fungal community in *Coptis chinensis* Franch. Continuous cropping fields. PLoS One 13:e0193811
Souza R, Ambrosini A, Passaglia LM (2015) Plant growth-promoting bacteria as inoculants in agricultural soils. Genet Mol Biol 38:401–419
Sun XG, Tang M (2013) Effect of arbuscular mycorrhizal fungi inoculation on root traits and root volatile organic compound emissions of *Sorghum bicolor*. S Afr J Bot 88:373–379
Sundar SK, Palavesam A, Parthipan B (2011) AM fungal diversity in selected medicinal plants of Kanyakumari District, Tamil Nadu, India. Ind J Microbiol 5:259–265
Tamilarasi S, Nanthakumar K, Karthikeyan K, Lakshmanaperumalsamy P (2008) Diversity of root associated microorganisms of selected medicinal plants and influence of rhizomicroorganisms on the antimicrobial property of *Coriandrum sativum*. J Environ Biol 29:127–134
Tang M, Xue S, Yang HP (2004) Vesicular arbuscular mycorrhizal (VAM) fungi of xerophyte in Gansu. J Yunnan AgricUniv 19:638–642
Teixeira da Silva JA, Egamberdieva D (2013) Plant-growth promoting rhizobacteria and medicinal plants. Recent progress in medicinal plants 38:26–42
Thombre SS, Kalamkar SS, Shaikh MN, Torawane SD, Mokat DN (2016) Studies on rhizosphere fungi and allelopathic potential of *Santalum album* L. Biosci Discov 7:158–161
Tian XY, Zhang CS (2017) Illumina-based analysis of endophytic and rhizosphere bacterial diversity of the coastal halophyte *Messerschmidia sibirica*. Front Microbiol 8:2288
Tiwari S, Lata C, Chauhan PS, Nautiyal CS (2016) *Pseudomonas putida* attunes morphophysiological, biochemical and molecular responses in *Cicer arietinum* L. during drought stress and recovery. Plant Physiol Biochem 99:108–117
Toussaint JP, Smith A, Smith E (2007) Arbuscular mycorrhizal fungi can induce the production of phytochemicals in sweet basil irrespective of phosphorus nutrition. Mycorrhiza 17:291–297
Vacheron J, Renoud S, Muller D, Babalola OO, Prigent-Combaret C (2015) Alleviation of abiotic and biotic stresses in plants by *Azospirillum*. In: Handbook for *Azospirillum*. Springer, Berlin, pp 333–365
Wahid OAA, Mehana TA (2000) Impact of phosphate-solubilizing fungi on the yield and phosphorus-uptake by wheat and faba bean plants. Microbiol Res 155:221–227
Wang S, Tang M, Niu ZC, Zhang HQ (2008) Relationship between AM fungi resources of rare medicinal plants and soil factors in Lishan Mountain. Acta Bot Bor-Occi Sin 28:355–361
Wang Y, Wang M, Li Y, Wu A, Huang J (2018) Effects of arbuscular mycorrhizal fungi on growth and nitrogen uptake of *Chrysanthemum morifolium* under salt stress. PLoS One 13:e0196408
Wei LH, Shao Y, Wan JW, Feng H, Zhu H, Huang HW, Zhou YJ (2014) Isolation and characterization of a rhizobacterial antagonist of root-knot nematodes. PLoS One 9:e85988
Whang KS, Lee JC, Lee HR, Han SI, Chung SH (2014) *Terriglobus tenax* sp. nov., an exopolysaccharide-producing *Acidobacterium* isolated from rhizosphere soil of a medicinal plant. Int J Syst Evol Microbiol 64:431–437
Xu Z, Xu QY, Zheng ZH, Huang YJ (2012) *Kribbella amoyensis sp.nov.*, isolated from rhizosphere soil of a pharmaceutical plant, *Typhonium giganteum*. Engl Int J Syst Evol Microbiol 62:1081–1085
Yang AN, Lu L, Wu CX, Xia MM (2011) Arbuscular mycorrhizal fungi associated with Huangshan Magnolia (*Magnolia cylindrica*). J Med Plant Res 5:4542–4548

Yang L, Chen ML, Shao AJ, Yang G (2012) Discussion on applications and mechanisms of biocontrol microoganisms used for controlling medicinal plant soil-borne diseases. China J Chin Mater Med 37:3188–3192

Zhang SS, Jin YL, Zhu WJ, Tang JJ, Hu SJ, Zhou TS, Chen X (2010) Baicalin released from *Scutellaria baicalensis* induces autotoxicity and promotes soil borne pathogens. J Chem Ecol 36:329–338

Zhang YQ, Chen J, Liu HY, Zhang YQ, Li WJ, Yu LY (2011a) *Geodermatophilus ruber* sp. nov., isolated from rhizosphere soil of a medicinal plant. Int J Syst Evol Microbiol 61:190–193

Zhang ZY, Lin WX, Yang YH, Chen H, Chen XJ (2011b) Effects of consecutively monocultured *Rehmannia glutinosa* L. on diversity of fungal community in rhizospheric soil. Agric Sci China 10:1374–1384

Zhang HY, Xue QH, Shen GH, Wang DS (2013) Effects of actinomycetes agent on ginseng growth and rhizosphere soil microflora. J Appl Ecol 24:2287–2293

Zhang TY, Yu Y, Zhu H, Yang SZ, Yang TM, Zhang MY, Zhang YX (2018) *Absidia panacisoli* sp. nov., isolated from rhizosphere of *Panax notoginseng*. Int J Syst Evol Microbiol 68:2468–2472

Zhao Z, Zhang X, Tan Z, Guo J, Zhu H (2013) Isolation and identification of cultivable myxobacteria in the rhizosphere soils of medicinal plants. Acta Microb Sin 53:657–668

Zubek S, Blaszkowski J (2009) Medicinal plants as hosts of arbuscular mycorrhizal fungi and dark septate endophytes. Phytochem Rev 8:571–580

Zubek S, Blaszkowski J, Mleczko P (2011) Arbuscular mycorrhizal and dark septate endophyte associations of medicinal plants. Acta Soc Bot Pol 80:285–292

Chapter 15
Insight to Biotechnological Advances in the Study of Beneficial Plant-Microbe Interaction with Special Reference to *Agrobacterium tumefaciens*

Pankaj Kumar and Dinesh Kumar Srivastava

Abstract *Agrobacterium tumefaciens* is the most scientifically investigated and proved cellular organism that has a natural capability to transfer genetic material between the kingdoms of life, from prokaryotes to eukaryotes. Recent advancement in biotechnological tools and techniques helps us to understand plant-microbe interactions because plants are closely associated with microorganisms that influence plants' overall fitness. With the rapid decline of natural resources and continuous increase in the population of developing countries, there is an utmost need for new tools and technologies that supplement the agriculture system and provide novel opportunities for ensuring global food and nutritional security. Till now, *Agrobacterium* has been widely used as a vector to genetically transform the plants with agronomically important traits. Apart from the role of *Agrobacterium tumefaciens* in plant genetic engineering, it also served plant biologist to scientifically investigate and reveal basic biological processes such as regulation of gene expression, gene identification and mapping, cell-cell recognition, cell-to-cell transport mechanism, nuclear import, and recombination mechanism and to study mutagenesis within plant cells. In this chapter, we mainly emphasized on importance of a systems understanding of plant-microbe interactions, with a special reference to *Agrobacterium tumefaciens*—as natural plant genetic engineer, signal transduction and host immune response, quorum sensing and quenching, plant genes involved in susceptibility/resistance, factor affecting *Agrobacterium* plant transformation, recent advances/application in plant biology research, and omics approaches for better understanding plant-microbe interaction complexity.

P. Kumar
CSIR-Institute of Himalayan Bioresource Technology, Palampur, Himachal Pradesh, India

D. K. Srivastava (✉)
Department of Biotechnology, Dr Yashwant Singh Parmar University of Horticulture and Forestry, Solan, Himachal Pradesh, India

A. Varma et al. (eds.), *Plant Microbe Symbiosis*,
https://doi.org/10.1007/978-3-030-36248-5_15

15.1 Introduction

Plants in their natural characteristic surroundings/habitats are encompassed by a large number of microorganisms that directly and indirectly interact with plants. Some microbes interact with plants in a mutually beneficial manner or otherwise colonize the plant only for their own benefits. In addition, microbes can also indirectly affect plants by severely changing their growth environment conditions. By using recent advancement in biotechnological approaches, molecular genetics and system biology approaches also provide insight to understand the true nature of plant-microbe interaction, basic biological processes, regulated gene expression, signal transduction mechanism, and plant immune signaling network and significantly offer novel strategies to augment quantitatively and qualitatively crop productivity and protection in an environment-friendly manner (Ishaq 2017; Pathak et al. 2018). Today, in the global world, the major challenges in agricultural biotechnology are to increase crop productivity and crop protection in adverse environmental conditions, manage resistance to pests and disease and herbicide tolerance, improve genetic engineering technologies to enhance public perception, and improve harvest index and nutrient cycling in agricultural ecosystems (Kossmann 2012; Prasad et al. 2018). Detailed scientific investigation on beneficial plant-microbe interactions possibly helps to develop microbial inoculants which can be used as plant strengtheners, phytostimulators, biopesticides, and biofertilizers. Advancement in understanding plant biology, novel genetic resources, and modification process and omics technologies revolutionized our concepts of sustainable food production, cost-effective alternative energy strategies, and novel biomaterial production that significantly contribute to revolutionize our agriculture system under changing climatic and environmental conditions (Moshelion and Altman 2015).

Agrobacterium tumefaciens gene tranfer method still prevailing as one of the oldest and most widely used method for genetic manipulation of various economically and horticulturally important monocot and dicot plants by optimized efficient and reliable tissue culture protocol, in planta transformation procedure, and more recently *Agrobacterium* T-DNA-derived nano-complex method for recalcitrant crop species (Ziemienowicz 2014; Kumar et al. 2018a). However, various factors of bacterial, host, and environmental origin affect the transformation efficiency which needs to be critically optimized and addressed. Till date, many economically important crop species or elite varieties have been developed using transgenic technology by the use of *Agrobacterium tumefaciens*-mediated gene transfer techniques. Genetic transformation of plants using *Agrobacterium*-mediated gene transfer is a key technique for plant molecular breeding to introduce agronomically important desirable characteristics/traits into the existing plant genome while preserving genetic identity of plants. The broad compensation of transgenic crops for a society to resolve the food and nutritional security issues has been well recognized and is addressed by added benefits such as tolerance to various biotic and abiotic stresses, herbicide tolerance, virus resistance, higher nutritional value, and enhancement of the fruit shelf life (Shukla et al. 2018). As the biotechnological advancement in the area of plant biotechnology and plant genetic engineering technologies, biotech crops/genetically modified crops are considered as the fastest adopted crop technology in current modern agriculture era, as per the latest update by the International Service for the Acquisition of Agri-biotech Applications (ISAAA). In 2017, biotech crops were planted on 189.8 million hectares area by 24 countries, and from the initial plantation of 1.7 million hectares in 1996, there was an ~112-fold increase in 2017. Soybean, maize, cotton, and canola are the most planted biotech crops in 2017. Biotech soybean occupied an area of 94.1 million hectares followed by biotech maize (59.7 million hectares), biotech cotton (24.1 million hectares), and biotech canola (10.2 million hectares) globally in 2017. Apart from soybean, maize, cotton, canola, and alfalfa, the following biotech crops, i.e., papaya, potato, sugar beet, eggplant, squash, and apple, were also planted in different countries. There was a 9% increase in the global market value of biotech crops in 2017 than from 2016, i.e., US$17.2 billion (ISAAA 2017).

15.2 *Agrobacterium tumefaciens*: Natural Plant Genetic Engineer

Agrobacterium tumefaciens is also called a natural plant genetic engineer because of its natural ability to transfer genetic material between the kingdoms of life, from prokaryotes to eukaryotes. *Agrobacterium tumefaciens* is a rod-shaped, motile, and Gram-negative soil bacterium having tumor-inducing (Ti) plasmid, which causes crown gall disease in plants (Tzfira and Citovsky 2008; Gelvin 2009, 2010; Ziemienowicz 2014). Ti plasmid maintains most of the Ti-plasmid genes in transcriptionally inactive state; their expression is activated by the presence of suitable plant host. *Agrobacterium* enters into the plants through wounded tissue, and phenolic compound acetosyringone and monosaccharides are released by wounded plant cell, which activate the expression of virulence genes, although these genes are silent until induced. Vir A is a sensor protein, and its periplasmic domain binds to a phenolic compound such as acetosyringone to activate cytoplasmic protein kinase domain leading to Vir A autophosphorylation. For phosphonyl group transfer from Vir A to Vir G genes (transcriptional regulator), Vir G genes activate all the other virulence genes by binding to a 12 bp DNA element called the Vir box located upstream of the Vir operon. *Agrobacterium* chromosomal virulence genes (*chv E*) bind to a certain plant-derived sugar and also promote autophosphorylation of Vir A and enhance its signaling. The Vir region is ~40 kb, and it includes several distinct operons. The first step in T-DNA transfer includes the action of single-stranded endonucleases that cleave T-DNA at the border sequence. The *Vir D* genes (Vir D_1, helicase activity; Vir D_2, endonucleases activity) and Vir D_2 site-specific endonucleases recognize 25 bp T-DNA border sequence, and Vir D_2 nicks only the bottom strand at each border sequence and releases the single-stranded T-DNA molecule. Vir D_2 remains covalently attached to $5'$ end of T-DNA throughout the transfer to host nucleus and is thought to protect T-DNA from exonucleolytic attack. Vir B operon encodes 11 proteins that along with Vir D_4 forms type IV secretion system (T4SS). The T-DNA-Vir D_2 complex moves into the plant cell through a large protein complex called type IV secretion system (T4SS) that spans in the bacterial outer and inner membranes and possibly the plant cell wall. T-DNA-Vir D_2 and other proteins in the Vir E region move through T4SS into the plant cytoplasm. A possible explanation is that Vir D_2 and Vir E_2 contribute to T-DNA movement into the nucleus because it has a nuclear localization signal that directs the complex to plant nucleus. A plant importin α (also known as karyopherin) also contributes to the movement of T-DNA complex into the nucleus. Once the T-DNA complex enters the nucleus, it is thought to attach to the plant nuclear DNA through interaction with nucleosomal proteins. The exact mechanism of DNA insertion into the plant genome is not fully resolved, but it occurs via a process known as nonhomologous or illegitimate recombination (Anand et al. 2008; Christie et al. 2014; Srivastava et al. 2016; Willig et al. 2018).

Agrobacterium tumefaciens as a natural plant genetic engineer is used as a plant transformation vector, which requires four major key steps that are disarming process (bacterium had to be made nonpathogenic), introduction of selected desirable gene(s) along with selectable markers genes into the T-DNA region, in vitro manipulation of large-size tumor-inducing (Ti) plasmid, and efficient method development for regeneration of whole plants from transformed cells. Due to the natural genetic engineering ability of *Agrobacterium tumefaciens*, it has been successfully commercially exploited to create transgenic plants using plant genetic engineering approaches to enhance crop productivity, nutritional value, and crop protection by increased tolerance to biotic and abiotic stresses and also helps in the reduction of harmful agrochemicals (Gelvin 2010; Mine et al. 2014; Christie et al. 2014). *Agrobacterium*-mediated gene transfer also conferred improved growth and grain yield in rice plant by enhancing the biosynthesis of an iron chelator at low iron bioavailability conditions (Takahashi et al. 2001). Using *Agrobacterium*-mediated transformation for the production of herbicide-tolerant crops such as soybean, canola, cotton, maize, sugar beet, and rapeseed oil resulted in the reduction of harmful agrochemicals (Slater et al. 2008; Jube and Borthakur 2009; Ziemienowicz 2014; Srivastava et al. 2016; Kumar and Srivastava 2016).

15.3 Signal Transduction and Host Immune Response

Agrobacterium is the mostly studied and scientifically investigated phytopathogen till date and proved for its ability to genetically transform host plants by transferring and integrating T-DNA of its tumor-inducing (Ti) plasmid or a foreign desirable gene such as agronomically important traits (Lacroix and Citovsky 2013). Gohlke and Deeken (2014) have reported how plants respond to *Agrobacterium tumefaciens* causative agent of crown gall disease. *Agrobacterium tumefaciens* recognition and signal transduction pathway responses to plant-derived signaling molecules for understanding complexity of *Agrobacterium*-plant interaction have been largely studied over several decades (Subramoni et al. 2014). For transferring and integrating its T-DNA of tumor-inducing (Ti) plasmid, *Agrobacterium* perceives and recognizes plant-derived signals to activate its virulence genes, which are responsible for T-DNA transfer into the plant nucleus. The expression of genes inside the plant hosts leads to the production of plant growth regulators, i.e., auxins {indole-3-acetic acid (IAA) and cytokinin}, and opines which resulted in tumor formation. *Agrobacterium* makes use of opines as sole nutrient sources as well as signaling molecule to activate quorum sensing (QS) for increased virulence and opine metabolism.

For active defense response in plants/host immune response against microbial pathogens, signal transduction pathway plays a central role (Hwang et al. 2015a). Recognition of *Agrobacterium* as pathogen by host plants resulted in plant defense response elicitation. However, plants do not have any specialized immune cells, but showed the sophisticated innate defense system. Plants have various receptor molecules which recognize conserved pathogen-associated molecular patterns (PAMPs). Pathogen-associated molecular patterns are structurally and sequence characterized, highly conserved within divergent classes of pathogens, such as bacteria/fungi, and also often found in nonpathogenic microbes that's why they are also called microbe-associated molecular patterns (MAMPs), which elicit sets of defense mechanisms (Ausubel 2005). The most characterized pathogen-associated molecular patterns are peptides derived from flagellin (flg22) and the elongation factor *EF-Tu* (*elf18*) (Gómez-Gómez et al. 2001). Flagellin (particularly flagellin peptide flg22) is recognized by a specific receptor kinase (*FLS2*) and induces the expression of various defense-related genes to prompt resistance against pathogenic bacteria (Chinchilla et al. 2006). *EF-Tu* is the most abundant protein present in many bacteria and upon cell membrane integrity disruption released into the extracellular spaces (Kunze et al. 2004; Zipfel et al. 2006; Nicaise et al. 2009). However, *Agrobacterium* flagellin protein (N-terminal conserved domain) is not recognized by the host plant and fails to induce defense response and does not trigger an immune response (Zipfel et al. 2004; Yuan and Williams 2014).

Agrobacterium has developed numerous strategies to evade the host's immune responses. Ditt et al. (2001) have investigated the plant response to *Agrobacterium tumefaciens* infection and its transformation in inoculated *Ageratum conyzoides* plant cultures by comparing cDNA amplified fragment length polymorphism (AFLP) pattern. From amplified cDNA fragment pattern, differential regulated gene expression was studied after co-cultivation (24 and 48 h) with *Agrobacterium.* To confirm the cDNA-AFLP differential pattern sequence, similarities were also carried out which revealed the role of amplified genes in the signal transduction of plant defense response. However, various scientific findings in a number of host species such as *Arabidopsis thaliana*, rice, and rye grass indicated that *Agrobacterium*-induced defenses limit transformation efficiency (Ditt et al. 2001; Zipfel et al. 2006; Taj et al. 2014; Zhang et al. 2013). To detect pathogenic and nonpathogenic microbes, extra- and intracellular receptors on host cell trigger signal transduction mechanism resulting in immediate physiological changes such as production of damaging reactive oxygen species (ROS), Ca^{2+} fluxes (secondary messenger molecule), extracellular alkalization, and hypersensitive response to localized programmed cell death (Boller and Felix 2009). Initial signal transduction events to regulate these responses modulate defense gene expression in mitogen-activated protein kinase (MAPK) pathway (Asai et al. 2002). Upregulation of hormone (salicylic acid, jasmonate, and ethylene)-specific defense gene is also being expressed and induces systemic defenses in plants (Jones and Dangl 2006; Spoel and Dong 2012; Yuan and Williams 2014; Hwang et al. 2015b).

15.4 Quorum Sensing and Quenching

Agrobacterium tumefaciens induces crown gall formation generally on a wide range of dicotyledonous plants and pathogenicity determinants of this bacterium mostly borne on tumor-inducing (Ti) plasmid. The conjugative transfer of the plasmid genes and expression of genes are regulated by quorum sensing (QS) (Dessaux and Faure 2018). Based upon accumulation of an acyl homoserine lactone (AHL), *A. tumefaciens* regulates the expression of genes in a population density-dependent manner, such kind of regulations were named as quorum sensing, and signals are called quorum signals, autoinducers, or quoromones. When *A. tumefaciens* infects the host plants, then quorum sensing is induced, and it varies among different strains, but here it is being discussed about the octopine-utilizing agrobacterial strains (Zhang et al. 2002). Transfer of T-DNA into the plant cells leads to the production of octopine which is recognized by *A. tumefaciens* transcriptional regulator named as octopine catabolism regulator (OccR). Once recognized, transcription activation of octopine catabolism operon (consists of genes involved in octopine uptake and catabolism) takes place. In this operon, *traR* gene encodes a transcription factor bound to acyl homoserine lactone (AHL) as quorum signals. Acyl homoserine lactone activates an operon particularly *traI* gene (encodes an enzyme for the synthesis of AHL) and then binds more avidly to *traR*, forming a positive feedback regulatory loop. Various other *traR* regulated genes involved in Ti-plasmid replication and conjugal transfer (Yuan and Williams 2014). Recent investigations have shown that plants are capable of recognizing bacterial quormone signals and responding to these signals through their various defense mechanisms and developmental stages, and thus in initiating quormone degradation processes, this process is collectively called quorum quenching (Dessaux and Faure 2018). Quorum quenching involves the exclusion of the quorum signal from the environment. Acyl homoserine lactone (AHL) lactonase encoded by *attM* of *Agrobacterium* is believed to degrade and quench the quormone signal. Plant defense response compounds such as salicylic acid and gamma-Aminobutyric acid which accumulate in crown gall tumors caused by *Agrobacterium tumefaciens* induced the expression of a*ttM*, which contributed quorum quenching to the plant and bacterium. Dessaux and Faure (2018) reported that quorum-sensing elements/signal molecule in conjunction with quorum quenching (QQ) activates a complex-integrated "go/no go system" that finely controls the Ti-plasmid transfer in response to various environmental cues. This finely tuned go/no go system of quorum sensing and quenching permits the bacteria to sense its presence or absence in a gall, in a decaying or living tumor, and in stressed plant tissues.

15.5 Plant Genes Involved in Susceptibility/Resistance to *Agrobacterium* Transformation

Gelvin (2010) and Yuan and Williams (2014) have reported different identified proteins such as *hat* (hypersusceptible to *Agrobacterium* transformation) and *rat* (resistant to *Agrobacterium* transformation) genes, which conferred hypersensitivity and resistance to *Agrobacterium* transformation. Resistant to *Agrobacterium* transformation; *rat* genes encode a protein that interacted with transferred DNA (T-DNA) integration and a VirB-interacting protein that facilitate contact between the type IV secretion systems (T4SSs) are large protein complexes contains a channel through which proteins or protein-DNA complexes can be translocated, which traverse the cell envelope of many bacteria and the host cell. Zhu et al. (2003) and Gelvin (2010) also discussed about various other identified genes involved in bacterial attachment/biofilm formation (arabinogalactan protein *AtAGP17*, cellulose synthase-like *CslA-09* and *CslB-05*, and plant defense reaction proteins), cytoplasmic trafficking (microtubules/kinesin, actin, myosin, and cyclophilin), nuclear targeting [importin α, importin β/transportin, CAK2M kinase, protein phosphatase 2C (*PP2C*), *VIP1*, caspase, GALLS interacting protein (*GIPa*)], transgene expression (histones H2A, H3-11, and H4), and susceptibility to transformation (*Myb* transcription factor). Comprehensive investigation of the plant's contributions to the infection process/disease procedure can help to identify and distinguish strategies/methods to protect plants, such as walnut (*Juglans regia*), grapevines, and almond (*Prunus dulcis*), that are vulnerable to crown gall disease and furthermore prompt more reliable and proficient plant transformation methods, especially for important monocotyledonous crops (Yuan and Williams 2014).

15.5.1 Factor Affecting Agrobacterium *Plant Transformation*

An efficient, successful *Agrobacterium* plant transformation using desirable gene (s) for agronomically important traits largely depends upon the optimization of various factors which determine the rate of transformation frequency in different crop species (Ziemienowicz 2014; Kumar and Srivastava 2016; Kumar et al. 2018a). Various factors which are known to affect the *Agrobacterium* plant transformation frequency include plant species, explant type (hypocotyl, cotyledon, root, leaf, and stem), agrobacterial strain (LBA4404, EHA101, C58, AGL1), vector plasmid (pCAMBIA, pGreen, pGA, pCG, pGPTV, Bi-BAC), culture medium compositions (micro- and macronutrient concentration, sugars, plant growth regulators), preculturing time, co-cultivation period (1–5 days; mostly: 24, 48, 60, 72 h), temperature of co-cultivation range 19–30 °C (optimal temp. dicots, 19–20 °C; monocots, 24–25 °C), pH of co-cultivation medium (acidic pH: 5.2, 5.5, 5.6, 5.8, or 6.0) and concentration of bacterial inoculum ($1{\sim}10^6$–$1{\sim}10^{10}$ cfu/ml), antibiotics (cefotaxime, carbenicillin, and kanamycin), selectable markers [neomycin phosphotransferase (*nptII*), phosphinothricin acetyltransferase (*pat*), and hygromycin phosphotransferase (*hpt*)], and effect of different inducers like acetosyringone. Extensive work to study the effects of all these factors has been carried out by various researchers during the *Agrobacterium* plant genetic transformation studies (Kavitah et al. 2010; Sood et al. 2011; Ziemienowicz 2014; Verma et al. 2014; Kumar et al. 2015, 2017, 2018a, b, c; Kumar and Srivastava 2015, 2016; Srivastava et al. 2016; Gambhir et al. 2017; Parmar et al. 2017; Shaunak et al. 2018).

Guo et al. (2012) reported that plant transformation efficiency in tomato varies depending on the selection marker and thus on the concentration of selective agent such as kanamycin and carbenicillin. Ziemienowicz (2014) reported the timentin effect on in vitro shoot regeneration and its use for the suppression of *Agrobacterium tumefaciens* bacterial growth in *Agrobacterium*-mediated genetic transformation of tobacco (*Nicotiana tabaccum*) and concluded that timentin may be an alternative antibiotic for the effective suppression of *A. tumefaciens* in genetic transformation. Kumar et al. (2017) studied the effects of antibiotics kanamycin and cefotaxime to determine the aptness of kanamycin resistance as a selectable marker and for cefotaxime in controlling excessive agrobacterial growth during *Agrobacterium* genetic transformation studies in broccoli (*Brassica oleracea* L. var. *italica* cv. Solan Green Head) using cultured hypocotyl, cotyledon, leaf, and petiole tissues. To find out the minimum concentration of kanamycin required for the selection of putative transformed cells during *Agrobacterium* transformation, increasing doses of kanamycin (10, 20, 30, 40, and 50 mg/l) to leaf and petiole explants were given. Leaf and petiole explants exhibited decreased in fresh weight as kanamycin concentration increased, resulting in full or partial inhibition of shoot regeneration. The nontransformed tissues did not survive on the selective medium containing kanamycin. A significant or nonsignificant negative correlation occurred between kanamycin concentration and explant fresh weight over the time. Cefotaxime also affects the plant regeneration potential and transformation efficiency. Similar findings were reported by Sharma (2010, 2014), Husaini (2010), Sharma et al. (2011), Aggarwal (2011), Ahmad et al. (2012), and Gambhir et al. (2017). Therefore, selection and identification of transgenic events from non transgenics are crucial steps of genetic transformation and kanamycin and cefotaxime proved as an effective selective agents in improving selection and transgenic regenerants efficiency.

In vitro manipulation of cultured explants was reported to improve the competency of plant cells for efficient T-DNA delivery and the subsequent recovery of transformed plant cells; however, it is largely species-dependent (Ziemienowicz 2014). Opabode (2006) had observed increased T-DNA delivery in rice explants when treated with sucrose, whereas desiccation improved the T-DNA delivery, and stable transformation was recorded in sugarcane, maize, wheat, and soybean. Cardoza and Stewart (2003) and Opabode (2006) also observed that preculturing condition, co-cultivation time, and *Agrobacterium* density affect T-DNA delivery and integration in canola. Bacterial culture (*Agrobacterium*) density higher than $1–10^{10}$ cfu/ml usually damaged the cultured plant cells/tissues and resulted in poor transformation efficiency (Ziemienowicz 2014; Kumar and Srivastava 2016).

Different selectable marker genes such as *nptII* (neomycin phosphotransferase-II), *hpt* (*hygromycin phosphotransferase*), and *pat* (phosphinothricin acetyltransferase) under the control of constitutive promoters, i.e., CaMV35S, work efficiently for the selection of desirable transgene expression in *Agrobacterium*-transformed cells from the nontransformed cells in various crop species (Kumar and Srivastava 2016; Kumar et al. 2017). But use of antibiotic selectable marker genes(s) has been strongly opposed, criticized, and facing problems among consumer acceptance for genetically modified crops, due to biosafety concerns. Moreover, selectable marker gene(s) apart from transformed cell selections does not have any functional relevance inside the plant cell, and the constitutive expression of antibiotic marker gene-encoded proteins might affect the plant metabolism (Parmar et al. 2017). So as to increase the consumer acceptance for genetically modified crops and to minimize the biosafety concerns, recent advent in molecular biotechnology led to the development of clean-gene technology or marker-free transgenic technology. Different strategies have been used by various researchers to eliminate the selection marker gene for marker-free transgenic plant development such as co-transformation, homologous recombination, use of multi-auto-transformation vectors, transposition system, and site-specific recombinations (Gleave et al. 1999; Puchta 2000; Tuteja et al. 2012).

15.6 Recent Advances/Application in Plant Biology Research

Agrobacterium tumefaciens-mediated transformation has been remarkably considered as an indispensable tool in plant biology research for economically and horticulturally important crop plants. Significant findings have been achieved in the past using *Agrobacterium tumefaciens*-mediated transformation method for studies of gene expression, gene function, and protein localization. For gene identification and gene mapping studies using reverse genetics approaches, transferred DNA (T-DNA) has also been successfully employed to generate mutants. Its random insertion and integration into the genome resulted in the interruption of gene with a known DNA sequence and can be used for gene identification. Initially large-scale T-DNA insertion libraries were developed in model crop plants, i.e., *Arabidopsis* and rice (*Oryza sativa*), in late 1990 (Yuan and Williams 2014). The unifying properties of *A. tumefaciens* strains to transfer T-DNA to fungi and yeast also extend this technique to generate insertion mutant libraries. Till now, *Agrobacterium*-mediated transformation remains the only genetic tool available in many medicinally important fungal species (Yuan and Williams 2014).

Recently, in spite of the numerous ongoing improvements in plant biotechnology and development of efficient alternative gene transfer techniques, *Agrobacterium tumefaciens* still prevails as the oldest and widely used successful cellular organism for plant transformation. *Agrobacterium tumefaciens* is an indispensable tool in the targeted gene delivery of selected desirable gene(s) such as agronomically important traits which are not naturally present in the plant genotypes/wild germplasm with high efficiency for the production of transgenic plants. For the production of biotic (pest resistant, bacterial, fungal, and virus resistant) and abiotic stress (heat, cold, flood, and heavy metals)-tolerant plants, resistance to herbicide, increasing nutritional status (vitamins, lipids, amino acids, and proteins), recombinant therapeutic (protein, vaccine, antibodies) production, introduction of new traits (flower color modification, fruit ripening), metabolic pathway regulations (production of sugars, essential oil, nutrient accumulation), and bioremediation processes (pollution control), numbers of targeted desired gene(s) of interest available incorporated into host plants using *Agrobacterium tumefaciens*-mediated transformation procedure (Srivastava et al. 2016; Kumar and Srivastava 2016; Parmar et al. 2017). However, regardless of the significant achievement of *Agrobacterium*-mediated genetic transformation in a number of plant species ranging from model plants, cereal crops, medicinal plants, woody species, nuts and fruit crops, ornamental plants, turf grasses, legume crops, vegetable crops, and forest plant species using plant genetic engineering techniques, still some limitations of host range and development of efficient regeneration protocols are there, which need to be addressed.

For efficient reliable, high-throughput *Agrobacterium* transformation, in planta transformation procedures have been optimized by Feldmann and Marks (1987) in the model plant *A. thaliana* initially. Application of this developed method relies on minimized labor cost and expenses, does not depend upon tissue culture regeneration which is time-consuming, requires a lot of standardization and optimization procedures, and limits down the mutagenesis rate and somaclonal variations during in vitro regenerated cultures. Another highly efficient transformation procedure has also been developed by advancement in the technique such as floral dip method in *Arabidopsis* (Clough and Bent 1998) and vacuum infiltration (Bechtold and Pelletier 1998) in cereal crops, i.e., rice and maize (Chumakov et al. 2006) and wheat (Mamontova et al. 2010). Successful in planta transformation method has been introduced to design time- and labor-efficient methods for the production of transgenic plants by various researchers to transform a number of plant/crop species such as safflower (Rao and Rohini 1999), radish (Curtis 2003), *Brassica napus* (oil seed crop) (Wang et al. 2003), mulberry (Ping et al. 2003), cotton (Keshamma et al. 2008), and groundnut (Rohini and Rao 2008).

Recent strategy using *Agrobacterium* T-DNA-derived nano-complex has been used to transform recalcitrant crop plant species (Ziemienowicz 2014). Chugh et al. (2009) used this technique in preparing in vitro nano-complex that consisted of *Agrobacterium* transferred DNA (T-DNA) region, virulence protein ($VirD_2$), and single-stranded DNA-binding protein (RecA) then successfully delivered to wheat (triticale microspores) using Tat2 cell-penetrating peptide. This recently used novel approach resulted in single transgene copy integration events into the genome of triticale plants and also prevented DNA degradation. Chugh et al. (2009) also observed integration of intact copies of the transgene and expression in all the regenerated transgenic plants, when triticale microspores were transfected with the prepared nano-complex. Using *Agrobacterium* nano-complex, the mediated method of transgene delivery has been proven highly efficient particularly in the crop species which are difficult to transform by other methods (Ziemienowicz et al. 2012). As cell-penetrating peptides were shown to transfer their DNA and proteins to monocot as well as dicot plants, this approach will be remarkably highly valuable for plant biology research (Chen et al. 2007; Chugh and Eudes 2008; Chugh et al. 2010).

15.7 Omics Approaches for Understanding Plant-Microbe Interaction Complexity

Omics-based approaches such as genomics, transcriptomics, proteomics, metabolomics, ionomics, and phenomics have accelerated the biological research and give insight to reveal the molecular mechanisms of plant-microbe interaction, insect resistance to pesticides, and plant tolerance to herbicides for better pest management, saving the time and expenses of producing better-quality food crops that are resistant to various stresses and exhibit a high nutritional value (Willig et al. 2018). Imam et al. (2016) have reported the use of various "omics" tools to understand beneficial and pathogenic effects of microbes and crop improvement with the recent advancement made in sequencing technologies. Interpreting plant-microbe interactions is a promising aspect to comprehend the advantages and the pathogenic effect of microbes using various "microbiomics" approaches which has impressively fast-tracked the ongoing research in biological sciences. Omics sciences enable a systems biology approach toward understanding the complex interfaces between genes, proteins, and metabolites within the resulting phenotype (Van Emon 2016). Plant-microbe interactions with special reference to *Agrobacterium tumefaciens*, which is an alpha-proteobacterium of the family *Rhizobiaceae*, affected a wide range of plants and act as a natural genetic engineer that makes it of great concern to the agriculture industry (Schenk et al. 2008). With the rapid change in climatic conditions, use of omics approaches and information on plant-pathogen interaction also broaden our ideas and perspective of a wide range of interaction (Torto-Alalibo et al. 2009; Knief 2014). Presently proteomics in blend with bioinformatics and computational biology approaches are generally utilized strategy to translate plant-pathogen association by the isolation, characterization, and expression profiling of the entire set of proteins inside a cell under specific conditions and time (Jayaraman et al. 2012; Imam et al. 2016). Proteomics approaches also have been applied to study protein-protein interactions involved in plant defense reactions and map protein modification which determines the difference between a wild-type and genetically modified organism. Similarly, metabolite-based approach, i.e., metabolomics, is also used to determine the nutritional difference between traditional and genetically modified crops, to identify plant defense metabolites, and to study differences at molecule/metabolite level between the healthy and diseased plant. So utilization of interventions of omics science and technologies significantly augments crop productivity, crop protection, and crop management practices in modern agriculture (Willig et al. 2018; Pathak et al. 2018; Singh et al. 2019).

15.8 Conclusion

Plant-associated microorganisms fulfill important function of plant health and growth and remarkably contributed to provide disease resistance, enhance stress tolerance, aid nutrient availability and uptake, and promote biodiversity. With biotechnological advancement, significant progress has been made in the past and ongoing for better understanding of plant-microbe interactions so that better strategies can be implemented and the problem of global food and nutrition security can be resolved. *Agrobacterium*-mediated transformation is still a prevailing important biotechnological tool for the transfer of desirable characteristics to crop plants and supplements the classical plant breeding practices to enhance agricultural production. Despite the significant achievement of *Agrobacterium*-mediated genetic transformation in plant genetic engineering, some problems still need to be addressed such as transformation of recalcitrant economically important crops, host range, development of efficient plant regeneration protocol, and introduction of multiple stacked traits, transgene stability, and inheritance in further generation without gene loss. Better understanding of host-pathogen interactions and detailed investigation of plant proteins involved in assisting T-DNA delivery into the host plant genome, elucidation of *Agrobacterium* signaling pathways and regulatory mechanism, and identification and characterization of plant genes and proteins essential for *Agrobacterium*-mediated transformation will provide new insights useful for plant genetic engineering. In the twenty-first century, there is an urge for the implementation of a systems bio-agriculture integrated approach to achieve significant improvement in agriculture to solve the issue of global food and nutritional security.

Acknowledgments The senior author (PK) thankfully acknowledges the award of National Postdoctoral Fellowship, Science and Engineering Research Board, Department of Science and Technology, Government of India, New Delhi, India.

References

Aggarwal G (2011) Studies on *Agrobacterium*-mediated insect resistance gene transfer studies in Himalayan poplar (*Populus ciliata* Wall.) and molecular analysis of regenerated plantlets. Ph.D. Thesis, Dr Y.S. Parmar University of Horticulture and Forestry, Nauni, Solan

Ahmad MZ, Hussain I, Muhammad A, Ali S, Ali GM, Roomi Z, Zia MA, Ijaz A (2012) Factor affecting *Agrobacterium* mediated transformation of rice chitinase gene in *Solanum tuberosum* L. Afr J Biotechnol 11:9716–9723

Anand A, Uppalapati SR, Ryu CM, Allen SN, Kang L, Tang Y et al (2008) Salicylic acid and systemic acquired resistance play a role in attenuating crown gall disease caused by *Agrobacterium tumefaciens*. Plant Physiol 146:703–715

Asai T, Tena G, Plotnikova J, Willmann MR, Chiu WL, Gomez-Gomez L et al (2002) MAP kinase signalling cascade in *Arabidopsis* innate immunity. Nature 415:977–983

Ausubel FM (2005) Are innate immune signaling pathways in plants and animals conserved? Nat Immunol 6:973–979

Bechtold N, Pelletier G (1998) In planta *Agrobacterium*-mediated transformation of adult *Arabidopsis thaliana* plants by vacuum infiltration. In: Martinez-Zapater JM, Salinas J (eds) *Arabidopsis* protocols. Humana, Totowa, pp 259–266

Boller T, Felix GA (2009) Renaissance of elicitors: perception of microbe-associated molecular patterns and danger signals by pattern-recognition receptors. Annu Rev Plant Biol 60:379–406

Cardoza V, Stewart CN (2003) Increased *Agrobacterium*-mediated transformation and rooting efficiencies in canola (*Brassica napus* L.) from hypocotyl segment explants. Plant Cell Rep 21:599–604

Chen CP, Chou JC, Liu BR, Chang M, Lee HJ (2007) Transfection and expression of plasmid DNA in plant cells by an arginine-rich intracellular delivery peptide without protoplast preparation. FEBS Lett 581:1891–1897

Chinchilla D, Bauer Z, Regenass M, Boller T, Felix G (2006) The *Arabidopsis* receptor kinase *FLS2* binds *flg22* and determines the specificity of flagellin perception. Plant Cell 18:465–476

Christie PJ, Whitaker N, Gonzalez-Rivera C (2014) Mechanism and structure of the bacterial type IV secretion systems. Biochim Biophys Acta 1843:1578–1591

Chugh A, Eudes F (2008) Cellular uptake of cell-penetrating peptides pVEC and transportan in plants. J Pept Sci 14:477–481

Chugh A, Amundsen E, Eudes F (2009) Translocation of cell-penetrating peptides and delivery of their cargoes in triticale microspores. Plant Cell Rep 28:801–810

Chugh A, Eudes F, Shim YS (2010) Cell-penetrating peptides: nanocarrier for macromolecule delivery in living cells. IUBMB Life 62:183–193

Chumakov MI, Rozhok NA, Velikov VA, Tyrnov VS, Volokhina IV (2006) In planta transformation of maize through inoculation of *Agrobacterium* into the silks. Genetika 42:1083–1088

Clough SJ, Bent AF (1998) Floral dip: a simplified method for *Agrobacterium*-mediated transformation of *Arabidopsis thaliana*. Plant J 16:735–743

Curtis IS (2003) The noble radish: past, present and future. Trends Plant Sci 8:305–307

Dessaux Y, Faure D (2018) Quorum sensing and quorum quenching in *Agrobacterium*: a "go/no go system"? Genes 9:210

Ditt RF, Nester EW, Comai L (2001) Plant gene expression response to *Agrobacterium tumefaciens*. Proc Natl Acad Sci USA 98:10954–10959

Feldmann KA, Marks MD (1987) *Agrobacterium*-mediated transformation of germinating seeds of *Arabidopsis thaliana*: a non-tissue culture approach. Mol Gen Genet 208:1–9

Gambhir G, Kumar P, Srivastava DK (2017) Effect of antibiotic sensitivity on different cultured tissues and its significance in genetic transformation of cabbage *Brassica oleracea*. Biosci Biotechnol Res Comm 10:652–661

Gelvin SB (2009) *Agrobacterium* in the genomics age. Plant Physiol 150:1665–1676

Gelvin SB (2010) Plant proteins involved in *Agrobacterium*-mediated genetic transformation. Annu Rev Phytopathol 48:45–68

Gleave AP, Mitra DS, Mudge SR, Morris BA (1999) Selectable marker-free transgenic plants without sexual crossing: transient expression of *cre* recombinase and use of a conditional lethal dominant gene. Plant Mol Biol 40:223–235

Gohlke J, Deeken R (2014) Plant responses to *Agrobacterium tumefaciens* and crown gall development. Front Plant Sci 5:155

Gómez-Gómez L, Bauer Z, Boller T (2001) Both the extracellular leucine-rich repeat domain and the kinase activity of *FSL2* are required for flagellin binding and signaling in *Arabidopsis*. Plant Cell 13:1155–1163

Guo M, Zhang YL, Meng ZJ, Jiang J (2012) Optimization of factors affecting *Agrobacterium*-mediated transformation of Micro-Tom tomatoes. Gen Mol Res 11:661–671
Husaini AM (2010) Pre- and post-agroinfection strategies for efficient leaf disk transformation and regeneration of transgenic strawberry plants. Plant Cell Rep 29:97–110
Hwang EE, Wang MB, Bravo JE, Banta LM (2015a) Unmasking host and microbial strategies in the *Agrobacterium*-plant defense tango. Front Plant Sci 6:200
Hwang H, Gelvin SB, Lai E (2015b) *Agrobacterium* biology and its application to transgenic plant production. Front Plant Sci 6(1–3):265
Imam J, Singh PK, Shukla P (2016) Plant microbe interactions in post genomic era: perspectives and applications. Front Microbiol 7:1488
ISAAA (2017) Global status of commercialized Biotech/GM crops: 2017. ISAAA Brief No. 53. ISAAA, Ithaca
Ishaq SL (2017) Plant-microbial interactions in agriculture and the use of farming systems to improve diversity and productivity. AIMS Microbiol 3:335–353
Jayaraman D, Forshey KL, Grimsrud PA, Ane JM (2012) Leveraging proteomics to understand plant–microbe interactions. Front Plant Sci 3:44
Jones JDG, Dangl JL (2006) The plant immune system. Nature 444:323–329
Jube S, Borthakur D (2009) Development of an *Agrobacterium*-mediated transformation protocol for the tree-legume *Leucaena leucocephala* using immature zygotic embryos. Plant Cell Tissue Organ Cult 96:325–333
Kavitah G, Taghipour F, Huyop F (2010) Investigation of factors in optimizing *Agrobacterium* mediated gene transfer in *Citrullus lanatus* cv round dragon. J Biol Sci 10:209–216
Keshamma E, Rohini S, Rao KS, Madhusudhan B, Kumar MU (2008) Tissue culture-independent in planta transformation strategy: an *Agrobacterium tumefaciens*-mediated gene transfer method to overcome recalcitrance in cotton (*Gossypium hirsutum* L.). J Cotton Sci 12:264–272
Knief C (2014) Analysis of plant microbe interactions in the era of next generation sequencing technologies. Front Plant Sci 5:216
Kossmann J (2012) Grand challenges in plant biotechnology. Front Plant Sci 3:61
Kumar P, Srivastava DK (2015) High frequency organogenesis in hypocotyl, cotyledon, leaf and petiole explants of broccoli (*Brassica oleracea* L. var. *italica*), an important vegetable crop. Physiol Mol Biol Plants 21(2):279–285
Kumar P, Srivastava DK (2016) Biotechnological advancement in genetic improvement of broccoli (*Brassica oleracea* L. var. *italica*), an important vegetable crop. Biotechnol Lett 38 (7):1049–1063
Kumar P, Gambhir G, Gaur A, Srivastava DK (2015) Molecular analysis of genetic stability in in vitro regenerated plants of broccoli (*Brassica oleracea* L. var. *italica*). Curr Sci 109 (8):1470–1475
Kumar P, Gaur A, Srivastava DK (2017) *Agrobacterium* – mediated insect resistance gene (cry1Aa) transfer studies pertaining to antibiotic sensitivity on cultured tissues of broccoli (*Brassica oleracea* L. var. *italica*): an important vegetable crop. Int J Veg Sci 23:523–535
Kumar P, Gambhir G, Gaur A, Thakur AK, Sharma KC, Srivastava DK (2018a) Development of transgenic broccoli with *cryIAa* gene for resistance against diamondback moth (*Plutella xylostella*). 3 Biotech 8(7):299
Kumar P, Thakur AK, Srivastava DK (2018b) Genetic engineering approaches for abiotic stress tolerance in broccoli: recent progress. In: Akula R, Gill SS (eds) Metabolic adaptations in plants during abiotic stress. Taylor & Francis (CRC), New York, pp 363–367
Kumar P, Dhiman K, Srivastava DK (2018c) Morphogenic potential of different explants of broccoli (*Brassica oleracea* L. var. *italica*): important "nutrient rich" vegetable, using Thidiazuron. In: Ahmad N, Faisal M (eds) Thidiazuron: from urea derivative to plant growth regulator. Springer, Singapore, pp 373–392
Kunze G, Zipfel C, Robatzek S, Niehaus K, Boller T, Felix G (2004) The N terminus of bacterial elongation factor *Tu* elicits innate immunity in *Arabidopsis* plants. Plant Cell 16:3496–3507

Lacroix B, Citovsky V (2013) The roles of bacterial and host plant factors in *Agrobacterium*-mediated genetic transformation. Int J Dev Biol 57(6–8):467–481

Mamontova EM, Velikov VA, Volokhina IV, Chumakov MI (2010) *Agrobacterium*-mediated in planta transformation of maize germ cells. Russ J Genet 46:501–504

Mine A, Sato M, Tsuda K (2014) Toward a systems understanding of plant–microbe interactions. Front Plant Sci 5(1–9):423

Moshelion M, Altman A (2015) Current challenges and future perspectives of plant and agricultural biotechnology. Trends Biotechnol 33(6):337–342

Nicaise V, Roux M, Zipfel C (2009) Recent advances in PAMP-triggered immunity against bacteria: pattern recognition receptors watch over and raise the alarm. Plant Physiol 150:1638–1647

Opabode J (2006) *Agrobacterium*-mediated transformation of plants: emerging factors that influence efficiency. Biotechnol Mol Biol Rev 1:12–20

Parmar N, Singh KH, Sharma P, Singh L, Kumar P, Nanjundan J, Khan YJ, Chauhan DK, Thakur AK (2017) Genetic engineering strategies for biotic and abiotic stress tolerance and quality enhancement in horticultural crops: a comprehensive review. 3 Biotech 7:239

Pathak RK, Baunthiyal M, Pandey D, Kumar A (2018) Augmentation of crop productivity through interventions of omics technologies in India: challenges and opportunities. 3 Biotech 8:454

Ping LX, Nogawa M, Nozue M, Makita M, Takeda M, Bao L, Kojima M (2003) In planta transformation of mulberry trees (*Morus alba* L.) by *Agrobacterium tumefaciens*. J Ins Biotechnol 72:177–184

Prasad R, Gill SS, Tuteja N (2018) Crop improvement through microbial biotechnology. Elsevier, Amsterdam. ISBN 9780444639882. https://www.elsevier.com/books/crop-improvement-through-microbialbiotechnology/prasad/978-0-444-63987-5

Puchta H (2000) Removing selectable marker genes: taking the shortcut. Trends Plant Sci 5:273–274

Rao SK, Rohini VK (1999) Gene transfer into Indian cultivars of safflower (*Carthamus tinctorius* L.) using *Agrobacterium tumefaciens*. Plant Biotechnol 16:201–206

Rohini VK, Rao KS (2008) A novel in planta approach to gene transfer for legumes. In: Kirti PB (ed) Handbook of new technologies for genetic improvement of legumes. CRC, New York, pp 273–286

Schenk PM, McGrath KC, Lorito M (2008) Plant–microbe and plant-insect interactions meet common grounds. New Phytol 179:251–255

Sharma C (2010) Studies on *Agrobacterium*- mediated insect resistance gene transfer in tomato (*Lycopersicon esculentum* Mill.). Ph.D. Thesis. Dr Y.S. Parmar University of Horticulture and Forestry, Nauni, Solan

Sharma P (2014) Studies on chitinase gene transfer in tomato (*Solanum lycopersicum* L.) and molecular analysis of transgenic plantlets. Ph.D. Thesis. Dr Y.S. Parmar University of Horticulture and Forestry, Nauni. In: Solan. H.P., India

Sharma C, Srivastava DK, Aggarwal G (2011) Effect of cefotaxime with kanamycin on regeneration efficiency and *Agrobacterium* growth in tomato plants. J Plant Sci Res 27:227–230

Shaunak I, Kumar P, Gaur A, Sharma S, Srivastava DK (2018) *Agrobacterium*-mediated gene transfer using binary vector in lettuce (*Lactuca sativa* L.). Agric Res J 55(3):443–450

Shukla M, Al-Busaidi KT, Trivedi M, Tiwari RK (2018) Status of research, regulations and challenges for genetically modified crops in India. GM Crops Food 04:1–16

Singh D, Raina TK, Kumar A, Singh J, Prasad R (2019) Plant microbiome: a reservoir of novel genes and metabolites. Plant Gene 18:100177. https://doi.org/10.1016/j.plgene.2019.100177

Slater A, Scott NW, Fowler MR (2008) The genetic manipulation of herbicide tolerance. In: Plant biotechnology: the genetic manipulation of plants, 2nd edn. Oxford University Press, Oxford, pp 54–74

Sood P, Bhattacharya A, Sood A (2011) Problems and possibilities of monocot transformation. Biol Plant 55:1–15

Spoel SH, Dong XN (2012) How do plants achieve immunity? Defence without specialized immune cells. Nat Rev Immunol 12:89–100

Srivastava DK, Kumar P, Sharma S, Gaur A, Gambhir G (2016) Genetic engineering for insect resistance in economically important vegetable crops. In: Ahmad N, Anis M (eds) Plant tissue culture: propagation, conservation and crop. Springer, New York, pp 343–378

Subramoni S, Nathoo B, Klimov E, Yuan Z (2014) *Agrobacterium tumefaciens* responses to plant-derived signaling molecules. Front Plant Sci 5(1–12):322

Taj G, Giri P, Tasleem M, Kumar A (2014) MAPK signaling cascades and transcriptional reprogramming in plant-pathogen interactions. In: Gaur RK, Sharma P (eds) Approaches to plant stress and their management. Springer, Berlin, pp 297–316

Takahashi M, Nakanishi H, Kawasaki S, Nishizawa NK, Mori S (2001) Enhanced tolerance of rice to low iron availability in alkaline soils using barley nicotianamine aminotransferase genes. Nat Biotechnol 19:466–469

Torto-Alalibo T, Collmer CW, Gwinn-Giglio M (2009) The plant-associated microbe gene ontology (PAMGO) consortium: community development of new gene ontology terms describing biological processes involved in microbe-host interactions. BMC Microbiol 9(Suppl 1):S1

Tuteja N, Verma S, Sahoo RK, Raveendar S, Reddy INBL (2012) Recent advances in development of marker-free transgenic plants: regulation and biosafety concern. J Biosci 37:167–197

Tzfira T, Citovsky V (eds) (2008) *Agrobacterium*, from biology to biotechnology. Springer, New York

Van Emon JM (2016) The omics revolution in agricultural research. J Agric Food Chem 64 (1):36–44

Verma H, Kumar P, Gambhir G, Srivastava DK (2014) *Agrobacterium*-mediated transformation of broccoli. Crop Improv 41(2):139–147

Wang WC, Menon G, Hansen G (2003) Development of a novel *Agrobacterium*-mediated transformation method to recover transgenic *Brassica napus* plants. Plant Cell Rep 22:274–281

Willig CJ, Duan K, Zhang ZJ (2018) Transcriptome profiling of plant genes in response to *Agrobacterium tumefaciens*-mediated transformation. Curr Top Microbiol Immunol 418:319–348. https://doi.org/10.1007/82_2018_115

Yuan Z, Williams M (2014) A really useful pathogen, *Agrobacterium tumefaciens*. Plant Cell. https://doi.org/10.1105/tpc.112.tt1012

Zhang RG, Pappas KM, Brace JL, Miller PC, Oulmassov T, Molyneaux JM, Anderson JC, Bashkin JK, Winans SC, Joachimiak A (2002) Structure of a bacterial quorum-sensing transcription factor complexed with pheromone and DNA. Nature 417:971–974 (Erratum. Nature 476:240)

Zhang C, Xie Q, Anderson RG, Ng G, Seitz NC, Peterson T et al (2013) Crosstalk between the circadian clock and innate immunity in *Arabidopsis*. PLoS Pathog 9:e1003370

Zhu Y et al (2003) Identification of *Arabidopsis* rat mutants. Plant Physiol 132:494–505

Ziemienowicz A (2014) *Agrobacterium*-mediated plant transformation: factors, applications and recent advances. Biocatal Agric Biotechnol 3(4):95–102

Ziemienowicz A, Shim YS, Matsuoka A, Eudes F, Kovalchuk I (2012) A novel method of transgene delivery into triticale plants using the Agrobacterium transferred DNA-derived nano-complex. Plant Physiol 158:1503–1513

Zipfel C, Robatzek S, Navarro L, Oakeley EJ, Jones JDG, Felix G et al (2004) Bacterial disease resistance in *Arabidopsis* through flagellin perception. Nature 428:764–767

Zipfel C, Kunze G, Chinchilla D, Caniard A, Jones JDG, Boller T et al (2006) Perception of the bacterial PAMP *EF-Tu* by the receptor EFR restricts *Agrobacterium*-mediated transformation. Cell 125:749–760

Chapter 16
Amelioration of Salt Stress Tolerance in Plants by Plant Growth-Promoting Rhizobacteria: Insights from "Omics" Approaches

Kavya Bakka and Dinakar Challabathula

Abstract Soil salinity is one of the major abiotic stresses known to drastically reduce agricultural productivity. Prolonged salinity stress in glycophytic plants may cause oxidative damage to the cells, thereby causing cell death. Although salt-tolerant crops can be produced by genetic engineering by introducing novel transgenes or by altering the expression levels of the existing genes, substantial enhancement of crop productivity is questionable, and the introduction of genetically modified transgenic plants into the ecosystem is not well received. Breeding for environmental stress tolerance in plants is also challenging, time consuming and cost intensive. Alternative to the above mentioned, the identification and usage of beneficial rhizobacteria are efficient, cost-effective approaches that have been successfully employed in various crops to improve their growth, yield and tolerance to salt stress. These beneficial plant growth-promoting rhizobacteria are naturally occurring soil bacteria that rapidly colonize plant roots and benefit plants by various mechanisms. These bacteria are able to survive in high-salt concentrations of the soil due to their inherent capability to accumulate some of the important compatible osmolytes required for maintaining intracellular osmotic homeostasis or possess the transporters that help them survive under high-salt conditions among other adaptive mechanisms. These soil bacteria grow luxuriously under high-salt conditions and possess plant growth-promoting and protecting traits that are responsible for facilitating plant growth and survival under high-salt conditions in the soil. In this chapter, we summarize the salinity stress responses in plants in terms of physiological, biochemical and molecular mechanisms followed by the plant growth-promoting rhizobacteria-mediated stress amelioration phenomenon. We describe the role of 'omics' approaches in generating comprehensive information essential for better understanding of plant growth promotion by plant growth-promoting rhizobacteria.

K. Bakka · D. Challabathula (✉)
Department of Life Sciences, School of Life Sciences, Central University of Tamil Nadu, Thiruvarur, Tamil Nadu, India
e-mail: dinakarc@cutn.ac.in

A. Varma et al. (eds.), *Plant Microbe Symbiosis*,
https://doi.org/10.1007/978-3-030-36248-5_16

16.1 Introduction

Agricultural productivity can be severely hampered by both abiotic and biotic stress conditions. Since plants are sessile they are continuously exposed to varied abiotic stresses such as drought, temperature fluctuations, salinity, high light intensity, flooding and nutrient starvation (Bailey-Serres and Voesenek 2008; Meena et al. 2017; Yang and Guo 2018a, b). With the rise in the area of land affected by high salt concentrations due to low precipitation, weathering of native rocks, poor cultivation practices, etc., salinity has become one of the most obstructive environmental factors for optimum crop productivity (Jamil et al. 2011; Hanin et al. 2016). Additionally, the salinization of soil has been constantly increasing due to steep changes in the ecosystem and increased human intervention wherein increased release of salt from natural sources leads to salt accumulation in the downstream reservoirs or lakes or similar water bodies which, in turn, increases the salinity of the land that gets irrigated by these water sources (Pitman and Lauchli 2002; Ilangumaran and Smith 2017). The salinity stress is further aggravated by the occurrence of natural catastrophes like drought and floods. Depletion of water table by increased irrigation, increased farming, unsustainable agricultural practices, deforestation, poor drainage systems and loss of top soil water by increased evaporation positively influence the accumulation of salt in the soil layers, thereby causing salinization of soils. Over time this leads to formation of salt seepages, salt lakes, salt marshes, marine sediments and scalds in the landscape (Pitman and Lauchli 2002; Rengasamy 2002; Ilangumaran and Smith 2017). Accumulation of salt in topsoil causes salt stress in plants with shallow root systems. While most of the crop plants are sensitive to salinity stress caused by high concentration of salts in the soil, the level of salt stress tolerance varies from one species to another (Bartels and Dinakar 2013; Flowers and Colmer 2015). Over the years, there has been a steady increase in the loss of the crop yields due to salinity stress which needs urgent intervention in the form of adaptations, changes and reforms in cultivation practices. This includes better resource management, developing resistant varieties and identifying plant growth–enhancing biological methods. Among the low cost and environmental friendly salinity stress management methods, microorganisms with properties to enhance plant growth promotion could be the most promising option (Acosta-Motos et al. 2017; Negrão et al. 2017; Etesami and Maheshwari 2018). The rhizosphere of soil serves as a nutrient-rich habitat, which harbors a wide variety of beneficial or harmful bacteria for the plant. The beneficial bacteria are known to improve plant growth, survival, productivity and stress tolerance by employing varied mechanisms. They have exceptional genetic flexibility, genetic diversity, ability to interact with variety of crop plants, tolerance to environmental stress factors and potential to produce plant growth-promoting hormones (Dodd and Perez-Alfocea 2012; Hanin et al. 2016; Meena et al. 2017). For the systematic application of beneficial microorganisms to enhance plant stress tolerance capacity, it is essential to understand the various direct and indirect mechanisms through which these microorganisms colonize the rhizosphere of plants and the signaling events involved therein. In this chapter, the salinity stress responses in plants are described and the mechanisms involved in plant growth promotion by variety of rhizobacteria are explained. The role of 'omics' approaches to understand plant growth promotion by rhizobacteria is discussed.

16.2 Physiological and Molecular Responses of Plants to Salinity Stress

The adverse effects of salinity on plant development, growth and metabolism has forever been affecting the global agricultural sector. All plant species do not respond to high salt stress in a similar way. Based on the salt stress tolerance limit, plants are being classified as salt sensitive (glycophytes) and salt tolerant (halophytes). While a majority of the crop species are glycophytes lacking the genetic architecture for salinity tolerance, halophytes are endowed with a genetic makeup and mechanisms to survive under high-saline conditions including sequestration of Na^+ ions into the cell vacuoles and utilizing them as an osmoticum to maintain osmotic balance (Blumwald 2000; Tester and Davenport 2003; Mishra and Tanna 2017). Halophytes employ different mechanisms to deal with salt stress and have evolved a number of adaptive traits that allow them to germinate, grow and complete their life cycle under high-saline conditions (Flowers et al. 1977; Chen et al. 2018). Acclimation or tolerance to salinity stress in plants requires restriction of Na^+ uptake across the plasma membrane by reduced influx and increased efflux, facilitating Na^+ and Cl^- sequestration into the vacuole, regulating the compatible osmolyte production and generation as well as accumulation of antioxidants to quench/scavenge reactive oxygen species (ROS) (Bartels and Dinakar 2013; Flowers et al. 2015; Liang et al. 2018; Kapoor et al. 2019). The Na^+/H^+ antiporter catalyzes the exchange of sodium ions for hydrogen ions across the membranes, regulating sodium levels in the cytoplasm and, thereby, maintaining intracellular ion homeostasis (Blumwald 2000; Brini and Masmoudi, 2012; Bartels and Dinakar 2013; Assaha et al. 2017). The existence of Na^+/H^+ antiporter for the sequestration of the ions into the vacuole is not only observed in halophytes but also demonstrated in moderately salt-tolerant glycophytes like *Beta vulgaris* and *Arabidopsis thaliana* (Blumwald and Poole 1985; Tester and Davenport 2003). These mechanisms are essential to maintain cellular homeostasis during salinity stress and to avoid damage to subcellular structures.

Salt tolerance in plants is quite variable depending on several factors like salt content in soil, water availability in soil, climatic conditions, physiological make-up of the plant and extent of exposure to salt conditions. There also exists a huge variability in salt tolerance between different species and within the same species depending on the inheritance of acquired and inherent genetic traits. Salt tolerance of a plant is usually calculated by comparing its growth (biomass, root and shoot length), development and duration of survival in the presence and absence of salt in the soil (Munns et al. 2002; Negrão et al. 2017). The soil salinity when reaches its threshold level or more, nutrient absorption, plant growth and development are directly affected which, in turn, leads to decreased yield of the crop. Several plants continue to grow at a modified rate and adapt to salt stress by restricting their growth. The adaptation strategies employed by these plants include controlling the water loss through stomata, metabolic adjustment, toxic ion homeostasis, upregulated antioxidant defenses and osmotic adjustment (Bohnert and Sheveleva 1998). Glycophytes that are sensitive to the presence of salt in the soil can tolerate very low levels of salt (less than 40 mM NaCl) and constitute the majority of plants. These plants cannot resist high salinity stress and, hence, eventually succumb to stress and die (Abogadallah 2010). Contrary to these plants, halophytes can tolerate high levels of salt (up to 500 mM NaCl or more) and avoid salt stress by salt excretion. The salt is absorbed by the plant but is then excluded by specialized structures like salt glands or precipitation in leaves (Flowers 1985; Flowers and Colmer 2008; Munns and Tester 2008). In an experimental study, two varieties of *Vicia faba* that differed in salinity tolerance have been compared for their accumulation of Na^+ and Cl^- ions. While the salt-sensitive Nura variety accumulated high amounts of Na^+ and Cl^- ions during salinity stress, the salt-tolerant line 1487/7 accumulated lower concentrations of Na^+ and Cl^- without showing any toxicity symptoms (Tavakkoli et al. 2010). These kinds of studies are important to decipher the differences between salt-tolerant and -sensitive varieties.

High salinity due to high Na^+ and Cl^- concentrations in soil causes hyperionic and hyperosmotic stress, thereby causing drastic decrease in photosynthesis, increased ROS associated with membrane damage and cell death. High salt concentrations in soil also impose drought stress in plants due to increased water evaporation and retention of excess amounts of salts in soil, thereby affecting plant growth and development (Mudgal et al. 2010). However, halophytic plants perform well under high-salinity conditions without any symptoms of damage. Hyperaccumulation of Na^+ and Cl^- ions by roots or leaves is harmful for plant osmotic regulation. The Na^+ and Cl^- ions in the saline soil move into the roots and shoots through transpiration flux and, if not regulated, result in high ionic accumulation in aerial parts of the plant causing extensive damage to tissues (Teakle and Tyerman 2010). Sodium ions are easily absorbed to enter the xylem through the roots and accumulate in the shoot (Tester and Davenport 2003; Plett and Moller 2010). The accumulation of Cl^- ions is detrimental to the proper functioning of chloroplasts and mitochondria. A significant reduction in photosynthetic rate due to increased chlorophyll degradation because of increased leaf Cl^- concentration under salinity stress was observed in *Vicia faba* plants (Tavakkoli et al. 2010). Halophytes are better protected than glycophytes and are known to accumulate more Na^+ and Cl^- in chloroplasts than glycophytes. While higher Na^+ concentrations in chloroplasts of halophytic plants such as *Mesembryanthemum crystallinum* and *Suaeda maritima* are associated with elevated photosynthetic performance, higher Cl^- concentrations in the chloroplasts of halophytic plants such as *Avicennia marina* and *Limonium vulgare* are known to enhance electron transport and oxygen evolution during salinity stress (Preston and Critchley 1986; Flowers et al. 2015). Halophytic plants possess high chloroplast number per cell, thereby maintaining optimal PSI and PS II activity along with higher enzymatic activity of the enzymes involved in CO_2 fixation during salinity stress (Bose et al. 2017).

The responses of glycophytic plants to salinity stress can be differentiated into different phases. During the first phase, when the plant is exposed to salinity stress above the threshold levels, the roots experience osmotic stress, water shortage and loss of cell turgor, thereby triggering an immediate response of cessation in plant growth (Shavrukov 2013). This phenomenon is temporary and the plant can be recovered from this phase over time if the salinity decreases. However, prolonged exposure to salt stress elicits the second phase wherein the plant experiences ion toxicity. At this point if the plant fails to exclude the ions by excretion or exclusion, the ions accumulate in the cell cytoplasm, thereby leading to irreversible damage to the cell organelles and cells (Munns 2002; Munns and Tester 2008). During prolonged exposure to salinity, cell elongation and cell division rates are drastically reduced in the root and shoot primordia of the plant leading to complete cessation of growth and development (Bartels and Sunkar 2005; Munns and Tester 2008). Lateral root formation is completely abolished which is a characteristic phenotype of salt stress. Additionally, stunted shoot and leaf death are also observed during extended exposure. The presence of excess salts within the cells leads to salt aggregate formation followed by disturbance in vesicular trafficking and lowering of availability of metabolic enzymes within the cell (Baral et al. 2015).

Salt stress affects phytohormone signaling and other long-distance growth-regulating mechanisms (Iqbal et al. 2014). For salt stress adaptation and for developing resistance, a tightly regulated alteration of inherent signaling pathways is essential in plants. Phytohormone signaling is dependent on positive and negative stimuli from parameters that regulate cellular homeostasis like the osmotic balance, ion concentrations, etc. The signaling involves a crosstalk between the five prominent phytohormones: auxin, gibberellin, abscisic acid, ethylene and cytokinin. These hormones are known to play a vital role in the ability of plants to acclimatize to salt stress conditions by mediating changes in growth, development and nutrient allocation (Fahad et al. 2015; Verma et al. 2016a). Many transcription factors and kinases along with other protein regulators are known to influence phytohormone signaling (Verma et al. 2016a). Alterations in phytohormone signaling control stomatal conductivity. Due to increased osmotic stress caused by high salinity, stomata become flaccid leading to reduced transpiration rates, thereby showing an immediate effect on plant metabolism, especially on carbohydrate metabolism. The reduced growth rate, leaf senescence and stomatal closure lead to reduced rates of photosynthesis. The blockage in the growth of plant primordia leads to lesser storage space for synthesized carbohydrates which accumulate and cause feedback inhibition of photosynthesis (Paul and Foyer 2001; Suwa et al. 2006). In tobacco plants, the effect of salinity stress on stem, which is a major sink tissue, appeared within few hours of stress treatment while the effect on photosynthesis rate was observed after a lag period of 3–4 days post stress treatment, suggesting that the effect of salinity on sink tissue is earlier than on the source activity. Additionally, the salinity stress also reduced the carbon export rate leading to increased leaf sugar concentration and Na^+ accumulation in all plant tissues (Suwa et al. 2006). Accumulation of sugars and compatible solutes is a general response towards salinity stress in plants. Positive correlation between sugar accumulation and tolerance towards osmotic stress induced by salinity is reported in experiments with transgenic plants (Abebe et al. 2003; Giri 2011; Li et al. 2011; Udawat et al. 2017). Low-molecular weight osmolytes like trehalose, mannitol, sorbitol, glycerol, proline, glycine betaine, etc., which are essential for cellular osmoregulation and for stabilizing desiccated cell organelles, are accumulated during osmotic stress induced by salinity stress to protect plants from toxic effects (Rhodes et al. 2002; Bartels and Sunkar, 2005).

ROS are usually generated and accumulated at low levels during normal conditions essentially as a byproduct from photosynthesis, respiration and photorespiration wherein it plays a major role in signaling cascades of several cellular and intercellular events. Constitutive activity and expression of various antioxidant enzymes have been reported in several plants suggesting that plants are equipped with the enzymes controlling ROS levels under normal physiological conditions (Abogadallah 2010). Studies have shown that salinity stress causes a steep increase in the production of cellular ROS leading to drastic decrease in the photosynthesis rate along with degradation of chlorophyll, thereby affecting cellular metabolism and growth of plants (Bartels and Dinakar 2013; Ismail et al. 2014; Choudhury et al. 2017). In order to mitigate the effect of ROS accumulation, increased antioxidants along with upregulation of antioxidant enzyme activities is generally observed (Jithesh et al. 2006; Sofo et al. 2015; Caverzan et al. 2016; Kapoor et al. 2019).

Plants respond to stress by regulating multiple signaling pathways and activate transcription of the genes that confer salinity tolerance. Salinity tolerance in plants is a very complex process involving numerous pathways and genes involved in accumulation of compatible solutes, upregulation of protective proteins, antioxidants and suppression of energy-consuming reactions (Apse and Blumwald 2002; Nikalje et al. 2017; Chen et al. 2018; Yang and Guo 2018a, b). The stress is initially perceived as a signal by the cell membrane receptors resulting in the activation of the signaling cascade and generation of secondary signaling molecules. Calcium, ROS and inositol phosphate are secondary messengers which are known to modulate salinity stress responses. These signaling molecules result in the induction of expression of stress-responsive genes which are either directly or indirectly involved in stress tolerance (Chinnusamy et al. 2004; Park et al. 2016). In *A. thaliana*, ion homeostasis during moderate salt stress is mediated mainly by the salt overly sensitive (SOS) signaling pathway in which excess Na^+ and high osmolarity are sensed by the receptors present at the plasma membrane, thereby inducing an increase in cytosolic calcium. This is sensed by SOS3 which in turn activates SOS2. The activated SOS3–SOS2 protein complex phosphorylates SOS1, a plasma membrane-localized Na^+/H^+ antiporter functioning in the efflux of Na^+ ions. Similarly, AtNHX1, a tonoplast localized Na^+/H^+ antiporter, is involved in the control of vacuolar osmotic potential in *A. thaliana* (Apse et al. 1999; Shi et al. 2000; Shi et al. 2002; Yang et al. 2009; Ji et al. 2013).

Breeding and production of transgenic plants are considered as feasible approaches to produce salt-tolerant plants, and hence efforts have been made to enhance the salinity tolerance of economically important plants by traditional breeding as well as by biotechnological approaches. However, success in these approaches is not progressive as expected due to restricted knowledge with regard to genetic traits and due to polygenic inheritance that promotes salinity tolerance in plants (Silva and Gerós 2009; Tang et al. 2015). Genetic engineering to produce transgenically modified plants is a tedious and expensive process as it may require manipulation of several genes that play a crucial role in alleviation of salinity stress (Roy et al. 2014; Nongpiur et al. 2016). Usage of plant growth-promoting rhizobacteria (PGPR) is a promising method to induce salinity stress tolerance in plants (Dimpka et al. 2009; Dodd and Perez-Alfocea 2012; Shrivastava and Kumar 2015; Ilangumaran and Smith 2017; Backer et al. 2018). Plants growing in environmentally stressed regions are supported by the beneficial actions of microbes present in and around the plant which is termed as the phytomicrobiome (Singh et al. 2019). It has been observed that the phytomicrobiome has a huge impact on plant metabolism and physiology and usually supports the adaptation of the plant to abiotic stress factors (Dimpka et al. 2009; Smith et al. 2015; Quiza et al. 2015).

16.3 Rhizospheric Bacteria to Mitigate Salt Stress in Plants

Halophytic and glycophytic plants use different strategies to cope with ion toxicity induced by salinity stress. While halophytes seem to take up sodium and sequester it into the vacuole, glycophytes usually limit sodium uptake or transport sodium to old leaves as an alternative way to extrude sodium out of plants. In halophytes, the high osmotic potential in vacuoles is balanced by accumulating compatible solutes in the cytoplasm. Plants, in order to combat oxidative damage induced by osmotic stress during salinity stress, should acquire effective ROS scavenging ability, should effectively balance the osmotic microenvironment of the cell by accumulating compatible solutes, should balance ion uptake and synthesize specific protective proteins to protect itself from damaging effects of salinity stress. Because of the existence of the above-mentioned characteristics, halophytes are used as model plants to decipher the mechanisms of salinity tolerance in plants (Ali and Yun 2017; Mishra and Tanna 2017; Numan et al. 2018). If any of the above-mentioned mechanisms are induced in plants by rhizospheric bacteria or by symbiosis of rhizospheric bacteria with plant roots leading to salinity tolerance in plants with growth promotion, that bacteria can be considered as PGPR with potential to alleviate salt stress tolerance in plants, even though the beneficial effects of the rhizobacteria on the plant growth may be direct or indirect (Lugtenberg and Kamilova 2009; Backer et al. 2018; El-Esawi et al. 2018a; Sarkar et al. 2018). While some of the PGPR are not physically associated with the roots of plants, majority of them are found colonizing the root surfaces and surviving in spaces in between the root hairs and the rhizodermal layers of roots (Gray and Smith 2005; Desbrosses et al. 2009; Vacheron et al. 2013). Some of the bacteria can further invade intercellular spaces of the host tissues and survive as endophytes establishing a mutually beneficial association. Symbiotic rhizospheric and endophytic bacteria are extremely useful for not only salinity stress tolerance in plants but also for increasing agricultural production under changing environmental conditions. A better understanding of the underlying morphological, physiological, biochemical and molecular mechanisms of bacterially mediated salinity stress tolerance is important to mitigate salinity stress in plants (Gouda et al. 2018; Lata et al. 2018). Several of the root colonizing nonpathogenic bacteria are known to release inorganic nutrients from the organic reserves to sustain rapid plant growth with increased tolerance to salinity stress (Sarkar et al. 2018). The bacteria belonging to the genera *Acetobacter, Achromobacter, Acinetobacter, Alcaligenes, Arthrobacter, Azoarcus, Azospirillum, Azotobacter, Bacillus, Beijerinckia, Burkholderia, Derxia, Enterobacter, Gluconacetobacter, Herbaspirillum, Klebsiella, Methylobacterium, Microbacterium, Ochrobactrum, Paenibacillus, Pantoae, Pseudomonas, Rhizobium, Rhodococcus, Staphylococcus, Serratia, Stenotrophomonas, Streptococcus, Variovorax* and *Zoogloea* have been extensively studied for their plant growth-promoting traits (Babalola 2010; Dodd and Perez-Alfocea 2012).

16.4 Role of PGPR in Salinity Stress Tolerance and Plant Growth Promotion

Many factors influence the signaling events during plant-microbe interactions. PGPR contribute to the general well-being and enhanced salt tolerance of the plants by regulating several physiological pathways and signaling networks in the plant. A continuous symbiotic relationship exists between the plants and the microbes present in the rhizospheric environment wherein the roots of the plants release nutrient exudates enriching the rhizospheric zone essential for signaling and regulating the communication in between beneficial microbes and plant, and PGPR in turn impart beneficial influence on the welfare of the plant (Lugtenberg and Kamilova 2009). The plants release a variety of phenolics, flavonoids and organic acids from the roots thereby acting as chemical signals to attract bacteria for efficient root colonization (Badri et al. 2009).

The beneficial effects of PGPR involve boosting of key physiological processes such as water and nutrient uptake, increased photosynthesis, coordinated source-sink signaling, etc., thereby promoting the overall growth and development of the plant (Fig. 16.1; Ilangumaran and Smith 2017). Additionally, PGPR support nutrient assimilation by increasing plant nitrogen fixation capacity, production of different phytohormones like auxins and cytokinins, mineralization and decomposition of organic matter, improved bioavailability of different mineral nutrients like iron and phosphorous (Sharma et al. 2013; Jin et al. 2014; Kuan et al. 2016), biological control of the plant diseases through induction of systemic resistance, production of antibiotics (Beneduzi et al. 2012; Chowdhury et al. 2015), degrading organic pollutants, reducing metal toxicity of contaminated soils (Janssen et al. 2015; Weyens et al. 2015) and improving plant growth in salt-affected soils (Ilangumaran and Smith 2017; Numan et al. 2018). Due to these beneficial effects, PGPR offer an attractive way to replace chemical fertilizers, pesticides and additional chemical

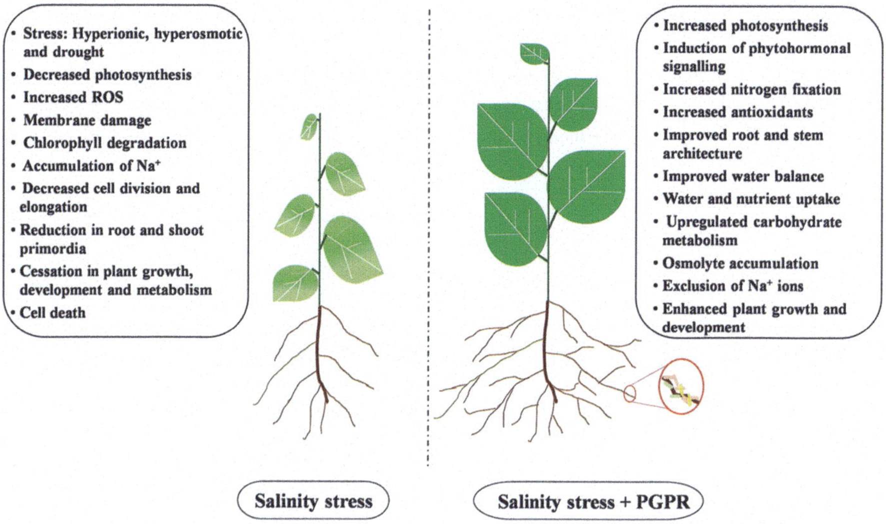

Fig. 16.1 Schematic representation of PGPR-mediated salinity stress tolerance in glycophytic plants. Changes observed in glycophytic plants during salinity stress in the presence and absence of PGPR are shown

supplements which may have deleterious effects in the natural environment. Improvement of salinity tolerance in plants using PGPR is becoming a promising approach to overcome the negative effects of salt stress on plant growth. PGPR-mediated salinity stress tolerance and growth promotion observed in different plant species due to activation of various mechanisms are listed in Table 16.1.

The mechanism by which PGPR bring about amelioration of salt stress in plants is a controlled multistep process involving crosstalk between the plant and the rhizobacteria wherein complex signaling events bring about changes in hormone signaling, thereby affecting the physiology of the plant. PGPR can be isolated from the rhizosphere of the plants growing in salt stress conditions and characterized in vitro for the growth-promoting features in plants subjected to salinity stress. A wide variety of culturable PGPR can be efficiently identified using this approach and can be tested for their efficacy in promoting plant growth under salt stress in laboratory conditions (Islam et al. 2016; Panwar et al. 2016a, b; Verma et al. 2016a, b). *Bacillus* species isolated from rhizospheric soils of different halophytic plants showed high phosphate solubilization potential along with the production of bacteriocin and siderophore, and upon co-inoculation on maize under salinity stress, plant growth is increased (Ullah and Bano 2015). These kinds of studies are important to identify and unravel the importance of salt-tolerant PGPR inhabiting the rhizosphere of halophytic plants.

PGPR are known to improve salinity tolerance in plants by numerous mechanisms which include activation of antioxidant enzymes, enhanced synthesis of antioxidant metabolites, ion homeostasis, polyamine biosynthesis, biosynthesis of compatible solutes and osmoprotectants, modulation in phytohormones and

Table 16.1 PGPR-mediated salinity stress tolerance and growth promotion observed in different plant species

Bacterial species	Plant species	Observed effects on alleviating salt stress	Reference
Arthrobacter protophormiae SA3 *Dietzia natronolimnaea* STR1	*Triticum aestivum*	– Enhanced photosynthetic efficiency – Upregulation of abiotic stress tolerance genes – Regulating ethylene signaling	Barnawal et al. (2017)
Pseudomonas fluorescens 002 (P.f.002)	*Zea mays*	– Increased growth in primary, lateral and seminal roots	Zerrouk et al. (2016)
Serratia sp. SL-12	*Triticum aestivum*	– Enhanced IAA production – Increased solubilized inorganic phosphate – Improved ion homeostasis – Accumulation of osmolytes	Singh and Jha (2016)
Bacillus megaterium	*Zea mays*	– Higher root hydraulic conditions	Marulanda et al. (2010)
Bacillus amyloliquefaciens SN13	*Oryza sativa*	– Transcription of genes involved in stress response mechanisms	Nautiyal et al. (2013)
Arthrobacter sp. SU18	*Triticum aestivum*	– Increased growth and dry biomass	Upadhyay and Singh (2015)
B. aquimaris SU44, SU8		– Accumulation of sugars	
B. subtilis SU47		– Reduction of Na^+ content in leaves and increased crop yield	
Azospirillum brasilense, Pantoea dispersa	*Capsicum annuum*	– Higher dry weight – Improved stomatal conductivity – Reduced Cl^- and increased NO_3^- concentrations, higher K^+:Na^+ ratios	del Amor and Cuadra-Crespo (2012)
Azotobacter strains C5, C9	*Zea mays*	– Improved K^+:Na^+ ratios – Increased photosynthetic pigments – Increased polyphenol content	Rojas-Tapias et al. (2012)
Burkholderia phytofirmans PsJN	*Arabidopsis thaliana*	– Proline accumulation – Upregulation of ABA signaling, ROS	Pinedo et al. (2015)

(continued)

Table 16.1 (continued)

Bacterial species	Plant species	Observed effects on alleviating salt stress	Reference
		scavenging and detoxification genes – Altered expression of ion homeostasis genes	
Bacillus subtilis GB03	*Puccinellia tenuiflora* (halophyte)	– Enhanced and selective absorbance of K^+ – Reduced uptake of Na^+ in roots and reduced transport of Na^+ from roots to shoots	Niu et al. (2016)
Bacillus amyloliquefaciens SQR9	*Zea mays*	– Enhanced chlorophyll content and growth – Increased soluble sugars – Increased glutathione content and peroxidase/catalase activity – Reduced Na^+ levels	Chen et al. (2016)
Enterobacter sp. EJ01	*Arabidopsis thaliana, Solanum lycopersicum*	– Production of IAA and ACC deaminase – Increased expression of salt stress responsive and proline biosynthesis genes – Enhanced ROS scavenging	Kim et al. (2014)
Achromobacter piechaudii	*Solanum lycopersicum*	– Increased water use efficiency (WUE) – Protection of photosynthesis – Increased fresh and dry weights	Mayak et al. (2004)
Bacillus subtilis GB03	*Arabidopsis thaliana*	– Tissue-specific regulation of high-affinity K^+ transporter (HKT1) – Lower Na^+ accumulation	Zhang et al. (2008)
Rhizobium phaseoli M1, M6, M9 and *Pseudomonas* sp. (coinoculation)	*Vigna radiata*	– Improved seedling growth and nodulation	Ahmad et al. (2011)
Exiguobacterium oxidotolerans STR36	*Bacopa monnieri*	– Higher proline levels – Higher yield – Decreased lipid peroxidation	Bharti et al. (2013)
Erwinia persicinus RA2	*Solanum lycopersicum*	– Improved fruit quality and growth	Shen et al. (2012)

(continued)

Table 16.1 (continued)

Bacterial species	Plant species	Observed effects on alleviating salt stress	Reference
Pseudomonas aureantiaca TSAU22, *Pseudomonas extremorientalis* TSAU6 and *Pseudomonas extremorientalis* TSAU20	*Triticum aestivum*	– Alleviates salinity-induced seed dormancy – Increased seedling root growth	Egamberdieva (2009)
Pseudomonas putida UW4	*Solanum lycopersicum*	– Enhanced growth and increased chlorophyll concentration – Altered expression of chloroplast import apparatus genes	Yan et al. (2014)
Variovorax paradoxus 5C-2	*Pisum sativum*	– Increased K^+ uptake by root and transport to shoot – Decreased stomatal resistance and xylem balancing pressure – Improved photosynthetic efficiency	Wang et al. (2016)
Enterobacter sp. UPMR18	*Abelmoschus esculentus*	– Induction of ROS-scavenging enzyme activity – Improved germination, growth and chlorophyll content	Habib et al. (2016)
Pseudomonas sp., *Flavobacterium* sp. and *Enterobacter* sp.	*Zea mays*	– Increased plant growth and yield – More N, P and K uptake – Improved K^+:Na^+ ratios	Nadeem et al. (2009)
Paraburkholderia phytofirmans PsJN	*Arabidopsis thaliana*	– Less Na^+ in leaves – Growth promotion	Ledger et al. (2016)
Pseudomonas simiae strain AU	*Glycine max*	– Upregulation of induced systemic tolerance (IST) proteins – Enhanced proline and chlorophyll content – Improved K^+:Na^+ ratios	Vaishnav et al. (2015)
Bacillus megaterium BOFC15	*Arabidopsis thaliana*	– Increased photosynthetic capacity – Higher plant biomass – Improved root architecture	Zhou et al. (2016)

(continued)

Table 16.1 (continued)

Bacterial species	Plant species	Observed effects on alleviating salt stress	Reference
Azospirillum brasilense strain Cd and *Rhizobium* sp. (coinoculation)	*Phaseolus vulgaris*	– Improved root branching – Enhanced production of root exudates by plant	Dardanelli et al. (2008)
Arthrobacter sp. and *Bacillus* sp.	*Capsicum annuum*	– Altered expression of stress-inducible genes	Sziderics et al. (2007)
Pantoea dispersa	*Cicer arietinum*	– Higher growth and yield – Reduced Na^+ and increased K^+ uptake – Improved leaf water and chlorophyll content – Reduced membrane damage	Panwar et al. (2016a, b)
Dietzia natronolimnaea STR1	*Triticum aestivum*	– Upregulation of stress-responsive and ABA-signaling cascade genes – Tissue-specific responses of ion transporters – Enhanced expression of antioxidant enzymes – Higher proline content	Bharti et al. (2016)
Pseudomonas putida H-2-3	*Glycine max*	– Increased shoot length and biomass – Higher chlorophyll content – Higher ROS scavenging activity – Lower ABA and salicylic acid levels, increased jasmonic acid levels	Kang et al. (2014)
Pseudomonas putida Rs-198	*Gossypium* sp.	– Increased germination rate – Increased Mg^+, K^+, Ca^+ absorption – Decreased uptake of Na^+ – Increased endogenous IAA and decreased ABA content	Yao et al. (2010)

(continued)

Table 16.1 (continued)

Bacterial species	Plant species	Observed effects on alleviating salt stress	Reference
Pseudomonas mendocina	*Lactuca sativa*	– Increased water-soluble carbohydrates – Increased soil aggregate stabilization	Kohler et al. (2006)
Halomonas variabilis (*HT1*) and *Planococcus rifietoensis* (RT4)	*Cicer arietinum*	– Increased soil aggregate formation – Improved plant growth	Qurashi and Sabri (2012)
Enterobacter sp. MN17 and *Bacillus* sp. MN 54	*Chenopodium quinoa*	– Improved plant-water relations – Reduced Na^+ uptake	Yang et al. (2016)

regulation of the uptake and transport of ions (Zhang et al. 2008; Dimpka et al. 2009; Dodd and Perez-Alfocea 2012; Bharti et al. 2016; Numan et al. 2018). Maize seedlings coinoculated with the PGPR strain *Bacillus amyloliquefaciens* SQR9 showed enhanced salinity stress tolerance with increased chlorophyll, improved peroxidase and catalase activities along with enhanced glutathione content (Chen et al. 2016). Increased antioxidant activities were also observed in plants such as *Capsicum annum, Pisum sativum, Zea mays, Glycine max,* etc., upon inoculation with PGPR strains such as *Microbacterium oleivorans* KNUC7074, *Brevibacterium iodinum* KNUC7183, *Rhizobium massiliae* KNUC7586, *Planomicrobium* sp. MSSA-10, *Serratia liquefaciens* KM4, *Bacillus firmus* SW5 and subjected to salinity stress (Hahm et al. 2017; El-Esawi et al. 2018a, b; Shahid et al. 2018). In another study, PGPR strain *Burkholderia phytofirmans* PsJN treatment enhanced salt stress tolerance levels in salt-sensitive *A. thaliana* plants with increased accumulation of proline and showed high expression of genes related to abscisic acid signaling (Pinedo et al. 2015).

The ability of PGPR to produce plant hormones is one of the most important mechanisms by which many rhizobacteria are known to promote plant growth (Prasad et al. 2015). PGPR can either synthesize plant hormones or supply them externally to plant roots during symbiosis, or alternatively they can induce their production in plant roots by intricate signaling events (Dodd and Perez-Alfocea 2012; Backer et al. 2018). Several of the PGPR are known to trigger phytohormones such as indole 3- acetic acid (IAA), abscisic acid, cytokinins, gibberellins, brassinosteroids and ethylene for root and shoot invigoration, thereby promoting plant growth and development (Perrig et al. 2007). The induction of hormones is important for plant adaptation and survival during salinity stress conditions (Perrig et al. 2007; Sureshbabu et al. 2016; Gouda et al. 2018). Auxin or indole-3-acetic acid (IAA) production in PGPR and its utilization by plant roots is well studied as it directly influences root growth and lateral root proliferation in plants. Most of the PGPR cannot produce endogenous tryptophan (IAA precursor) and absorb it from plant root exudates. Within the bacterial cell, IAA is synthesized from tryptophan which when released in the rhizosphere is absorbed by the plant. The combined IAA levels produced by the plant and absorbed from rhizosphere determine the total IAA response in the plant. A tightly controlled signaling exists between plant and bacteria which determines the amounts of tryptophan release and IAA absorption by plant (Spaepen et al. 2007). Although tryptophan-independent pathways for IAA biosynthesis exist in plants, evidences for the existence of tryptophan-independent pathways for IAA biosynthesis in PGPR are not clear (Ahmed and Hasnain 2014; Kasahara 2016). PGPR strains such as *Rhizobium*, *Microbacterium* and *Mycobacterium* are known to produce high amounts of IAA. However, many PGPR are also known to produce auxin-like bioactive compounds which are involved in plant growth promotion (Thakuria et al. 2004; Egamberdieva 2009).

Abscisic acid (ABA) is a stress hormone whose levels are known to increase in roots, xylem sap and shoots during salinity stress, thereby decreasing the transpiration rate, increasing the stomatal closure and finally decreasing the net photosynthetic rate. However, reports suggest that upon inoculation of plants with PGPR during salinity stress, ABA levels are reduced in the plants leading to increased photosynthesis rate, growth and biomass (Yao et al. 2010; Dodd and Perez-Alfocea 2012; Barnawal et al. 2017). PGPR are also known to influence the ABA-signaling pathway by upregulating ABA-responsive genes and stress-responsive genes (Bharti et al. 2016). These in turn are known to modulate the SOS pathway, antioxidant production, osmolyte production and upregulate the genes encoding the ion transporters, thereby alleviating salt stress symptoms in salt stress-affected plants.

Production of cytokinin is a common trait that is observed in the majority of PGPR, and increased salinity stress tolerance along with plant growth is positively correlated with cytokinin production by PGPR (Dodd and Perez-Alfocea 2012). *Bacillus megaterium* UMCV1, a PGPR strain isolated from the rhizosphere of *Phaseolus vulgaris*, has cytokinin receptors which are involved in plant growth promotion. Inoculation with *B. megaterium* caused a significant increase in fresh weights of shoots and roots of *A. thaliana* and *P. vulgaris* (López-Bucio et al. 2007; Ortiz-Castro et al. 2008). Using cytokinin signaling mutants of *A. thaliana* lacking one, two or three of the putative cytokinin receptors named as *CRE1, AHK2, AHK3* and *RPN12*, it was shown that the plant growth promotion by *B. megaterium* is reduced in single- and double-mutant combinations of the cytokinin receptors. The triple cytokinin receptor *CRE1-12/AHK2-2/AHK3-3* knockout with retarded growth of the primary root and arrested shoot development was insensitive to the inoculation and does not show any growth promotion and root developmental responses. These results indicate that root cytokinin receptors play an important role in plant growth promotion by *B. megaterium* (Ortiz-Castro et al. 2008).

A small number of PGPR have been identified to produce gibberellins which are known to improve soil fertility and thus help indirectly in the promotion of plant growth. These PGPR regulate the plant hormone either by direct synthesis of gibberellins or by deconjugation of glucosyl gibberellins or by changing the inactive gibberellins into active gibberellic acid residues (Piccoli et al. 1999; Cassán et al. 2001). Several strains such as *Bacillus cereus*, *B. macroides*, and *B. pumilus* are known to produce a variety of gibberellins such as GA5, GA8, GA34, GA44 and GA53. Upon inoculation of these bacterial strains to *Capsicum annuum*, plant growth promotion along with increased endogenous gibberellins levels was observed (Joo et al. 2004, 2005). The growth promotion due to increased gibberellins from the PGPR is also observed in a variety of plants such as cucumber, tomato, maize, etc., upon inoculation with PGPR strains such as *Acinetobacter calcoaceticus*, *Promicromonospora* sp. SE188, *Pseudomonas*, etc., (Kang et al. 2014). Brassinosteroids are natural hormones important for plant growth and development. The brassinosteroid-deficient-of-signaling mutants of *Arabidopsis* with shortened root phenotypes are used to study signal transduction and growth promotion by PGPR. Results revealed that the elicitation of plant growth promotion by PGPR in vitro involved brassinosteriod signaling (Ryu et al. 2005; Fahad et al. 2015).

The ethylene levels are known to increase in leaves during stress conditions thereby triggering senescence. Although plant growth is compromised during stress conditions, ethylene is one of the important hormones which is known to regulate plant adaptation to stress conditions. Increased ethylene levels inhibit the transcription of auxin-responsive genes, thereby restricting the plant growth. Several PGPR secrete 1-aminocyclopropane-1-carboxylase (ACC) deaminase for restricting ethylene biosynthesis in plants thereby promoting plant growth under salinity conditions. The presence of high ethylene levels in the plant are known to inhibit growth and even cause cell death (Singh et al. 2015). In the plant, the enzyme ACC deaminase is responsible for cleaving the ethylene precursor ACC into ammonia and α-ketobutyrate, thereby decreasing ethylene levels in the cell. The possession of this enzyme by rhizobacteria serves as an indicator for plant growth promotion (Ali and Kim 2018; Sarkar et al. 2018). Strains of *Pseudomonas fluorescens* and *Enterobacter* sp. are known to produce ACC deaminase and when inoculated to maize plants showed a significant increase in maize yield in salt-affected soil with higher K^+/Na^+ ratios and higher nitrogen, phosphorous and potassium uptakes (Nadeem et al. 2009).

16.5 Omics Approaches to Address the Alleviation of Salt Stress by PGPR: Transcriptome, Proteome and Metabolome Analyses

Technical advances utilizing 'omics' approaches have made it possible to analyze the changes in gene expression/protein levels and metabolite accumulation on a larger scale. Transcriptomic, proteomic and metabolomic approaches aim to provide complete information on the changes in transcripts/proteins/metabolites that occur in plants upon inoculation with PGPR, giving an overview of the metabolic status of the plant (Ilangumaran and Smith 2017). These approaches together with comparative studies with PGPR inoculated and non-inoculated plants subjected to salinity stress will provide novel insights on the molecular processes that are regulated by PGPR in host plants to protect the plant against damaging effects of salinity stress. These approaches are important for the identification of novel proteins/metabolites

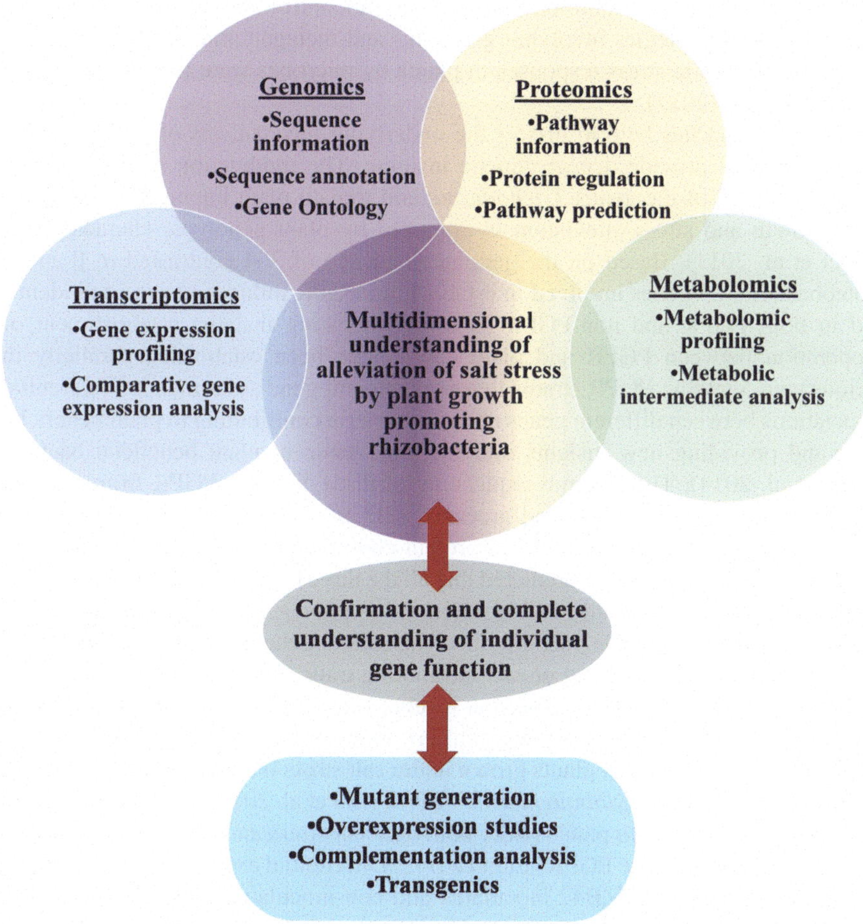

Fig. 16.2 Diagrammatic representation of different 'omics' approaches that can be used to understand the molecular mechanisms of PGPR-mediated salinity stress tolerance in plants. Integration of the data helps in understanding the molecular mechanisms and identification of genes required for salinity tolerance

that are accumulated in response to inoculation with PGPR during salinity stress in plants (Fig. 16.2). Genomics and metagenomics are two other approaches that are included under the omics approaches. Advances in genome sequencing technologies made large amount of genomic data, gene expression profiles, tools and other useful resources accessible; hence, it is easier to identify the genes necessary for breeding and stress alleviation. Genomics-based technologies are important components in crop improvement programs. Metagenomics is the culture-independent application of genomics techniques giving information about the distribution of microbial communities prevalent in a specific natural habitat and possessing traits such as plant growth promotion, antibiotic production, biocontrol and xenobiotic degradation. The omics strategies involving genomics and metagenomics employed in the mitigation of abiotic stress responses in plants by microbes were reviewed recently by Meena et al. (2017).

Omics approaches help to dissect the underlying mechanisms of plant-microbe interactions and provide deeper genetic insights. The modulation of rhizosphere-competent fungi of the genus *Trichoderma* and tomato plant interaction leading to plant growth and stress alleviation is impacted by plant genotypic characteristics (Tucci et al. 2011). Based on the genome sequence of 304 contrasted α, β and γ proteobacteria, 23 genes involved in establishing PGPR influence have been identified in both PGPR and non-PGPR bacteria, indicating that the establishment of cooperation between PGPR and plant could have been established parallelly in various taxa, yielding PGPR strains that use different gene assortments. Preferential associations between different genes occur in bacteria contributing to plant beneficial traits and providing new insights into the emergence of plant beneficial bacteria (Bruto et al. 2014). The genome sequencing of three different PGPR from coconut (CPCRI-1), cocoa (CPCRI-2) and arecanut (CPCRI-3) encoded 4056 (CPCRI-1), 4637 (CPCRI-2) and 4286 (CPCRI-3) protein-coding genes, respectively. The functional annotation of the genes predicted that all the three bacteria encoded the proteins needed for mineral phosphate solubilization, siderophores, acetoin, butanediol, ACC deaminase, chitinase, phenazine, 4-hydroxybenzoate, trehalose and quorum sensing molecules supporting the plant growth-promoting traits (Gupta et al. 2014).

To identify the underlying molecular mechanisms that allow plants to respond to PGPR and exhibit salinity tolerance, transcriptome analysis has been performed in shoot tissues of *A. thaliana* plants grown under salt stress inoculated with or without PGPR strain *Bacillus amyloliquefaciens* FZB42 (Liu et al. 2017). Growth promotion is observed in *A. thaliana* plants under both non-salt stress and salt stress conditions upon inoculation with the PGPR strain FZB42. Differential expression was observed in 1461 genes between FZB42 inoculated and non-inoculated shoot samples under control conditions without salt stress. Upon imposition of salt stress, 1288 genes were differentially expressed out of which 1024 were upregulated and 264 were downregulated. Six genes involved in the synthesis of the photosynthesis antenna proteins pathway were upregulated under salinity stress, suggesting the importance of FZB42 in enhancing plant photosynthesis. Major gene expression changes were observed in the transcripts related to photosynthesis, auxin biosynthesis, ROS scavenging, Na^+ translocation and biosynthesis of osmoprotectants such as trehalose and proline. Additionally, usage of hormone mutants demonstrated that FZB42 might induce salinity tolerance in *A. thaliana* plants by activating ethylene and jasmonic acid signaling (Liu et al. 2017). In another study, a carotenoid-producing salt-tolerant PGPR, *Dietzia natronolimnaea* STR1, was found to protect wheat plants from salinity stress by modulating the transcriptional machinery (Bharti et al. 2016). The effect of salt stress on the expression of stress-responsive genes, genes related to abscisic acid metabolism, genes related to SOS pathway, expression pattern of transcription factors TaWRKY10, TaWRKY17 and TaMYB33, ion transporters and antioxidant levels were analyzed in wheat seedlings exposed to salt stress in the presence and absence of STR1 strain. The expression studies confirmed the involvement of ABA signaling leading to the induction of transcription factors thereby increasing the expression of stress-related genes. Enhanced expression of *TaST*, a salt stress-induced gene, associated with promoting salinity tolerance, was observed in PGPR-inoculated plants in comparison to uninoculated control plants. While tissue-specific responses were observed for the genes encoding ion transporters *TaNHX1*, *TaHAK*, and *TaHKT1*, modulation in SOS pathway-related genes (*SOS1* and *SOS4*) was observed in PGPR-treated wheat plants under salinity stress. Enhanced antioxidant gene expression along with high proline content and stress-related gene expression cumulatively contributed to increased salinity stress tolerance of PGPR-inoculated wheat plants (Bharti et al. 2016). A combination of transcriptome and proteome analyses in *A. thaliana* plants inoculated with *Bacillus megaterium* revealed the importance of jasmonic acid metabolism for protecting *Arabidopsis* plants from salinity stress. The adjustment of jasmonic acid metabolism through the upregulation of jasmonil-isoleucine-12-hydroxylase (CYP94B3), lipoxygenase 4 and allene-oxide cyclase in inoculated plants under salinity stress is observed. Proteomic analysis revealed increased Rubisco spot volumes along with the alteration of antioxidant defense system through the upregulation of monodehydroascorbate reductase and ATP synthase in salt-stressed *A. thaliana* plants treated with *B. megaterium* (Erice et al. 2017). *Arabidopsis thaliana* plants inoculated with *Pseudomonas putida* (MTCC5279) resulted in significant increase in growth compared to non-inoculated control with upregulation of genes involved in the maintenance of genome integrity, phytohormone biosynthesis, amino acid synthesis, ABA signaling, suppression of ethylene biosynthesis, Ca^{+2} signaling and induced systemic resistance, thus giving a hint about the metabolic status of the plant during symbiosis with PGPR (Srivastava et al. 2012). These kinds of studies are important to unravel the molecular mechanisms that are important for PGPR-mediated salinity stress tolerance in plants.

Integrated transcriptome, proteome and metabolome analyses in *Arabidopsis* roots and shoots in response to PGPR strain *Paenibacillus polymyxa* E681 revealed a positive correlation between transcript expressions with protein abundance. While the proteins involved in amino acid metabolism, antioxidant defense, stress response, photosynthesis and plant hormone biosynthesis were upregulated, five proteins including three carbohydrate metabolism proteins, one amino acid metabolism related protein and one protein with unknown function were downregulated. Metabolite analysis revealed significantly higher levels of tryptophan, indole-3-acetonitrile (IAN), IAA and camalexin in the *P. polymyxa*-treated plants suggesting that PGPR promote plant growth by inducing metabolism (Kwon et al. 2016).

Proteome and metabolome analyses of *Pisum sativum* plants inoculated with *Didymella piondes* revealed enhanced pisatin biosynthesis as well as accumulation of amino acid and tricarboxylic acid cycle intermediates in the inoculated samples (Desalegn et al. 2016). Several PGPR are known to trigger systemic resistance in plants along with growth promotion by activating various cellular defense responses that are effective against multiple pathogens. The root-colonizing *Pseudomonas fluorescens* strain (Pf.SS101) enhanced resistance in *Arabidopsis* against several bacterial pathogens, including *Pseudomonas syringae* pv. tomato (Pst) and the insect pest *Spodoptera exigua* by activating salicylic acid-regulated genes. Genome-wide transcriptomic and untargeted metabolomic analyses showed that in roots and leaves of *Arabidopsis* plants treated with *P. fluorescens*, around 1910 genes and 50 metabolites were differentially regulated when compared to uninoculated plants (van de Mortel et al. 2012).

Use of transcriptomics, proteomics and metabolomics approaches for PGPR-plant interaction studies provides effective and reliable data that can be compared with other plant species. Identifying the genes, proteins and metabolites that are expressed/accumulated during rhizobacteria-plant interaction and studying their regulation during salinity stress conditions will provide insights on the genes and pathways that are induced/regulated in plants during plant-PGPR interactions, paving a path for generating better varieties for salinity stress tolerance.

16.6 Future Perspectives

Rhizosphere, the narrow zone surrounding the plant roots, harbors numerous microorganisms including rhizospheric bacteria that exhibit metabolic capability and flexibility with enormous potential to mitigate abiotic stress in plants. The interaction of beneficial rhizospheric bacterial population with plants is an integral and essential part of the ecosystem wherein rhizospheric bacteria are natural plant partners with an inherent capacity to promote growth in plants, modulate local and systemic mechanisms and to offer defense under adverse abiotic stress conditions. Since plants are sessile and are sensitive to salinity stress, it is essential to gain deeper insights into the bacterial-mediated salinity stress-mitigating mechanisms to obtain crops with high growth and productivity under salinity stress conditions. Genomics, transcriptomics, proteomics and metabolomics approaches are useful to study the interaction of plants with microbes and their external environment, thereby generating multidimensional understanding about the changes that occur in a plant cell during salinity stress. Integration of the data obtained from the 'omics' approaches is critical for gaining novel insights on molecular mechanisms that are operative in plants during rhizobacteria-mediated salinity stress tolerance. Although the usage of certain rhizobacteria for plant growth promotion and alleviation of salt stress symptoms in plants is not novel, the current knowledge and understanding about PGPR is still very limited. A comprehensive experimentation about halotolerant rhizobacteria inhabiting different soil types and different agricultural crops is needed at the moment to screen out active novel beneficial rhizobacteria with plant-beneficial metabolites. Currently, the majority of PGPR are tested for their salt stress alleviation either under laboratory conditions or in green house conditions; however, their efficacy should be evaluated under field conditions with different soil types. More investigations with holistic approaches are needed to analyze and assess the role of active PGPR on crop growth under various abiotic environmental stresses like salinity and drought. Since host genotypes and abiotic factors may influence the composition of plant microbiomes, detailed studies are needed to experimentally elucidate the underlying mechanisms of bacterial community assembly and the beneficial effects of bacterial population on plant growth. The diversity of the root microbiota within a phylogenetic framework of hosts should be explored in order to identify selectable traits that are required by bacteria to find a suitable host plant and to protect them under salinity stress conditions.

Acknowledgments Dinakar Challabathula acknowledges the grant from SERB, and Kavya Bakka acknowledges SERB-NPDF funding.

References

Abebe T, Guenzi AC, Martin B, Cushman JC (2003) Tolerance of mannitol accumulating transgenic wheat to water stress and salinity. Plant Physiol 131:1748–1755

Abogadallah GM (2010) Antioxidative defense under salt stress. Plant Signal Behav 5:369–374

Acosta-Motos JR, Ortuno MF, Bernal-Vicente A, Diaz-Vivancos P, Sanchez-Blanco MJ, Hernandez JA (2017) Plant responses to salt stress: adaptive mechanisms. Agronomy 7:18

Ahmad M, Zahir ZA, Asghar HN, Asghar M (2011) Inducing salt tolerance in mung bean through co-inoculation with rhizobia and plant-growth-promoting rhizobacteria containing 1-aminocyclopropane-1-carboxylate deaminase. Can J Microbiol 57:578–589

Ahmed A, Hasnain S (2014) Auxins as one of the factors of plant growth improvement by plant growth promoting rhizobacteria. Pol J Microbiol 63:261–266

Ali S, Kim WC (2018) Plant growth promotion under water: decrease of waterlogging-induced ACC and ethylene levels by ACC deaminase-producing bacteria. Front Microbiol 9:1096

Ali A, Yun DJ (2017) Salt stress tolerance; what do we learn from halophytes? J Plant Biol 60:431–439

Apse MP, Blumwald E (2002) Engineering salt tolerance in plants. Curr Opin Biotechnol 13:146–150

Apse MP, Aharon GS, Snedden WA, Blumwald E (1999) Salt tolerance conferred by overexpression of a vacuolar Na^+/H^+ antiport in *Arabidopsis*. Science 285:1256–1258

Assaha DVM, Ueda A, Saneoka H, Al-Yahyai R, Yaish MW (2017) The role of Na^+ and K^+ in salt stress adaptation in glycophytes. Front Physiol 8:509

Babalola OO (2010) Beneficial bacteria of agricultural importance. Biotechnol Lett 32:1559–1570

Backer R, Rokem JS, Ilangumaran G, Lamont J, Praslickova D, Ricci E, Subramanian S, Smith DL (2018) Plant growth promoting rhizobacteria: context, mechanisms of action, and roadmap to commercialization of biostimulants for sustainable agriculture. Front Plant Sci 9:1473

Badri DV, Weir TL, van der Lelie D, Vivanco JM (2009) Rhizosphere chemical dialogues: plant-microbe interactions. Curr Opin Biotechnol 20:642–650

Bailey-Serres J, Voesenek LA (2008) Flooding stress: acclimations and genetic diversity. Ann Rev Plant Biol 59:313–339

Baral A, Shruthi KS, Mathew MK (2015) Vesicular trafficking and salinity responses in plants. IUBMB Life 67:677–686

Barnawal D, Bharti N, Pandey SS, Pandey A, Chanotiya CS, Kalra A (2017) Plant growth promoting rhizobacteria enhance wheat salt and drought stress tolerance by altering endogenous phytohormone levels and TaCTR1/TaDREB2 expression. Physiol Plant 161:502–514

Bartels D, Dinakar C (2013) Balancing salinity stress responses in halophytes and non-halophytes: a comparison between *Thellungiella* and *Arabidopsis thaliana*. Funct Plant Biol 40:819–831

Bartels D, Sunkar R (2005) Drought and salt tolerance in plants. CRC Crit Rev Plant Sci 24:23–58

Beneduzi A, Ambrosini A, Passaglia LMP (2012) Plant growth promoting rhizobacteria (PGPR): their potential as antagonists and biocontrol agents. Genet Mol Biol 35:1044–1051

Bharti N, Yadav D, Barnawal D, Maji D, Kalra A (2013) *Exiguobacterium oxidotolerans*, a halotolerant plant growth promoting rhizobacteria, improves yield and content of secondary metabolites in *Bacopa monnieri* (L.) Pennell under primary and secondary salt stress. World J Microbiol Biotechnol 29:379–387

Bharti N, Pandey SS, Barnawal D, Patel VK, Kalra A (2016) Plant growth promoting rhizobacteria *Dietzia natronolimnaea* modulates the expression of stress responsive genes providing protection of wheat from salinity stress. Sci Rep 6:34768

Blumwald E (2000) Sodium transport and salt tolerance in plants. Curr Opin Cell Biol 12:431–434

Blumwald E, Poole RJ (1985) Na^+/H^+ antiport in isolated tonoplast vesicles from storage tissue of Beta vulgaris. Plant Physiol 78:163–167

Bohnert HJ, Sheveleva E (1998) Plant stress adaptations-making metabolism move. Curr Opin Plant Biol 1:267–274

Bose J, Munns R, Shabala S, Gilliham M, Pogson B, Tyerman SD (2017) Chloroplast function and ion regulation in plants growing in saline soils. Lessons from halophytes. J Exp Bot 68:3129–3143

Brini F, Masmoudi K (2012) Ion transporters and abiotic stress tolerance in plants. ISRN Mol Biol 2012:927436

Bruto M, Pringent-Combaret C, Muller D, Moenne-Loccoz Y (2014) Analysis of genes contributing to plant beneficial functions in plant growth promoting rhizobacteria and related proteobacteria. Sci Rep 4:6261

Cassán F, Bottini R, Schneider G, Piccoli P (2001) *Azospirillum brasilense* and *Azospirillum lipoferum* hydrolyze conjugates of GA_{20} and metabolize the resultant aglycones to GA_1 in seedlings of rice dwarf mutants. Plant Physiol 125:2053–2058

Caverzan A, Casassola A, Brammer SP (2016) Antioxidant responses of wheat plants under stress. Genet Mol Biol 39:1–6

Chen L, Liu Y, Wu G, Veronican Njeri K, Shen Q, Zhang N, Zhang R (2016) Induced maize salt tolerance by rhizosphere inoculation of *Bacillus amyloliquefaciens* SQR9. Physiol Plant 158:34–44

Chen M, Yang Z, Liu J, Zhu T, Wei X, Fan H, Wang B (2018) Adaptation mechanism of salt excluders under saline conditions and its applications. Int J Mol Sci 19:E3668

Chinnusamy V, Schumaker K, Zhu JK (2004) Molecular genetic perspectives on cross-talk and specificity in abiotic stress signaling in plants. J Exp Bot 55:225–236

Choudhury FK, Rivero RM, Blumwald E, Mittler R (2017) Reactive oxygen species, abiotic stress and stress combination. Plant J 90:856–867

Chowdhury SP, Hartmann A, Gao XW, Borriss R (2015) Biocontrol mechanism by root-associated *Bacillus amyloliquefaciens* FZB42. Front Microbiol 6:780

Dardanelli MS, Fernández de Córdoba FJ, Espuny MR, Rodríguez Carvajal MA, Soria Díaz ME, Gil Serrano AM, Okon Y, Megias M (2008) Effect of *Azospirillum brasilense* coinoculated with rhizobium on *Phaseolus vulgaris* flavonoids and Nod factor production under salt stress. Soil Biol Biochem 40:2713–2721

del Amor FM, Cuadra-Crespo P (2012) Plant growth-promoting bacteria as a tool to improve salinity tolerance in sweet pepper. Funct Plant Biol 39:82–90

Desalegn G, Turetschek R, Kaul HP, Wienkoop S (2016) Microbial symbionts affect *Pisum sativum* proteome and metabolome under *Didymella pinodes* infection. J Proteome 143:173–187

Desbrosses G, Contesto C, Varoquaux F, Galland M, Touraine B (2009) PGPR-Arabidopsis interactions is a useful system to study signaling pathways involved in plant developmental control. Plant Signal Behav 4:321–323

Dimpka C, Weinand T, Asch F (2009) Plant-rhizobacteria interactions alleviate abiotic stress conditions. Plant Cell Environ 32:1682–1694

Dodd IC, Perez-Alfocea F (2012) Microbial amelioration of crop salinity stress. J Exp Bot 63:3415–3428

Egamberdieva D (2009) Alleviation of salt stress by plant growth regulators and IAA producing bacteria in wheat. Acta Physiol Plant 31:861–864

El-Esawi MA, Alaraidh IA, Alsahli AA, Alamri SA, Ali HM, Alayafi AA (2018a) *Bacillus firmus* (SW5) augments salt tolerance in soybean (*Glycine max* L.) by modulating root system architecture, antioxidant defense systems and stress responsive genes expression. Plant Physiol Biochem 132:375–384

El-Esawi MA, Alaraidh IA, Alsahli AA, Alzahrani SM, Alayafi AA, Ahmad M (2018b) *Serratia liquefaciens* KM4 improves salt stress tolerance in maize by regulating redox potential, ion homeostasis, leaf gas exchange and stress related gene expression. Int J Mol Sci 19:E3310

Erice G, Ruiz-Lozano JM, Zamarreno AM, Garcia-Mina JM, Aroca R (2017) Transcriptomic analysis reveals the importance of JA-Ile turnover in the response of Arabidopsis plants to plant growth promoting rhizobacteria and salinity. Environ Exp Bot 143:10–19

Etesami H, Maheshwari DK (2018) Use of plant growth promoting rhizobacteria (PGPRs) with multiple plant growth promoting traits in stress agriculture: action mechanisms and future prospects. Ecotoxicol Environ Safety 156:225–246

Fahad S, Hussain S, Matloob A, Khan FA, Khaliq A, Saud S, Hassan S, Shan D, Khan F, Ullah N, Faiq M, Khan MR, Tareen AK, Khan A, Ullah A, Ullah N, Huang J (2015) Phytohormones and plant responses to salinity stress: a review. Plant Growth Regul 75:391–404

Flowers TJ (1985) Physiology of halophytes. Plant Soil 89:41–56

Flowers TJ, Colmer TD (2008) Salinity tolerance in halophytes. New Phytol 179:945–963

Flowers TJ, Colmer TD (2015) Plant salt tolerance: adaptations in halophytes. Ann Bot 115:327–331

Flowers TJ, Troke PF, Yeo AR (1977) The mechanism of salt tolerance in halophytes. Annu Rev Plant Physiol 28:89–121

Flowers TJ, Munns R, Colmer TD (2015) Sodium chloride toxicity and the cellular basis of salt tolerance in halophytes. Ann Bot 115:419–431

Giri J (2011) Glycinebetaine and abiotic stress tolerance in plants. Plant Signal Behav 6:1746–1751

Gouda S, Kery RG, Das G, Paramithiotis S, Shin HS, Patra JK (2018) Revitalization of plant growth promoting rhizobacteria for sustainable development in agriculture. Microbiol Res 206:131–140

Gray EJ, Smith DL (2005) Intracellular and extracellular PGPR: commonalities and distinctions in the plant-bacterium signaling processes. Soil Biol Biochem 37:395–412

Gupta A, Gopal M, Thomas GV, Manikandan V, Gajewski J, Thomas G, Seshagiri S, Schuster SC, Rajesh P, Gupta R (2014) Whole genome sequencing and analysis of plant growth promoting bacteria isolated from the rhizosphere of plantation crops coconut, cocoa and arecanut. PLoS One 9:e104259

Habib SH, Kausar H, Saud HM (2016) Plant growth-promoting rhizobacteria enhance salinity stress tolerance in okra through ROS-scavenging enzymes. Biomed Res Int 2016:6284547

Hahm MS, Son JS, Hwang YJ, Kwon DK, Ghim SY (2017) Alleviation of salt stress in pepper (*Capsicum annum* L.) plants by plant growth promoting rhizobacteria. J Microbiol Biotechnol 27:1790–1797

Hanin M, Ebel C, Ngom M, Laplaze L, Masmoudi K (2016) New insights on plant salt tolerance mechanisms and their potential use for breeding. Front Plant Sci 7:1787

Ilangumaran G, Smith DL (2017) Plant growth promoting rhizobacteria in amelioration of salinity stress: a systems biology perspective. Front Plant Sci 8:1768

Iqbal N, Umar S, Khan NA, Khan MIR (2014) A new perspective of phytohormones in salinity tolerance: regulation of proline metabolism. Environ Exp Bot 100:34–42

Islam S, Akanda AM, Prova A, Islam MT, Hossain MM (2016) Isolation and identification of plant growth promoting rhizobacteria from cucumber rhizosphere and their effect on plant growth promoting and disease suppression. Front Microbiol 6:1360

Ismail A, Takeda S, Nick P (2014) Life and death under salt stress: same players, different timing. J Exp Bot 65:2963–2979

Jamil A, Riaz S, Ashraf M, Foolad MR (2011) Gene expression profiling of plants under salt stress. CRC Crit Rev Plant Sci 30:435–458

Janssen J, Weyens N, Croes S, Beckers B, Meiresonne L, Van Peteghem P, Carleer R, Vangronsveld J (2015) Phytoremediation of metal contaminated soil using willow: exploiting plant-associated bacteria to improve biomass production and metal uptake. Int J Phytoremediation 17:1123–1136

Ji H, Pardo JM, Batelli G, Van Oosten MJ, Bressan RA, Li X (2013) The salt overly sensitive (SOS) pathway: established and emerging roles. Mol Plant 6:275–286

Jin CW, Ye YQ, Zheng SJ (2014) An underground tale: contribution of microbial activity to plant iron acquisition via ecological processes. Ann Bot 113:7–18

Jithesh MN, Prashanth SR, Sivaprakash KR, Parida AK (2006) Antioxidative response mechanisms in halophytes: their role in stress defence. J Genet 85:237–254

Joo GJ, Kim YM, Lee IJ, Song KS, Rhee IK (2004) Growth promotion of red pepper plug seedlings and the production of gibberellins by *Bacillus cereus*, *Bacillus macroides,* and *Bacillus pumilus*. Biotechnol Lett 26:487–491

Joo GJ, Kim YM, Kim JT, Rhee IK, Kim JH, Lee IJ (2005) Gibberellins-producing rhizobacteria increase endogenous gibberellins content and promote growth of red peppers. J Microbiol 43:510–515

Kang SM, Radhakrishnan R, Khan AL, Kim MJ, Park JM, Kim BR, Shin DH, Lee IJ (2014) Gibberellin secreting rhizobacterium, *Pseudomonas putida* H-2-3 modulates the hormonal and stress physiology of soybean to improve the plant growth under saline and drought conditions. Plant Physiol Biochem 84:115–124

Kapoor D, Singh S, Kumar V, Romero R, Prasad R, Singh J (2019) Antioxidant enzymes regulation in plants in reference to Reactive Oxygen Species (ROS) and Reactive Nitrogen Species (RNS). Plant Gene 19:100182. https://doi.org/10.1016/j.plgene.2019.100182

Kasahara H (2016) Current aspects of auxin biosynthesis in plants. Biosci Biotechnol Biochem 80:34–42

Kim K, Jang YJ, Lee SM, Oh BT, Chae JC, Lee KJ (2014) Alleviation of salt stress by *Enterobacter* sp. *EJ*01 in tomato and *Arabidopsis* is accompanied by up-regulation of conserved salinity responsive factors in plants. Mol Cells 37:109–117

Kohler J, Caravaca F, Carrasco L, Roldan A (2006) Contribution of *Pseudomonas mendocina* and *Glomus intraradices* to aggregate stabilization and promotion of biological fertility in rhizosphere soil of lettuce plants under field conditions. Soil Use Manag 22:298–304

Kuan KB, Othman R, Abdul Rahim K, Shamsuddin ZH (2016) Plant growth-promoting rhizobacteria inoculation to enhance vegetative growth, nitrogen fixation and nitrogen remobilisation of maize under greenhouse conditions. PLoS One 11:e0152478

Kwon YS, Lee DY, Rakwal R, Baek SB, Lee JH, Kwak YS, Seo JS, Chung WS, Bae DW, Kim SG (2016) Proteomic analyses of the interaction between the plant growth promoting rhizobacterium *Paenibacillus polymyxa* E681 and *Arabidopsis thaliana*. Proteomics 16:122–135

Lata R, Chowdhury S, Gond SK, White JF Jr (2018) Induction of abiotic stress tolerance in plants by endophytic microbes. Lett Appl Microbiol 66:268–276

Ledger T, Rojas S, Timmermann T, Pinedo I, Poupin MJ, Garrido T, Richter P, Tomayo J, Donoso R (2016) Volatile-mediated effects predominate in *Paraburkholderia phytofirmans* growth promotion and salt stress tolerance of *Arabidopsis thaliana*. Front Microbiol 7:1838

Li M, Li Y, Li H, Wu G, Nasholm T (2011) Overexpression of *AtNHX5* improves tolerance to both salt and drought stress in *Broussonetia papyrifera* (L.) vent. Tree Physiol 31:349–357

Liang W, Ma X, Wan P, Liu L (2018) Plant salt tolerance mechanism: a review. Biochem Biophys Res Comm 495:286–291

Liu S, Hao H, Lu X, Zhao X, Wang Y, Zhang Y, Xie Z, Wang R (2017) Transcriptome profiling of genes involved in induced systemic salt tolerance conferred by *Bacillus amyloliquefaciens* FZB42 in *Arabidopsis thaliana*. Sci Rep 7:10795

López-Bucio J, Campos-Cuevas JC, Hernández-Calderón E, Velásquez-Becerra C, Farías-Rodríguez R, Macías-Rodríguez LI, Valencia-Cantero E (2007) *Bacillus megaterium* rhizobacteria promote growth and alter root-system architecture through an auxin-and ethylene-independent signaling mechanism in *Arabidopsis thaliana*. Mol Plant Microbe Interact 20:207–217

Lugtenberg B, Kamilova F (2009) Plant growth promoting rhizobacteria. Ann Rev Microbiol 63:541–556

Marulanda A, Azcon R, Chaumont F, Ruiz-Lozano JM, Aroca R (2010) Regulation of plasma membrane aquaporins by inoculation with a *Bacillus megaterium* strain in maize (*Zea mays* L.) plants under unstressed and salt-stressed conditions. Planta 232:533–543

Mayak S, Tirosh T, Glick BR (2004) Plant growth-promoting bacteria confer resistance in tomato plants to salt stress. Plant Physiol Biochem 42:565–572

Meena KK, Sorty AM, Bitla UM, Choudhary K, Gupta P, Pareek A, Singh DP, Prabha R, Sahu PK, Gupta VK, Singh HB, Krishnani KK, Minhas PS (2017) Abiotic stress responses and microbe-mediated mitigation in plants: the omics strategies. Front Plant Sci 8:172

Mishra A, Tanna B (2017) Halophytes: potential resources for salt stress tolerance genes and promoters. Front Plant Sci 8:829

Mudgal V, Madaan N, Mudgal A (2010) Biochemical mechanisms of salt tolerance in plants. Int J Bot 6:136–143

Munns R (2002) Comparative physiology of salt and water stress. Plant Cell Environ 25:239–250

Munns R, Tester M (2008) Mechanisms of salinity tolerance. Ann Rev Plant Biol 59:651–681

Munns R, Husain S, Rivelli AR, James RA, Condon AGT, Lindsay MP, Lagudah ES, Schachtman DP, Hare RA (2002) Avenues for increasing salt tolerance of crops, and the role of physiologically based selection traits. Plant Soil 247:93–105

Nadeem SM, Zahir ZA, Naveed M, Arshad M (2009) Rhizobacteria containing ACC-deaminase confer salt tolerance in maize grown on salt-affected fields. Can J Microbiol 55:1302–1309

Nautiyal CS, Srivastava S, Chauhan PS, Seem K, Mishra A, Sopory SK (2013) Plant growth-promoting bacteria *Bacillus amyloliquefaciens* NBRISN13 modulates gene expression profile of leaf and rhizosphere community in rice during salt stress. Plant Physiol Biochem 66:1–9

Negrão S, Schmöckel SM, Tester M (2017) Evaluating physiological responses to salinity stress. Ann Bot 119:1–11

Nikalje GC, Nikam TD, Suprasanna P (2017) Looking at halophytic adaptation to high salinity through genomics landscape. Curr Genomics 18:542–552

Niu SQ, Li HR, Pare PW, Aziz M, Wang SM, Shi HZ, Li J, Han QQ, Guo SQ, Li J, Guo Q, Ma Q, Zhang JJ (2016) Induced growth promotion and higher salt tolerance in the halophyte grass *Puccinellia tenuiflora* by beneficial rhizobacteria. Plant Soil 407:217–230

Nongpiur RC, Singla-Pareek SL, Pareek A (2016) Genomics approaches for improving salinity stress tolerance in crop plants. Curr Genomics 17:343–357

Numan M, Bashir S, Khan Y, Mumtaz R, Shinwari ZK, Khan AL, Khan A, AL-Harrasi A (2018) Plant growth promoting bacteria as an alternative strategy for salt tolerance in plants: a review. Microbiol Res 209:21–32

Ortiz-Castro R, Valencia-Cantero E, Lopez-Bucio J (2008) Plant growth promotion by *Bacillus megaterium* involves cytokinin signaling. Plant Signal Behav 3:263–265

Panwar M, Tewari R, Gulati A, Nayyar H (2016a) Indigenous salt-tolerant rhizobacterium *Pantoea dispersa* (PSB3) reduces sodium uptake and mitigates the effects of salt stress on growth and yield of chickpea. Acta Physiol Plant 38:278

Panwar M, Tewari R, Nayyar H (2016b) Native halo-tolerant plant growth promoting rhizobacteria *Enterococcus* and *Pantoea* sp. improve seed yield of mungbean (*Vigna radiata* L.) under soil salinity by reducing sodium uptake and stress injury. Physiol Mol Biol Plants 22:445–459

Park HJ, Kim WY, Yun DJ (2016) A new insight of salt stress signaling in plant. Mol Cells 39:447–459

Paul MJ, Foyer CH (2001) Sink regulation of photosynthesis. J Exp Bot 52:1383–1400

Perrig D, Boiero M, Masciarelli O, Penna C, Ruiz O, Cassán F, Luna M (2007) Plant growth promoting compounds produced by two agronomically important strains of *Azospirillum brasilense*, and implications for inoculant formulation. Appl Microbiol Biotechnol 75:1143–1150

Piccoli P, Masciarelli O, Bottini R (1999) Gibberellin production by *Azospirillum lipoferum* cultured in chemically defined medium as affected by oxygen availability and water status. Symbiosis 27:135–146

Pinedo I, Ledger T, Greve M, Poupin MJ (2015) *Burkholderia phytofirmans* PsJN induces long term metabolic and transcriptional changes involved in *Arabidopsis thaliana* salt tolerance. Front Plant Sci 23:466

Pitman MG, Lauchli A (2002) Global impact of salinity and agricultural ecosystems. In: Lauchli A, Luttage U (eds) Salinity: environment – plants – molecules. Kluwer Academic, Amsterdam, pp 3–20

Plett DG, Moller IS (2010) Na^+ transport in glycophytic plants: what we know and would like to know. Plant Cell Environ 33:612–626

Prasad R, Kumar M, Varma A (2015) Role of PGPR in soil fertility and plant health. In: Egamberdieva D, Shrivastava S, Varma A (eds) Plant Growth-Promoting Rhizobacteria (PGPR) and medicinal plants. Springer, Cham, pp 247–260

Preston C, Critchley C (1986) Differential effects of K^+ and Na^+ on oxygen evolution activity of photosynthetic membranes from two halophytes and spinach. Funct Plant Biol 13:491–498

Quiza L, St-Arnaud M, Yergeau E (2015) Harnessing phytomicrobiome signaling for rhizosphere microbiome engineering. Front Plant Sci 6:507

Qurashi AW, Sabri AN (2012) Bacterial exopolysaccharide and biofilm formation stimulate chickpea growth and soil aggregation under salt stress. Braz J Microbiol 43:1183–1191

Rengasamy P (2002) Transient salinity and subsoil constraints to dryland farming in Australian sodic soils: an overview. Aust J Exp Agric 42:351–361

Rhodes D, Nadolska-Orczyk A, Rich P (2002) Salinity, osmolytes and compatible solutes. In: Läuchli A, Lüttge U (eds) Salinity: environment – plants – molecules. Springer, Dordrecht, pp 181–204

Rojas-Tapias D, Moreno-Galván A, Pardo-Díaz S, Obando M, Rivera D, Bonilla R (2012) Effect of inoculation with plant growth-promoting bacteria (PGPB) on amelioration of saline stress in maize (*Zea mays*). Appl Soil Ecol 61:264–272

Roy SJ, Negrao S, Tester M (2014) Salt resistant crop plants. Curr Opin Biotechnol 26:115–124

Ryu CM, Hu CH, Locy RD, Kloepper JW (2005) Study of mechanisms for plant growth promotion elicited by rhizobacteria in *Arabidopsis thaliana*. Plant Soil 268:285–292

Sarkar A, Pramanik K, Mitra S, Soren T, Maiti TK (2018) Enhancement of growth and salt tolerance of rice seedlings by ACC deaminase-producing *Burkholderia* sp. MTCC 12259. J Plant Physiol 231:434–442

Shahid M, Akram MS, Khan MA, Zubair M, Shah SM, Ismail M, Shabir G, Basheer S, Aslam K, Tariq M (2018) A phytobeneficial strain *Planomicrobium* sp. MSSA-10 triggered oxidative stress responsive mechanisms and regulated the growth of pea plants under induced saline environment. J Appl Microbiol 124:1566–1579

Sharma SB, Sayyed RZ, Trivedi MH, Gobi TA (2013) Phosphate solubilizing microbes: sustainable approach for managing phosphorus deficiency in agricultural soils. Springerplus 2:587

Shavrukov Y (2013) Salt stress or salt shock: which genes are we studying? J Exp Bot 64:119–127

Shen M, Kang YJ, Wang HL, Zhang XS, Zhao QX (2012) Effect of plant growth-promoting rhizobacteria (PGPRs) on plant growth, yield, and quality of tomato (*Lycopersicon esculentum* Mill.) under simulated seawater irrigation. J Gen Appl Microbiol 58:253–262

Shi H, Ishitani M, Kim C, Zhu JK (2000) The *Arabidopsis thaliana* salt tolerance gene SOS1 encodes a putative Na^+/H^+ antiporter. Proc Natl Acad Sci USA 97:6896–6901

Shi H, Quintero FJ, Pardo JM, Zhu JK (2002) The putative plasma membrane Na^+/H^+ antiporter SOS1 controls long distance Na^+ transport in plants. Plant Cell 14:465–477

Shrivastava P, Kumar R (2015) Soil salinity: a serious environmental issue and plant growth promoting bacteria as one of the tools for its alleviation. Saudi J Biol Sci 22:123–131

Silva P, Gerós H (2009) Regulation by salt of vacuolar H^+-ATPase and H^+- pyrophosphatase activities and Na^+/H^+ exchange. Plant Signal Behav 4:718–726

Singh RP, Jha PN (2016) Alleviation of salinity-induced damage on wheat plant by an ACC deaminase-producing halophilic bacterium *Serratia* sp SL-12 isolated from a salt lake. Symbiosis 69:101–111

Singh RP, Shelke GM, Kumar A, Jha PN (2015) Biochemistry and genetics of ACC deaminase: a weapon to "stress ethylene" produced in plants. Front Microbiol 6:937

Singh D, Raina TK, Kumar A, Singh J, Prasad R (2019) Plant microbiome: a reservoir of novel genes and metabolites. Plant Gene 18:100177. https://doi.org/10.1016/j.plgene.2019.100177

Smith DL, Praslickova D, Ilangumaran G (2015) Inter-organismal signaling and management of the phytomicrobiome. Front Plant Sci 6:722

Sofo A, Scopa A, Nuzzaci M, Vitti A (2015) Ascorbate peroxidase and catalase activities and their genetic regulation in plants subjected to drought and salinity stresses. Int J Mol Sci 16:13561–13578

Spaepen S, Vanderleyden J, Remans R (2007) Indole-3-acetic acid in microbial and microorganism-plant signaling. FEMS Microbiol Rev 31:425–448

Srivastava S, Chaudhry V, Mishra A, Chauhan PS, Rehman A, Yadav A, Tuteja N, Nautiyal CS (2012) Gene expression profiling through microarray analysis in *Arabidopsis thaliana* colonized by *Pseudomonas putida* MTCC5279, a plant growth promoting rhizobacterium. Plant Signal Behav 7:235–245

Sureshbabu K, Amaresan N, Kumar K (2016) Amazing multiple function properties of plant growth promoting rhizobacteria in the rhizosphere soil. Int J Curr Microbiol Appl Sci 5:661–683

Suwa R, Nguyen NT, Saneoka H, Moghaieb R, Fujita K (2006) Effect of salinity stress on photosynthesis and vegetative sink in tobacco plants. Soil Sci Plant Nutrition 52:243–250

Sziderics AH, Rasche F, Trognitz F, Sessitsch A, Wilhelm E (2007) Bacterial endophytes contribute to abiotic stress adaptation in pepper plants (*Capsicum annuum* L.). Can J Microbiol 53:1195–1202

Tang X, Mu X, Shao H, Wang H, Brestic M (2015) Global plant-responding mechanisms to salt stress: physiological and molecular levels and implications in biotechnology. CRC Crit Rev Biotechnol 25:425–437

Tavakkoli E, Rengasamy P, McDonald G (2010) High concentrations of Na^+ and Cl^- ions in soil solution have simultaneous detrimental effects on growth of faba bean under salinity stress. J Exp Bot 61:4449–4459

Teakle NL, Tyerman SD (2010) Mechanisms of Cl^- transport contributing to salt tolerance. Plant Cell Environ 33:566–589

Tester M, Davenport R (2003) Na^+ tolerance and Na^+ transport in higher plants. Ann Bot 91:503–5027

Thakuria D, Talukdar NC, Goswami C, Hazarika S, Boro RC, Khan MR (2004) Characterization and screening of bacteria from the rhizosphere of rice grown in acidic soils of Assam. Curr Sci 86:978–985

Tucci M, Ruocco M, De Masi L, De Palma M, Lorito M (2011) The beneficial effect of *Trichoderma* spp. on tomato is modulated by the plant genotype. Mol Plant Pathol 12:341–354

Udawat P, Jha RK, Mishra A, Jha B (2017) Overexpression of a plasma membrane localized SbSRP-like protein enhances salinity and osmotic stress tolerance in transgenic tobacco. Front Plant Sci 8:582

Ullah S, Bano A (2015) Isolation of plant growth-promoting rhizobacteria from rhizospheric soil of halophytes and their impact on maize (*Zea mays* L.) under induced soil salinity. Can J Microbiol 61:307–313

Upadhyay SK, Singh DP (2015) Effect of salt-tolerant plant growth-promoting rhizobacteria on wheat plants and soil health in a saline environment. Plant Biol 17:288–293

Vacheron J, Desbrosses G, Bouffaud ML, Touraine B, Moënne-Loccoz Y, Muller D, Legendre L, Wisniewski-Dyé F, Prigent-Combaret C (2013) Plant growth-promoting rhizobacteria and root system functioning. Front Plant Sci 4:356

Vaishnav A, Kumari S, Jain S, Varma A, Choudhary DK (2015) Putative bacterial volatile-mediated growth in soybean (*Glycine max* L. *Merrill*) and expression of induced proteins under salt stress. J Appl Microbiol 119:539–551

van de Mortel JE, de Vos RCH, Dekkers E, Pineda A, Guillod L, Bouwmeester K, van Loon JJA, Dicke M, Raaijmakers JM (2012) Metabolic and transcriptomic changes induced in *Arabidopsis* by the rhizobacterium *Pseudomonas fluorescens* SS101. Plant Physiol 160:2173–2188

Verma V, Ravindran P, Kumar PP (2016a) Plant hormone mediated regulation of stress responses. BMC Plant Biol 16:86

Verma P, Yadav AN, Khannam KS, Kumar S, Saxena AK, Suman A (2016b) Molecular diversity and multifarious plant growth promoting attributes of bacilli associated with wheat (*Triticum aestivum* L.) rhizosphere from six diverse agro-ecological zones of India. J Basic Microbiol 56:44–58

Wang QY, Dodd IC, Belimov AA, Jiang F (2016) Rhizosphere bacteria containing 1-aminocyclopropane-1-carboxylate deaminase increase growth and photosynthesis of pea plants under salt stress by limiting Na^+ accumulation. Funct Plant Biol 43:161–172

Weyens N, Beckers B, Schellingen K, Ceulemans R, Van der Lelie D, Newman L, Taghavi S, Carleer R, Vangronsveld J (2015) The potential of the Ni-resistant TCE degrading *Pseudomonas putida* W619-TCE to reduce phytotoxicity and improve phytoremediation efficiency of poplar cuttings on A Ni-TCE co-contamination. Int J Phytoremediation 17:40–48

Yan JM, Smith MD, Glick BR, Liang Y (2014) Effects of ACC deaminase containing rhizobacteria on plant growth and expression of Toc GTPases in tomato (*Solanum lycopersicum*) under salt stress. Botany 92:775–781

Yang Y, Guo Y (2018a) Elucidating the molecular mechanisms mediating plant salt-stress responses. New Phytol 217:523–539

Yang Y, Guo Y (2018b) Unraveling salt stress signaling in plants. J Integr Plant Biol 60:796–804

Yang Q, Chen ZZ, Zhou XF, Yin HB, Li X, Xin XF, Hong XH, Zhu JK, Gong Z (2009) Overexpression of SOS (salt overly sensitive) genes increases salt tolerance in transgenic Arabidopsis. Mol Plant 2:22–31

Yang AZ, Akhtar SS, Iqbal S, Amjad M, Naveed M, Zahir ZA, Jacobsen SE (2016) Enhancing salt tolerance in quinoa by halotolerant bacterial inoculation. Funct Plant Biol 43:632–642

Yao LX, Wu ZS, Zheng YY, Kaleem I, Li C (2010) Growth promotion and protection against salt stress by *Pseudomonas putida* Rs-198 on cotton. Eur J Soil Biol 46:49–54

Zerrouk IZ, Benchabane M, Khelifi L, Yokawa K, Ludwig-Muller J, Baluska F (2016) A *Pseudomonas* strain isolated from date-palm rhizospheres improves root growth and promotes root formation in maize exposed to salt and aluminum stress. J Plant Physiol 191:111–119

Zhang H, Kim MS, Sun Y, Dowd SE, Shi H, Paré PW (2008) Soil bacteria confer plant salt tolerance by tissue specific regulation of the sodium transporter HKT1. Mol Plant Microbe Interact 21:737–744

Zhou C, Ma Z, Zhu L, Xiao X, Xie Y, Zhu J, Wang J (2016) Rhizobacterial strain *Bacillus megaterium* BOFC15 induces cellular polyamine changes that improve plant growth and drought resistance. Int J Mol Sci 17:976

Chapter 17
Plant Microbial Ecology as a Potential Option for Stress Management in Plants

Deepkamal Jha, Shweta Kulshreshtha, and Sunita Chauhan

Abstract Plants provide different products required for sustaining life. There are many stress factors that affect plant growth and development adversely such as drought, salinity, heavy metal toxicity, osmotic potential, water and cold stress. These stresses produce their own effects individually or in combination. For example, water stress leads to drought conditions and increases salt concentration in the soil that eventually results in reduced growth, low productivity and, ultimately, plant death. Therefore, it is necessary to induce tolerance in the plant to different stress factors. The role of microbes in improving plant growth under stress conditions is well established in the literature. Here, in this chapter, the role of microbes and their mechanisms in inducing tolerance are mentioned. Detailed insight into the mechanisms involved in inducing tolerance to stress conditions will help us in making a suitable strategy for developing stress-resistant plant varieties that can cope up with stress conditions.

17.1 Introduction

Plants are *metaorganisms* that have a well-defined microbiome and close symbiotic relationships with different microorganisms (Mendes et al. 2013). Soil consists a pool of microscopic life forms including fungi, protozoa, algae and actinomycetes, and, of these, bacteria are the most common (Ho et al. 2017). Plants interact with a large number of these microbes in manners that have essential consequences for their optimal growth and wellness (Powell and Klironomos 2007). Microbes living in close association with plant roots compete for water, nutrients and space (Hartmann et al. 2009). Microbial activities help plants in nutrient uptake and endow protection against pathogen attacks (Berendsen et al. 2012). Plant-microbe interactions assist numerous transformations in the rhizosphere, for example, nutrient cycling, especially, carbon and nitrogen sequestration, which influence different aspects of

---

D. Jha · S. Kulshreshtha (✉)
Amity Institute of Biotechnology, Amity University Rajasthan, Jaipur, India

S. Chauhan
Kumarappa National Handmade Paper Institute, Jaipur, India

A. Varma et al. (eds.), *Plant Microbe Symbiosis*,
https://doi.org/10.1007/978-3-030-36248-5_17

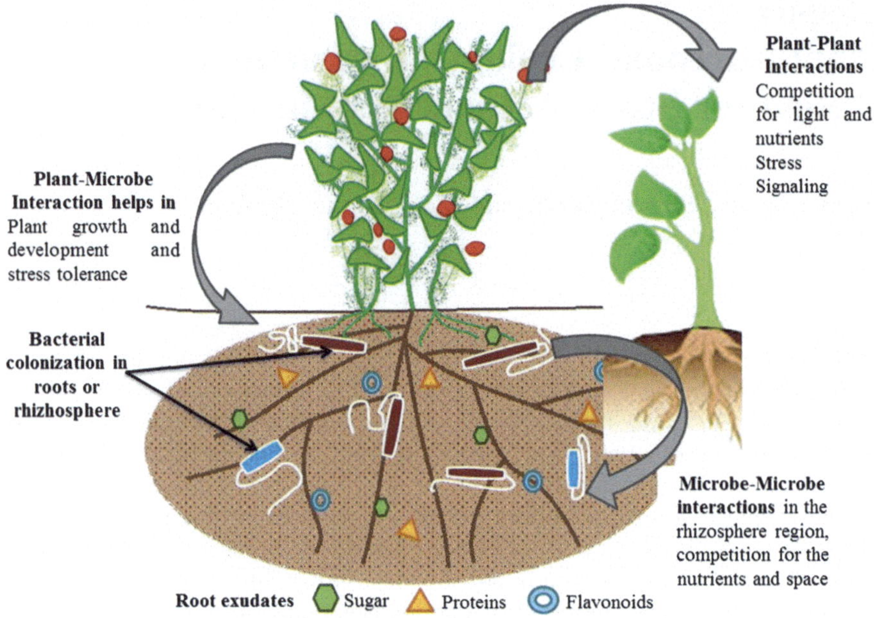

Fig. 17.1 Rhizosphere hub of plant-microbe interactions (modified from Singh and Chauhan (2017))

ecosystem functioning (Velmourougane et al. 2017). In the rhizosphere hub, interactions are not limited only between soil and microbes but also occur between microbe-microbe, plant-microbe and plant-plant (as shown in Fig. 17.1). These interactions may be favorable, unfavorable or neutral (Singh and Chauhan 2017). Plant community dynamics are steered by the microbial regulation of soil resource partitioning and inhibition of other host symbionts. Thus, the phenotype and ecology of the plant can be detected by the influence of symbiotic microbes on the environment and the competition for soil resources. The roles of both plant-related microbes and the host in the functioning of the ecosystem have been described; however, comprehensive mechanisms are not known clearly. Plants are immobile and they have co-evolved with microorganisms and procured various mechanisms that modulate the end result of their interactions (Oldroyd 2013). Active roots continuously synthesize, secrete and accumulate a broad spectrum of compounds into the soil such as H^+ ions, enzymes, mucilage and carbon-carrying primary and secondary metabolites, which are collectively called as *root exudates*. These root exudates provide enough essential factors to support the growth and maintenance of soil microbiota (Campbell and Greaves 1990; Faure et al. 2009). Microorganisms in the rhizosphere were 100 times higher compared to that of soil. Consequently, uptake efficiency or root surface area of the host plant increases to uptake more nutrients and water (Powell and Klironomos 2007). Plant growth-promoting rhizobacteria (PGPR) live in close association with root in the rhizosphere region (Prasad et al. 2015). They stimulate plant growth by the generous mutualism mentioned above (Peñuelas et al. 2014; Sirari et al. 2016). Besides, they can produce a multiplex blend of volatile substances that are unique among bacterial species and other closely related species (Garbeva et al. 2014). A few of these volatiles can suppress disease by stimulating induced systemic resistance (ISR) (Yi et al. 2013) or antagonizing phytopathogens, insects, nematodes and other inhibiting microbes.

Crop plants, in general, are constantly subjected to several abiotic and biotic stress factors from the time of planting up to harvesting, transport, storage and their consumption. Agriculture already being one of the profoundly unprotected sectors to climate change gets severely influenced by abiotic and biotic stresses; subsequently, the effects of these stresses have been an important reason for the development of static crop production (Grover et al. 2011). These stresses not only exert detrimental effects on crop health but also they affect worldwide crop production, which results in millions of dollars of losses (Suzuki et al. 2014). These factors influence crops qualitatively and quantitatively and also exert critical pressure on land and water resources. There is the possibility that changing climate and unevenness in environmental conditions, stresses like drought, floods, rains, frost damages and heat waves may rapidly increase in the future. The most recent report by the UN stated that there are 7.3 billion people—and that number may hit 9.7 billion by 2050 (UN DESA 2015). This growth, alongside rising per capita incomes in developing countries, is always driving up global food demand. In order to meet this increasing future food demand, a scientifically mediated agricultural revolution promising enhanced yield and better nutritive qualities is the need of the hour.

Abiotic and biotic stresses are considered as major obstacles to achieve this task. Consequently, to manage these stresses, simple, low-cost and less time-consuming biological modification plans are required. Microbes live in close association with plants in the rhizosphere and provide several benefits to the plant along with the power of facing different abiotic and biotic stresses like endurance to extreme conditions, genetic diversity, ubiquity and, thus,, can play a pivotal role in these aspects.

Beneficial and positive interactions among plants, microbes and nutrients induce favorable soil environment, shaping the premise of good agricultural practices, and increase crop production. A healthy soil is prerequisite for plant productivity. Plant-microbe interactions are crucial for the plant in terms of promotion of growth and maintenance, bio-control and stress management (Prasad et al. 2018). Besides plant stress management, they can be used as critical models for stress tolerance, adaptation and response mechanisms. These can be transferred to plants in order to resist climate change caused by plant stresses (Grover et al. 2011). Now, questions are raised on plant and microbial interactions and the mechanism of inducing tolerance. How does the plant microbiome affect plant growth, yield and host survival in various associations? How do plants tolerate adverse environmental and/or biological conditions? The solution to these questions will help us in finding out the solution to plant stresses and their economic impacts. Therefore, the aim of this chapter is to focus on plant stresses and the tolerance to stress provided by microbes. This will further help in developing future sustainable crop production and protection strategies.

17.2 What is Stress?

The original concept of stress for living organisms was developed by Selye (1936) as the agent which produces both stress and stress-specific action and can act as a stressor. The responses may be stressor specific or non-specific general ones. Levitt (1980) defined stress as "any environmental factor potentially unfavourable to living organisms." Later on, Larcher (1987) described plant stress as a "state in which increasing demands made upon a plant lead to an initial destabilization of functions, followed by normalization and improved resistance." It was also explained that the result may be permanent damage or even death if the limits of tolerance are exceeded and the adaptive capacity is overworked.

Lichtenthaler (1996) extended the stress concept of plants by including both the positive and negative elements in the form of eu-stress and dis-stress. Eu-stress was regarded as an activating, stimulating stress and a positive element for plant development whereas dis-stress was termed as a severe stress that negatively affects the plant and causes damage. A mild stress may activate cell metabolism besides increasing the physiological activity of a plant. Such mild stimulating stresses are favorable for a plant and never cause any damaging effect even at a long duration. Thus, stress is actually a dose-dependent matter, for instance, any herbicide, a stressor, can stimulate plant metabolism and plant growth at fairly low concentrations. Similarly, low doses of a xenobiotic can have the opposite effect of higher doses. However, at a concentration 100 or 1000 times higher, the same xenobiotics will prove to be damaging to the plant by inducing early senescence and finally leading to death if the stressor is not removed. Such damaging stressor concentrations and all high doses of stress constraints are negative for the physiology and development of plants, and thus represent a true stress in the sense of a dis-stress. Overall, true stress is reflected when the certain threshold of a stressor is exceeded beyond a limit that can no longer be compensated by the plant. Thus, the applicability of the "stressor dose-stress effect relationship" seems to be obvious but more research is required in this area.

17.3 Stress in Plants

Plants are the principal source of food for a large part of the world population as well as for the animals that are utilized for meat and milk production. Climate and environmental components are the main determinants of the geographical distribution of plant species and also the types of crops to be planted and their yields (Mosa et al. 2017). Environmental components such as abiotic and biotic stressors play a crucial role in the growth and productivity of plants. Any unfavorable alterations in these components can result in halted plant development, decline in crop yields and loss of function. Thus, plants need to endure these unfavorable alterations so as to minimize their impact. Predominantly, acquaintance of the plant with unfavorable conditions, either biological or environmental, is known as "plant stress." Larcher (1987) reported plant stress as "a state in which increasing demands made upon a plant prompt the desensitization of functions, succeeded by normalization and enhanced resistance." In the event when the thresholds of tolerance are over reached and the adaptive capacity is surpassed, permanent impairment or even death may ensue. Consequently, stress can be regarded as a temporary state and the extent of the damage which is relative to the strength and duration of the adverse conditions. There are a few other different meanings of stress that were formulated later by plant researchers. For example, stress was characterized as "any unfavourable condition or substance that affects or obstructs a plant's metabolism and development." On the other hand, it was additionally described as "a condition caused by components that have a tendency to revise equilibrium." Despite the differences among the several definitions of plant stress, they are altogether focused on describing the alterations in the conditions that influence the plant, the plant response to change and the level of damage that is introduced by the changed condition.

17.4 Types of Plant Stresses

Plant stresses can be characterized into following types in compliance with various factors as stated below:

1. Based on the cause of stress, it can be characterized into "abiotic stresses" that are caused by non-living components such as changes in salinity and temperatures, drought; and "biotic stresses" that are caused by living organisms such as microbes, pathogens and other plants.
2. Based on the effect of the stress, it can be characterized into eu-stress and dis-stress. Eu-stress shows positive effects while dis-stress shows negative effects of stress on plants. The balance between sensitivity and tolerance decides the effect of the stress.
3. Based on the duration of the stress, it can be characterized into "short-term stresses" where the plant can overcome the stress and "long-term stresses" that result in considerable and irreversible damages (Kranner et al. 2010).
4. Based on the number of interacting factors, stresses can be grouped into three categories, i.e., single, multiple individual and combined stresses. A single stress represents only one stress factor affecting plant growth and development. Multiple stresses represent the impact of two or more stresses occurring at different time periods without any overlap (multiple individual) or occurring concurrently with at least some degree of overlap between them (combined). The co-occurrence of drought and heat stresses during summer is an example of a combined abiotic stress whereas a bacterial and fungal pathogen attacking a plant at the same time represents a case of combined biotic stress. When bacteria and fungi attack the plant at different times then it is considered as multiple individual biological stress.

Table 17.1 Types of stress/stress factors

Natural stress factors		
Abiotic factors	Biotic factors	Anthropogenic factors
• High irradiance (photoinhibition, photooxidation) • Heat (increased temperature) • Low temperature (chilling) • Sudden and late frost • Water shortage (desiccation problems) • Natural mineral deficiency (e.g. nitrogen shortage) • Long rainy periods	• Insects • Viral • Fungal • Bacterial • Pathogens	• Herbicides, pesticides, fungicides • Air pollutants (e.g. SO_2, NO, NO_2, NO_x) • Ozone (O_3) and photochemical smog • Formation of highly reactive oxygen species (O_2, radicals $O_2^{\bullet-}$ and $OH^{\bullet}$, H_2O_2) • Photooxidants (e.g. peroxyacylnitrates) • Acid rain, acid fog, acid morning dew • Acid pH of soil and water • Mineral deficiency of the soil, often induced by acid rain • Oversupply of nitrogen (dry and wet NO_3 deposits) • Heavy metal load (lead, cadmium, etc.) • Overproduction of NH_4^+ in breeding stations • Uncoupling of electron transport • Increased UV radiation (UV-B and UV-A) • Increased CO_2, global climate change

5. Plant stresses can further be characterized into "internal stresses" that originate from within the plant and "external stresses" that exist outside the plant. External and internal stresses are generally referred to as stress factors and stresses, respectively.

In the simplest manner, stress can be divided into three categories: abiotic, biotic and anthropogenic. Abiotic stress has a huge impact on growth and, consequently, it is responsible for severe losses in the field. The resulting growth reductions can reach >50% in most plant species due to abiotic stress. Biotic stress is an additional challenge inducing strong pressure on plants and adding to the damage through pathogen or herbivore attack. In the present scenario, anthropogenic stress factors are also very important for the overall development of a plant. All the three categories of stress/stress factors are enlisted in Table 17.1.

17.5 Abiotic and Biotic Plant Stresses

As specified in the aforesaid paragraph, the key determinant of plant development and proliferation is growth in optimal conditions. However, plants are exposed to a vast range of biotic and abiotic stress factors which deviates the growth conditions from the ideal point to the stressed point. Abiotic stresses are the outcome of alterations in non-biological components, i.e., nutritional or environmental components. It also determines geographical distribution of plants. Consequently, induce responses from the plant to re-establish normal conditions or to minimize the deleterious effects of these changes. Among the different environmental conditions, salinity, cold and drought intensely affect plants and cause heavy economic losses (Beck et al. 2007). On the other hand, biotic stresses are the consequence of interactions between plant and other living organisms like bacteria, viruses, fungi, nematodes and herbivores that result in either partial or critical damages. The plant may or may not cope with these damages. Biotic stress affects ecosystem nutrient cycling along with plant population dynamics. Environmental conditions are responsible for controlling the occurrence of pathogens and pests. For example, dispersal of pathogens increases with increase in the temperature extremes. There is drop in the defense capability of plants which makes them susceptible to pathogen attack due to abiotic stress factors.

The co-evolution of plants and pathogens results in the development of plant defense mechanisms. Whenever plants are exposed to adverse environmental conditions or attacked by pathogens, they need to maintain balance between their growth and defense requirements which reduces the crop yield. Concerning food security, worldwide research is required to develop crops with high yields and efficiency to survive during unfavorable abiotic (Mosa et al. 2017) and biotic stress situations. In the near-term, research should be carried out to find out a solution for both biotic and abiotic stresses in order to reduce harmful effects on plants, whether these may be the result of only one stress or the combination of stresses.

17.6 Different Phases Induced by Stress

Normally, plants remain in a certain standard situation of physiology, i.e., the optimum limits set by the nutrient, light, water and carbon dioxide supply conditions of the niche. Stressors or complex stress events lead to three stress response phases: (1) response phase, (2) restitution phase and (3) end phase. On removing the stressors, they enter into the regeneration phase (phase 4) only in cases of mild damage. These four phases are given below:

17.6.1 Response Phase

This phase occurs at the beginning of stress and is represented by an alarm reaction such as deviation from functional norm, reflecting a decrease in several physiological functions, viz., photosynthesis, transport of metabolites or uptake and translocation of ions, thereby leading to a decline in vitality and an excess of catabolism over anabolic processes (Lichtenthaler 1988). During this alarm phase, most plants activate their stress coping mechanisms such as acclimation of metabolic fluxes, activation of repair processes and long-term metabolic and morphological adaptations.

17.6.2 Restitution Phase

The general alarm phase induces hardening of the plants by establishing a new physiological standard, which is an optimum stage of physiology under the impact of the stressor and corresponds to the plants' "resistance maximum." The restitution phase consists of three stages of resistance, i.e., adaptation processes, repair processes and reactivation processes like hardening of plants.

17.6.3 End Phase

Under long-term stress or a severe stress dose, the stress-coping mechanism of plants is overloaded thereby leading to the end phase. The end phase is also known as the stage of exhaustion or long-term stress. It occurs when the stress intensity is too high and the adaptation capacity is overloaded, leading to chronic disease or death of the plant.

17.6.4 Regeneration Phase

If the stressors are removed at a critical time prior to the initiation of the senescence processes, the plants regenerate and adapt new physiological standards, i.e., the regeneration phase. It is characterized by partial or full regeneration of the physiological functions of a plant when the stressor is removed and the damage has not been too high to replace.

The time and stage of exhaustion at which stressors are removed from the plant play a very crucial role in determining the new physiological standard of the plant. However, a continuous stress does not mean that the damage must happen; nevertheless, plants will orient themselves within the range set by the resistance minimum and resistance maximum, and in such cases, damage symptoms are not detectable (Wang et al. 2003).

17.7 Plant-Microbe Interactions for Stress Management in Plants

Environmental stresses influence agricultural outputs by hindering rhizosphere functioning. A healthy plant rhizosphere is not only helpful in nutrient and water uptake, but also imparts sustained benefits to microbial diversity, which ultimately encourages plant health (Vimal et al. 2017). Plants endure several types of stresses either by self-adaptation mechanisms or by microbes like mycorrhizal fungi and plant growth-

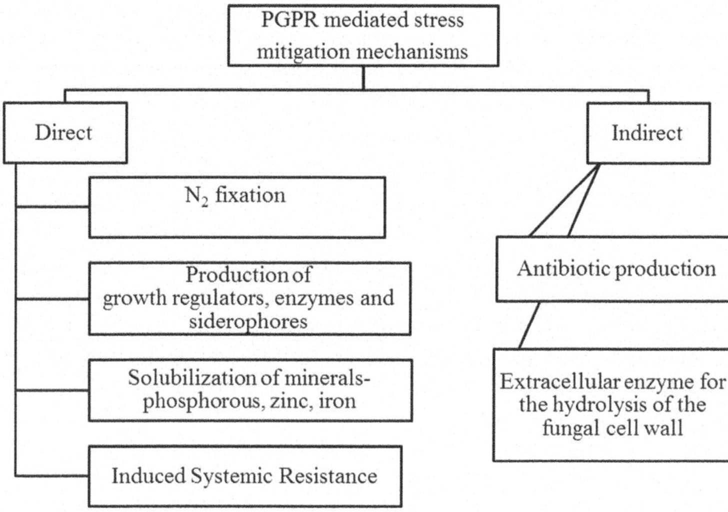

Fig. 17.2 Plant growth-promoting rhizobacteria-mediated mitigation of abiotic and biotic stresses

promoting rhizobacteria (Nadeem et al. 2014; Hamilton et al. 2016). Microbes with their intrinsic genetic and metabolic capabilities play an important role to mitigate abiotic stresses in plants. A number of microbes promote plant responses via managing the level of several antioxidant enzymes, defense proteins, phytohormones and polysaccharides, for instance, rhizobacteria-induced drought tolerance and management (Pandey et al. 2016). These methodologies make plants able to endure these environmental stress conditions.

Plant growth and crop productivity can be improved under biotic and abiotic stresses by using plant growth-promoting rhizobia (PGPR) which helps in alleviating the effect of these stresses on plants. The mechanisms include both biochemical and physiological changes. The processes in making plants resistant against stresses are not completely understood; however, research has demonstrated that a few mechanisms directly or indirectly (Fig. 17.2) stimulate plant growth and protection from environmental stress conditions (Singh and Chauhan 2017).

Directly PGPR can control stress of plants by imitating production of plant hormones or increasing the availability of minerals and nitrogen in the soil for a prolonged period of time. For example, the leguminous symbiont *Rhizobium* fixes environmental nitrogen to the soil and makes it available to plants in the form of amino acids. Indirect PGPR-mediated stress mitigation mechanisms include the production of different antibiotic compounds, siderophore or volatiles (2,3-butanediol and acetoin) or induction of plant-mediated induced systemic resistance (ISR) (Saharan and Nehra 2011). Endophytic and/or mycorrhizal fungi can interact with several plant species and, thereby, develop adaptation mechanisms of these plants to a multitude of environmental stresses (Rodriguez et al. 2008). The fungal networks improve water availability and phosphorous uptake under water-scarce conditions (Barnawal et al. 2014). Besides, the fungi are found helpful as they spread around the root zones ensuring plant protection against different pathogens and, therefore, act as possible bio-control agents (Fabbro and Prati 2014).

Beneficial microbes are associated with plant roots symbiotically and non-symbiotically and promote plant health via a broad spectrum of mechanisms. These mechanisms include producing secondary metabolites, controlling pathogen attacks and increasing the tolerance against abiotic and biotic stresses. There are numerous cross-talks between plants and microbes amid their interactions which improve our understanding on physiological methods related with roots, chemical molecules produced by them, signaling between root and microbes and potential defense mechanisms molecules involved in providing stress resistance to plants. The process of managing various stress conditions in different plants particularly incorporates remediation through the production of growth-promoting hormones, metabolites, organic volatiles and enzymes.

17.8 Mechanisms of Plant-Microbe Interactions in Rhizosphere

Plant-microbe associations, either positive or negative, occur in and around the rhizosphere through diverse mechanisms. A number of the widely reported mechanisms (Table 17.2) in plant-microbe associations include quorum sensing, plant or microbial signaling and production of volatile compounds.

Table 17.2 Mechanisms of plant-microbe interactions in rhizosphere

Mechanism	Compounds involved	References
Quorum sensing	Altered oligopeptides, *N*-acyl homoserine lactones	Uroz et al. (2009), Crépin et al. (2012)
Volatiles	Acetoin, 2-amino acetophenone, 2-pentylfuran, 2,3-butanediol,13-tetradecadien-1-ol,2-methyl-n-1-tridecene and 2-butanone	Park et al. (2015), Audrain et al. (2015), Schmidt et al. (2015), Kai et al. (2016)
Plant-mediated signalling	Salicylic acid, ethylene and jasmonic acid	Stam et al. (2014), Lebeis et al. (2015), Rosier et al. (2016)
Plant hormones	Auxins, cytokinin, abscisic acid, gibberellin	Pieterse et al. (2012), Giron et al. (2013)

17.9 Plant-Microbe Interactions for Alleviation of Abiotic Stresses

The interaction of mycorrhizal fungi, growth-promoting bacteria and other microbes with certain plants can assist the plants combat several abiotic stresses and save them from dying. The occurrence of *Azotobacter*, *Bradyrhizobium*, *Rhizobium*, *Azospirillum*, *Pseudomonas*, *Bacillus* and many plant-associated microbes has been investigated from saline and alkaline areas, acid soils, eroded hill slopes and desert ecosystems of India (Tilak et al. 2005; Selvakumar et al. 2009; Upadhyay et al. 2009). These microbes support host plants under different abiotic stress conditions. A variety of mechanisms have been proposed along with microbial-induced stress tolerance in plants against drought, salinity, heavy metal toxicity, etc., which are discussed here.

17.9.1 Drought as Stress

The growth and productivity of plants are affected by a water stress condition called drought. Drought is the result of low rainfall, high intensity of light and other extreme environmental conditions like high salinity, and high or low temperatures. It causes many changes in the plant physiological condition and molecular traits along with morphological and biochemical changes. Plants developed different mechanisms and adaptation conditions in order to avoid drought stress such as modified molecular mechanisms that regulate expression of genes, productions of phytohormones, closure of stomata, modified root morphology and maintenance of osmotic potential. Two regulatory genes, i.e., GUDK and HYR in rice, when expressed, resulted in high yield in normal as well as drought conditions due to increase in photosynthesis. The production of abscisic acid in the root and its translocation to the leaves lead to stomatal closure and further drought adaptation. Abscisic acid regulates water deficit stress by controlling various molecular events and activates anion channels which lead to the depolarization of the plasma membrane of guard cells (Levchenko et al. 2005; Negi et al. 2008). Drought resistance is also induced by modifications in the root system through microRNA miR393. The overexpression of the *sucrose:fructan-6-fructosyltransferase* (*6-SFT*) gene from *Psathyrostachys huashanica* in tobacco and the trehalose-6-phosphate phosphatase gene *OsTPP1* in rice also induces drought tolerance in plants. Besides chemical mediators, there are many microbes which confer drought tolerance in the plant. These are enlisted in Table 17.3.

Table 17.3 Plant-microbial operations against drought tolerance in plants

Plant	Microbes	Mechanisms	References
Sunflower	*Rhizobium* sp.	Soil aggregation through EPS	Alami et al. (2000)
Wheat	*Azospirillum* sp.	Improved water relations	Creus et al. (2004)
Pepper and tomato	*Achromobacter piechaudii* ARV8	Degradation of the ethylene precursor ACC by ACC deaminase	Mayak et al. (2004)
Pea	*Variovorax paradoxus*	Synthesis of 1-aminocyclopropane-1-carboxylate (ACC) deaminase	Dodd et al. (2004)
Tomato	*Azospirillum brasilense*	Produces nitric oxide, a signaling molecule in IAA inducing pathway	Creus et al. (2005) Molina-Favero et al. (2008)
Sorghum	AM fungi	Improved water relation	Cho et al. (2006)
Rice	Brome mosaic virus	–	Márquez et al. (2007)
Trifolium	*Bacillus megaterium* and *Glomus* sp.	IAA and proline production	Marulanda et al. (2007)
Common bean	*Rhizobium tropici* with *Paenibacillus polymyxa* strains (DSM 36) and Loutit (L)	Increased nodulation and nitrogen content	Figueiredo et al. (2008)
Pea	*Pseudomonas* sp.	Decreased ethylene production	Arshad et al. (2008)
Lettuce	*Pseudomonas mendocina* and *Glomus intraradices*	Improved antioxidant status	Kohler et al. (2008)
Sunflower	*Pseudomonas putida* P45	Soil aggregation through EPS	Sandhya et al. (2009)
Maize	*Azospirillum lipoferum*	Increased accumulation of soluble sugar, free amino acids and proline	Bano et al. (2013)
Soybean	*Pseudomonas putida* H-2-3	Lowers the level of abscisic acid and salicylic acid, increases the level of jasmonate	Kang et al. (2014)
Wheat	*Bacillus thuringiensis* AZP2	Production of volatile organic compounds	Timmusk et al. (2014)
Arabidopsis	*Azospirillum brasilense* sp. 245	Decreased stomatal conductance	Cohen et al. (2015)
Indian mustard	*Trichoderma harzianum*	Enhanced accumulation of antioxidants and osmolytes and decreased Na^+ uptake	Ahmad et al. (2015)
Chickpea	*Pseudomonas putida* MTCC5279 (RA)	Osmolyte accumulation, ROS scavenging ability and stress-responsive gene expression	Tiwari et al. (2016)

(continued)

Table 17.3 (continued)

Plant	Microbes	Mechanisms	References
Medicago truncatula	*Sinorhizobium medicae*	Root nodulation and increased nutrient acquisition	Staudinger et al. (2016)
Brassica oxyrrhina	*Pseudomonas libanensis* TR1 and *Pseudomonas reactans* Ph3R3	Decreased concentrations of proline and malondialdehyde in leaves	Ma et al. (2016a, b)
Rice	*Trichoderma harzianum*	Upregulation of dehydrin, malonialdehyde and aquaporin genes	Pandey et al. (2016)
Maize and sorghum	*Pseudomonas variovencis* XiU1297 and *Luteibacter* sp. XiU1292, *Acinetobacter inoffii* XiU12138	Produces 1-aminocyclopropane-1-carboxylate (ACC) deaminase and ethylene	Mull et al. (2017)

17.9.2 Salinity as Stress

Another devastating abiotic stress is soil salinity which leads to reduction in cultivable land area for crop cultivation, quality and productivity. Saline soil is commonly characterized as one in which the electrical conductivity (EC) of the saturation extract (EC_e) around the root surpasses 4 dSm^{-1} (~40 mM NaCl) at 25 °C. The yield of most crop plants tends to cease at this EC_e; however, a few crops tend to provide low yield (Kohler et al. 2010). It has been assessed that worldwide salinity of soil is increasing at a rate of 10% annually due to different reasons, including weathering of native rocks, high surface evaporation, low precipitation, poor agricultural practices and irrigation with saline water (Kohler et al. 2010). Salinity affects nearly all aspects of plant development by forcing ion toxicity, oxidative and osmotic stress and nutrient (Zn, N, K, Ca, Fe, and P) deficiency on plants (Vimal et al. 2018a). Rhizosphere microorganisms may enhance plant performance under extreme salinity and, thus, enhance yield directly or indirectly (Vimal et al. 2018b). These microbes must be targeted to manage the problem of soil salinization via a variety of mechanisms like controlling the sodium ion efflux in the root, modulating the transcriptional machinery responsible for salinity tolerance, enhancing production of IAA and 1-aminocyclopropane-1-carboxylate (ACC) deaminase and many other mechanisms as mentioned in Table 17.4.

17.9.3 Heavy Metal Toxicity as Stress

The term "heavy metals" refers to any metallic element that is a positive-charged ion, high density and is toxic beyond the acceptable limits as given by EU Regulation1881/2006/EU. Generally, it implies to the group of metals and metalloids with

Table 17.4 Plant-microbial operations against salinity tolerance in plants

Plant	Microbes	Mechanisms	References
Rice	*Escherichia coli*	High net catalytic efficiency for trehalose formation	Garg et al. (2002)
Wheat	*B. amylolequifaciens*, *B. insolitus*, *Microbacterium* sp., *P. syringae*	Restricted Na^+ influx	Ashraf et al. (2006)
Maize	*Azospirillum*	Amino acid and proline production	Hamdia et al. (2004)
Arabidopsis	*Xanthomonas campestris pv. Vesicatoria*	Pathogen-induced gene encoding RAV (related to ABI3/VP1) transcription factor	Sohn et al. (2006)
Vitis vinifera, Capsicum annuum	*Burkholderia*, *Arthrobacter* and *Bacillus*	Increased accumulation of proline	Barka et al. (2006)
Rice	*Scytonema*	Gibberellic acid and extracellular products	Rodriguez et al. (2006)
Lotus	*Glomus intraradices BAFC*3108	Decreased root and shoot Na^+ accumulation and enhanced root K^+ concentrations	Sannazzaro et al. (2006)
Groundnut	*Pseudomonas fluorescens*	Synthesis of 1-aminocyclopropane-1-carboxylate (ACC) deaminase	Saravanakumar and Samiyappan (2007)
Soybean	*Glomus etunicatum*	Increased root but decreased shoot proline concentrations	Sharifi et al. (2007)
Lettuce	*Glomus intraradices* BEG121	Reduced concentration of abscisic acid	Aroca et al. (2008)
Arabidopsis thaliana	*Bacillus subtilis* GB03	Tissue-specific regulation of sodium transporter HKT1	Zhang et al. (2008)
Gossypium hirsutum	*Pseudomonas putida* Rs-198	Prevented abscisic acid accumulation in seedlings	Yao et al. (2010)
Lettuce	*Pseudomonas mendocina* with *Glomus mosseae*	Enhanced plant biomass	Kohler et al. (2010)
Mung bean (*Vigna radiate*)	*Rhizobium* and *Pseudomonas*	Synthesis of 1-aminocyclopropane-1-carboxylate (ACC) deaminase	Ahmad et al. (2011)
Canola and maize	*Pseudomonas putida* UW4	Modulation of plant protein differential expression and 1-aminocyclopropane-1-carboxylate (ACC) deaminase activity	Cheng et al. (2012)
Wheat	*Azospirillum* sp. with *Piriformospora indica*	Improved proline accumulation	Zarea et al. (2012)
Capsicum annuum	*Azospirillum brasilense* and *Pantoea dispersa*	High stomatal conductance and photosynthesis	del Amor and Cuadra-Crespo (2012)

(continued)

Table 17.4 (continued)

Plant	Microbes	Mechanisms	References
Groundnut	*Brachybacterium saurashtrense* (JG-06), *Brevibacterium casei* (JG-08) and *Haererohalobacter* (JG-11)	Higher K^+/Na^+ ratio and higher Ca^{2+}, PO_4^- and N_2 content	Shukla et al. (2012a, b)
Rice	*Bacillus amyloliquefaciens* NBRISN13 (SN13)	Modulating differential transcription in a set of at least 14 genes	Nautiyal et al. (2013)
Rice GJ-17	*Pseudomonas pseudoalcaligenes* and *Bacillus pumilus*	Reduced ROS toxicity and lipid peroxidation and superoxide dismutase activity	Jha and Subramanian (2014)
Barley and oats	*Pseudomonas* sp. and *Acinetobacter* sp.	Enhanced production of IAA and 1-aminocyclopropane-1-carboxylate (ACC) deaminase	Chang et al. (2014)
'Micro-tom' tomato	*Streptomyces* strain PGPA39	Reduction in leaf proline content	Palaniyandi et al. (2014)
Pea	*Arthrobacter protophormiae* with *Glomus mosseae*	Reduced proline content and lipid peroxidation, increased pigment activity	Barnawal et al. (2014)
Barley	*Hartmannibacter diazotrophicus* E19	ACC deaminase activity and lower ethylene content	Suarez et al. (2015)
Lettuce seeds	*Azospirillum*	Lowers browning intensity	Fasciglione et al. (2015)
Soybean	*Pseudomonas koreensis* strain AK-1	Reduction in Na^+ level and increase in K^+ level	Kasotia et al. (2015)
Soybean	*Pseudomonas simiae*	4-nitroguaiacol and quinoline promote seed germination	Vaishnav et al. (2016)
Wheat	*Dietzia natronolimnaea* STR1	Modulates the transcriptional machinery responsible for salinity tolerance	Bharti et al. (2016)
Rice	*Curtobacterium albidum* SRV4	Enhancement in antioxidative enzymatic activities CAT, SOD, POX and APX and K^+ uptake	Vimal et al. (2018a)
Wheat	*Bacillus* sp. (JG3) and *Pseudomonas* sp. (JG7)	–	Vimal et al. (2018b)

atomic density greater than water or greater than 4 g/cm^3. A huge amount of heavy metals is present in soil and aquatic ecosystems in polluted sites and industrial areas. Plants are usually sensitive to both the lack and overabundant availability of some heavy metal ions. Contamination of agricultural soil by heavy metals has become a serious environmental concern due to their extensive occurrence and acute and chronic toxic effects on plants. Heavy metals are required by plants in trace amounts for their growth and development. However, it is toxic for the plant beyond the tolerable limits. Heavy metal toxicity varies with specific plant species, chemical form of the metal and its concentration, soil pH and composition. A number of heavy metals are bio-accumulative and, therefore, cannot be easily degraded in the environment and persist for a longer time. They enter in the food chain through uptake at primary producer levels (plants) and then accumulate at consumer levels (Wang et al. 2007). Rhizosphere hub bacteria have been found in many studies assisting plants under heavy metal stress via a variety of mechanisms as mentioned in Table 17.5.

17.9.4 Cold Stress in Plants

Low- and high-temperature stresses are very crucial for crop production as plants show maximum growth rate an optimum diurnal range of temperatures. Plants experience cold or chilling stress at temperatures from 0 °C to 15 °C (Vos et al. 2012). Plants exposed to cold stress show various phenotypic symptoms that include poor germination, stunted seedlings, yellowing of leaves, withering and reduced tillering. The major adverse effect of cold stress in plants is plasma membrane damage as a consequence of ice formation in plant tissues that causes dehydration. There is a need for developing frost-tolerant crops, and therefore mechanisms of tolerance to cold must be studied. Soil-associated bacteria have been proven to be promising options for improving crops against cold tolerance. The microbes and their mechanisms of inducing tolerance to cold stress are depicted in Table 17.6.

17.9.5 Heat Stress of Plants

Heat stress often occurs where the temperature is high for prolonged durations of time. The prolonged exposure to heat causes permanent damage to plant growth and affects their function. Also, high temperatures of soil and air, at day or night, affect the rate of reproductive development. High temperature causes direct effects by raising tissue temperatures that can result in damage to components of the leaf essential for photosynthesis. It also produces indirect effects by increasing the rate of evaporation and premature death of plants. As a resistance mechanism to heat, the plant develops resistance to water with low evaporative efficiency. Rhizosphere hub bacteria have been found to regulate high temperatures in plant under stressed conditions. A variety of mechanisms have been proposed to describe microbial-induced high-temperature tolerance in plants as given in Table 17.7.

Table 17.5 Plant-microbial operations against heavy metal toxicity in plants

Plant	Microbes	Mechanisms	References
Indian mustard (Ni, Pb, Zn)	*Kluyvera ascorbata* SUD165	No increase of metal uptake	Burd et al. (1998)
Sorghum vulgare (Cu)	*Glomus caledonium*, *Glomus claroideum*, *Glomus mosseae*	Increased heavy metal absorption rate	Gonzalez-Chavez et al. (2002)
Astragalus sinicus (Cd)	*Mesorhizobium huakuii* subsp. rengei B3	Increased ability of cells to bind Cd^{2+}	Sriprang et al. (2003)
Brassica juncea (Ni)	*Bacillus subtilis* SJ-101 (Ni)	Facilitates Ni accumulation	Zaidi et al. (2006)
Trifolium repens	*Brevibacillus brevis* with *Glomus mosseae*	Reduced metal acquisition	Vivas et al. (2006)
Tomato (Ni, Cd)	*Methylobacterium oryzae*, *Burkholderia* sp.	Reduced uptake and translocation	Madhaiyan et al. (2007)
Maize (Cu, Zn, Pb, Cd)	*Glomus caledonium*, *Gigaspora margarita*, *Gigaspora decipiens*, *Scutellospora gilmorei*	Increased P uptake	Wang et al. (2007)
Thlaspi praecox (Cd, Pb, Zn)	*Phialophora verrucosa*, *Rhizoctonia* sp., *Penicillium brevi compactum*, *Rhodotorula aurantiaca*	Enhanced heavy metal uptake by plant roots	Pongrac et al. (2009)
Chrysopogon zizanioides (Pb)	*Glomus mosseae*	Heavy metal uptake by plant roots and its translocation to plant shoots	Punamiya et al. (2010)
Alphitonia neocaledonica, *Cloezia artensis* (Ni)	*Glomus etunicatum*	Reduced heavy metal concentration in roots and shoots	Amir et al. (2013)
Arabidopsis thaliana (Cd)	*Pseudomonas putida* UW4	Synthesis of ACC deaminase and IAA	Glick (2014)
Pinus densiflora, *Quercus variabilis* (Cu)	*Pisolithus* sp., *Cenococcum geophilum*, *Laccaria laccata*	Reduced heavy metal accumulation in shoots	Zong et al. (2015)
Solanum nigrum (Cd)	*Glomus versiforme*	Enhanced phosphatase activity, higher heavy metal uptake	Liu et al. (2015)
Cymbopogon citratus (Pb)	*Rhizophagus clarus*	Enhanced productivity of essential oils from plants	Lermen et al. (2015)

17.9.6 Osmotic Stress to Plants

Osmotic stress conditions are induced by salinity, drought and low-temperature conditions which decrease the plant growth and crop productivity. In water-deficit conditions, the status of plant water is controlled by stress signaling and sensing

Table 17.6 Plant-microbial operations against cold stress in plants

Plant	Microbe(s)	Mechanisms	References
Grapevine	*Burkholderia phytofirmans* PsJN	Synthesis of ACC deaminase	Barka et al. (2006)
Canola	*P. putida*	Synthesis of ACC deaminase	Chang et al. (2007)
Wheat	*Methylobacterium phyllosphaerae* strain IARI-HHS2–67	Improved nutrient uptake	Verma et al. (2015)
Tomato	*Arthrobacter, Flavobacterium, Flavimonas, Pedobacter*, and *Pseudomonas*	Improved plant height, root length	Subramanian et al. (2016)

Table 17.7 Plant-microbial operations against heat stress in plants

Plant	Microbe(s)	Mechanisms	References
Sorghum	*Pseudomonas* sp. AMK-P6	Induction of heat shock proteins	Ali et al. (2009)
Wheat	*Pseudomonas putida* strain AKMP7	Reduced the activity of enzymes SOD, APX and CAT	Ali et al. (2011)
Wheat	*Bacillus amyloliquefaciens, Azospirillum brasilense*	Reduced regeneration of reactive oxygen species	El-Daim et al. (2014)
Dichanthelium lanuginosum, Tomato	*Curvularia proturberata* isolate Cp4666D	Colonization of roots	de Zelicourt et al. (2013)
Arabidopsis	*Paraphaeosphaeria quadriseptata*	Induction of HSP 90	McLellan et al. (2007)

Table 17.8 Plant-microbial operations against osmotic stress in plants

Plant	Microbe(s)	Mechanisms	References
Phaseolus vulgaris	*Glomus intraradices* BEG 123	Increased active solute transport through roots	Aroca et al. (2007)
Pepper	*Arthrobacter* sp., *Bacillus* sp.	IAA and proline production	Sziderics et al. (2007)
Maize	*Bacillus megaterium*	High hydraulic conductance, increased root expression of two ZmPIP isoforms	Marulanda et al. (2010)

which lead to rapid changes in gene expression. During osmotic stress conditions, stomatal closure is controlled by the production of abscisic acid and turgor pressure. To resist osmotic stress, the plant produces proline in high amounts, which preserves membrane integrity and acts as a scavenger of reactive oxygen radicals. Snf1-related protein kinase 2 helps in the maintenance of plant water status and plays a pivotal role in tolerance to osmotic stress. A number of soil microbes have been noted for their roles in osmotic stress regulation and management. Microbial-induced tolerance in plants against osmotic stress is mentioned in Table 17.8.

17.9.7 Alleviation of Biotic Stresses

Apart from abiotic stresses, plants are susceptible to various sorts of pathogens including fungi, viruses, bacteria, and nematodes and herbivores. Plants grown under biotic stress conditions have a tendency to screen and interact more with beneficial microbes. The machinery of biocontrol activity of rhizospheric microbes is either through direct inhibition (killing) or by restraining establishment of pathogens (Pieterse et al. 2012). Several plant-related microbes have been found to assist their hosts in critical pathogen attacks which are enlisted in Table 17.9.

17.10 OMICS Approach of Stress Management in Fields

Since microbial interactions with plants are an integral part of the living ecosystem, they are believed to be natural partners that modulate local and systemic mechanisms in plants to offer defense under adverse external conditions. Work on plant-microbe interactions at biochemical, physiological and molecular levels established that microbial associations largely direct plant responses toward stresses. Under the continuous pressure of increasing climatic changes, plant-microbe relationships can be defined and interpreted in terms of protection against abiotic stresses. Multi-omics approaches involving genomics, transcriptomics, proteomics, metabolomics and phenomics. Integrated studies on the interaction of plants with microbes and their external environment generate multi-layered information that can be helpful in finding out real-changes happening within cells. These approaches together give big data to decide the outcome and possibilities and further knowledge for implementation in the fields (Fig. 17.3).

Advances in the field of bioinformatics and its tools help in acquiring data at the multi-omics level, which further improve our knowledge about the microorganisms present in rhizosphere area and their interconnections with the other microbes and root systems of plants. These tools are also useful in acquiring information about the role of microbes related to plant stress. Meta-omics approaches of bioinformatics including metagenomics, metatranscriptomics and metaproteomics are also used nowadays to determine the role of microbes and their roles in stress management in plants. Thus, multi-omics approaches on specific plant-microbe-abiotic stress system are proved to be very useful to resolve many facts related to precise mechanisms of stress tolerance/mitigation in crop plants.

Table 17.9 Plant-microbial operations against biotic stress tolerance in plants

Host plants	Beneficial microbes	Diseases/ pathogens	Mechanism/effect	References
Tobacco	*Trichoderma harzianum, Glomus mosseae*	*Ralstonia solanacearum*	Enhanced systemic resistance	Yuan et al. (2016)
Senecio vernalis, Senecio inaequidens, Inula conyza, Conyza canadensis, Solidago virgaurea, Solidago gigantean	*Glomus intraradices, Glomus mosseae, Glomus geosporum, Glomus claroideum, Glomus etunicatum*	*Pythium ultimum*	Inhibited pathogen	Fabbro and Prati (2014)
Helianthus tuberosus	*Trichoderma harzianum, Glomus clarum*	*Sclerotium rolfsii*	Reduced disease incidence	Sennoi et al. (2013)
Lycopersicon esculentum	*Glomus mosseae*	*Meloidogyne incognita, Pratylenchus penetrans*	Reduced disease development	Vos et al. (2012)
Lycopersicon esculentum	*Glomus mosseae, Glomus intraradices*	*Phytophthora nicotianae*	Resistance to pathogen invasion	Lioussanne et al. (2009)
Phaseolus vulgaris	*Glomus intraradices, Glomus mosseae, Glomus clarum, Gigaspora gigantean*	*Fusarium solani*	Enhanced phenolic content and defensive enzyme activities	Al-Askar and Rashad (2010)
Hordeum vulgare	*Glomus mosseae*	*Gaeumannomyces graminis*	Inhibited pathogens with dense mycorrhizal colonization	Castellanos-Morales et al. (2011)
Phoenix dactylifera	*Glomus monosporus, Glomus deserticola, Glomus clarum*	*Fusarium oxysporum*	Improved defensive enzyme activities	Jaiti et al. (2007)
Cabbage	*Paenibacillus* sp.	Black rot (*Xanthomonas campestris*)	ISR	Ghazalibiglar et al. (2016)
Cucumber	*Bacillus subtilis, Pseudomonas, Fluorescens, Azotobacter chroococcum,*	Cucumber mosaic cucumovirus (CMV)	Production of pathogen-related protein	El-Borollosy and Oraby (2012)

Mustard	–	*Pseudomonas putida*	Phosphate solubilization, IAA and siderophores production	Ahemad and Khan (2012a, b)
Panax ginseng	*Bacillus amyloliquefaciens*HK34	Root diseases (*Phytophthora cactorum*)	ISR	Lee et al. (2015)
Rice	*Bacillus* sp.	Bacterial leaf blight (*Xanthomonas oryzae*)	Increased accumulation of phenylal-anine, ammonia lyase, peroxidase and polyphenol oxidase	Chithrashree et al. (2011)
Pepper	*Brevibacterium iodinum* KUDC1716	Gray leaf spot dis-ease (*Stemphylium lycopersici*)	Production of pathogen-related protein	Son et al. (2014)
Elaeis guineensis	*Pseudomonas aeruginosa UPMP3* with *Glomus intraradices*, *Glomus clarum*	*Ganoderma* basal stem rot disease	Reduced epidemic rates of disease	Sundram et al. (2015)
Papaya	*Pseudomonas* sp. with AM fungi	*Fusarium oxysporum*	Reduced pathogen colonization	Hernández-Montiel et al. (2013)

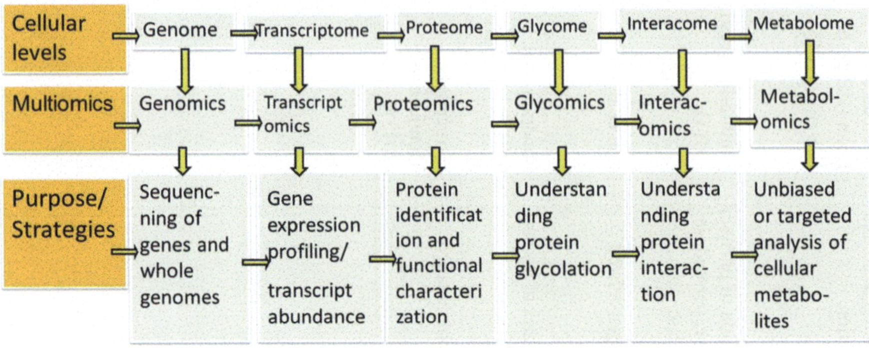

Fig. 17.3 Multi-omics approaches in understanding plant stress mechanisms

17.11 Conclusion

Microbes can be utilized positively to alter the growth capabilities of plants and to make them more tolerant against abiotic and biotic stresses that will occur substantially and frequently with progressing climate change. Permanent alleviation of plant stress demands a complex range of associations between plants and microbes; understanding these associations may serve better to improve their stress management mechanisms.

The development and improvement of stress-tolerant crop varieties are a time-consuming endeavor whilst microbial inoculation to alleviate stresses in plants could be a more economical and eco-friendly alternative which would be accessible in shorter time. In coming times, intensive research is needed on field assessment and the use of potential microorganisms. Concerns over environmental issues give microbial biocontrol a compelling perspective. Therefore, application of naturally occurring soil microbes instead of detrimental chemicals can give an exceptionally promising substitute for alleviating plant stress conditions.

References

Ahemad M, Khan MS (2012a) Effect of fungicides on plant growth promoting activities of phosphate solubilizing *Pseudomonas putida* isolated from mustard (*Brassica compestris*) rhizosphere. Chemosphere 86(9):945–950. https://doi.org/10.1016/j.chemosphere.2011.11.013

Ahemad M, Khan MS (2012b) Alleviation of fungicide-induced phytotoxicity in greengram [*Vigna radiata* (L.) Wilczek] using fungicide-tolerant and plant growth promoting *Pseudomonas* strain. Saud J Biol Sci 19:451–459

Ahmad P, Hashem A, Abd-Allah EF, Alqarawi AA, John R, Egamberdieva D, Gucel S (2015) Role of *Trichoderma harzianum* in mitigating NaCl stress in Indian mustard (Brassica juncea L) through antioxidative defense system. Front Plant Sci. https://doi.org/10.3389/fpls.2015.00868

Ahmad M, Zahir ZA, Asghar HN, Asghar M (2011) Inducing salt tolerance in mung bean through coinoculation with rhizobia and plant-growth-promoting rhizobacteria containing 1-aminocyclopropane-1-carboxylate deaminase. Can J Microbiol 57:578–589. https://doi.org/10.1139/W11-044

Alami Y, Wafa A, Marol C, Heulin T (2000) Rhizosphere soil aggregation and plant growth promotion of sunflowers by an exopolysaccharide-producing *Rhizobium* sp. strain isolated from sunflower roots. Appl Environ Microbiol 66:3393–3398

Al-Askar AA, Rashad YM (2010) Arbuscular mycorrhizal fungi: a biocontrol agent against common bean fusarium root rot disease. Plant Pathol J 9(1):31–38. https://doi.org/10.3923/ppj.2010.31.38

Ali SZ, Sandhya V, Grover M, Kishore N, Rao LV, Venkateswarlu B (2009) *Pseudomonas* sp. strain AKM-P6 enhances tolerance of sorghum seedlings to elevated temperatures. Biol Fertil Soils 46:45–55

Ali SZ, Sandhya V, Grover M, Linga VR, Bandi V (2011) Effect of inoculation with a thermotolerant plant growth promoting *Pseudomonas putida* strain AKMP7 on growth of wheat (*Triticum* spp.) under heat stress. J Plant Interact 6:239–246. https://doi.org/10.1080/17429145.2010.545147

Amir H, Lagrange A, Hassaïne N, Cavaloc Y (2013) Arbuscular mycorrhizal fungi from new Caledonian ultramafic soils improve tolerance to nickel of endemic plant species. Mycorrhiza 23 (7):585–595. https://doi.org/10.1007/s00572-013-0499-6

Aroca R, Porcel R, Ruiz-Lozano JM (2007) How does arbuscular mycorrhizal symbiosis regulate root hydraulic properties and plasma membrane aquaporins in *Phaseolus vulgaris* under drought, cold or salinity stresses? New Phytol 173(4):808–816. https://doi.org/10.1111/j.1469-8137.2006.01961.x

Aroca R, Vernieri P, Ruiz-Lozano JM (2008) Mycorrhizal and non-mycorrhizal *Lactuca sativa* plants exhibit contrasting responses to exogenous ABA during drought stress and recovery. J Exp Bot 59(8):2029–2041. https://doi.org/10.1093/jxb/ern057

Arshad M, Shaharoona B, Mahmood T (2008) Inoculation with *Pseudomonas* spp. containing ACC-deaminase partially eliminates the effects of drought stress on growth, yield, and ripening of pea (*Pisum sativum* L.). Pedosphere 18:611–620. https://doi.org/10.1016/s1002-0160(08)60055-7

Ashraf M, Hasnain S, Berge O (2006) Effect of exo-polysaccharides producing bacterial inoculation on growth of roots of wheat (*Triticum aestivum* L.) plants grown in a salt-affected soil. Int J Environ Sci Technol 3:43–51. https://doi.org/10.1007/BF03325906

Audrain B, Farag MA, Ryu CM, Ghigo JM (2015) Role of bacterial volatile compounds in bacterial biology. FEMS Microbiol Rev 39(2):222–233. https://doi.org/10.1093/femsre/fuu013

Bano Q, Ilyas N, Bano A, Zafar N, Akram A, Hassan FU (2013) Effect of Azospirillum inoculation on maize (*Zea mays* L.) under drought stress. Pak J Bot 45(S1):13–20

Barka EA, Nowak J, Clément C (2006) Enhancement of chilling resistance of inoculated grapevine plantlets with a plant growth-promoting *Rhizobacterium, Burkholderia phytofirmans* strain PsJN. Appl Environ Microbiol 72:7246–7252. https://doi.org/10.1128/AEM.01047-06

Barnawal D, Bharti N, Maji D, Chanotiya CS, Kalra A (2014) ACC deaminase-containing *Arthrobacter protophormiae* induces NaCl stress tolerance through reduced ACC oxidase activity and ethylene production resulting in improved nodulation and mycorrhization in *Pisum sativum*. J Plant Physiol 171:884–894. https://doi.org/10.1016/j.jplph.2014.03.007

Beck EH, Fettig S, Knake C, Hartig K, Bhattarai T (2007) Specific and unspecific responses of plants to cold and drought stress. J Biosci 32:501–510

Berendsen RL, Pieterse CM, Bakker PA (2012) The rhizosphere microbiome and plant health. Trends Plant Sci 17(8):478–486. https://doi.org/10.1016/j.tplants.2012.04.001

Bharti N, Pandey SS, Barnawal D, Patel VK, Kalra A (2016) Plant growth promoting rhizobacteria *Dietzia natronolimnaea* modulates the expression of stress responsive genes providing protection of wheat from salinity stress. Sci Rep 6:34768

Burd GI, Dixon GD, Glick BR (1998) A plant growth-promoting bacterium that decreases nickel toxicity in seedlings. Appl Environ Microbiol 64:3663–3668

Campbell R, Greaves MP (1990) Anatomy and community structure of the rhizosphere. Rhizosphere 11–34

Castellanos-Morales V, Keiser C, Navarro RC, Grausgruber H, Glauninger J, Garrido JMC, Steinkellner S, Sampedro I, Hage-Ahmed K, Illana A (2011) The bioprotective effect of AM root colonization against the soil-borne fungal pathogen *Gaeumannomyces graminis* var. tritici in barley depends on the barley variety. Soil Biol Biochem 43(4):831–834. https://doi.org/10.1016/j.scienta.2015.02.029

Chang WS, van de Mortel M, Nielsen L, de Guzman GN, Li X, Halverson LJ (2007) Alginate production by *Pseudomonas putida* creates a hydrated microenvironment and contributes to biofilm architecture and stress tolerance under water-limiting conditions. J Bacteriol 189 (22):8290–8299. https://doi.org/10.1128/JB.00727-07

Chang P, Gerhardt KE, Huang XD, Yu XM, Glick BR, Gerwing PD, Greenberg BM (2014) Plant growth-promoting bacteria facilitate the growth of barley and oats in salt-impacted soil: implications for phytoremediation of saline soils. Int J Phytoremediation 16 (7–12):1133–1147. https://doi.org/10.1080/15226514.2013.821447

Cheng Z, Woody OZ, McConkey BJ, Glick BR (2012) Combined effects of the plant growth-promoting bacterium *Pseudomonas putida* UW4 and salinity stress on the *Brassica napus* proteome. Appl Soil Ecol 61:255–263

Chithrashree UAC, Nayaka BS, Reddy MS, Srinivas C (2011) Plant growth-promoting rhizobacteria mediate induced systemic resistance in rice against bacterial leaf blight caused by *Xanthomonas oryzae* pv. oryzae. Biol Control 59(2):114–122. https://doi.org/10.1016/j.biocontrol.2011.06.010

Cho K, Toler H, Lee J, Ownley B, Stutz JC, Moore JL, Augé RM (2006) Mycorrhizal symbiosis and response of sorghum plants to combined drought and salinity stresses. J Plant Physiol 163 (5):517–528. https://doi.org/10.1016/j.jplph.2005.05.003

Cohen AC, Bottini R, Pontin M, Berli FJ, Moreno D, Boccanlandro H, Travaglia CN, Piccoli PN (2015) Azospirillum brasilense ameliorates the response of *Arabidopsis thaliana* to drought mainly via enhancement of ABA levels. Physiol Plant 153:79–90. https://doi.org/10.1111/ppl.12221

Crépin A, Barbey C, Cirou A et al (2012) Biological control of pathogen communication in the rhizosphere: a novel approach applied to potato soft rot due to *Pectobacterium atrosepticum*. Plant Soil. (2012 358:27. https://doi.org/10.1007/s11104-011-1030-5

Creus CM, Sueldo RJ, Barassi CA (2004) Water relations and yield in Azospirillum-inoculated wheat exposed to drought in the field. Can J Bot 82:273–281. https://doi.org/10.1139/b03-119

Creus CM, Graziano M, Casanovas EM, Pereyra MA, Simontacchi M, Puntarulo S, Barassi CA, Lamattina L (2005) Nitric oxide is involved in the *Azospirillum brasilense*-induced lateral root formation in tomato. Planta 221(2):297–303

de Zelicourt A, Al-Yousif M, Hirt H (2013) Rhizosphere microbes as essential partners for plant stress tolerance. Mol Plant 6:242–245

del Amor FM, Cuadra-Crespo P (2012) Plant growth-promoting bacteria as a tool to improve salinity tolerance in sweet pepper. Funct Plant Biol 39:82–90. https://doi.org/10.1071/FP11173

Dodd IC, Belimov AA, Sobeih WY, Safronova VI, Grierson D, Davies WJ (2004) Will modifying plant ethylene status improve plant productivity in water-limited environments ? In: Proceedings for the 4th International Crop Science Congress, vol 26, Brisbane, Australia

El-Borollosy AM, Oraby MM (2012) Induced systemic resistance against cucumber mosaic cucumovirus and promotion of cucumber growth by some plant growth-promoting rhizobacteria. Ann Agric Sci 57:91–97

El-Daim IAA, Bejai S, Meijer J (2014) Improved heat stress tolerance of wheat seedlings by bacterial seed treatment. Plant Soil 379:337–350

Fabbro CD, Prati D (2014) Early responses of wild plant seedlings to arbuscular mycorrhizal fungi and pathogens. Basic Appl Ecol 15:534–542

Fasciglione G, Casanovasa EM, Quillehauquya V, Yommi AK, Goñi MG, Roura SI, Barassi CA (2015) Azospirillum inoculation effects on growth, product quality and storage life of lettuce plants grown under salt stress. Sci Hort 195:154–162
Faure D, Vereecke D, Leveau JHJ (2009) Molecular communication in the rhizosphere. Plant Soil 321(1–2):279–303. https://doi.org/10.1007/s11104-008-9839-2
Figueiredo MVB, Burity HA, Martínez CR, Chanway CP (2008) Alleviation of drought stress in the common bean (*Phaseolus vulgaris* L.) by co-inoculation with *Paenibacillus polymyxa* and *Rhizobium tropici*. Appl Soil Ecol 40(1):182–188. https://doi.org/10.1016/j.apsoil.2008.04.005
Garbeva P, Hordijk C, Gerards S, de Boer W (2014) Volatiles produced by the mycophagous soil bacterium *Collimonas*. FEMS Microbiol Ecol 87: 639–649. https://doi.org/10.1111/1574-6941.12252
Garg AK, Kim JK, Owens TG, Ranwala AP, Choi YD, Kochian LV, Wu RJ (2002) Trehalose accumulation in rice plants confers high tolerance levels to different abiotic stresses. Proc Natl Acad Sci USA 99(25):15898–15903. https://doi.org/10.1073/pnas.252637799
Ghazalibiglar H, Hampton JG, de Jong EZ, Holyoake A (2016) Is induced systemic resistance the mechanism for control of black rot in *Brassica oleracea* by a *Paenibacillus* sp.? Biol Control 92:195–201. https://doi.org/10.1016/j.biocontrol.2015.10.014
Giron D, Frago E, Glevarec G, Pieterse CM, Dicke M (2013) Cytokinins as key regulators in plant–microbe–insect interactions: connecting plant growth and defence. Funct Ecol 27(3):599–609. https://doi.org/10.1111/1365-2435.12042
Glick BR (2014) Bacteria with ACC deaminase can promote plant growth and help to feed the world. Microbiol Res 169(1):30–39. https://doi.org/10.1016/j.micres.2013.09.009
Gonzalez-Chavez C et al (2002) Copper sorption and accumulation by the extraradical mycelium of different *Glomus* spp. (arbuscular mycorrhizal fungi) isolated from the same polluted soil. Plant Soil 240:287–297
Grover M, Ali SZ, Sandhya V, Rasul A, Venkateswarlu B (2011) Role of microorganisms in adaptation of agriculture crops to abiotic stresses. World J Microbiol Biotechnol 27:1231–1240
Hamdia MAES, Shaddad MAK, Doaa MM (2004) Mechanisms of salt tolerance and interactive effects of *Azospirillum brasilense* inoculation on maize cultivars grown under salt stress conditions. Plant Growth Regul 44:165–174
Hamilton CE, Bever JD, Labbe J, Yang X, Yin H (2016) Mitigating climate change through managing constructed-microbial communities in agriculture. Agric Ecosyst Environ 216:304–308. https://doi.org/10.1016/j.agee.2015.10.006
Hartmann A, Schmid M, van Tuinen D, Berg G (2009) Plant-driven selection of microbes. Plant Soil 321:235–257
Hernández-Montiel LG, Rueda-Puente EO, Cordoba-Matson MV, Holguín-Peña JR, Zulueta-Rodríguez R (2013) Mutualistic interaction of rhizobacteria with arbuscular mycorrhizal fungi and its antagonistic effect on *Fusarium oxysporum* in *Carica papaya* seedlings. Crop Prot 47:61–66. https://doi.org/10.1016/j.cropro.2013.01.008
Ho, YN, Mathew DC, Huang CC (2017) Plant-microbe ecology: interactions of plants and symbiotic microbial communities. In: Plant ecology - traditional approaches to recent trends. InTech, London. https://doi.org/10.5772/intechopen.69088
Jaiti F, Abdelillah M, Hadrami IE (2007) Effectiveness of arbuscular mycorrhizal fungi in the protection of date palm (*Phoenix dactylifera* L.) against bayoud disease. Plant J Plant Mol Biol 71:166–173
Jha Y, Subramanian RB (2014) PGPR regulate caspase-like activity, programmed cell death, and antioxidant enzyme activity in paddy under salinity. Physiol Mol Biol Plants 20(2):201–207. https://doi.org/10.1007/s12298-014-0224-8
Kai M, Effmert U, Piechulla B (2016) Bacterial-plant-interactions: approaches to unravel the biological function of bacterial volatiles in the Rhizosphere. Front Microbiol 7:108. https://doi.org/10.3389/fmicb.2016.00108
Kang SM, Radhakrishnan R, Khan AL, Kim MJ, Park JM, Kim BR, Shin DH, Lee IJ (2014) Gibberellin secreting rhizobacterium, *Pseudomonas putida* H-2-3 modulates the hormonal and

stress physiology of soybean to improve the plant growth under saline and drought conditions. Plant Physiol Biochem 84:115–124. https://doi.org/10.1016/j.plaphy.2014.09.001
Kasotia A, Varma A, Choudhary DK (2015) *Pseudomonas*-mediated mitigation of salt stress and growth promotion in *Glycine max*. Agric Res 4:31–41
Kohler J, Hernández JA, Caravaca F, Roldán A (2008) Plant-growth-promoting rhizobacteria and arbuscular mycorrhizal fungi modify alleviation biochemical mechanisms in water-stressed plants. Funct Plant Biol 35:141–151
Kohler J, Caravaca F, Roldán A (2010) An AM fungus and a PGPR intensify the adverse effects of salinity on the stability of rhizosphere soil aggregates of *Lactuca sativa*. Soil Biol Biochem 42 (3):429–434. https://doi.org/10.1071/FP0721
Kranner I, Minibayeva FV, Beckett RP, Seal CE (2010) What is stress? Concepts, definitions and applications in seed science. New Phytol 188:655–673. https://doi.org/10.1111/j.1469-8137.2010.03461.x
Larcher W (1987) Streß bei Pflanzen. Naturwissenschaften 74:158–167
Lebeis SL, Paredes SH, DS L, Breakfield N, Gehring J, McDonald M, Malfatti S, Glavina del Rio T, Jones CD, Tringe SG, Dangl JL (2015) Plant microbiome. Salicylic acid modulates colonization of the root microbiome by specific bacterial taxa. Science 349(6250):860–864. https://doi.org/10.1126/science.aaa8764
Lee BD, Dutta S, Ryu H, Yoo SJ, Suh DS, Park K (2015) Induction of systemic resistance in *Panax ginseng* against *Phytophthora cactorum* by native *Bacillus amyloliquefaciens* HK34. J Ginseng Res 39:213–220
Lermen C, Mohr FBM, Alberton O (2015) Growth of Cymbopogon citratus inoculated with mycorrhizal fungi under different levels of lead. Sci Hortic 186:239–246. https://doi.org/10.1016/j.scienta.2015.02.029
Levchenko V, Konrad KR, Dietrich P, Roelfsema MR, Hedrich R (2005) Cytosolic abscisic acid activates guard cell anion channels without preceding Ca2+ signals. Proc Natl Acad Sci USA 102:4203–4208
Levitt J (1980) Responses of plants to environmental stresses, vol 1. Academic, New York
Lichtenthaler HK (1988) In vivo chlorophyll fluorescence as a tool for stress detection in plants. In: Lichtenthaler HK (ed) Applications of chlorophyll fluorescence. Kluwer Academic, Dordrecht, the Netherlands, pp 129–142
Lichtenthaler HK (1996) Vegetation stress: an introduction to the stress concept in plants. J Plant Physiol 148:4–14
Lioussanne L, Jolicoeur M, St-Arnaud M (2009) Role of the modification in root exudation induced by arbuscular mycorrhizal colonization on the intraradical growth of *Phytophthora nicotianae* in tomato. Mycorrhiza 19:443–448
Liu H, Yuan M, Tan S, Yang X, Lan Z, Jiang Q, Ye Z, Jing Y (2015) Enhancement of arbuscular mycorrhizal fungus (*Glomus versiforme*) on the growth and Cd uptake by Cd-hyperaccumulator *Solanum nigrum*. Appl Soil Ecol 89:44–49. https://doi.org/10.1016/j.apsoil.2015.01.006
Ma Y, Rajkumar M, Zhang C, Freitas H (2016a) Beneficial role of bacterial endophytes in heavy metal phytoremediation. J Environ Manag 174:14–25. https://doi.org/10.1016/j.jenvman.2016.02.047
Ma Y, Rajkumar M, Zhang C, Freitas H (2016b) Inoculation of *Brassica oxyrrhina* with plant growth promoting bacteria for the improvement of heavy metal phytoremediation under drought conditions. J Hazard Mater 320:36–44. https://doi.org/10.1016/j.jhazmat.2016.08.009
Madhaiyan M, Poonguzhali S, Sa T (2007) Metal tolerating methylotrophic bacteria reduces nickel and cadmium toxicity and promotes plant growth of tomato (*Lycopersicon esculentum* L.). Chemosphere 69:220–228
Márquez LM, Redman RS, Rodriguez RJ, Roossinck MJ (2007) A virus in a fungus in a plant: three-way symbiosis required for thermal tolerance. Science 315:513–515
Marulanda A, Porcel R, Barea JM, Azcón R (2007) Drought tolerance and antioxidant activities in lavender plants colonized by native drought-tolerant or drought-sensitive *Glomus* species. Microb Ecol 54(3):543–552. https://doi.org/10.1007/s00248-007-9237-y

Marulanda A, Azcón R, Chaumont F, Ruiz-Lozano JM, Aroca R (2010) Regulation of plasma membrane aquaporins by inoculation with a *Bacillus megaterium* strain in maize (*Zea mays* L.) plants under unstressed and salt-stressed conditions. Planta 232(2):533–543. https://doi.org/10.1007/s00425-010-1196-8

Mayak S, Tirosh T, Glick BR (2004) Plant growth-promoting bacteria confer resistance in tomato plants to salt stress. Plant Physiol Biochem 42:565–572

McLellan CA, Turbyville TJ, Wijeratne EMK, Kerschen A, Vierling E, Queitsch C, Whitesell L, Gunatilaka AAL (2007) A Rhizosphere fungus enhances Arabidopsis thermotolerance through production of an HSP90 inhibitor. Plant Physiol 145:174–182. https://doi.org/10.1104/pp.107.101808

Mendes R, Garbeva P, Raaijmakers JM (2013) The rhizosphere microbiome: significance of plant beneficial, plant pathogenic, and human pathogenic microorganisms. FEMS Microbiol Rev 37:634–663

Molina-Favero C, Creus CM, Simontacchi M, Puntarulo S, Lamattina L (2008) Aerobic nitric oxide production by *Azospirillum brasilense* Sp245 and its influence on root architecture in tomato. Mol Plant Microbe Interact 21(7):1001–1009. https://doi.org/10.1094/MPMI-21-7-1001

Mosa KA, Ismail A, Helmy M (2017) Introduction to plant stresses. Plant stress tolerance. Springer, Cham, pp 1–19. https://doi.org/10.1007/978-3-319-59379-1_1

Mull JGC, De La Riva GADL, Vargas-Sámano CD, Pérez-Machado G, Agüero-Chapin G (2017) Plant growth promoting bacteria isolated from a mexican natural ecosystem induce water stress resistance in maize and sorghum plants. J Microb Biochem Technol 9:209–219

Nadeem SM, Ahmad M, Zahir ZA, Javaid A, Ashraf M (2014) The role of mycorrhizae and plant growth promoting rhizobacteria (PGPR) in improving crop productivity under stressful environments. Biotechnol Adv 32(2):429–448. https://doi.org/10.1016/j.biotechadv.2013.12.005

Nautiyal CS, Srivastava S, Chauhan PS, Seem K, Mishra A, Sopory SK (2013) Plant growth-promoting bacteria *Bacillus amyloliquefaciens* NBRISN13 modulates gene expression profile of leaf and rhizosphere community in rice during salt stress. Plant Physiol Biochem 66:1–9. https://doi.org/10.1016/j.plaphy.2013.01.020

Negi J, Matsuda O, Nagasawa T, Oba Y, Takahashi H, Kawai-Yamada M, Uchimiya H, Hashimoto M, Iba K (2008) CO2 regulator SLAC1 and its homologues are essential for anion homeostasis in plant cells. Nature 452:483–486. https://doi.org/10.1038/nature06720

Oldroyd GE (2013) Speak, friend, and enter: signalling systems that promote beneficial symbiotic associations in plants. Nat Rev Microbiol 11:252–263. https://doi.org/10.1038/nrmicro2990

Palaniyandi SA, Damodharan K, Yang SH, Suh JW (2014) *Streptomyces* sp. strain PGPA39 alleviates salt stress and promotes growth of 'Micro Tom' tomato plants. J Appl Microbiol 117(3):766–773. https://doi.org/10.1111/jam.12563

Pandey V, Ansari MW, Tula S, Yadav S, Sahoo RK, Shukla N, Bains G, Badal S, Chandra S, Gaur AK, Kumar A, Shukla A, Kumar J, Tuteja N (2016) Dose-dependent response of *Trichoderma harzianum* in improving drought tolerance in rice genotypes. Planta 243(5):1251–1264. https://doi.org/10.1007/s00425-016-2482-x

Park YS, Dutta S, Ann M, Raaijmakers JM, Park K (2015) Promotion of plant growth by *Pseudomonas fluorescens* strain SS101 via novel volatile organic compounds. Biochem Biophys Res Commun 461(2):361–365. https://doi.org/10.1016/j.bbrc.2015.04.039

Peñuelas J, Asensio D, Tholl D, Wenke K, Rosenkranz M, Piechulla B, Schnitzler JP (2014) Biogenic volatile emissions from the soil. Plant Cell Environ 37(8):1866–1891. https://doi.org/10.1111/pce.12340

Pieterse CM, Van der Does D, Zamioudis C, Leon-Reyes A, Van Wees SC (2012) Hormonal modulation of plant immunity. Annu Rev Cell Dev Biol 28:489–521. https://doi.org/10.1146/annurev-cellbio-092910-154055

Pongrac P, Sonjak S, Vogel-Mikuš K, Kump P, Nečemer M, Regvar M (2009) Roots of metal hyperaccumulating population of *Thlaspi praecox* (brassicaceae) harbour arbuscular mycorrhizal and other fungi under experimental conditions. Int J Phytoremediation 11(4):347–359. https://doi.org/10.1080/15226510802565527

Powell J, Klironomos J (2007) The ecology of plant-microbial mutualisms. In: Soil microbiology, ecology and biochemistry, 3rd edn. Academic, New York, pp 257–281

Prasad R, Kumar M, Varma A (2015) Role of PGPR in soil fertility and plant health. In: Egamberdieva D, Shrivastava S, Varma A (eds) Plant Growth-Promoting Rhizobacteria (PGPR) and medicinal plants. Springer, Cham, pp 247–260

Prasad R, Gill SS, Tuteja N (2018) Crop improvement through microbial biotechnology. Elsevier, Amsterdam. ISBN 9780444639882. https://www.elsevier.com/books/crop-improvement-through-microbialbiotechnology/prasad/978-0-444-63987-5

Punamiya P, Datta R, Sarkar D, Barber S, Patel M, Das P (2010) Symbiotic role of *Glomus mosseae* in phytoextraction of lead in vetiver grass [*Chrysopogon zizanioides* (L.)]. J Hazard Mater 177 (1–3):465–474. https://doi.org/10.1016/j.jhazmat.2009.12.056

Rodríguez AA, Stella AM, Storni MM, Zulpa G, Zaccaro MC (2006) Effects of cyanobacterial extracellular products and gibberellic acid on salinity tolerance in *Oryza sativa* L. Saline Syst 2 (7). https://doi.org/10.1186/1746-1448-2-7

Rodriguez RJ, Henson J, Van Volkenburgh E, Hoy M, Wright L, Beckwith F, Kim YO, Redman RS (2008) Stress tolerance in plants via habitat-adapted symbiosis. ISME J 2(4):404–416. https://doi.org/10.1038/ismej.2007.106

Rosier A, Bishnoi U, Lakshmanan V, Sherrier DJ, Bais HP (2016) A perspective on inter-kingdom signaling in plant-beneficial microbe interactions. Plant Mol Biol 90(6):537–548. https://doi.org/10.1007/s11103-016-0433-3

Saharan BS, Nehra V (2011) Plant growth promoting rhizobacteria: a critical review. Life Sci Med Res 21:1–30

Sandhya V, Ali SZ, Grover M, Reddy G, Venkateswarlu BSSS (2009) Alleviation of drought stress effects in sunflower seedlings by the exopolysaccharides producing *Pseudomonas putida* strain GAP-P45. Biol Fertil Soils 46(1):17–26. https://doi.org/10.1007/s00374-009-0401-z

Sannazzaro AI, Ruiz OA, Albertó EO, Menéndez AB (2006) Alleviation of salt stress in *Lotus glaber* by *Glomus intraradices*. Plant Soil 285:279–287. https://doi.org/10.1007/s11104-006-9015-5

Saravanakumar D, Samiyappan R (2007) ACC deaminase from *Pseudomonas fluorescens* mediated saline resistance in groundnut (*Arachis hypogea*) plants. J Appl Microbiol 102:1283–1292

Schmidt R, Cordovez V, de Boer W, Raaijmakers J, Garbeva P (2015) Volatile affairs in microbial interactions. ISME J 9(11):2329–2335. https://doi.org/10.1038/ismej.2015.42

Selvakumar G, Joshi P, Mishra PK, Bisht JK, Gupta HS (2009) Mountain aspect influences the genetic clustering of psychrotolerant phosphate solubilizing pseudomonads in the Uttarakhand Himalayas. Curr Microbiol 59:432–438. https://doi.org/10.1007/s00284-009-9456-1

Selye H (1936) A syndrome produced by diverse nocuous agents. Nature 138:32

Sennoi R, Singkham N, Jogloy S, Boonlue S, Saksirirat W, Kesmala T, Patanothai A (2013) Biological control of southern stem rot caused by *Sclerotium rolfsii* using *Trichoderma harzianum* and arbuscular mycorrhizal fungi on Jerusalem artichoke (*Helianthus tuberosus* L.). Crop Prot 54:148–153. https://doi.org/10.1016/j.cropro.2013.08.011

Sharifi M, Ghorbanli M, Ebrahimzadeh H (2007) Improved growth of salinity-stressed soybean after inoculation with salt pre-treated mycorrhizal fungi. J Plant Physiol 164:1144–1151

Shukla N, Awasthi RP, Rawat L, Kumar J (2012a) Biochemical and physiological responses of rice (*Oryza sativa* L.) as influenced by *Trichoderma harzianum* under drought stress. Plant Physiol Biochem 54:78–88. https://doi.org/10.1016/j.plaphy.2012.02.001

Shukla PS, Agarwal PK, Jha B (2012b) Improved salinity tolerance of *Arachis hypogaea* (L.) by the interaction of halotolerant plant-growth-promoting rhizobacteria. J Plant Growth Regul 31 (2):195–206

Singh A, Chauhan PS (2017) Ecological significance of soil-associated plant growth-promoting biofilm-forming microbes for stress management. In: Ahmad I, Husain FM (eds) Biofilms in plant and soil health. Wiley, Hoboken, NJ, p 291

Sirari K, Kashyap L, Mehta CM (2016) Stress management practices in plants by microbes. In: Microbial inoculants in sustainable agricultural productivity. Springer, New Delhi, pp 85–99. https://doi.org/10.1111/pce.12340

Sohn KH, Lee SC, Jung HW, Hong JK, Hwang BK (2006) Expression and functional roles of the pepper pathogen-induced transcription factor RAV1 in bacterial disease resistance, and drought and salt stress tolerance. Plant Mol Biol 61:897–915. https://doi.org/10.1007/s11103-006-0057-0
Son JS, Sumayo M, Hwang YJ, Kim BS, Ghim SY (2014) Screening of plant growth-promoting rhizobacteria as elicitor of systemic resistance against gray leaf spot disease in pepper. Appl Soil Ecol 73:1–8. https://doi.org/10.1016/j.apsoil.2013.07.016
Sriprang R, Hayashi M, Ono H, Takagi M, Hirata K, Murooka Y (2003) Enhanced accumulation of Cd2+ by a *Mesorhizobium* sp. transformed with a gene from *Arabidopsis thaliana* coding for phytochelatin synthase. Appl Environ Microbiol 69:1791–1796
Stam JM, Kroes A, Li Y, Gols R, van Loon JJ, Poelman EH, Dicke M (2014) Plant interactions with multiple insect herbivores: from community to genes. Annu Rev Plant Biol 65:689–713. https://doi.org/10.1146/annurev-arplant-050213-035937
Staudinger C, Mehmeti-Tershani V, Gil-Quintana E, Gonzalez EM, Hofhansl F, Bachmann G, Wienkoop S (2016) Evidence for a rhizobia-induced drought stress response strategy in *Medicago truncatula*. J Proteome 136:202–213. https://doi.org/10.1016/j.jprot.2016.01.006
Suarez C, Cardinale M, Ratering S, Steffens D, Jung S, Montoya AMZ, Geissler-Plaum R, Schnell S (2015) Plant growth-promoting effects of *Hartmannibacter diazotrophicus* on summer barley (*Hordeum vulgare* L.) under salt stress. Appl Soil Ecol 95:23–30. https://doi.org/10.1016/j.apsoil.2015.04.017
Subramanian P, Kim K, Krishnamoorthy R, Mageswari A, Selvakumar G, Sa T (2016) Cold stress tolerance in psychrotolerant soil bacteria and their conferred chilling resistance in tomato (*Solanum lycopersicum* mill.) under low temperatures. PLoS One 11(8):e0161592. https://doi.org/10.1371/journal.pone.0161592
Sundram S, Meon S, Seman IA, Othman R (2015) Application of arbuscular mycorrhizal a fungus with *Pseudomonas aeruginosa* UPMP3 reduces the development of *Ganoderma* basal stem rot disease in oil palm seedlings. Mycorrhiza 25:387–397. https://doi.org/10.1007/s00572-014-0620-5
Suzuki N, Rivero RM, Shulaev V, Blumwald E, Mittler R (2014) Abiotic and biotic stress combinations. New Phytol 203:32–43. https://doi.org/10.1111/nph.12797
Sziderics AH, Rasche F, Trognitz F, Sessitsch A, Wilhelm E (2007) Bacterial endophytes contribute to abiotic stress adaptation in pepper plants (*Capsicum annuum* L.). Can J Microbiol 53 (11):1195–1202. https://doi.org/10.1139/W07-082
Tilak KVBR, Nandanavanam R, Pal KK, Johri BN (2005) Diversity of plant growth and soil health supporting bacteria. Curr Sci 89:136–150
Timmusk S, El-Daim IAA, Copolovici L, Tanilas T, Kännaste A, Behers L, Nevo E, Seisenbaeva G, Stenström E, Niinemets Ü (2014) Drought-tolerance of wheat improved by rhizosphere bacteria from harsh environments: enhanced biomass production and reduced emissions of stress volatiles. PLoS One 9(5):e96086. https://doi.org/10.1371/journal.pone.0096086
Tiwari S, Lata C, Chauhan PS, Nautiyal CS (2016) *Pseudomonas putida* attunes morphophysiological, biochemical and molecular responses in *Cicer arietinum* L. during drought stress and recovery. Plant Physiol Biochem 99:108–117. https://doi.org/10.1016/j.plaphy.2015.11.001
UN DESA (2015) World population prospects: the 2015 revision, key findings and advance tables. United Nations Department of economic and social affairs. Population Division Working Paper No. ESA/P/WP 241
Upadhyay SK, Singh DP, Saikia R (2009) Genetic diversity of plant growth promoting rhizobacteria isolated from rhizospheric soil of wheat under saline condition. Curr Microbiol 59(5):489–496. https://doi.org/10.1007/s00284-009-9464-1
Uroz S, Dessaux Y, Oger P (2009) Quorum sensing and quorum quenching: the Yin and Yang of bacterial communication. Chembiochem 10(2):205–216. https://doi.org/10.1002/cbic.200800521
Vaishnav A, Kumari S, Jain S, Varma A, Tuteja N, Choudhary DK (2016) PGPR-mediated expression of salt tolerance gene in soybean through volatiles under sodium nitroprusside. J Basic Microbiol 56:1274–1288. https://doi.org/10.1002/jobm.201600188

Velmourougane K, Saxena G, Prasanna R (2017) Plant-microbe interactions in the rhizosphere: mechanisms and their ecological benefits. In: Singh DP, Singh HB, Prabha R (eds) Plant-microbe interactions in agro-ecological perspectives. Springer, Singapore, pp 193–219

Verma P, Yadav AN, Khannam KS, Panjiar N, Kumar S, Saxena AK, Suman A (2015) Assessment of genetic diversity and plant growth promoting attributes of psychrotolerant bacteria allied with wheat (*Triticum aestivum*) from the northern hills zone of India. Ann Microbiol 65:1885–1899

Vimal SR, Singh JS, Arora NK, Singh S (2017) Soil-plant-microbe interactions in stressed agriculture management: a review. Pedosphere 27:177–192. https://doi.org/10.1016/S1002-0160(17)60309-6

Vimal SR, Patel VK, Singh JS (2018a) Plant growth promoting *Curtobacterium albidum* strain SRV4: an agriculturally important microbe to alleviate salinity stress in paddy plants. Ecol Indic. https://doi.org/10.1016/j.ecolind.2018.05.014

Vimal SR, Gupta J, Singh JS (2018b) Effect of salt tolerant *Bacillus* sp. and *Pseudomonas* sp. on wheat (*Triticum aestivum* L.) growth under soil salinity: a comparative study. Microbiol Res 9 (1). https://doi.org/10.4081/mr.2018.7462

Vivas A, Barea JM, Biró B, Azcón R (2006) Effectiveness of autochthonous bacterium and mycorrhizal fungus on *Trifolium* growth, symbiotic development and soil enzymatic activities in Zn contaminated soil. J Appl Microbiol 100:587–598. https://doi.org/10.1111/j.1365-2672.2005.02804.x

Vos CM, Tesfahun AN, Panis B, DDe W, Elsen A (2012) Arbuscular mycorrhizal fungi induce systemic resistance in tomato against the sedentary nematode *Meloidogyne incognita* and the migratory nematode *Pratylenchus penetrans*. Appl Soil Ecol 61:1–6. https://doi.org/10.1016/j.apsoil.2012.04.007

Wang WX, Vinocur B, Altman A (2003) Plant responses to drought, salinity and extreme temperatures: towards genetic engineering for stress tolerance. Planta 218:1–14. https://doi.org/10.1007/s00425-003-1105-5

Wang FY, Lin XG, Yin R (2007) Role of microbial inoculation and chitosan in phytoextraction of Cu, Zn, Pb and Cd by *Elsholtzia splendens*-a field case. Environ Pollut 147:248–255

Yao L, Wu Z, Zheng Y, Kaleem I, Li C (2010) Growth promotion and protection against salt stress by *Pseudomonas putida* Rs-198 on cotton. Eur J Soil Biol 46:49–54. https://doi.org/10.1016/j.ejsobi.2009.11.002

Yi H-S, Yang JW, Ryu CM (2013) ISR meets SAR outside: additive action of the endophyte *Bacillus pumilus* INR7 and the chemical inducer, benzothiadiazole, on induced resistance against bacterial spot in field-grown pepper. Front Plant Sci 4:122. https://doi.org/10.3389/fpls.2013.00122

Yuan S, Li M, Fang Z, Liu Y, Shi W, Pan B, Wu K, Shi J, Shen B, Shen Q (2016) Biological control of tobacco bacterial wilt using *Trichoderma harzianum* amended bioorganic fertilizer and the arbuscular mycorrhizal fungi *Glomus mosseae*. Biol Control 92:164–171. https://doi.org/10.1016/j.biocontrol.2015.10.013

Zaidi S, Usmani S, Singh BR, Musarrat J (2006) Significance of *Bacillus subtilis* strain SJ-101 as a bioinoculant for concurrent plant growth promotion and nickel accumulation in *Brassica juncea*. Chemosphere 64:991–997

Zarea M, Hajinia S, Karimi N, Goltapeh EM, Rejal F, Varma A (2012) Effect of *Piriformospora indica* and *Azospirillum* strains from saline or non-saline soil on mitigation of the effects of NaCl. Soil Biol Biochem 45:139–146. https://doi.org/10.1016/j.soilbio.2011.11.00

Zhang H, Kim MS, Sun Y, Dowd SE, Shi H, Paré PW (2008) Soil bacteria confer plant salt tolerance by tissue-specific regulation of the sodium transporter HKT1. Mol Plant Microbe Interact 21(6):737–744. https://doi.org/10.1094/MPMI-21-6-0737

Zong K, Huang J, Nara K, Chen Y, Shen ZG, Lian C (2015) Inoculation of ectomycorrhizal fungi contributes to the survival of tree seedlings in a copper mine tailing. J For Res 20(6). https://doi.org/10.1007/s10310-015-0506-1

The manufacturer's authorised representative in the EU is Springer Nature Customer Service Centre GmbH, Europaplatz 3, 69115 Heidelberg, Germany. If you have any concerns regarding our products, please contact ProductSafety@springernature.com

Printed and bound by CPI Group (UK) Ltd, Croydon, CR0 4YY
30/09/2026
02225503-0011